# Global Europe: Legal and Policy Issues of the EU's External Action

Volume 4

**Editors-in-Chief**



**Advisory Editors**

The *Global Europe Book Series: Legal and Policy Issues of the EU's External Action Series* is a peer-reviewed book series that explores the legal and policy issues pertaining to the EU's global actorness. The editors of the book series welcome edited volumes as well as monographs exploring the EU's identity as an international actor. The coverage extends to submissions identifying the internal and external factors that may challenge the EU's capacity to exercise value-based global leadership on crucial issues affecting European citizens as well as those from other parts of the world; critically reflecting on whether the external projection of the EU as a virtuous normative power comports with its practice on the ground as well as addressing the descriptive, normative and conceptual challenges that complement the ever-expanding global reach of EU law.

Ramses A. Wessel ·
Jamile Bergamaschine Mata Diz ·
Júlia Péret Tasende Társia · Saide Esra Akdogan
Editors

# EU External Relations Law and Sustainability

## The EU, Third States and International Organizations

ASSER PRESS

Springer

*Editors*
Ramses A. Wessel
Faculty of Law
University of Groningen
Groningen, The Netherlands

Júlia Péret Tasende Társia
Direito Público
Universidade Federal de Minas Gerais
Belo Horizonte, Brazil

Jamile Bergamaschine Mata Diz
Department of Public Law
Federal University of Minas Gerais
(UFMG)
Belo Horizonte, Brazil

Saide Esra Akdogan
Law Group
Wageningen University and Research
(WUR)
Wageningen, The Netherlands

ISSN 2666-4828 ISSN 2666-4836 (electronic)
Global Europe: Legal and Policy Issues of the EU's External Action
ISBN 978-94-6265-654-3 ISBN 978-94-6265-655-0 (eBook)
https://doi.org/10.1007/978-94-6265-655-0

Published by T.M.C. ASSER PRESS, The Hague, The Netherlands www.asserpress.nl
Produced and distributed for T.M.C. ASSER PRESS by Springer-Verlag Berlin Heidelberg



This T.M.C. ASSER PRESS imprint is published by the registered company Springer-Verlag GmbH, DE, part of Springer Nature.
The registered company address is: Heidelberger Platz 3, 14197 Berlin, Germany

# Contents

# Chapter 1
# Introduction: EU External Relations Law and Sustainability

**Saide Esra Akdoğan, Júlia Péret T. Társia, Jamile B. Mata Diz and Ramses A. Wessel**

Amidst growing global environmental challenges, the importance of sustainability in international relations has escalated, positioning the European Union (EU) as a pivotal player in promoting global sustainable development. This book, entitled *"EU External Relations Law and Sustainability: The EU, Third States and International Organizations"* delves deeply into the integration of sustainability within the European Union's external action, reflecting its increasing prominence on the international political agenda and its ambition to contribute to reaching the UN's Sustainable Development Goals (SDGs).[1] The aim of the book is to consolidate diverse legal perspectives and discussions about the EU's approach to sustainability, both from inside the EU, but in particular also from the outside. It chronicles the EU's strategic efforts to weave sustainability into its external relations and examines its role in shaping international standards through policy initiatives and practical implementations.

As multilateral relations intensify in a globalized world, the European Union's role in international relations has similarly expanded, along with its attempts to

[1] See United Nations (UN) General Assembly, 'Transforming Our World: The 2030 Agenda for Sustainable Development', UN Doc/A/RES/70/1, 25 Sept. 2015.

S. E. Akdoğan (✉)
Wageningen University and Research (WUR), Wageningen, The Netherlands
e-mail: s.e.akdogan@rug.nl

J. P. T. Társia · J. B. M. Diz
Federal University of Minas Gerais, Belo Horizonte, Brazil

J. P. T. Társia
Pompeu Fabra University, Barcelona, Spain

R. A. Wessel
University of Groningen, Groningen, The Netherlands
e-mail: r.a.wessel@rug.nl

 
R. A. Wessel et al. (eds.), *EU External Relations Law and Sustainability*, Global Europe: Legal and Policy Issues of the EU's External Action 4,
https://doi.org/10.1007/978-94-6265-655-0_1

export its values and objectives worldwide. Among these objectives, as delineated in Articles 3(5) and 21 TEU, and Articles 11 and 191 TFEU, is the promotion of sustainable development. Thus, sustainability concerns form a key part of the design and execution of nearly all EU internal and external policies, ranging from the financial and industrial sectors to food security and imports.

In recent decades, with the signing of several international treaties on environmental protection such as the Kyoto Protocol and the Paris Accords, the EU has intensified its external action in the environmental area, particularly in the trade sector. The renewed attention to sustainability questions (also driven by the SDGs), has also entered the Free Trade Agreements (FTAs) that are the key instruments for the Union to regulate its trade relations with third states.[2]

The legal basis for EU FTAs is primarily derived from the European Union's exclusive competence over trade policy, as outlined in the Treaty on the Functioning of the European Union (TFEU). Specifically, Articles 206 and 207 of the TFEU grant the EU the authority to negotiate and conclude trade agreements with third countries on behalf of its member states. These days, non-trade values seem to have become a standard element of FTAs concluded by the EU in the form of so-called 'Trade and Sustainable Development' (TSD) chapters.[3] The Court of Justice of the EU (CJEU) has, in principle, confirmed this broader scope. Thus, in relation to sustainability provisions in FTA's, the CJEU in Opinion 2/15 concerning the EU's FTA with Singapore, clearly held that 'the objective of sustainable development [...] forms an integral part of the common commercial policy'.[4] As will be outlined in this book, sustainability as such also needs to be seen in broad terms. Thus, the new generation of FTAs may include provisions on environmental and labor standards, but the list seems to be getting longer and may also include issues such as climate change or gender equality.

The impact of this normative role of the Union should not be underestimated,[5] also given the large number of FTA's and the dependence of many countries around the world on access to the EU's internal market. FTA's have thus become part of the Union's extraterritorial regulatory reach, famously coined the 'Brussels Effect'.[6] At this moment, the EU lists 76 FTAs in force, 25 FTAs adopted, 6 FTAs in the pipeline, and 22 FTAs on hold.[7] Indeed, countries not affected by the EU's trade policy are hard to find. The normative, or standard-setting role flows from the Union's own

---

[2] See in general Bosse-Platière and Rapoport 2019; Kübek, forthcoming.

[3] See Marín Durán 2020. The 2008 EU-CARIFORUM Economic Partnership Agreement is often mentioned as the first FTA to include legally binding environmental and labour provisions (O.J. 2008, L 289/3, Articles 183–196). The first FTA that included a full-fledged TSD Chapter was the 2010 FTA with Korea (O.J. 2011, L 127/6).

[4] Opinion 2/15 EU-Singapore FTA, EU:C:2017:376, para 147. See further Kübek (n 5) 64.

[5] See in general Neuman, Wessel and de Zee 2024; as well as Odermatt and Wessel 2025 (forthcoming).

[6] Bradford 2017.

[7] See for an overview: https://policy.trade.ec.europa.eu/eu-trade-relationships-country-and-region/negotiations-and-agreements_en.

Treaty brief to 'export' its own values (Articles 3(5) and 21 TEU in particular) and has now become part and parcel of the Commission's policies in this regard.[8]

Despite the clear Treaty obligations, the EU's global actions to enhance environmental, labor and social standards through trade have also been criticized. There are two main points of critique. First, TSD Chapters have been widely condemned for their lack of effectiveness on the ground. Against that background, secondly, critical questions about the extent to which the EU should (aim to) interfere in the internal affairs of third states through its external action emerged.

In 2022, the European Commission unveiled a 'new approach' to trade and sustainability.[9] In particular, the Commission has broadened its 'sustainability' agenda and intends to tailor it more closely to the situation in partner countries. Moreover, the Commission seeks to improve the enforcement of TSD commitments, including by moving from a 'cooperative' (i.e. legally non-binding) to a 'sanctions-based' approach for individual commitments. At around the same time, the Commission furthermore proposed a series of unilateral trade instruments that essentially make access to the EU internal market conditional on compliance with environmental and/or human rights standards. Examples include the Deforestation Regulation,[10] the proposed Corporate Due Diligence Directive,[11] and the CBAM.[12] In view of the EU's increased focus on the export and enforcement of global sustainability norms and the resulting (ambition of) the EU to exercise its jurisdiction extraterritorially, the above-mentioned second point of critique has re-gained prominence.

---

[8] See Commission Communication, 'Trade for al—Towards a more responsible trade and investment policy', COM(2015)497 final, at p. 15; Commission, 'Strategic Plan 2016–2020l—DG Trade', 22 Dec. 2017. Cf. also Y. Kaspiarovich and R.A. Wessel, 'The Role of Values in EU External Relations: A Legal Assessment of the EU as a Good Global Actor', in E. Fahey and I. Mancini (eds.), *Understanding the EU as a Good Global Actor: Ambitions, Values and Metrics* (Cheltenham, Edward Elgar Publishing, 2022) 92–106. More specifically on the treaty brief the EU has in this regard: E. Kassoti and R.A. Wessel, 'The Normative Effect of Article 3(5) TEU: Observance and Development of International Law by the European Union', in P. García Andrade (ed.), *Interacciones entre el Derecho de la Unión Europea y el Derecho Internacional Público* (Tirant lo Blanch, 2023) 19–46.

[9] European Commission, 'Commission unveils new approach to trade agreements to promote green and just growth', Press Release, 22 June 2022. See, in more detail, Communication from the Commission to the European Parliament, the Council, the European Economic and Social Committee and the Committee of the Regions, 'The power of trade partnerships: together for green and just economic growth', COM(2022) 409 final. See Kübek and Wessel 2024. Parts of this introduction draw from that publication.

[10] Regulation (EU) 2023/1115 of the European Parliament and of the Council of 31 May 2023 on the making available on the Union market and the export from the Union of certain commodities and products associated with deforestation and forest degradation and repealing Regulation (EU) No 995/2010 [2023] OJ L 150/206.

[11] Proposal for a DIRECTIVE OF THE EUROPEAN PARLIAMENT AND OF THE COUNCIL on Corporate Sustainability Due Diligence and amending Directive (EU) 2019/1937, COM/2022/71 final.

[12] Regulation (EU) 2023/956 of the European Parliament and of the Council of 10 May 2023 establishing a carbon border adjustment mechanism [2023] OJ L 130/52.

As the European Union's external policy grows and expands to new regions and countries, its sustainability concerns are also exported in the form of sanitary regulations and sustainable investment regulations, among others.

The EU's Green Deal has been extensively debated in the media and specialized literature, as it manifests the Union's commitment to sustainable development through multi-sectoral policies, aiming at a 50% reduction of the block's greenhouse gas emissions by 2030 and achieving carbon neutrality by 2050. The new Green Deal also highlights the debate surrounding 'carbon leakage', as the EU reinforces the importance of emissions reduction in cooperation with third states, and enforces a 'carbon border adjustment mechanism' (CBAM) on imported products, reinforcing its role as a champion for climate change mitigation and adaptation on the international stage.

While attention to the sustainability dimensions of the EU's external activities in specific areas such as trade and investment remains high, the various legal consequences of the increase in both the EU's external action and its sustainability concerns have not been brought together coherently until now. The current literature on the subject is scarce and scattered, making it challenging for scholars and practitioners to form a cohesive picture of how sustainability concerns are deeply ingrained in the EU's external action, not only in Free Trade Agreements but also in investment policies and in its diplomatic relations with neighboring states.

With these developments in mind, this book investigates the sustainability dimension of the EU's external action, responding to the increasing calls for integrating 'sustainability' into the EU's relations with third states and international organizations. As the European Union's external activities expand, the importance of sustainability in these engagements is underscored by numerous provisions in the EU Treaties, mandating that environmental protection requirements must be integrated into the definition and implementation of the Union's policies, particularly to promote sustainable development and mainstream environmental aspects throughout its policies.

This book represents the first instance where environmental law and external relations experts have come together to study this field of EU law comprehensively. It aims to become a reference work for EU law experts, as environmental concerns increasingly permeate the Union's external policy. Although attention to the sustainability dimensions of the EU's external activities, such as trade and investment, is high on the political and academic agenda, the various legal consequences have historically not been unified coherently.

With the contributions of experts from around the globe, this book offers a multilateral and diverse perspective on a subject that is, by nature, complex. The outside and inside perspectives presented provide comprehensive insights into the complexities of external relations and sustainability concerns, from carbon emissions to water resources. The contributors to this volume have been selected based on their specific expertise in this field. The result is a comprehensive set of chapters that together cover the many aspects of this new and increasingly important area of research.

As Editors, we hope that *EU External Relations Law and Sustainability* will serve as an essential academic resource for scholars, policymakers, and practitioners interested in the intersections of law, sustainability, and EU external relations.

## References

A. Bradford (2017), The Brussels Effect: How the European Union Rules the World. Oxford University Press, Oxford

I. Bosse-Platière and C. Rapoport (2019) (ed.), The Conclusion and Implementation of EU Free Trade Agreements: Constitutional Challenges. Edward Elgar Publishing, Cheltenham

E. Fahey (2017), The Global Reach of EU Law. Routledge, London

G. Kübek (forthcoming), The European Union's External Trade Agreements in C. Herrmann and J. Chaisse (eds.), The International Law of Economic Integration. Oxford University Press, Oxford

G. Marín Durán (2020), Sustainable Development Chapters in EU Free Trade Agreements: Emerging Compliance Issues (2020) 57 CMLR 1031-1068

M. Neuman, R.A. Wessel and T. de Zee (2024) (eds.), The EU in a Globalising World. T.M.C. Asser Press, The Hague

J. Odermatt and R.A. Wessel (2025) (forthcoming) (eds.), Research Handbook on EU Law and Global Challenges. Edward Elgar Publishing, Cheltenham

Gesa Kübek and Ramses A. Wessel (2024), Governing Sustainability through Trade in EU External Relations: The "New Approach" and its Challenges, in J. Bergamaschine Mata Diz (ed.), Trade and Sustainable Development: The Foreign Relations of the European Union

**Saide Esra Akdogan** is a lecturer and researcher at Wageningen University and Research (WUR).

**Júlia Péret T. Társia** is a PhD Candidate at the Federal University of Minas Gerais.

**Jamile B. Mata Diz** is a Professor of International Public Law and Coordinator of the Jean Monnet Center of Excellence of the Federal University of Minas Gerais.

**Ramses A. Wessel** is a Professor of European Law of the University of Groningen.

# Part I
# Principles and Policies in EU External Relations

# Chapter 2
# Worldwide Expansion of the EU Precautionary Principle Through Free Trade Agreements

**Bruno Immanuel Striebel and Nuno Cunha Rodrigues**

## Contents



**Abstract** Due to the freezing of multilateralism in international trade, a change can be perceived in the external trade policy of the EU. In recent years, the EU has signed different bilateral and regional agreements (Free Trade Agreements—FTAs) with which it has tried to position itself as a key player in the international trading system. By doing so it also aims to relaunch a multilateral approach to world trade. Until 2006, EU bilateral agreements principally served non-economic purposes (neighborhood and development objectives), while EU economic interests were served by multilateral agreements. Since then, FTAs have largely been justified on economic interests. However, they also serve to promote EU values and principles, including the precautionary one. The "precautionary principle" ensures that the EU

B. I. Striebel (✉)
Higher Regional Court Karlsruhe and Institute for German and European Corporate and Economic Law, Heidelberg University, Heidelberg, Germany
e-mail: bruno.striebel@igw.uni-heidelberg.de

N. Cunha Rodrigues
University of Lisbon, Lisbon, Portugal
e-mail: nunorodrigues@fd.ulisboa.pt



R. A. Wessel et al. (eds.), *EU External Relations Law and Sustainability*, Global Europe: Legal and Policy Issues of the EU's External Action 4,
https://doi.org/10.1007/978-94-6265-655-0_2

can continue to protect health and the environment even if this affects trade, in situations where scientific information is not conclusive or insufficient. This principle has been included in several recent FTAs concluded by the EU, namely the CETA; JETA and the EU/Mercosur agreement. Still, in some cases, the implementation of the precautionary principle has led to the creation of deadlocks in the negotiations of bilateral agreements—as in the case of TTIP—or to its implementation—as in the case of the EU/Mercosur. This Chapter aims to analyze the role of the precautionary principle in FTAs concluded or negotiated by the EU and whether the EU has been trying to expand its understanding thereof through FTAs.

**Keywords** EU Law · Precautionary principle · Free Trade Agreements · Trade and sustainable development · Worldwide expansion of EU law · Environmental law · Common Commercial Policy

## 2.1 The Precautionary Principle—History and Meaning

Looking at the history of the precautionary principle, a striking development from a widely unknown peculiarity of environmental law to a cornerstone of EU environmental, health and food security law becomes visible.[1]

The principle originates from German environmental law ("Vorsorgeprinzip"), where it emerged when the social-liberal coalition, which came to power in 1969, introduced different laws and programmes that contributed to the development of modern environmental legislation.[2] From there it found its way into international public law, while still being only related to environmental issues.[3]

The core idea of the principle was already reflected, although not named as such, in the 1984 Ministerial Declaration of the International Conference on the Protection of the North Sea, stipulating that states may take action without having to wait for proof of harmful effects.[4] The preamble of the Montreal Protocol from 1987 refers to "precautionary measures"[5] and principle 15 of the Rio Declaration from 1992 (Rio Declaration) provides that States shall apply a precautionary approach in order to protect the environment.[6]

---

[1] De Sadeleer 2006, p. 1.

[2] Arndt 2009, p. 13f.; Kahl in Streinz Article 191 TFEU para 77ff.

[3] Kahl in Streinz Article 191 TFEU Rn. 77ff.

[4] Laowonsiri 2010, p. 568.

[5] Montreal Protocol on Substances that Deplete the Ozone Layer. Concluded at Montreal on 16 September 1987.

[6] Principle 15, Rio Declaration from 1992, A/CONF.151/26 (Vol. I): "In order to protect the environment, the precautionary approach shall be widely applied by States according to their capabilities. Where there are threats of serious or irreversible damage, lack of full scientific certainty shall not be used as a reason for postponing cost-effective measures to prevent environmental degradation"; The preceding Report on the United Nations Conference on the Human Environment, Stockholm 1972 did not contain any comparable provision.

In the case of EU-primary law, it was the Single European Act in 1987 that initially incorporated provisions on environmental law into the treaties, yet without an explicit reference to the precautionary principle.[7]

Five years later, the Maastricht Treaty amended Title XVI and finally provided that the Community Policy on the environment shall be based on the precautionary principle.[8] As of 2009 and the Treaty of Lisbon entering into force, it has been constituted in Article 191 para 2 TFEU.

Although the treaties only refer to the precautionary principle in the title on environmental policy, its scope is much broader. The CJEU explicitly applied it in the domain of public health and there is secondary legislation based on the precautionary principle in different legal areas.[9] The Court of First Instance (CFI) even considered it to be a general principle of EU law.[10]

## 2.2 The Interpretation and Application of the Precautionary Principle by the CJEU and the Commission

### *2.2.1 Introduction*

The precautionary principle is not universally defined and there are differences in the interpretation of its scope and presuppositions in national, EU and international law.[11]

For the purpose of this chapter, these differences shall be set aside and only the aspects relevant to analyzing the role of the precautionary principle in FTAs shall be highlighted. Yet, it is necessary to consider in which situations the precautionary principle can be applied and what presuppositions have to be met under EU law.

Accordingly, a brief look shall be taken at how the Commission and the CJEU interpret it. Furthermore, it shall also be considered where the principle can be found in WTO law and how it is interpreted in this domain. The aim of this chapter is not to provide an exhaustive analysis of case law or academic literature dealing with the precautionary principle.

---

[7] Article 25 Single European Act, OJ 29.6.87 L 169/1.

[8] Article 130r para 2 TEU, OJ 29.7.92 C 191, 29/07/1992 P.

[9] Pursuant to Article 1(3) Regulation (EC) No 1907/2006 of the European Parliament and of the Council of 18 December 2006 (REACH), “its provisions are underpinned by the precautionary principle”. There is much more secondary law implementing the precautionary principle, for example Directive 2011/92/EU of the European Parliament and of the Council of 13 December 2011 on the assessment of the effects of certain public and private projects on the environment.

[10] Cases T-74, 76, 83-85, 132, 137 and 141/00 Artegodan [2002], para 181ff; Craig and De Búrca 2020, p. 143.

[11] Robert-Cuendet 2020, p. 101; Arndt 2009, pp. 2, 26.

### 2.2.2 The European Commission Communication

In 2000, the Commission unveiled a Communication on the precautionary principle with the aim to inform of the manner of its application.[12] According to the Commission, the principle plays a role both in risk assessment and risk management.[13] It is relevant "only in the event of a potential risk, even if this risk cannot be fully demonstrated or quantified or its effects determined because of the insufficiency or inclusive nature of the scientific data".[14]

The Communication sets out two steps that have to be taken, in order to take recourse to the principle. Firstly, a potentially negative effect of a phenomenon has to be identified.[15] Earlier, the Commission was more restrictive in that regard, requiring a threat of severe and irreversible damages, following the wording of the Rio Declaration.[16]

Secondly, a risk assessment must be carried out, evaluating the available scientific data relevant to the risk, and resulting in a conclusion on the possibility and severity of a hazard's impact on health or the environment.[17] If it is impossible to determine with certainty the risk in question, because of the insufficiency of the data, their inconclusive or imprecise nature, action based on the precautionary principle may be triggered.[18] In doing so, the regulator can decide on whether to act or not to act and, if it opts to act, on the nature of the action ultimately taken. This decision shall be guided by an assessment of the potential consequences of inaction and the existing uncertainties.[19] The potential actions range from legally binding measures and the funding of research, to the information of the public about possible adverse effects of a product.[20]

Any measure adopted must be proportionate, non-discriminatory and consistent,[21] which is nothing specific to measures based on the precautionary principle. The principle of proportionality is a general principle of EU law[22] and the adoption of measures will in most cases coincide with the infringement of fundamental freedoms or basic rights that can only be justified if they are proportionate and non-discriminatory. Afterwards, the regulator must examine whether there are scientific developments because the measure should only be maintained as long as the scientific data is still inadequate, inconsistent or inconclusive.[23]

---

[12] Communication from the Commission on the precautionary principle, COM (2000) 1 final, p. 8.

[13] COM (2000) 1 final, p. 12.

[14] COM (2000) 1 final, p. 13.

[15] COM (2000) 1 final, p. 13.

[16] Arndt 2009, p. 80, 82.

[17] COM (2000) 1 final, p. 13.

[18] COM (2000) 1 final, p. 14.

[19] COM (2000) 1 final, p. 16.

[20] COM (2000) 1 final, p. 15.

[21] COM (2000) 1 final, p. 17.

[22] Arndt 2009, p. 266; Craig and De Búrca 2020, p. 142.

[23] COM (2000) 1 final, p. 17, 19.

### 2.2.3 *The Jurisprudence of the CJEU*

The CJEU first addressed the issues of scientific uncertainty and precaution in the 1980s,[24] but it was only at the end of the 1990s that the precautionary principle—without being expressly referred to as such—was defined in two preliminary rulings.[25]

The CJEU derived from Article 130r (1) and (2) of the EC treaty, according to which the policy on the environment pursues the objective of protecting human health and aims at a high level of protection, where there is uncertainty as to the existence or extent of risks to human health, the institutions may take protective measures without having to wait until the reality and seriousness of those risks become fully apparent. From then on, the judges in Luxembourg elaborated the jurisprudence and began to directly refer to the precautionary principle.[26]

To apply the precautionary principle according to the CJEU, a two-stage analysis must be conducted.[27] Potentially[28] negative consequences for health or the environment must be identified and a comprehensive assessment of that risk, based on the most reliable scientific data available and the most recent results of international research, has to be conducted.[29]

If this risk assessment reveals scientific uncertainties regarding the existence or the extent of a risk, and the likelihood of considerable harm persists were the negative eventuality to occur, a Member State can adopt restrictive measures based on the precautionary principle provided that they are objective, non-discriminatory and proportionate.[30] The competent authority may even be obliged by the precautionary principle to give the protection of health or the environment precedence over economic interests.[31] With regard to judicial review, the Commission has wide discretion: only a manifestly inappropriate risk-management measure would be annulled.[32]

To understand how the precautionary principle can be invoked by regulators under EU law, two judgements of the CJEU shall be reviewed with a focus on the role the principle played in the reasoning.

---

[24] C-174/82 Sandoz [1983], para 17; Bándi 2020, p. 45.

[25] C-180/96 United Kingdom v Commission [1998], para 99f.; C-157/96 National Farmers' Union and Others [1998], para 64.

[26] C-236/01 Monsanto [2003], para 111.

[27] Kingston et al. 2017, p. 95.

[28] A purely hypothetical consideration does not suffice, C-192/01 Commission v. Denmark [2003], para 49; EFTA Court, E-3/00—EFTA Surveillance Authority v. Norway, para 29; De Sadeleer 2006, p. 10.

[29] C-77/09 Gowan [2010] para 75; C-333/08 Commission v France [2010] para 91; EFTA Court, E-3/00 EFTA Surveillance Authority v. Norway [2001], para 30.

[30] C-192/01 Commission v. Denmark [2003], para 49; EFTA Court, E-3/00—EFTA Surveillance Authority v. Norway [2001], para 29; C-77/09 Gowan [2010] para 76; C-333/08 Commission v France [2010] para 92; C-78/16 Pesce u.a. [2016], para 48.

[31] Cases T-74, 76, 83-85, 132, 137 and 141/00 Artegodan [2002], para 184; Bándi 2020, p. 46.

[32] C-78/16 Pesce u.a. [2016], para 49; Kingston et al. 2017, p. 95.

#### 2.2.3.1 National Farmers Union (C-157/96)

A request for a preliminary ruling lodged in 1996, concerned the validity of a decision by the Commission prohibiting the export of live bovine animals or meat thereof from the territory of the United Kingdom. The Decision was taken in light of uncertainties concerning the transmissibility of BSE to humans.[33] The National Farmers Union argued that the decision was not proportionate. Amongst other reasons, it brought forward that there was no evidence whether an export ban to third countries was needed to protect public health.[34]

The CJEU first—due to the given discretion—limited the extent of its jurisdiction over the measure stating that its legality is only affected if it is manifestly inappropriate in relation to the pursued objective.[35]

It then considered the great uncertainty of the risks when the contested decision was adopted.[36] Without expressly taking recourse to the precautionary principle the CJEU acknowledged that the institutions may take protective measures without having to wait until the reality and seriousness of those risks become fully apparent.[37] It was furthermore considered that the Commission itself had—in the preamble of the decision—recognized that there was a need for further information and that the contested decision had to be reviewed.[38] Consequently, the Court found that the measure in question was not manifestly inappropriate.[39]

#### 2.2.3.2 Preliminary Ruling in Monsanto (C-236/01)

In *Monsanto*, the main proceedings concerned an action of annulment against an Italian decree suspending the trade-in and use of certain transgenic products on a precautionary basis.[40] Article 12 of Regulation 258/97 concerning the placing on the market of novel foods or novel food ingredients essentially provided that where a Member State has detailed grounds for considering that the use of a food endangers human health or the environment, it may temporarily restrict or suspend the trade-in and use of the food in its territory.[41] The Regulation did not contain any explicit reference to the precautionary principle. The applicants maintained that

[33] Preamble of Commission Decision of 27 March 1996 on emergency measures to protect against bovine spongiform encephalopathy, OJ 1996 L 078/47.

[34] C-157/96 National Farmers' Union and Others [1998], para 47.

[35] C-157/96 National Farmers' Union and Others [1998], para 61.

[36] C-157/96 National Farmers' Union and Others [1998], para 62.

[37] C-157/96 National Farmers' Union and Others [1998], para 63.

[38] C-157/96 National Farmers' Union and Others [1998], para 65.

[39] C-157/96 National Farmers' Union and Others [1998], para 67.

[40] C-236/01 Monsanto [2003] para 40.

[41] Article 12 of Regulation (EC) No 258/97 of the European Parliament and of the Council of 27 January 1997 concerning novel foods and novel food ingredients (no longer in force). The succeeding Regulation (EU) 2015/2283 of the European Parliament and of the Council of 25 November 2015 makes direct reference to the precautionary principle in Recital 20, as well as in Article 12, according

the conditions for applying Article 12, the safeguard clause, were not met since the decree was not founded on detailed grounds as there was a lack of available scientific information, that the novel foods were dangerous for human health or the environment.[42]

The CJEU interpreted the presupposition of "detailed grounds" in the light of the precautionary principle and found that protective measures may be taken based on Article 12, even if a full risk assessment is impossible because of the inadequate nature of the available scientific data in the particular circumstances.[43] Subsequently adopted measures would require that the preceding risk assessment of the national authorities provides specific evidence which makes it possible to conclude that the measure is necessary to avoid novel foods, with potential risks to human health being offered on the market.[44]

The precautionary principle thus finds expression in the risk assessment preceding the actual measure and in answering the question whether the adopted restriction is necessary. The example illustrates quite well how the precautionary principle is applied in interpreting secondary law concerning the question whether it allows the national administration to enact regulatory measures protecting human health on ambiguous scientific grounds. In the end, its application broadened the regulator's possibility to adopt a measure with negative effects on certain affected traders.

## 2.3 International Trade Law and the Precautionary Principle

Shifting the focus to international trade law, it must first be noted that the precautionary principle has not been referred to by the GATT 1947 which was adopted almost 30 years before the principle's emergence in environmental law. Article XX of the GATT 1947 on general exceptions only allowed for necessary measures to protect human, animal or plant life or health.[45]

However, there are other WTO agreements relating to the trading in goods adopted after the principle became prominent, namely the Agreement on the Application of Sanitary and Phytosanitary Measures (SPS Agreement) and the Agreement on Technical Barriers to Trade (TBT Agreement) which entered into force with the establishment of the WTO on January 1, 1995.

---

to which the Commission must explicitly consider the principle in its decision on the authorization of a novel food.

[42] C-236/01 Monsanto [2003] para 86.

[43] C-236/01 Monsanto [2003] para 112.

[44] C-236/01 Monsanto [2003] para 113.

[45] Although this does not expressly require a risk-assessment, there must still be scientific evidence for an existing risk. It is moreover disputed whether Article XX GATT allows for precautionary measures. Arndt 2009, p. 364.

### 2.3.1 The SPS Agreement

The SPS Agreement has the objective of harmonizing sanitary and phytosanitary measures to minimize their negative effects on trade.[46] Article 2, being the general provision, stipulates that the Members may adopt SPS measures[47], inter alia, as long as they are necessary and based on scientific principles, complying with the requirements in Articles 3 to 6.[48] With regard to the presupposition of Article 2(2) whereas measures shall be based on scientific principles, Article 5(1) specifies that a risk assessment has to be conducted, which can then be used as a basis for regulatory actions, providing information for the decision making.[49] Article 5 provides detailed requirements for the risk assessment, which is strictly applied and interpreted by the Appellate Body significantly limiting the right of the Members to determine their own level of protection.[50]

The measures adopted have to be based on the risk-assessment meaning that the results must reasonably support it.[51] The substantial link between the risk assessment and the risk management narrows down the possibility of deviating from the results of the assessment.[52]

The precautionary principle is reflected, though not expressly mentioned, in Article 5(7) of the SPS Agreement, allowing for provisional measures in cases where relevant scientific evidence is insufficient.[53]

The "insufficiency criterion" delimits Article 5(1) and Article 5(7) from each other.[54] According to the Appellate Body, the application of Article 5(7) is not triggered by scientific uncertainty but rather the insufficiency of scientific evidence.[55] Due to this rigid interpretation of Article 5(7), the EC—unsuccessfully—invoked the precautionary principle in the ambit of Article 5(1) and (2) to override the strict requirements for the risk assessment and circumvent the limits of Article 5(7).[56]

---

46 See the Preamble as well as Article 3(1) of the SPS Agreement.

47 The term "sanitary or phytosanitary measure" is defined in Annex A—Definitions (1) and includes any measure protecting human, animal, or plant life from certain risks arising for example from the entry or spread of pests and diseases or from additives and toxins in foods, beverages or feedstuffs.

48 Seibert-Fohr in Wolfrum et al. (2007), Article 2 SPS, para 1.

49 Stoll/Strack in Wolfrum et al. (2007), Article 5 SPS para 12.

50 Arndt 2009, p. 349f.

51 Stoll/Strack in Wolfrum et al. (2007), Article 5 SPS, para 22.

52 Arndt 2009, p. 354.

53 Laowonsirei, Application of the Precautionary Principle, Max Planck UNYB 14 (2010) 567 (568).

54 Arndt 2009, p. 358.

55 Appellate Body, Japan—Apples, WT/DS245/AB/R v. 26.11.2003, para 184; Arndt 2009, p. 358.

56 This was brought forward by the EC in the *Hormones* case, arguing that the precautionary principle would constitute a general customary rule of international law, WT/DS26/AB/R v. 16.1.1998, paras 16, 120f.; Arndt 2009, p. 356.

### 2.3.2 The TBT Agreement

According to the preamble[57] of the TBT Agreement, it aims to facilitate trade by encouraging the development of international standards and conformity assessment systems, and by ensuring that technical regulations and standards do not create unnecessary obstacles to international trade. Article 2.2 TBT Agreement provides that technical regulations should not create unnecessary obstacles to trade and shall not be more trade-restrictive than necessary to fulfil a legitimate objective, taking account of the risks that non-fulfilment would create. Available scientific and technical information shall be considered in assessing the respective risk. The risk assessment under the TBT Agreement is of less importance than under the SPS Agreement: scientific information only has to be considered and there are no explicit requirements for it. Accordingly, the TBT Agreement does not refer to the precautionary principle.[58]

## 2.4 The Precautionary Principle in EU Free Trade Agreements

### 2.4.1 Introduction

While the question of whether FTAs pose a risk to the EU's precautionary principle has attracted considerable political interest,[59] there are, as of now, few scientific analyses scrutinizing the role of the precautionary principle in such agreements.[60]

It is not too far-fetched to say that on the one hand there is a significant tension between the precautionary principle allowing for the adoption of regulative measures despite scientific uncertainties, and the objective of FTAs on the other hand, aiming at the removal of barriers to trade stemming from different regulatory standards.[61] The fourth section of this chapter will therefore analyze the role and meaning of the precautionary principle in FTAs concluded by the EU. A special emphasis will be put on whether the EU actually expanded the principle and its understanding thereof through the conclusion of such bilateral agreements. For this purpose, the FTAs

---

57 Koebell in Wolfrum et al. (2007), Preamble TBT para 1.

58 Arndt 2009, p. 369f.

59 Especially in respect of the ongoing TTIP and CETA negotiations, different positions were taken by stakeholders on whether the FTAs endanger the precautionary principle. The Federation of German Industries (BDI) argued that the concerns that the precautionary principle would be undermined were unfounded, Federation of German Industries (BDI) 2016.

60 Robert-Cuendet 2020, pp. 99ff.; See also a study commissioned by food watch concluding that the precautionary principle is not sufficiently anchored in the CETA and TTIP texts, Stoll et al. 2009.

61 Robert-Cuendet 2020, p. 103.

concluded[62] by the EU from 2010 on will be explored as well as available textual proposals by the Commission for negotiations.[63]

Since the 1970's the EU[64] has concluded a multitude of FTAs and there are different agreements currently being negotiated or awaiting ratification.[65] Originally, the EU preferred a multilateral approach to world trade within the framework of the WTO. With the failure of the Doha Round and the stagnating legal framework of international trade, the focus shifted towards bilateral agreements.[66]

In the past, the main objective for the EU to conclude FTAs has been the abolition of tariff- and non-tariff barriers and in a broader sense the liberalization of (bilateral) trade. However, the focus on purely economic aspects shifted over time. With the Treaty of Lisbon entering into force in 2009, Article 207(1) TFEU provides that the common commercial policy shall be conducted in the context of the principles and objectives of the Union's external action, for example sustainable development, Article 21(2) lit. f) TEU.[67] With these changes in primary law and beginning with the FTA between the EU and Korea from 2010 (EU–Korea FTA) all the agreements concluded afterwards contain a Chapter on Trade and Sustainable Development (TSD chapter).[68] The development found its current pinnacle in the Commission's Communication from June 2022, which aims to improve the significance of the TSD chapters by providing for the possibility to apply trade sanctions, as a matter of last resort if obligations are not fulfilled.[69]

Although the precautionary principle is also relevant in other parts of FTAs, the only express reference to it can be found in these TSD chapters. Because of its wide-ranging application under EU law this sectoral approach is initially surprising. FTAs regularly also contain Chapters on SPS and TBT measures in which the precautionary principle is also of relevance.[70] It would thus have been a possibility to determine the principle in a general chapter governing the entire agreement. However, this is not typical for FTAs which regularly do not at all contain such an initial chapter

---

[62] FTAs which are already in force or such, where a political agreement on the text exists.

[63] The Article will not consider the 2008 EU CARIFORUM economic partnership agreement as well as the 2021 EU UK Trade and Cooperation Agreement as they do not correspond to the typical structure of new generation FTAs. Economic Partnership Agreement between the CARIFORUM States, of the one part, and the European Community and its Member States of the other part, OJ 2008, L 289/3; Agreement between the European Union and the European Atomic Energy Community, of the one part, and the United Kingdom of Great Britain and Northern Ireland, of the other part, OJ 2021, L 149/10.

[64] Respectively the EC.

[65] See for an overview of FTA currently being negotiated or in place: https://policy.trade.ec.europa.eu/eu-trade-relationships-country-and-region/negotiations-and-agreements_en, Accessed 3 February 2023; See also Weiß 2022, § 12 para 66ff.

[66] Rodrigues 2017, p. 135; Stoll et al. 2014, p. 388.

[67] C-2/15, EU Singapore Free Trade Agreement [2017], para 142.

[68] Durán 2020, p. 1034.

[69] COM (2022) 409 final, The power of trade partnerships: together for green and just economic growth" p. 11; The Communication has been endorsed by the European Parliament resolution of 6 October 2022.

[70] See 3. on the precautionary principle and international trade law.

determining principles or definitions of general application to the entire agreement.[71] As there is no general provision providing that measures may be adopted based on the precautionary principle, it is necessary to look into the different Chapters, where the principle (might) play a role: the TSD chapter as well as the Chapters on Sanitary and Phytosanitary Measures (SPS chapter) and on Technical Barriers to Trade (TBT-Chapter).

## *2.4.2 The Precautionary Principle within the Ambit of the TSD Chapter*

### 2.4.2.1 TSD Chapters in EU FTAs

TSD chapters promise to enhance environmental and labor standards being an expression of the EU's value-based trade policy.[72] Not only do they exemplify the wider range of objectives pursued by FTA's, but they are also the focal point for analyzing the role of the precautionary principle in the agreements. This is why it is necessary to briefly review the most important provisions in these TSD chapters.

The first Article regularly sets out the objectives of the TSD chapter, where the Parties recall certain international declarations and agreements, for example, the preamble of the WTO agreement, the Agenda 21 from 1992,[73] or the UN 2030 Agenda for Sustainable Development.[74]

The underlying principle of the TSD chapter is that every Party has the right to regulate and specify its level of environmental and labor protection.[75] This right is however limited in a threefold way. Firstly, there is an obligation to respect certain internationally recognized standards or agreements.[76] With regard to labor law, some agreements are listed,[77] whereas in the context of environmental law, most FTA refer

---

[71] Only the EU-Chile Advanced Framework Agreement following the political conclusion between the Parties, announced on 9 December 2022 (EU–Chile FTA), contains a general section with certain definitions of general application (Article 1.3) and setting out the objectives of the agreement (Article 1.2). "General Principles" governing the entire agreement are also stipulated by Article 1 of the Trade Agreement between the European Union and its Member States, of the one part, and Colombia and Peru, of the other part (EU–Colombia and Peru FTA), though the respective provisions are very broad making reference only to the rule of law, democratic principles and fundamental human rights, OJ 2012 L 35.

[72] Durán 2020, p. 1032.

[73] Article 12.1(1) Free Trade Agreement between the European Union and the Republic of Singapore, OJ 2019, L 294/3 (EU–Singapore FTA).

[74] Article 1(2) of the TSD chapter in the EU-Mercosur Trade Agreement, as agreed on the 28th of June 2019 (EU–Mercosur FTA).

[75] Article 12.2 EU–Singapore FTA.

[76] Article 12.2 EU–Singapore FTA.

[77] The ILO Declaration on Fundamental Principles and Right at Work and the follow up process.

to all agreements binding on a Party of the FTA.[78] Although this does not create new obligations, it establishes a minimum level of protection as regulatory acts adopted in the relevant fields may not fall within that standard. Furthermore, the respective multilateral agreements are made applicable on a bilateral basis,[79] and are subject to the dispute settlement mechanism of the TSD chapter.[80]

Secondly, the Parties commit to improving their labor and environmental laws and policies and strive towards high levels of protection. The imprecise and broad language of this "best-endeavor-clause" makes the provision hardly enforceable but may at least prevent a severe weakening of existing standards.[81]

Thirdly, a Party shall not waive or derogate from its environmental and labor laws in a manner affecting trade or investment between the Parties.[82] This "non-regression obligation" shall prevent a race to the bottom in environmental and labor regulations, resulting from competitive pressure to lift those standards.[83]

Therefore, the TSD chapters leave much room for the right to regulate and do not harmonize or create new obligations to improve environmental or labor law standards. Yet, they prohibit a deterioration from the status quo.

#### 2.4.2.2 The Article on Science and Technical Information

Within the TSD chapters, the precautionary principle is—if directly or indirectly referred to—always included in the Article on Science and Technical Information,[84] which is why it is necessary to take a closer look into that matter.

These Articles are part of every FTA containing a TSD chapter.[85] They pertain to the consideration of scientific information concerning regulatory measures aiming to

---

[78] Article 270 EU–Colombia and Peru FTA lists certain agreements whereas for example Article 12.6(2) EU–Singapore FTA stipulates that the Parties shall effectively implement, in their respective laws, regulations or other measures and practices in their territories, the multilateral environmental agreements to which they are a party.

[79] Durán 2020, p. 1036f.

[80] The general dispute settlement and mediation mechanisms are not applicable to the TSD chapters which provide special procedures and consultations for disputes arising. These are rather complex and do not contain any possibility to enforce the rules stipulated. See for example the consultation and expert review procedure in the TSD chapter of the EU-Peru and Colombia Trade Agreement, Marx et al. 2017, p. 23; Durán 2020, p. 1041.

[81] Durán 2020, p. 1038.

[82] Article 12.5 EU–Singapore FTA.

[83] Durán 2020, p. 1039.

[84] Article 278 EU–Colombia and Peru FTA; Article 24.8, Comprehensive Economic and Trade Agreement (CETA), between Canada, of the one part, and the European Union and its Member States, of the other part, OJ 2017, L 11/23; Article 13.1, Free Trade Agreement between the European Union and the Socialist Republic of Viet Nam (EU–Vietnam FTA), OJ 2020 L 186/1; Articles 12.5 and 12.9, EU–Singapore FTA; Article 16.9, Agreement between the European Union and Japan for an Economic Partnership (JETA), OJ 2018 L 330/3; Article 10(2) TSD chapter, EU-Mercosur FTA.

[85] Article 13.8 Free Trade Agreement between the European Union and its Member States, of the one part, and the Republic of Korea, of the other part (EU–Korea FTA), OJ 2011, L 127/63; Article

protect the environment or occupational health, provided that trade may be affected.[86] Looking at their substantive nature, a change can be perceived: Both the EU–Korea, as well as the EU–Colombia and Peru FTA provide that "The Parties recognize the importance to take into account scientific information".[87] The permissive wording does not create an actual obligation and the right to regulate is not constrained.

In contrast to that, all the FTAs concluded or negotiated afterwards stipulate that the Parties "shall take into account available scientific and technical information".[88] The verb "shall"—especially compared to the wording of the two preceding FTAs—indicates that this is an obligation and thus constitutes another limit on the "right to regulate".[89] This is also underlined by the fact that certain FTAs provide that, in cases of scientific uncertainty, the Parties *may* adopt measures based on the grounds of the precautionary principle".[90] *E contrario*, unless the precautionary principle is applicable the measure may not be adopted. It can therefore be concluded that the Articles on Science and Technical Information requiring the consideration of scientific information constitutes another limitation to the right to regulate within the TSD chapter.

---

278 EU–Colombia and Peru FTA; Article 24.8 CETA; Article 12.5 EU–Singapore FTA; Article 16.9 JETA; Article 13.1 EU–Vietnam FTA;

Article 11.2 EU-Mexico Agreement in principle announced on 21 April 2018 (EU–Mexico FTA); Article 10.2 EU–Mercosur FTA; Article 19.13 Text of the EU New Zealand Free Trade Agreement following the announcement of conclusion of the negotiations on 30 June 2022 and having completed legal revision (EU–New Zealand FTA);

Article 26.4 of the EU-Chile FTA; Article X.10(2), EU proposal for the TSD chapter of the EU-Australia FTA, tabled for discussion on 25 February 2019 (Proposal EU–Australia FTA); Article X.12, EU proposal for a legal text on trade and sustainable development in the EU–India FTA from 22 September 2022 (Proposal EU–India FTA);

Article X.11, EU proposal for a legal text on trade and sustainable development in the EU-Indonesia FTA from 29 May 2019, tabled for discussion with Indonesia (Proposal EU–Indonesia FTA).

[86] Only the EU–Korea FTA and the EU–Colombia and Peru FTA require an effect on trade, while the other FTA presuppose that the measure "may" affect trade.

[87] Article 13.8 EU–Korea FTA; Article 278 EU–Colombia and Peru FTA.

[88] Article 13.1 EU–Vietnam FTA; Article 24.8(1) CETA; Article 16.9 JETA; Article 12.5 and 12.9 EU–Singapore FTA; Article 19.13(1) EU–New Zealand FTA; Article 11(1) EU–Mexico FTA; Article 26.4(1) EU–Chile FTA; Article 10(1) TSD chapter EU–Mercosur FTA; Article X.10(1) Proposal EU–Australia FTA; Article X.12(1) Proposal EU–India FTA; Article X.11 Proposal EU–Indonesia FTA.

[89] In *EC—Hormones*, the Panel considered that the use of the word "should" instead of "shall" would indicate that Article 5(4) of the SPS-Agreement, whereas the objective of minimising negative trade effects when determining the appropriate level of protection, does not impose an obligation, EC–Hormones (US), WT/DS26/R/USA, para 8.166.

[90] Article 11.2 EU–Mexico FTA; Article 26.4 EU–Chile FTA; Article 10.2 EU–Mercosur FTA; Article X(2) Proposal EU–Australia FTA; Article X.12 Proposal EU–India FTA; Article X.11 Proposed EU–Indonesia FTA.

#### 2.4.2.3 The Role of the Precautionary Principle

Having seen that the requirement to take into account scientific information changed from a "best-endeavor-clause" to another, fourth[91] limitation to the right to regulate, the role and substance of the precautionary principle within the TSD chapters can now be analyzed. Doing so, a development becomes visible which is closely related to the changing nature of the Article on Science and Technical Information.

The 2010 EU–Korea FTA, the first agreement containing a TSD chapter but not posing an obligation to take into account scientific information, does not mention the precautionary principle.[92] Without an obligation to take into account scientific information, there is—from a legal perspective—no need to address the situation in which a regulative measure can be enacted despite insufficient information.[93] The EU–Colombia and Peru FTA[94] concluded in 2012 and provisionally applied since 2013 as well as CETA, signed in 2016 and provisionally applied since 2017 are already a step further.

Although they do not expressly refer to the "precautionary principle" they circumscribe its scope of application. It is acknowledged that when there is a threat of serious or irreversible damage, the lack of full scientific certainty should not be used as a reason for postponing protective measures.[95] Interestingly, to the EU–Colombia and Peru FTA, it is added that Peru interprets this provision against the background of Principle 15 of the Rio Declaration—the precautionary approach.[96]

From then on, all the FTAs concluded,[97] or such which are currently being negotiated[98] make direct reference to the precautionary principle in the Article on Science and Technical Information. Thus, the requirement to take into account scientific information and the FTAs referring to the precautionary principle evolved more or less simultaneously.

While the obligation to take into account scientific information limits the right to regulate, the precautionary principle constitutes an exception to that. Consequently, with taking into account scientific information becoming mandatory, the precautionary principle found its way into the FTAs as well. After the context, in which

---

[91] See 4.2.1.

[92] Article 13.8 EU–Korea FTA.

[93] Article 278 of the FTA between the EU and Colombia and Peru, provides that "while acknowledging that, where there are threats of serious or irreversible damage, the lack of full scientific certainty should not be used as a reason for postponing protective measures".

[94] The EU–Colombia and Peru FTA, in contrast to CETA, did not create an obligation to take into account scientific information, see 4.2.2.

[95] Article 278 EU–Colombia and Peru FTA; Article 24.8(2) CETA.

[96] Footnote 81 of the Agreement is stating that "Peru interprets this Article against the background of Principle 15 of the Rio Declaration on Environment and Development.

[97] Articles 12.5 and 12.9, EU–Singapore FTA; Article 13.1, EU–Vietnam FTA; Article 16.9, JETA; Article 10(2) TSD Chapter, EU-Mercosur FTA; Article 19.13(2), EU–New Zealand FTA; Article 26.4 EU–Chile FTA.

[98] Article X.10(2), Proposal EU–Australia FTA; Article X.12, Proposal EU–India FTA; Article X.11, Proposal EU–Indonesia FTA.

the precautionary principle is referred to within FTAs has been established, as well as the role it plays in this context, the meaning of the principle shall be analysed, namely the presuppositions under which it can be invoked and the limitations to that.

First of all, it must be considered that certain earlier FTAs, which already oblige the Parties to take into account scientific information, only refer to the principle without specifying or defining it at all.[99] The meaning of "precautionary principle" in this context would have to be interpreted on the grounds of Articles 31 and 32 of the Vienna Convention on the Law of Treaties.[100]

Opposed to that, certain textual proposals by the EU for currently negotiated FTAs and texts, where an agreement exists in principle, are more detailed in this regard. In general, two presuppositions must be met in order to invoke the principle. The first one is similar to all texts: there must be a threat or risk "of serious or irreversible damage" to the environment or occupational health,[101] adapting the wording of principle 15 of the Rio Declaration. With regard to the second presupposition, there are differences between the respective texts. Some require the "lack of full scientific certainty",[102] while others refer to cases in which "scientific evidence or information is insufficient or inconclusive".[103] The first alternative referring to "lack of full scientific certainty" is broader than the latter wording which resembles rather Article 5(7) SPS-Agreement. The texts of the EU–Mercosur, EU–New Zealand and EU–Chile FTA furthermore determine certain limits for measures adopted on the grounds of the precautionary principle: "[They] shall not be applied in a manner, which would constitute a means of arbitrary or unjustifiable discrimination or a disguised restriction to trade".[104] In addition, the text of the EU–New Zealand FTA provides that the measure must also be appropriate to prevent damage stemming from the respective risk.[105] While this already follows from fundamental rights and freedoms regarding the subordinate relation between an individual and the EU or a Member State,[106] it is different in the ambit of an FTA being a bilateral treaty between states.

Lastly, certain FTAs create obligations after a measure based on the precautionary principle has been adopted. Some require that the measures shall be subject to review

---

[99] Articles 12.5 and 12.9 EU–Singapore FTA; Article 16.9 JETA; Article 13.1 EU Vietnam FTA.

[100] Mitchell et al. (2014) p. 117.

[101] Article 24.8(2) CETA; Article 19.13(2) EU–New Zealand FTA; Article 11(2) EU–Mexico FTA; Article X.10(2), Proposal EU–Australia FTA; Article X.12, Proposal EU–India FTA; Article X.11, Proposal EU–Indonesia FTA. Article 10(2) TSD Chapter EU–Mercosur FTA as well as Article 26.4 EU–Chile FTA require a risk of serious environmental degradation, whereas with regard to occupational health, a mere risk suffices.

[102] Article 24.8(2) CETA (without expressly referring to the precautionary principle); Aricle. 11(2) TSD chapter EU–Mexico FTA; Article X.10(2) EU–Australia FTA; Article X.12(2) TSD Chapter EU–India FTA; Indonesia; Article 19.13(1) EU–New Zealand FTA.

[103] Article 10(2) TSD chapter EU–Mercosur FTA; Article 26.4(2) EU–Chile FTA.

[104] Article 10(4) TSD chapter Mercosur; Article 19.13(3) EU–New Zealand FTA; Article 26.4(4) TSD chapter EU Chile FTA.

[105] Article 19.13(3) EU–New Zealand FTA.

[106] Article 51 para 1 Charter of Fundamental Freedoms of the EU.

in the light of new scientific information.[107] The EU-Mercosur FTA as well as the EU-Chile FTA furthermore provide that when such a measure has an impact on trade or investment, the adopting Party may be requested to provide information indicating that scientific knowledge is insufficient or inconclusive and may request discussion of the matter in a TSD Sub-Committee.[108]

From 2010 on and following the EU–Korea FTA, the precautionary principle constantly gained importance in FTAs, especially since taking into account scientific information became mandatory in the ambit of the TSD chapters. Because of the growing importance, the respective parts of the agreements got more and more detailed, with notable differences between the individual FTAs. Considering that the EU Commission intends to make the TSD provisions enforceable by the application of trade sanctions as a matter of last resort, this development is likely to continue.

### *2.4.3 The Precautionary Principle and the SPS and TBT Chapters in FTA*

FTAs concluded or currently negotiated by the EU Commission have all included an SPS and a TBT chapter since 2010.[109]

The objectives of the SPS chapters are to enhance the implementation of the SPS Agreement, to facilitate trade by ensuring that SPS measures do not create unnecessary obstacles, and to protect human, animal and plant life or health.[110] Substantially, the Chapters merely refer to the SPS Agreement and the Parties reaffirm their rights and obligations thereunder.[111] Although some of the proposed texts refer to situations in which scientific information is insufficient, these parts only repeat the presuppositions already laid down by Article 5 of the SPS Agreement and therefore add nothing new to the regulatory framework.[112] This is surprising, considering

[107] Article 11(2) EU–Mexico FTA; Article 10(2) TSD Chapter EU–Mercosur FTA; Article 26.4(2) EU–Chile FTA.

[108] Article 10(3) TSD Chapter EU–Mercosur FTA; Article 26.4(3) EU–Chile FTA.

[109] Robert-Cuendet 2020, p. 107.

[110] See for example Article 6.2 EU–Vietnam FTA and Article 6.1 JETA.

[111] Article 5.4 EU–Korea FTA; Article 86 EU–Colombia and Peru FTA; Article 5.4 EU–Singapore FTA; Article 5.4 CETA; Article 6.4 JETA; Article 4 EU–Mercosur FTA; Article 6.4(1) EU–Vietnam FTA; Article 6(1-2) EU–Mexico FTA; Article 6.2(1) EU–New Zealand FTA; Article 6.2 EU–Chile FTA; Article X.4 Proposal EU–Australia FTA; Article X.3(1) Proposal EU–Indonesia FTA; Article X.3 Proposal EU–India FTA.

[112] Article X.6 of the proposed EU text for the EU–India FTA stipulates: "The Parties shall ensure that their SPS measures are based on a risk assessment in accordance with relevant provisions, including Article 5 of the SPS Agreement. In particular, the Parties recall that in cases where relevant scientific evidence is insufficient, a Party may provisionally adopt SPS measures based on available pertinent information, including that from the relevant international organizations. In such circumstances, the Party adopting the measure shall seek to obtain the additional information necessary for a more objective assessment of risk and review the measure accordingly within a reasonable period of time".

the interpretation of the precautionary principle in the SPS Agreement has been a controversial issue and it was the EU that failed in its efforts to establish that the principle allows for wider-ranging measures. It has thus been criticized that the EU had apparently conceded its former position.[113]

The same applies to the respective TBT chapters. They are intended to facilitate trade in goods between the parties by ensuring that for example technical regulations and standards do not create unnecessary obstacles to trade.[114] Just as in the SPS chapters, the Parties reaffirm their rights and obligations under the TBT Agreement and certain Articles thereof are expressly incorporated and made part of the FTA.[115] Here, the precautionary principle is not mentioned at all.

## 2.5 Conclusion—Expanding of the EU Precautionary Principle Through FTA

Having analyzed the role and meaning of the precautionary principle in (proposed) EU FTAs, it is time to finally look into whether the EU actually expanded the principle and its understanding thereof through these agreements.

With respect to the SPS and TBT Chapters, the answer is obvious. The agreements solely refer to the respective WTO agreements incorporating or reaffirming the obligations thereunder. Although the SPS Agreement reflects the precautionary principle to some extent in Article 5(7), the limitation to provisional measures in cases of insufficient scientific evidence makes it hardly possible to adopt precautionary measures as it would be possible under EU law. With regard to the TBT Agreement, it remains unclear whether precaution plays any role at all. Insofar, as the EU did not expand its understanding of the precautionary principle. Concerning the TSD chapter and the Article on Science and Technical Information, it is rather difficult to come to such an unambiguous conclusion.

To a certain extent, the FTAs reflect the extensive understanding of the precautionary principle under EU law, while other aspects rather relate to the stricter understanding prevalent in international law, for example under the Rio Declaration or the SPS Agreement.

The broader understanding is expressed by the fact that precautionary measures must not be provisional.[116] Secondly, the obligation to take into account scientific information is—compared to the required risk assessment under the SPS Agreement—significantly less restrictive and the measure adopted must not be based on the risk assessment. This leaves more leeway for precautionary measures, compared with the requirements under the SPS agreement. With regard to the presuppositions

[113] Stoll et al. 2016, p. 4.

[114] See for example Article 7.1 lit. a) JETA.

[115] Article 7.3 JETA.

[116] Yet, there is according to some FTAs an obligation to review the measure and to collect new information, which is comparable to EU law.

that have to be met in order to invoke the principle, requiring a threat of serious or irreversible damage to occupational health or the environment is adapting Principle 15 of the Rio Declaration instead of EU law according to which potentially negative effects are sufficient. The FTAs that refer to scientific uncertainty reflect the broader EU understanding, while the ones requiring insufficient or inconclusive information are rather oriented to the wording of Article 5(7) of the SPS Agreement. All in all, it can be summarized that the meaning of the precautionary principle as provided for in the TSD chapters of recent FTAs does in some ways differ from EU law, but in general is rather close to the interpretation by the Commission and the CJEU. The EU has expanded its understanding of the precautionary principle through FTAs. What stands out in particular, is the continuous change in the provisions concerning the principle, which are becoming more and more detailed. This process is likely to continue with the provisions in the TSD chapters possibly being subject to dispute settlement and sanctions in the future. Whether the EU will coin the principle according to its broader understanding, or whether it will prefer to proceed with the more restrictive interpretation prevalent in international law, will become apparent in the near future.

## References


Arndt B (2009) Das Vorsorgeprinzip im EU-Recht [The Precautionary Principle under EU Law]. Mohr Siebeck, Tübingen

Bándi G (2020) Principles of EU Environmental Law. In: Peeters M and Eliantonio M (eds.), Research Handbook on EU Environmental Law. Edward Elgar, Cheltenham / Northampton, pp 36–53

Craig P and De Búrca G (2020) EU Law. Oxford University Press, Oxford

De Sadeleer N (2006) The Precautionary Principle in EC Health and Environmental Law, European Law Journal, 12:139:172

Durán G (2020) Sustainable Development Chapters in EU Free Trade Agreements: Emerging Compliance Issues, CMLRev 57:1031:1068

European Parliament resolution of 6 October 2022 on the outcome of the Commission's review of the 15-point action plan on trade and sustainable development (2022/2692(RSP)). https://www.europarl.europa.eu/doceo/document/TA-9-2022-0354_EN.pdf Accessed 19 November 2022

EU-Chile Advanced Framework Agreement following the political conclusion between the Parties, 9 December 2022 (EU - Chile FTA). https://policy.trade.ec.europa.eu/eu-trade-relationships-country-and-region/countries-and-regions/chile/eu-chile-agreement/text-agreement_en Accessed 03 February 2023

EU-Mercosur Trade Agreement, as agreed on the 28th of June 2019 (EU - Mercosur FTA). https://circabc.europa.eu/ui/group/09242a36-a438-40fd-a7af-fe32e36cbd0e/library/63854154-7f3f-45d6-bfe6-53e330818fd0/details Accessed 17 November 2022

EU-Mexico Agreement announced on 21 April 2018 (EU - Mexico FTA). https://policy.trade.ec.europa.eu/eu-trade-relationships-country-and-region/countries-and-regions/mexico/eu-mexico-agreement/agreement-principle_en Accessed 7 February 2023

EU-New Zealand Free Trade Agreement following the announcement of conclusion of the negotiations on 30 June 2022 (EU - New Zealand FTA). https://policy.trade.ec.europa.eu/eu-trade-relationships-country-and-region/countries-and-regions/new-zealand/eu-new-zealand-agreement/text-agreement_en Accessed 7 February 2023

EU proposal for the TSD chapter of the EU-Australia FTA, tabled for discussion on 25 February 2019 (Proposal EU - Australia FTA). https://policy.trade.ec.europa.eu/eu-trade-relationships-country-and-region/countries-and-regions/australia/eu-australia-agreement/documents_en Accessed 7 February 2023

EU proposal for a legal text on trade and sustainable development in the EU - India FTA from 22 September 2022 (Proposal EU - India FTA). https://policy.trade.ec.europa.eu/eu-trade-relationships-country-and-region/countries-and-regions/india/eu-india-agreement_en Accessed 7 February 2023

EU proposal for a legal text on trade and sustainable development in the EU-Indonesia FTA from 29 May 2019 (Proposal EU - Indonesia FTA). https://policy.trade.ec.europa.eu/eu-trade-relationships-country-and-region/countries-and-regions/indonesia/eu-indonesia-agreement/documents_en Accessed 7 February 2023

Federation of German Industries (BDI) (2016) The Precautionary Principle in New EU Free Trade Agreements. https://english.bdi.eu/media/themenfelder/aussenwirtschaftspolitik/TTIP/positionen/BDI-Position_Vorsorgeprinzip_englisch.pdf Accessed 20 January 2023

ILO Declaration on Fundamental Principles and Right at Work and the follow up process. https://www.ilo.org/wcmsp5/groups/public/---ed_norm/---declaration/documents/normativeinstrument/wcms_716594.pdf Accessed 18 November 2022

Kingston S, Heyvaert V and Čavoški A (2017) European Environmental Law. Cambridge University Press, Cambridge

Laowonsiri A (2010) Application of the Precautionary Principle in the SPS Agreement. In: Von Bogdandy A and Wolfrum R (eds.), Max Planck Yearbook of United Nations Law. Nijhoff, Leiden / Boston, pp 565–623

Marx A, Ebert F, Hachez N, Wouters J (2017) Dispute Settlement in the Trade and Sustainable Development Chapters of EU TradeAgreements. <ghum.kuleuven.be/ggs/publications/books/final-report-9-february-def.pdf> Accessed 19 November 2022

Mitchell A, Voon T and Sheargold E (2014) Free Trade Agreements and Public International Law. In: Lester S, Mercurio B and Bartels L (eds.), Bilateral and Regional Trade Agreements. Cambridge University Press, Cambridge, pp. 115–141

Robert-Cuendet S (2020) Les approches de précaution dans les accords internationaux de libre-échange [Precautionary Approaches in International Free Trade Agreements]. In: Sève R (ed.), Le Principe de Précaution. Dalloz, Paris, pp 99–116

Rodrigues N Cunha (2017) The Use of Public Procurement as a Non-Tariff Barrier, Public Procurement Law Review, 26:135–164

Stoll P, Krüger H, Xu J (2014) Freihandelsabkommen und ihre Umweltschutzregelungen [Free Trade Agreements and their Environmental Protection Provisions], ZUR 25:387:396

Stoll P, Douma W, De Sadeleer N, Abel P (2009) CETA, TTIP and the EU precautionary principle. www.foodwatch.org/fileadmin/Themen/TTIP_Freihandel/Dokumente/2016-06-21_foodwatch-study_precautionary-principle.pdf Accessed 24 November 2022

Streinz R (2018) Beck'sche Kurzkommentare, EUV/AEUV [Beck Short Commentary TEU/TFEU]. C.H.Beck, München

Weiß W (2022) Vertragliche Handelspolitik der EU [Treaty-based trade policy of the EU]. In Von Arnauld A and Bungenberg M (eds.), Europäische Außenbeziehungen - Enzyklopädie Europarecht Band 12 [European External Relations - Encyclopaedia European Law Volume 12]

Wolfrum R, Stoll P, Seibert-Fohr A (2007), Max Planck Commentaries on World Trade Law: WTO - Technical Barriers and SPS Measures. Nijhoff, Leiden / Boston

**Bruno Immanuel Striebel** is a legal trainee at the Higher Regional Court Karlsruhe and a research assistant to Prof. Dr. Dr. h.c.mult. Müller-Graff, Institute for German and European Corporate and Economic Law, Heidelberg University.

**Nuno Cunha Rodrigues** is an Associate Professor at the Faculty of Law of the University of Lisbon and holder of a Jean Monnet Chair.

# Chapter 3
# How to be a Green Global Standard-Setter—The EU's Externalization of the Environmental Integration Principle

**Eva Pander Maat**

**Contents**



**Abstract** This Chapter examines internal environmental measures as tools for the European Union's (EU) external action. It notes that the EU, spurred by the environmental integration principle, has become increasingly coercive in externalizing internal environmental measures. Crucially, the external dimension of internal measures makes it relevant to appreciate the role of legislative institutions in constructing the EU as an external actor. The Chapter evaluates this role

E. Pander Maat (✉)
City, University of London, Northampton Square, London, UK
e-mail: eva.pander-maat@city.ac.uk

 
R. A. Wessel et al. (eds.), *EU External Relations Law and Sustainability*, Global Europe: Legal and Policy Issues of the EU's External Action 4,
https://doi.org/10.1007/978-94-6265-655-0_3

by assessing the legislative contribution to factors which enhance the potential of internal measures to set green global standards: timing and innovation, stringency and international law compatibility. The Chapter concludes that the Commission generally contributes most to the EU's externalization of the environmental integration principle on questions of timing and international law compatibility. The Parliament spearheads EU leadership by pushing its environmental ambition. The Council contributes least to establishing EU internal measures as green global standards. Overall, the legislative institutions are increasingly aware of and strategic on the external dimension of EU internal measures. Notwithstanding, the EU's effectiveness in this external dimension is not seldomly—and understandably—trumped by internal or non-environmental considerations. Consequently, the place of internal measures in EU external sustainability law will raise ever more legal questions in the years to come.

**Keywords** Environmental integration · Externalization · Global standard-setting · European Union

## 3.1 Introduction

The European Union (EU) was the first global actor to adopt an environmental integration principle in its domestic law.[1] Article 11 Treaty on the Functioning of the European Union (TFEU) now has the status of a general principle of EU law and is arguably more forceful and more specific than the general coherence clauses or the other integration clauses in the Treaties.[2] The principle has most evidently materialized in market standards.[3] For example, in the context of the Green Deal, the EU has adopted or amended sustainability product standards on eco-design; forestry due diligence; financial products; the energy market; and an emissions-based market access levy.[4]

Importantly, these internal legislative measures are not just of relevance within the EU. Instead, they are externalized outside its borders, as the EU has gathered the ever-increasing ambition to elevate its market standards to global examples. It is important to analyze the externalization of internal measures as part of EU external action because the latter extends beyond the Common Foreign and Security Policy (CFSP). It can be defined as "any form of interaction [...] between the European Union [...] and the outside world, typically third countries or international organizations [...], but also non-state actors and individuals".[5] This broader view of external

[1] Following the 1972 Stockholm Declaration, in the 1987 Single European Act.

[2] E.g. Articles 7 and 21(3) TEU; Articles 8, 9, 10, 12 and 13 TFEU.

[3] As mandated especially by Article 114(3) TFEU.

[4] Respectively, COM/2022/142 final; Regulation (EU) No 995/2010; Regulation (EU) 2020/852; Regulation (EU) 2023/1115; Directive 2018/2001/EU; Regulation 2023/956.

[5] Schunz et al. 2018, p. 15.

action falls within the seminal literature on EU actorness.[6] The externalization of internal measures was first noted in literature on EU governance in its neighboring countries[7] before it was picked up in political economy literature,[8] and, eventually, legal scholarship.[9] However, authors remain overwhelmingly focused on the external dimension and the effectiveness of externalization, such as the impact of EU internal measures on progress in bi- and multilateral fora.[10] This contribution therefore looks at the internal dimension of externalization. This focus can help understand how the EU constructs itself as an external actor and sheds light on the internal-external nexus in EU law, which remains understudied.[11]

To understand the internal dimension of externalization, it is especially relevant to differentiate between institutions. The study of the EU as an external actor has historically focused on purely external policy areas such as security. Accordingly, the institutional focus has been on the Council and the European External Action Service (EEAS). Insufficient attention has been paid to how the institutions in their capacity as internal legislators construct the EU's external actorness. This chapter hence helps illuminate the internal dynamics of the EU as an external actor via internal market externality by drawing on recent examples.

The Chapter is structured as follows. The second section explains how EU internal measures extend their reach over third-country actors with a view to setting global environmental standards. The third section identifies factors which contribute to this ambition. The factors for effective externalization are then applied in the fourth section, which analyzes how EU institutions approach them in the legislative process of environmental market standards.

## 3.2 How to Deploy Internal Market Measures to Externalize Environmental Norms

This section distinguishes the different ways in which internal market externality functions as a tool of EU external environmental policy. The externalization of internal measures relies on three main constituent characteristics: objectives, external reach and coercive effects. These characteristics are addressed in turn. First, the section assesses the EU's increasingly evident ambition to create transposable measures, or 'set a global standard' with internal legislation.[12] Second, it differentiates the variety of legal techniques deployed to extend the reach of internal measures

[6] Bretherton and Vogler 2006.

[7] Lavenex 2004, p. 11; Schimmelfennig and Sedelmeier 2005.

[8] Damro 2012; Vogel 2012; Drezner 2005.

[9] Bradford 2013; Hadjiyianni 2019; Jaremba 2020.

[10] Focusing on its participation in multilateral negotiations. See, Walker and Biedenkopf 2018; Groen and Niemann 2011.

[11] Schunz and Damro 2020, p. 123.

[12] Dupont et al. 2018, p. 120

over third-country actors. Third, to better understand the incentives for third-country actors to comply with or co-opt the EU's environmental norms, the section will seek to qualify the degree of coercion exerted by this external reach. Recent developments suggest a tendency towards more assertiveness in these techniques.

### *3.2.1 Objectives: How the EU Developed the Ambition to be a Global Standard-Setter*

Internal measures increasingly function to externalize the EU's environmental values outside of its borders. This has come to attract significant scholarly attention in the past decade, yielding the observation that the Union's environmental unilateralism is, at times, more successful than bi- or multilateral measures.[13] Scholarly accounts diverge, however, on the question of whether or not the EU is intentionally seeking to externalize its internal measures. Bradford is most tentative in identifying the Brussels Effect as "an ancillary and largely unintended by-product" of an internal regulatory agenda.[14] Damro posits that "While intentionality is a frequent part of externalization, there may also be an unintentional dimension."[15] Other authors are more accepting of the EU's intentions.[16] Hadjiyianni argues that internal measures with external effects 'are *legally designed to* unilaterally extend the territorial remit of EU legislation.'[17] Schunz only considers internal measures a foreign policy tool when adopted with the express purpose of exerting an external effect.[18] The thesis of 'contingent unilateralism' put forward by Scott and Rajamani is based on the premise that the EU strategically adopts ambitious unilateral market-based measures to exert pressure on multilateral environmental negotiations.[19] Similarly, Dupont, et al., see the externalization of internal measures as a way for the EU to 'manipulate utility calculations'.[20]

Although the central place of the externalization of internal measures in EU external environmental policy may have come about without a preconceived strategy, as early as 2007 the Commission posited that the EU is 'looked upon as the global standard-setter' in its internal market regulation.[21] More recently, the awareness that

[13] As stated by Wouters et al., 'efforts to lead by example have arguably been more successful.' Wouters et al. 2021, p. 354; Kulovesi 2012, p. 115.

[14] Bradford 2020, p. 7.

[15] Damro 2012, p. 690.

[16] Öberg 2020; Cremona 2017; Gstöhl 2008.

[17] In her theorization of IEMEIs. Hadjiyianni 2019, p. 14, [emphasis added].

[18] Schunz 2019, p. 345.

[19] Marín Durán 2020, p. 385; Scott and Rajamani 2012.

[20] Dupont et al. 2018, p. 120.

[21] 'A single market for 21st century Europe' COM(2007) 724 final (Brussels, 20 November 2007).

internal measures can be strategically deployed to enhance the Union's power on the global stage has become broadly shared within EU institutions.[22]

This development is illustrated by the case study of the Renewable Energy Directive (RED). The external dimension of the Directive was tentatively acknowledged in 2009 and has since evolved into an explicit ambition for global leadership. In its proposal for the first iteration of the Directive, the Commission noted that the 'issue of legal framework of third countries must be addressed.'[23] In the eventual legal text, the RED was devoted to the 'development of second and third-generation biofuels in the Community and worldwide'.[24] In the run-up to the 2018 recast, the Commission observes that the Directive has 'also had a spill-over effect worldwide, triggering the adoption of renewable energy policies outside the European Union and helping renewables towards becoming a cost-competitive energy source'.[25] The perceived effectiveness of the externalization of RED I proved fertile ground for the Commission's 'political ambition to be a global leader in renewables' with RED II.[26] In an unmistakable sign to third countries, these ambitions for technological and industrial leadership were codified in the Directive.[27] The Green Deal review is similarly meant to contribute to this leadership.[28] In literature, the RED—and especially its provisions on biofuels—are hence seen 'as an attempt to ensure internal environmental integration with a view to providing a good-practice example to other countries to inspire national action in partner countries or to influence ongoing international negotiations.'[29]

Parallel to the RED, the aim to become a 'global standard-setter' is now explicitly vocalized as integral to recent environmental legislation in the context of the Green Deal. Without prejudice to the fact that externalization may still partly take place unintentionally, this chapter therefore focuses on measures to which this global ambition was considered in their adoption. This includes the Carbon Border Adjustment

[22] Commission, 'Answers to the European Parliament: Questionnaire to the Commission-Designate Paolo Gentiloni, Commissioner-designate for Economy' (October 2020); Parliament, 'A WTO-compatible EU carbon border adjustment mechanism: European Parliament resolution of 10 March 2021 towards a WTO-compatible EU carbon border adjustment mechanism (2020/2043(INI))' OJ C 474/01 (Parliament CBAM resolution), preamble E.

[23] COM(2008) 19 final (Brussels, 23 January 2008).

[24] Recital 66 Directive 2009/28/EC.

[25] Impact Assessment, SWD(2016) 418 final PART 1/4 (Brussels, 30 November 2016), p. 19.

[26] Impact Assessment, SWD(2016) 418 final PART 1/4 (Brussels, 30 November 2016), p. 64.

[27] Recital 2 and 20 Directive 2018/2001.

[28] Although the Commission Proposal did not seek to codify it in the legal text. This variance in the Commission's and the Parliament's approach will be further set out in Section 4. Explanatory Memorandum to COM(2021) 557 final (Brussels, 14 July 2021), 1; Recital 2(a), Amendments adopted by the European Parliament on 17 January 2018 on the Proposal for a Directive of the European Parliament and of the Council on the promotion of the use of energy from renewable sources (recast) P8_TA (2018)0009.

[29] Kulovesi and Muñoz 2011, p. 879.

Mechanism (CBAM),[30] the sustainable finance Taxonomy,[31] and the Deforestation Regulation.[32] For example, the latter is meant to incentivize 'the transition to sustainable supply chains in all producing countries, within or beyond the EU.'[33] This is considered necessary to 'make the EU a credible global standard-setter.'[34]

### 3.2.2 *Qualifying External Effects: How the Scope of Internal Measures is Expanded*

The second constitutive component of the externalization of EU internal measures is their external reach. This reach is the primary embodiment of the external objectives discussed above. It forces third-country actors to take EU measures into account. The qualification 'internal' indicates that measures were adopted unilaterally, on a legal basis which does not fall within the Treaty chapter on EU external action.[35] Notwithstanding, internal measures can extend their reach by including actors or conduct outside of EU borders within their scope. This must be explained by a brief discussion of the *personal* and the *material* scope of measures, with reference to jurisdictional doctrine.

---

[30] Paolo Gentiloni, Commissioner for Economy, has stated that 'As an early mover, the EU can become a standard-setter.' High-Level Conference on Carbon Border Adjustments for Climate on March 23rd, 2021<https://www.youtube.com/watch?v=sh8HUcwj7Xs> at 31:45. See also European Parliament resolution of 10 March 2021 towards a WTO-compatible EU carbon border adjustment mechanism, Preamble N.

[31] The Commission has expressed the hope that the Taxonomy 'will make the EU a global leader in setting standards for sustainable finance.' European Commission, 'Press release: Sustainable Finance and EU Taxonomy: Commission takes further steps to channel money towards sustainable activities' (Brussels, 21 April 2021).

[32] Regulation (EU) 2023/1115 of the European Parliament and of the Council of 31 May 2023 on the making available on the Union market and the export from the Union of certain commodities and products associated with deforestation and forest degradation and repealing Regulation (EU) No 995/2010 [2023] OJ L 150 (Deforestation Regulation).

[33] Explanatory Memorandum to COM(2021) 706 final (Brussels, 17 November 2021), p. 5.

[34] Ibid.

[35] Pursuant to Article 4(1) TEU, the Union is only able to act where the Treaties have conferred upon it a legal competence to do so. The Treaty sections on external action are Title Five of the TEU (Articles 21–46) and Part Five of the TFEU (Articles 205–222), covering amongst others security, trade and development policy. It may moreover be noted that the objectives of EU environmental policy (Article 191 TFEU) and the inherently cross-border nature of environmental protection offers a solid basis for environmental measures with a scope which covers actors or conduct outside of EU borders.

#### 3.2.2.1 Personal Scope

The *personal* scope of measures pertains to the actors which are under a legal obligation to comply. This scope corresponds to what in jurisdictional doctrine is referred to as 'enforcement' or 'adjudication' jurisdiction.[36] Actors within the personal scope of EU measures are thus subject to the enforcement of EU law, for example by EU or Member State agencies, and, in case of non-compliance, can be called before EU or Member State courts.[37]

In several areas of environmental protection, the EU extends the personal scope of measures to any actor who wishes to participate in the EU market. The main EU chemicals (REACH) regulation, for example, applies to anyone producing chemicals within the EU, regardless of their place of establishment.[38] The sustainable finance Taxonomy similarly applies to non-EU undertakings who want to offer financial services on the EU market.[39] Beyond environmental policy, the infamous example of EU measures with an extended personal scope is that of data protection, as data processing managers or subcontractors who are not established within the Union but nevertheless process data of EU citizens must comply with the GDPR.[40]

It follows that an extended personal scope creates a situation where non-EU actors are directly subject to EU regulation. This is deemed intrusive under customary international law. However, as the conduct over which the jurisdiction is exerted is initiated within EU territory, it might be justified under the territoriality principle.[41]

#### 3.2.2.2 Material Scope

In a larger number of cases, the EU adopts measures of which the personal scope is limited to actors established within the EU, but which nonetheless have significant

---

[36] Ryngaert 2015a.

[37] The majority of EU law is applied by national courts, which are subject to the principle of sincere cooperation (Article 4(3) TEU). See Case 14/83 *Von Colson and Kaufmann v Land Nordrhein-Westfalen*, ECLI:EU:C:1984:153, p. 26.

[38] See Article 3(4) of Regulation (EC) No 1907/2006 of the European Parliament and of the Council of 18 December 2006 concerning the Registration, Evaluation, Authorisation and Restriction of Chemicals (REACH), establishing a European Chemicals Agency, amending Directive 1999/45/EC and repealing Council Regulation (EEC) No 793/93 and Commission Regulation (EC) No 1488/94 as well as Council Directive 76/769/EEC and Commission Directives 91/155/EEC, 93/67/EEC, 93/105/EC and 2000/21/EC; See also Biedenkopf 2018; Biedenkopf 2015, pp. 107–136.

[39] 'Frequently Asked Questions on the final reports of the Technical Expert Group on Sustainable Finance (TEG) on the EU Taxonomy and the EU Green Bond Standard (EU GBS)', p. 10.

[40] Article 3 of Regulation (EU) 2016/679 of the European Parliament and of the Council of 27 April 2016 on the protection of natural persons with regard to the processing of personal data and on the free movement of such data, and repealing Directive 95/46/EC (General Data Protection Regulation). Due to the seemingly borderless nature of the internet, the scope of the GDPR has been considered controversial from an international law perspective. See, amongst others, Van Alsenoy 2017; Tene and Wolf 2013.

[41] Dobson 2021, p. 35.

external effects on third-country actors by merit of their material scope. The *material* scope pertains to the conduct or circumstances covered by the measure. In jurisdictional doctrine, this constitutes 'prescriptive' jurisdiction, which is considered less intrusive than the aforementioned 'adjudicative' variant.[42] Notwithstanding, it may effectively force non-EU producers to whom the EU market is of relevance to take EU law into account.

An expansive material scope has come to be a regular occurrence in EU environmental protection. For example, the Timber Regulation requires EU importers of timber products to factor in the circumstances under which the products were produced.[43] In the case of climate change mitigation, non-EU conduct can include greenhouse gasses emitted outside of EU territory. For example, the Renewable Energy Directive (RED) defines sustainability criteria for imported biofuels.[44] These criteria include GHG emissions reductions for installations,[45] as well as environmental protection safeguards related to the land on which biofuels are produced.[46] Both criteria apply regardless of geographical location.[47] The CBAM is another example: strictly speaking, it only places EU importers under a legal obligation, namely to pay a levy over their imports.[48] However, because the levy is contingent upon the amount of non-EU emissions, non-EU producers are effectively subjected to EU legislation.[49] Moreover, if they do not submit their individual emissions data to a central EU register, a punitive average benchmark will apply.[50] Consequently, although non-EU actors do not technically fall within the personal scope of the CBAM, they are subjected to strong economic incentives to both submit their data—which incurs a significant administrative burden—and reduce their emissions.

In terms of customary international law, it is not as apparent how an expanded material scope can be justified. Although some regard the 'territorial trigger' of importation sufficient for an appeal on the territoriality principle,[51] this seems like a stretch of the doctrine, as both the actor (the non-EU producer) and the conduct (the non-EU emissions) are located outside the EU. The most opportune pathway to justify environmental measures with an expansive material scope appears to be based on the effects doctrine, which legitimizes jurisdiction by those affected by the effects of the conduct (i.e. the EU can regulate emissions which exacerbate the consequences of

---

[42] Ryngaert 2015b, pp. 54–55; Gerber 1984, p. 189.

[43] Regulation (EU) No 995/2010 of the European Parliament and of the Council of 20 October 2010 laying down the obligations of operators who place timber and timber products on the market.

[44] Directive (EU) 2018/2001 of the European Parliament and of the Council of 11 December 2018 on the promotion of the use of energy from renewable sources.

[45] Article 29(10) Directive (EU) 2018/2001.

[46] Article 29(6) and (7) Directive (EU) 2018/2001.

[47] Article 29(3), second paragraph, Directive (EU) 2018/2001.

[48] Article 5(1), Regulation (EU) 2023/956.

[49] Dero-Bugny and Motte-Baumvol 2021, p. 56.

[50] Annex IV, Point 4.1 Regulation 2023/956.

[51] Scott 2013, p. 90.

climate change within its borders),[52] or the universality principle, which articulates that the universality of the violated norm justifies any state to exercise jurisdiction.[53]

Although an expansive material scope may seem less intrusive to third-country actors, both the RED and the CBAM have faced fierce opposition and challenges under the law of the World Trade Organization (WTO law). Hence, to enhance our understanding of the impact of the externalization of internal measures on third-country actors, it is necessary to qualify the coercion they exert in greater detail.

#### 3.2.2.3 Qualifying Coercion: How Third-country Actors are Compelled to Comply

Having established that measures have an external reach, the question presents itself what the implications of this reach are for third-country actors. This will depend on other features of a measure's legal design. This chapter distinguishes between financial and regulatory conditionality and between negative and positive conditionality, which will be illustrated with examples.

#### 3.2.2.4 Very Strong Coercion: Financial, Negative Conditionality

Measures with negative *financial* conditionality make access to the EU market conditional to a monetary levy, of which the height is based on non-EU conduct. The CBAM is the first EU instrument which will create financial conditionality. Although it is not yet imposing its cost, the fact that it has been adopted is notable, since its infamous predecessor, the Aviation Directive, never entered into force due to international pressure.[54] The Directive would have required non-EU airlines which land at or depart from an EU airport to pay a price for the emissions of the whole of their flight, also those which were emitted outside of EU territory.[55]

The CBAM, which will price the embedded emissions of imports, was announced in 2019.[56] Again, the initiative was met with fierce backlash from international

52 Ahlstroem v Commission (Wood Pulp) [1993] ECR I-1307, paras 15–18.

53 Ryngaert 2015a, pp. 186–187.

54 The literature on the Aviation Directive saga is ominous. See, amongst many others, Scott and Rajamani (2013); Voigt (2012); Bogojević (2012).

55 Directive 2008/101/EC.

56 Communication from the Commission to the European Parliament, the Council, the European Economic and Social Committee and the Committee of the Regions, 'The European Green Deal' COM/2019/640 final (Brussels, 11 December 2019).

partners.[57] This time, however, the EU showed no signs of withdrawing its initiative.[58] The legislative proposal was presented in July 2021.[59] Moreover, despite their strongly dismissive initial response, global partners seem to have acquiesced to the inevitable adoption of a CBAM. The EU will thus start implementing financial negative conditionality very soon.

#### 3.2.2.5 Strong Coercion: Regulatory, Negative Conditionality

Measures with negative *regulatory* conditionality restrict access to the EU market based on compliance with rules pertaining to conduct outside of EU borders. Rather than a monetary levy, this type of conditionality only creates an administrative burden to ensure and assess whether or not the rules are complied with. Contrary to financial conditionality, the EU has imposed ample measures which create regulatory negative conditionality, amongst which the Seals Regulation,[60] the Timber Regulation,[61] the Illegal, Unreported and Unregulated (IUU) Fishing Regulation,[62] and the monitoring, reporting and verification of MRV Regulation.[63] Pursuant to the latter, for example, any ship which wishes to report to EU ports is required to report on the $CO_2$ emissions of its entire journey. This is meant to enhance transparency and, by extension, environmental norms in the heavily polluting maritime sector.

Such as financial conditionality, regulatory conditionality constitutes a cost of production for those non-EU actors hoping to access the EU market. In Dobson's account, they are hence jointly categorized.[64] However, whilst it is relevant to acknowledge the coercive effects of regulatory conditionality, the strong opposition against the Aviation Directive and the CBAM shows that financial conditionality is deemed considerably more controversial in the international arena. A potential explanation is the fact that their impact on non-EU actors is easier to quantify than a regulatory burden. It is therefore posited here that financial conditionality exerts

---

[57] Joint Statement issued at the conclusion of the 30th BASIC Ministerial Meeting on Climate Change hosted by India on 8th April 2021; 'John Kerry warns EU against carbon border tax' (*Financial Times*, 11 March 2021) <https://t.co/rCldsWbQrk?amp=1> accessed 12 March 2021.

[58] Camilla Hodgso, 'EU rebuffs US concerns over carbon border tax threat' (*Financial Times*, March 31st, 2021).

[59] Proposal for a Regulation of the European Parliament and the Council establishing a carbon border adjustment mechanism COM(2021) 564 final.

[60] Based on Article 114 TFEU.

[61] Based on Article 192 TFEU.

[62] Council Regulation (EC) No 1005/2008 of 29 September 2008 establishing a Community system to prevent, deter and eliminate illegal, unreported and unregulated fishing, amending Regulations (EEC) No 2847/93, (EC) No 1936/2001 and (EC) No 601/2004 and repealing Regulations (EC) No 1093/94 and (EC) No 1447/1999.

[63] Regulation (EU) 2015/757 of the European Parliament and of the Council of 29 April 2015 on the monitoring, reporting and verification of carbon dioxide emissions from maritime transport, and amending Directive 2009/16/EC.

[64] Dobson 2021, p. 50.

the highest degree of coercion. Its ascent in EU policy is hence notable—it signals a new, more assertive era in EU environmental externality.

It may be submitted that non-EU actors are not forced to comply with the CBAM, the MRV Regulation or other measures of negative conditionality. They are free to abandon the EU market in order to avoid compliance. Such a viewpoint, however, negates the reality that the EU currently constitutes the largest consumer market in the world. Whilst this may change over the coming decades, it is not realistic for smaller and larger enterprises alike to evade the EU market solely for regulatory reasons. The costs of compliance do not currently outweigh those of neglecting the EU market altogether.

#### 3.2.2.6 Moderate Coercion: Regulatory, Positive Conditionality

Whereas negative conditionality is a *condition sine qua non* for access to the EU market, positive conditionality rather creates a reward: compliance yields access to a privileged part of the EU market.[65] Otherwise put: negative conditionality functions as a stick, positive conditionality as a carrot. The carrot is often formed by a sustainability label. As these measures do not bar market access altogether, they cause less interference and are less coercive than negative conditionality.[66]

Examples of positive regulatory conditionality are the RED, the sustainable finance Taxonomy Regulation and the Green Bond Standard. The RED establishes criteria for the production and promotion of renewable energy which can be counted towards Member State targets.[67] Energy imports which do not fulfil the RED criteria are still free to enter the EU market. However, they may not count towards the percentage of Member States' energy mixes which the RED mandates must be provided by renewable sources (the target is currently 32% by 2030,[68] but is set to be increased to 42.5% in the context of the Green Deal and RePowerEU).[69] As such, the RED conditions access to the privileged part of the EU energy market designated for sustainable energy sources.

---

[65] This distinction is inspired by Smith 2003. Smith, however, does not yet consider market regulations as instruments of externalization, potentially because their use was not yet commonplace in the early 00's. Damro, who also cites Smith, does, however, recognize their role, adding that they are not sufficiently subject to academic analysis. Damro 2012, p. 691; Smith 2003, p. 57.

[66] Dobson 2021, p. 246.

[67] Initially, Directive 2009/28/EC; Recast, Directive (EU) 2018/2001 of the European Parliament and of the Council of 11 December 2018 on the promotion of the use of energy from renewable sources.

[68] Article 3(1) Directive 2018/2001.

[69] The initial Commission Proposal sought an increase to 40%; in the context of the RePowerEU plan, this was increased to 45%; the Parliament and the Council settled on 42.5%. General Secretariat of the Council, 'Letter sent to the Chair of the European Parliament Committee on Industry, Research and Energy (ITRE)' 10746/21 + ADD 1 (Brussels, 19 June 2023).

Similarly, the Taxonomy does not restrict non-compliant financial products from being freely traded on the EU market. However, they will not be eligible to be designated as 'environmentally sustainable'.[70] The Taxonomy is intended to be incorporated across EU and Member State legislation as a standard for green investments.[71] To the extent that this happens, it will condition access to the privileged, sustainable part of the EU financial market. Economic actors are incentivized to gain this access because sustainable finance products are expected to become more desirable on the financial market. However, an important difference between the Taxonomy and the RED is that the latter creates the privileged part of the market in the measure itself, by designating a legal threshold for the proportion of renewable energy in national energy mixes. By contrast, the coercive effects of the Taxonomy are contingent upon market developments or subsequent legal instruments. Its external reach will fundamentally change when a minimum threshold for Taxonomy-compliant financial products is codified. Although the EU had initially intended to use the Taxonomy as the sustainability standard for the Covid Recovery Fund, which requires Member States to spend at least 37% of the funds on investments supporting the green transition,[72] this did not happen. Similarly, the EU Green Bond Standard and Climate Benchmarks are not yet aligned with the Taxonomy,[73] leading to a patchwork of sustainable finance standards which will dilute the impact of a single instrument. It follows that the EU's sustainable finance instruments currently constitute a weaker form of positive conditionality, as opposed to the stronger conditionality in the RED. However, future regulatory developments might amend this.

This section has established that the EU now unabashedly articulates its ambition to be a global standard-setter. It explained how internal measures include non-EU actors and conduct within their scope and offered insight into the ways and degrees to which third-country actors may be compelled to comply with and co-opt the EU's example. Having identified these legal mechanisms, the chapter will shift to an outside-in perspective to identify components in the legal design of internal measures which contribute to the effectiveness of their externalization. This perspective facilitates the institutional analysis in the fourth section, which dissects the legislative institutions' approaches to each of these components.

---

[70] Article 2(1) Regulation (EU) 2020/852.

[71] Article 1(2)(a) Regulation (EU) 2020/852.

[72] Regulation (EU) 2021/241 of the European Parliament and of the Council of 12 February 2021 establishing the Recovery and Resilience Facility.

[73] Regulation (EU) 2019/2089 of the European Parliament and of the Council of 27 November 2019 amending Regulation (EU) 2016/1011 as regards EU Climate Transition Benchmarks, EU Paris-aligned Benchmarks and sustainability-related disclosures for benchmarks; Regulation (EU) 2019/2089 and Proposal for a Regulation of the European Parliament and the Council on European green bonds, COM/2021/391 final.

## 3.3 How to Ensure the Effectiveness of Environmental Externalization

This section explores the criteria for effective externalization. These criteria are meant to serve the analysis in the fourth section, which will assess how the EU's legislative institutions approach them in the adoption of internal measures meant to externalize the environmental integration principle. The section synthesizes the main scholarly findings on effective EU externalization and adds its own reflections based on recent case studies. The section first presents a categorization of the kinds of effectiveness criteria, after which it takes a closer look at the criteria which are subject to legislative decisions in turn.

### *3.3.1 Categories of Effectiveness Criteria*

The literature on EU externalization has been overwhelmingly concerned with effectiveness. Despite this attention, conflicting accounts remain. Numerous authors have made the case of the powerful global presence of the EU internal market *acquis*,[74] as well as the role of internal measures in its environmental leadership.[75] Counter perspectives are offered by those who either question the 'frequent overestimation' of the effectiveness of the EU's asserted leadership,[76] especially regarding its (in)ability to externalize norms and values.[77] Overall, the selection bias in either category is subject to scrutiny.[78]

Upon a closer look, three categories of criteria for effectiveness emerge. The first pertains to external factors, be they regulatory, political or economic, which are almost entirely out of the EU's control.[79] As per Bretherton and Vogler, this category essentially entails the question of whether or not the EU has an *opportunity* to act.[80] Examples of criteria within this category include other actors' resources,[81] their regulatory capacity and the institutional density at the multilateral level.[82]

The second category focuses on the EU's institutional and economic characteristics at a macro level, which are partly within the EU's control but at such a macro level that they go far beyond the design of individual measures. For example, the EU's large internal market and its relative accessibility to foreign firms is often considered

74 Anu Bradford 2020; Cremona 2017; Damro 2012; Gstöhl 2008.

75 Wurzel and Connelly 2016, p. 13; Parker and Karlsson 2015; Kulovesi 2012.

76 Newman and Posner 2015, p. 1321.

77 Lavenex 2022; Poletti and Sicurelli 2018.

78 Young 2015, p. 1240.

79 Newman and Posner 2015, p. 1320.

80 Bretherton and Vogler 2013, p. 378; Bretherton and Vogler 2006.

81 Quaglia 2014, p. 427.

82 Newman and Posner 2015, p. 1320.

a primary feature of its regulatory leverage. Importantly, this criterion is relative—it co-depends on the home market and the adjustment cost of the third country.[83] This relativity partly explains why the EU is comparatively less effective in exporting financial market standards—the barriers to enter the EU capital market are considerably more burdensome than those for goods.[84] Moreover, in the case of climate, leverage also includes the EU's share of global emissions—global actors with a large share of global emissions have more negotiating power than smaller emitters. From this perspective, the EU's ability to export climate norms is set to decrease as it approaches net zero. However, these economic and climate dynamics are 'not a sufficient predictor of regulatory bargaining leverage or influence'.[85] Institutional characteristics are pivotal as well. The EU's regulatory capacity to set and enforce norms,[86] including the amount of internal actors with an interest in administrative rule-making and externalization,[87] are frequently considered its most empowering feature as a global standard-setter.

The third category pertains to the characteristics of internal measures and will be the focus of this chapter because these characteristics are subject to decisions made in the legislative process. They include the timing of measures; their stringency, or coherence with the non-economic norms they seek to export; and their relation to international law. Each criterion is discussed in turn.

### *3.3.2 Timing and Innovation: The EU as a First-mover*

The timing of measures is relevant as it offers the EU a chance to benefit from a 'first-mover advantage'.[88] This chance is especially opportune where issues have been subject to little global regulation. Timing is hence also contingent upon the global regulatory landscape.[89] Notwithstanding, it remains up to the EU to make use of opportunities where regulatory gaps arise. It has done so successfully on a number of occasions, for example with its chemicals regulation and on forestry standards,[90]

[83] Bradford 2020, p. 27.

[84] Another explanation is the 'elasticity' of the market: it is much easier to externalize standards on markets which are geographically fixed, such as goods, than those which are more mobile, such as capital or maritime shipping. Bradford 2020, p. 32.

[85] Newman and Posner 2015, p. 1323.

[86] Bradford 2020, p. 32.

[87] Bradford 2020, p. 41.

[88] Quaglia 2014, p. 427; Nicolaïdis and Egan 2001, p. 457.

[89] Newman and Posner 2015, p. 1320.

[90] Regulation (EC) No 1907/2006 of the European Parliament and of the Council of 18 December 2006 concerning the Registration, Evaluation, Authorisation and Restriction of Chemicals (REACH); Regulation (EU) No 995/2010 of the European Parliament and of the Council of 20 October 2010 laying down the obligations of operators who place timber and timber products on the market.

both of which have been extensively externalized.[91] Another legal instrument which the EU has actively sought to externalize is its pioneering Emissions Trading System, albeit with variable results.[92] In the case of the Taxonomy, the EU was initially similarly a regulatory innovator—but it will be discussed below how this potential was foregone due to internal divisions.

The CBAM is perhaps the EU's most innovative environmental measure so far. It would be the first carbon border measure ever implemented. Accordingly, the EU has always hoped that it would set a global standard.[93] However, as noted, the EU's initiative was hardly welcomed by global partners. The frosty reception of the CBAM illustrates that timing is a fickle thing. The aim is to lead by example without being a preference outlier (i.e. failing to trail third countries along by treading too far ahead of the herd).[94] It may be posited that this balance was lost in the case of the 2009 Aviation Directive, which the EU never proceeded to implement. However, the pitfalls of being a preference outlier are predominantly observed in multilateral contexts, where consensus is achieved by voting.[95] Despite the initial backlash to the CBAM, a number of third countries are now considering similar measures.[96] It thus appears as if the EU's initiative was worth the leap, or its timing was more carefully considered this time around.

### *3.3.3 Stringency and Comprehensiveness: The EU as a Trailblazer*

A central conclusion of the literature on regulatory diffusion is that strict standards are most likely to set global examples.[97] The rationale behind this is that exporting companies will choose to comply with the most stringent regulatory framework because it would be too costly to comply with various different frameworks. This effect could have been witnessed in the case of the GDPR,[98] as well as on several environmental issues.[99] Importantly, stringency is also relevant from an internal, EU law perspective—it satisfies the need to observe the environmental integration

[91] Simintzis 2022, p. 57; Biedenkopf 2018; 2015, pp. 107–136; Wouters et al. 2015, p. 38; Zeitlin 2015, p. 148.

[92] Unger 2021; Borghesi et al. 2018.

[93] 'As an early mover, the EU can become a standard-setter.' Paolo Gentiloni, High-Level Conference on Carbon Border Adjustments for Climate on March 23rd, 2021 <https://www.youtube.com/watch?v=sh8HUcwj7Xs> at 31:45. See also Geneviève Pons, Europe Jacques Delors, speaking at EU Trade Policy Day (April 26, 2021) <html> at 14:01; European Parliament resolution of 10 March 2021 towards a WTO-compatible EU carbon border adjustment mechanism, Preambles N

[94] Renckens et al. 2017, pp. 1432–1433.

[95] Young 2014.

[96] Amongst others the US, Turkey and India.

[97] Vogel 1997; Bradford 2020, p. 37.

[98] De Ville and Gunst 2021.

[99] For example the aforementioned instruments on chemicals and timber.

principle.[100] Moreover, in some cases, stringency can compensate for the less ideal timing of measures. A good example of this is the RED. The US and Brazil were well ahead of the EU in creating a sustainable biofuels policy.[101] Nonetheless, the EU's standards were and remain the most stringent by comparison. Consequently, the RED has led to the enhancement of environmental protection regulation in various countries which are big importers to the EU.[102] A similar effect might take hold with the Taxonomy. The external relevance of stringency also enhances the importance of evaluating the extent to and manner in which EU legislative institutions contribute to the regulatory capacity and political will to set stricter environmental standards.[103]

However, stringency is a balance to strike: there is such a thing as being too stringent.[104] Four main pitfalls prevail. First, targets or standards which are overly ambitious in a global context can undermine the ambition of externalization.[105] This risk is closely interrelated with that of being a 'preference outlier' in terms of timing. A second risk is that of overly complex standards compromising the feasibility of non-EU compliance with EU standards. In the case of the Taxonomy, for example, the overcomplication of the classification system is said to impair its externalization.[106] A third factor is scope. An overly wide scope, covering sectors where implementation will be technically complex, similarly compromises effectiveness. Fourth and finally, the environmental stringency of standards must be balanced against their compatibility with international law. This final factor deserves its own subsection.

### *3.3.4 International Law Compatibility: The EU as Defender and Challenger of the Rule-Based Order*

Because EU environmental externality conditions market access, it is subject to international trade law. As it transcends borders, it is subject to customary international law on jurisdiction. Compatibility with these regimes is important for the perceived legitimacy of EU measures,[107] especially given the EU's Treaty commitments to the rule of law and international law,[108] as well as its self-proclaimed status as a 'credible supporter' of the multilateral rule-based trade order.[109]

---

100 Bretherton and Vogler 2013, p. 383.

101 Renckens et al. 2017, pp. 1432–1433.

102 Such as Argentina, Brazil, Indonesia, and Malaysia. Commission Staff Working Document SWD 2013 102 final, 17, 32.

103 Anu Bradford 2020, p. 32.

104 Drezner 2007.

105 Bäckstrand and Elgström 2013; Oberthür and Rabitz 2014.

106 Naomi O'Leary, 'The Brussels defect: when compromise goes wrong' (*Irish Times*, 10 February 2022).

107 Diez and Pace 2011, pp. 210–225.

108 Article 2 TEU.

109 COM(2021) 66 final, 6.

Due to its relative effectiveness, WTO law is of particular importance.[110] The WTO-compatibility of EU environmental externality is far from self-evident.[111] The RED has been subject to particularly strong resistance.[112] The Panel Report on the two most pressing challenges by Malaysia and Indonesia is yet to be finalized.[113] The CBAM can similarly be expected to be challenged under the WTO. The ambiguity of the WTO-compatibility of carbon border measures is, in fact, a primary reason why they have not yet been implemented. These tensions have been extensively explored in legal literature.[114] A synthesis of this literature is beyond the scope of this chapter. However, WTO limitations will be discussed where relevant to the analysis of the legislative process of internal measures in the subsequent section.

The relevance of international law compatibility is not equally relevant in all cases of environmental externalization. For example, positive conditionality without market thresholds, such as the Taxonomy, is considerably less likely to be hindered by WTO constraints. However, the difference might also partly be a matter of market elasticity and political economy. Third countries will be more likely to initiate proceedings where their domestic markets are considerably affected by EU internal measures.

## 3.4 To What Extent do the EU's Legislative Institutions Prioritize the Effectiveness of Environmental Externalization via Internal Measures?

This chapter so far has established that the EU adopts internal measures with the objective of externalizing the environmental integration principle. This external objective is more prominent than ever in the EU's legislative agenda. This prominence raises the question of whether it is prioritized in the legislative process. To what extent do the EU's legislative institutions act in accordance with the objective of externalization in drafting internal environmental measures? The previous section identified several criteria which facilitate the successful dissemination of internal measures to third-country jurisdictions. This section analyzes the approach of each legislative institution to these criteria in the legislative process. The analysis will illuminate the internal dynamics of the EU as an external actor in its capacity to externalize internal environmental measures.

---

[110] Although it is interesting to note that it was not the main focus in relation to the Aviation Directive. Howse 2012, pp. 29–30.

[111] Geraets and Natens 2014; Fishman and Obidzinski 2014.

[112] Stattman and Gupta 2015.

[113] WT/DS600/9, *European Union and Certain Member States—Certain Measures concerning Palm Oil and Oil Palm Crop-Based Biofuels—Communication from the Panel* (24 February 2023); WT/DS593/12, *European Union—Certain Measures concerning Palm Oil and Oil Palm Crop-Based Biofuels—Communication from the Panel* (24 February 2023).

[114] Mehling et al. 2019; Holzer 2014; Hillman 2013; Goh 2004.

The section shall first sketch the roles of the Commission, the Council and the Parliament in the EU's externalization of internal environmental measures. Subsequently, it will inquire how each institution fulfils this role with regards to the codification of the externalization ambition; the issue of timing; the environmental stringency of measures; and their international law compatibility.

### *3.4.1 The Legal Foundations of the Roles of the Legislative Institutions in the EU's Externalization of Internal Measures*

The institutional focus of the study of the EU as an external actor has been on the Council and the EEAS. This logically follows from the prevalence of unanimity voting in the Council in the CFSP and the role of the High Representative for Foreign and Security Policy (HRFSP). More specifically as regards the external dimension of environmental integration, which was identified as an emanation of the principle of coherence, Article 21(3) Treaty on European Union (TEU) singles out the Commission and Council, assisted by the HRFSP, to ensure coherence in the EU's external action.[115]

Importantly, however, this *lex specialis* is not applicable to the EU's externalization of internal measures. In this capacity, the EU's legislative institutions define its external actorness.[116] The Treaties contain a couple of indications of their role in externalizing environmental integration via internal measures. As per Article 13(1) TEU, the Union's entire institutional framework 'shall aim' to ensure coherence. Article 7 TFEU provides that this ought to take place 'in accordance with the principle of conferral of powers'. The first phrase of Article 13(2) TEU also articulates the responsibility of each institution to respect the principle of conferral,[117] whilst adding 'and in conformity with the procedures, conditions and objectives set out in [the Treaties].' Pursuant to these provisions, the analysis shall focus on the roles allocated to and the subdivision of powers between the institutions in the Ordinary Legislative Procedure (OLP).[118]

The Commission, the Council and the Parliament are the main actors in the OLP. The European Council does not have an official position in the process. However, it exerts influence over the Commission's legislative agenda by setting general policy priorities, as well as over the Council's position in cases when national divergences

[115] Franklin 2011.

[116] Since Article 11 TFEU refers both to the 'definition' as well as the 'implementation' of EU policy, executive institutions such as EU agencies also play a role in pursuing climate coherence. Nonetheless, as this chapter focuses on the adoption of legislative acts, it focuses on the legislative institutions. Marín Durán and Morgera 2012, p. 31.

[117] 'Each institution shall act within the limits of the powers conferred on it in the Treaties'.

[118] As the majority of the examples discussed in this chapter is adopted based on Articles 114(1), 192(1) and 194(2) TFEU, it shall focus on the main actors in the ordinary legislative procedure.

cannot be resolved at the Council level.[119] In this capacity, it may be noted that it has been increasingly interested in EU climate policies due to global leadership ambitions. In the interrelation between the three legislative institutions, the Commission's position is characterized by its exclusive right to draft legislative proposals.[120] Moreover, Article 114(3) TFEU allocates particular responsibility to the Commission in environmental integration in the internal market policy. As such, the Commission bears an obligation 'to take action where action otherwise would not have been taken'.[121] However, as per Article 114(3) TFEU, the European Parliament and the Council are similarly addressed to endeavor for environmental integration 'within their respective powers.' Their most prevalent co-legislative responsibility hence lies in integrating environmental protection objectives in its review of legislative proposals put before them by the Commission.[122] Importantly, this obligation is not just procedural.[123] Read in light of Article 13(1) TEU, it is evidently also a substantive one. The subsequent subsections shall assess how these roles are fulfilled in cases of environmental externalization.

### 3.4.2 Objectives

The EU has increasingly vocalized its ambition to be a global standard-setter with internal market regulation.[124] Generally, this ambition is most expressly prioritized by the Commission. By exception, in the case of the RED, the Parliament and the Council were the first actors to frame the measure in the context of EU leadership. As early as its first Report on the proposal for the 2009 iteration of the Directive, the leading ITRE committee cites existing and aspired EU leadership on renewables as its motivation for amendments.[125] The Parliament moreover sought in vain to codify the Directive's contribution to 'industrial leadership'.[126] Similarly, the Council noted as early as 2008 that 'the promotion of the leadership and competitiveness of European

---

119 Wurzel et al. 2019, p. 253.

120 Article 294(2) TFEU.

121 Which would mean just to check 'whether environmental concerns had been considered' by the Commission. Sjåfjell 2015, p. 60.

122 Article 294(3)-(15) TFEU.

123 Sjåfjell 2015, p. 60.

124 European Commission, 'Commission Staff Working Document: The external dimension of the Single Market Review', SEC(2007) 1519, p. 9.

125 It both states that 'Europe is world leader in renewable energy' and gives conditions to lead an industrial revolution on sun technologies. Committee on Industry, Research and Energy, 'Report on the proposal for a Directive of the European Parliament and of the Council on the promotion of the use of energy from renewable sources' A6-0369/2008 (Brussels, 26 September 2008).

126 Amendment 83, Committee on Industry, Research and Energy, 'Report on the proposal for a Directive of the European Parliament and of the Council on the promotion of the use of energy from renewable sources' A6-0369/2008 (Brussels, 26 September 2008).

industries in those energy technologies should be an important element.'[127] In the 2018 recast, however, the external effects of the Directive have been duly noted and become part of its objectives also to the Commission, which cites its "spill-over effect worldwide, triggering the adoption of renewable energy policies outside the European Union.'[128] The Commission is hence as bold as to refer to its 'political ambition to be global leader on renewables'.[129] This ambition culminated in two codifications of the Directive's contribution to EU leadership.[130]

The assertiveness of the Commission continues undoubtedly in the majority of Green Deal environmental externality. In the case of the Taxonomy, the Commission is the only institution who appears to have consciously considered the ambition of global standard-setting.[131] Neither of the co-legislators make reference to it in their approach to the instrument. The Commission's proposal for the Deforestation Regulation is framed in the objective 'for the EU to set new standards for sustainable growth and use its economic weight to shape international standards that are in line with EU environmental and climate ambitions.'[132] In the new proposal for an eco-design Regulation, the Commission even seeks to codify that:

> 'the Union internal market provides a critical mass that is able to influence global standards on product sustainability and product design. This Regulation should therefore play a significant role towards achieving several targets established under the United Nations' Sustainable Development Goals of the UN's 2030 Agenda for Sustainable Development 'Responsible consumption and production', both inside and outside the Union.'[133]

However, the cases of RED and CBAM demonstrate that external limitations have come to impact the EU's formulation of external objectives especially in cases which encounter resistance within the WTO. To be eligible for the environmental exception under Article XX(g) GATT,[134] it is pivotal for the measure to genuinely pursue environmental objectives.[135] The appearance of economic or third country policy change objectives are detrimental to this test. As noted, the RED has been and still is subject to several WTO challenges. Not unsurprisingly, the Commission has hence become significantly less vocal on its global ambition. In the Green Deal amendment, it solely notes that the Directive can 'contribute to the EU's technological

---

[127] Press Release: 2854th Council meeting Transport, Telecommunications and Energy C/08/45 (Brussels, 28 February 2008), p. 9.

[128] Impact Assessment, SWD(2016) 418 final PART 1/4 (Brussels, 30 November 2016), p. 19.

[129] Impact Assessment, SWD(2016) 418 final PART 1/4 (Brussels, 30 November 2016), p. 64.

[130] Recital 3 and 20 Directive 2018/2001.

[131] European Commission, 'Press release: Sustainable Finance and EU Taxonomy: Commission takes further steps to channel money towards sustainable activities' (Brussels, 21 April 2021) <html> accessed 2 November 2021

[132] Explanatory Memorandum to COM(2021) 706 final (Brussels, 17 November 2021), p. 63.

[133] Recital 8, COM(2022) 142 final (Brussels, 30 March 2022).

[134] Which is necessary if the measure violates core WTO provisions, likely in the case of CBAM.

[135] Appellate Body Report, *United States—Import Prohibition of Certain Shrimp and Shrimp Products*, WT/DS58/AB/R (November 6, 1998).

and industrial leadership'.[136] Only the Parliament still seeks to codify references to leadership.[137]

The case of CBAM similarly displays consideration of external limitations to its objectives. It is an interesting case since combating carbon leakage, its main objective, is a double-edged sword: it inherently serves both environmental and economic objectives.[138] Nonetheless, all legislative institutions expressly frame it as an environmental, instead of an economic measure.[139] Moreover, whilst the Commission notes EU leadership as a justification for the measure, as it 'must go hand in hand with bold domestic action,'[140] it steers away from formulating third-country policy change as an objective. By contrast, the co-legislators are considerably less cautious of the external dimension. The Parliament seeks to frame the CBAM as 'an effective measure to lower emissions in third countries.'[141] Both co-legislators moreover want to codify that the EU will labor for the establishment of a so-called climate club, 'in order to promote the implementation of ambitious climate policies in all countries and pave the way for global carbon pricing.'[142]

It follows that the Commission is the most vocal on the global ambition of environmental externality, but also most cautious of the need to balance this with international law compatibility. The co-legislators are less consistent in their articulation of global standard-setting objectives, but also less in tune with external limitations. These findings align with the Commission's role as the main initiator and architect of measures, as well as the EU's main representation at the WTO. It could also be posited that global leadership is cited as an argument to gather co-legislative support for environmental standards.

---

[136] Explanatory Memorandum to COM(2021) 557 final (Brussels, 14 July 2021), p. 1.

[137] Seeking to codify that the Directive is meant to contribute to 'opportunities for [...] continued European leadership'. Recital 2(a), P8_TA (2018)0009.

[138] Carbon leakage refers to the risk that businesses respond to climate measures by relocating their activities to other regions with laxer regulation, resulting in no net global emissions reductions and a loss of competitiveness.

[139] WTO-compatibility. Explanatory Memorandum, COM(2021) 564 final, 1.

[140] COM(2021) 564 final (Brussels, 14 July 2021), p. 3.

[141] Amendment 10, Amendments adopted by the European Parliament on 22 June 2022 on the proposal for a regulation of the European Parliament and of the Council establishing a carbon border adjustment mechanism (COM(2021)0564—C9-0328/2021—2021/0214(COD)) TA/2022/0248.

[142] The Council understands this as 'the parallel setting up of an alliance of countries with carbon pricing instruments or other comparable instruments.' Recital 54a, Council of the European Union, 'Draft regulation of the European Parliament and of the Council establishing a carbon border adjustment mechanism—General approach' 7226/22 (Brussels, 15 March 2022).

### 3.4.3 *Timing*

The matter of timing is most invariably the prerogative of the Commission because of its monopoly on submitting legislative proposals. The Commission does not have full freedom over its legislative agenda: it is meant to act within the 'general political directions and priorities' set out by the European Council,[143] which represents Member State *ex ante* interests in the EU's legislative agenda. Nonetheless, these directions will generally not be precise enough to evoke the legislative details which determine how a proposal cleverly navigates the global regulatory landscape, which remains up to the Commission.

The only sidenote to the Commission's prerogative is the Parliament's right to, by a majority of members, request the Commission to submit a legislative proposal 'on matters on which it considers that a Union act is required for the purpose of implementing the Treaties'.[144] Such requests are not binding and infrequent. However, the Parliament has used its powers to impact the timing of environmental externality by the adoption of Resolutions urging legislative action. Before the adoption of the Commission Proposal, it showed its strong support for and indicated its preferences on the design of a CBAM.[145] Previously, it signaled its prioritization of the place of palm oil in the RED.[146] More indirectly, the Parliament can indicate its policy priorities in debates and through the agendas of political groups. For example, the legislative fervor with which the Green Deal was presented by the Von der Leyen Commission is a result of the desire to honor the outcome of the parliamentary elections.[147]

The Council has little formal means of impacting the timing of the initiation of the legislative process. However, a collection of members have occasionally taken a stance on the urgency of a particular proposal.[148] It follows that although the co-legislators have considerably less influence on the timing of the proposal of measures, they can be seen to have used the powers at their disposal to speed up especially the CBAM.

Crucially, however, the co-legislators are pivotal to the speed of the legislative process, which determines the timing of the eventual *adoption* of measures. Institutional disagreement on environmental measures can cause significant delay, inhibiting the EU's ability to set global standards. This can be witnessed in the case of the Taxonomy. Whilst the Commission harbored the ambition that the Taxonomy

---

143 Article 15 TEU.

144 Article 225 TFEU. Another sidenote is the Citizen's Initiative, Article 11(4) TEU and Regulation 2011/211.

145 European Parliament resolution of 10 March 2021 towards a WTO-compatible EU carbon border adjustment mechanism (2020/2043(INI)).

146 European Parliament resolution of 4 April 2017 on palm oil and deforestation of rainforests (2016/2222(INI)).

147 And the failure of the Spitzenkandidat system. Jordan (ed) (2021), p. 368.

148 Gernot Blümel et al., 'To fight climate change, fight carbon leakage' (*Politico*, March 23, 2021).

'will make the EU a global leader in setting standards for sustainable finance',[149] the momentum to realize this ambition seems to have passed. This is due to internal divisions on the adoption of its delegated acts. When the Commission submitted the initial draft of the so-called Climate Delegated Act (CDA), 10 Member States threatened to block its adoption if gas were not eligible as an environmentally sustainable activity.[150] The Commission finally decided to wholly omit gas and nuclear energy from the main CDA.[151] Whereas the latter was delayed by a year, the separate CDA for gas and nuclear had a further year's delay.[152]

Due to the Council's resistance, the innovative potential of the EU Taxonomy has been surpassed by several jurisdictions, including China, Russia and South Korea. These countries completed the adoption of their sustainable investment taxonomies without much regard to the EU standard. In an apparent recognition of the fact that the EU Taxonomy will not be able to ensure its own externalization, the EU has cooperated with China to draft a 'global common ground taxonomy' (GCGT) in the International Platform on Sustainable Finance (IPSF).[153] The GCGT is intended to identify areas of overlap and divergence between the EU and the Chinese taxonomies,[154] and exemplifies the EU's inability to rely on the unilateral dissemination of its standards. Regardless of this momentum foregone, its Taxonomy may still bear some example-setting effects by merit of its stringency, discussed in the subsequent subsection.

---

[149] European Commission, 'Press release: Sustainable Finance and EU Taxonomy: Commission takes further steps to channel money towards sustainable activities' (Brussels, 21 April 2021) <html> accessed 2 November 2021.

[150] Frédéric Simon, 'Brussels postponed green finance rules after 10 EU states wielded veto' *Euractiv* (Brussels, 18 January 2021) <html> accessed 7 November 2021.

[151] Commission Delegated regulation (EU) 2021/2139 of 4 June 2021 supplementing Regulation (EU) 2020/852 of the European Parliament and of the Council by establishing the technical screening criteria for determining the conditions under which an economic activity qualifies as contributing substantially to climate change mitigation or climate change adaptation and for determining whether that economic activity causes no significant harm to any of the other environmental objectives.

[152] Commission Delegated Regulation (EU) 2022/1214 of 9 March 2022 amending Delegated Regulation (EU) 2021/2139 as regards economic activities in certain energy sectors and Delegated Regulation (EU) 2021/2178 as regards specific public disclosures for those economic activities (Text with EEA relevance).

[153] European Commission, 'International Platform on Sustainable Finance' <https://europa.eu/!CP69Y> accessed 7 November 2021.

[154] European Commission, 'Questions and Answers: Taxonomy Climate Delegated Act and Amendments to Delegated Acts on fiduciary duties, investment and insurance advice' (Brussels, 21 April 2021) <html> accessed 7 December 2021; 'Common Ground Taxonomy Table' <html> accessed 7 December 2021.

### 3.4.4 Stringency

The stringency and environmental ambition of EU standards is the second main factor which contributes to effective externalization. Contrary to the question of timing, the legislative institutions are equally able to pursue environmental stringency. Nonetheless, they all take their own approach. On balance, the Commission's approach is characterized by caution in balancing environmental ambition with implementability, whilst the Parliament relatively unabatedly pursues stringent standards and the Council is most apprehensive.[155] These approaches will now be discussed in turn.

The Commission sees environmental ambition as a key feature of the EU's ability to set global standards. At the same time, it seeks to keep standards 'implementable'.[156] As such, it most apparently expresses an awareness of the need to balance ambition with its practical pitfalls. The balancing act between ambition and feasibility, effectiveness and international law compliance is hence reflected in the Commission's approach to several features of the legal design of internal measures meant for externalization. At times, the balancing act falls in favor of environmental effectiveness. For example, as per the Commission's proposal, the price of the CBAM is based on actual emissions of products, instead of average benchmarks.[157] Whilst this significantly enhances the administrative burden for exporting producers, the Commission argues this is outweighed by environmental effectiveness.[158]

However, the Commission often prioritizes implementability. For example, whilst CBAM would be more environmentally effective as its sectoral scope increases, the Commission preferred a smaller scope limited to a number of ETS sectors.[159] Similarly, it proposed that indirect emissions—which are considerable, but more technically complex to calculate—will only be covered 'after the end of a transition period and upon further assessment'.[160] It took a similar position on the question of including indirect land use change (ILUC) emissions in the RED. These emissions take place where croplands are transformed to produce crops for biofuels, which forces food and feed production for other purposes to relocate.[161] ILUC can also cause biodiversity loss and deforestation.[162] When the Commission fulfilled its obligation to review the future inclusion of ILUC emissions in the RED,[163] it concluded that this

---

155 In governance literature, this role is framed as 'entrepreneurial activities' on environmental policy. Jordan 2021, p. 360.

156 Explanatory Memorandum to COM(2021) 706 final (Brussels, 17 November 2021), 5.

157 Article 32, COM(2021) 564 final; Article 7 Regulation 2023/956.

158 COM (2021) 564 final, p. 9.

159 Sectors are selected by their high emissions, vulnerability to carbon leakage and the moderate complexity of calculating embedded emissions. Preamble 26–37; Article 2, COM(2021) 564 final.

160 Recital 17, COM(2021) 564 final.

161 IPCC (2020).

162 IPCC (2020).

163 Recital 85, Directive 2009/28/EC.

'would require major industrial adjustment which does not seem achievable [..*and*..] would not take into account the limits of the modeling'.[164]

Finally, in some cases the Commission chooses not to prioritize environmental effectiveness for internal, rather than external reasons. For example, in its proposal the CBAM will temporarily apply alongside free allowances under the ETS, of which the environmental effectiveness has been frequently scrutinized.[165] Especially contentious is the aforementioned separate CDA in the Taxonomy, which classifies gas and nuclear energy as 'environmentally sustainable'.[166] The Commission proposed to place the emissions threshold below which gas can be considered environmentally sustainable at 270g CO2e/kWh, or 550kgCO2e/kWH over a 20 year average.[167] This was controversial as the threshold for all other activities is set at 100 CO2e/kWh.[168] Moreover, there are no guarantees on the future emissions reductions which can be claimed by installations with emissions over 550kgCO2e/kW. This appears at odds with the three Taxonomy provisions which compel the Commission to account of the long- and short-term impacts and the life-cycle assessment of activities,[169] crucial to maintaining environmental effectiveness.[170]

On all aforementioned counts, the Commission's environmental ambition is outraced by the Parliament, which is less mindful of implementability, international law compliance or the preference outlier risk. For example, it proposed a considerably wider scope for the CBAM.[171] It also pursues greater certainty of a pathway towards the inclusion of all ETS sectors,[172] as well as indirect emissions.[173] Moreover, it sought to limit the use of default values for emissions calculations in electricity.[174] For all other sectors, it proposed that the default value which applies in the absence of product-specific data is based not on the sectoral average of the

---

164 COM, SWD(2012)343 final, p. 8.

165 Article 31, COM(2021) 564 final.

166 Commission Delegated Regulation (EU) 2022/1214 of 9 March 2022 amending Delegated Regulation (EU) 2021/2139 as regards economic activities in certain energy sectors and Delegated Regulation (EU) 2021/2178 as regards specific public disclosures for those economic activities (Text with EEA relevance).

167 Annex I to C(2022) 631 / 3, Section 4.29, 3.1 (b)(1).

168 Commission Delegated Regulation (EU) 2021/2139 of 4 June 2021 supplementing Regulation (EU) 2020/852 of the European Parliament and of the Council by establishing the technical screening criteria for determining the conditions under which an economic activity qualifies as contributing substantially to climate change mitigation or climate change adaptation and for determining whether that economic activity causes no significant harm to any of the other environmental objectives (Text with EEA relevance).

169 Article 19(1)(b) and (g) TR.

170 EU Platform on Sustainable Finance, 'Response to the Complementary Delegated Act' (21 January 2022), p. 2.

171 Also including organic chemicals, plastics, and hydrogen.

172 By committing the Commission to submit a legislative proposal to this end by 2030. Amendments 9, p. 22.

173 Amendment 15, TA/2022/0248.

174 Amendment 30, TA/2022/0248.

exporting country, but on the 10% worst-performing installations.[175] In the RED, the Parliament pursues the highest national targets.[176] It has moreover initiated the EU's ambition on mandatory targets for advanced biofuels, which are reputed to have a better environmental impact,[177] and was the only institution to favor including ILUC in emissions accounting.[178] Barring some detailed exceptions—certain GHG emissions savings criteria,[179] and the minimum threshold for the applicability of GHG emissions savings for heating, electricity and cooling—[180]the Parliament's environmental ambition thus outruns that of its co-legislators in the vast majority of cases. This was most evident in the case of the Taxonomy. The co-rapporteurs of the TR,[181] as well as the chairs of the ECON/ENVI committees that objected to the CDA on gas and nuclear.[182] Although the absolute majority necessary to reject the act in plenary was just missed,[183] one member of Parliament is now pursuing legal action.[184]

Of the three institutions, the Council tends to be most tentative on the environmental ambition of measures. This has been attributed to Member State concerns on the economic implications of environmental policy.[185] For example, the Council sought to subject the application of the CBAM to a threshold of EUR150, under

[175] Amendment 181, TA/2022/0248.

[176] In 2018, Parliament wanted the cap to be at 35%, but it ended up at 32%. Article 3(1) Directive 2018/2001; https://www.europarl.europa.eu/doceo/document/TA-8-2018-0009_EN.html?redirect.

[177] Article 3(4)(e) Directive 2009/28/EC, as amended by Article 2(2)(b)(iv) Directive 2015/1513.

[178] Renckens et al. 2017, p. 1438.

[179] Article 29(10)(d) Directive 2018/2001.

[180] Whereas the Commission proposed to lower this threshold from 20MW to 5MW, the Parliament favoured 7.5MW. Recital 35, Amendments adopted by the European Parliament on 14 September 2022 on the proposal for a Directive of the European Parliament and of the Council amending Directive (EU) 2018/2001 of the European Parliament and of the Council, Regulation (EU) 2018/1999 of the European Parliament and of the Council and Directive 98/70/EC of the European Parliament and of the Council as regards the promotion of energy from renewable sources, and repealing Council Directive (EU) 2015/652 (COM(2021)0557—C9-0329/2021—2021/0218(COD)) P9_TA(2022)0317.

[181] Bas Eickhout and Sirpa Pietikäinen, 'Remarks regarding the technical screening criteria for fossil gas' (21 January 2022) <https://groenlinks.nl/sites/groenlinks/files/2022-01/letter-rapporteurs-taxonomy.pdf> accessed 5 February 2022.

[182] Committee on Economic and Monetary Affairs and Committee on the Environment and Public Health, 'Letter to Commissioner Mairead McGuinness: Concerns about the Commission procedure for the adoption of the draft Complementary Delegated Act covering certain additional energy activities pertaining to nuclear energy and gas' Ref. D(2022)1563 (18 January 2022).

[183] The Parliament was 75 MEPs short of the required absolute majority to reject the delegated act. European Parliament, 'Minutes of proceedings Results of roll-call votes—Annex', PE 734.819 (Brussels, 6 July 2022).

[184] Case T-628/22 *Repasi v Commission.*

[185] Skjærseth 2018.

which consignments would not have to apply for allowances.[186] Moreover, it envisioned further apprehension in the extension to indirect emissions.[187] In the case of the RED, it introduced further leniency in the sustainability criteria for waste.[188] It also took the least ambitious position on the cap on conventional biofuels.[189] In the 2022 amendment, it did not concur with the Commission and the Parliament's positions to heighten the RED target to 45%.[190] Similarly, it seeks several flexibilities. For example, if a Member State were to set a lower cap, it would be rewarded with the option of lowering its overall target for renewables in transport.[191] Furthermore, it seeks to limit the application of the proposed forest biomass sustainability criteria.[192] Overall, however, it is worth noting that the Council can be strongly internally divided on environmental ambition. This appeared very evidently in the case of the inclusion of gas and nuclear in the Taxonomy: opposite the vocal proponents, several Member States publicly spoke out against it,[193] some of which have since threatened legal action against the outcome.[194]

[186] Article 2(2a), Council of the European Union, 'Draft regulation of the European Parliament and of the Council establishing a carbon border adjustment mechanism—General approach' 7226/22 (Brussels, 15 March 2022).

[187] Replacing 'may' with 'could'. Recital 34.

[188] Article 26(1)(c), 'Meeting Document: Proposal for a Directive of the European Parliament and of the Council on the promotion of the use of energy from renewable sources (recast)—Third informal trilogue' (17 May 2018, Brussels); Article 29(1), second and third subparagraph, Directive 2018/2001.

[189] Article 3(4)(d) Directive 2009/28/EC, as amended by Article 2(2)(b)(iv) Directive 2015/1513.

[190] Article 1(2)(a), Permanent Representatives Committee, 'Proposal for a Directive of the European Parliament and of the Council amending Directive (EU) 2018/2001 of the European Parliament and of the Council, Regulation (EU) 2018/1999 of the European Parliament and of the Council and Directive 98/70/EC of the European Parliament and of the Council as regards the promotion of energy from renewable sources, and repealing Council Directive (EU) 2015/652—General approach' 10488/22 (Brussels, 24 June 2022).

[191] General Secretariat of the Council, proposal for a directive of the European Parliament and of the Council on the promotion of the use of energy from renewable sources (recast)' 15893/17 (Brussels, 20 December 2017).

[192] Only to areas 'very rich in biodiversity', Article 1(18)(b) Council of the European Union, 'General approach' Interinstitutional File 2021/0218(COD) (Brussels, 24 June 2022).

[193] Especially Austria, Denmark, Germany, Luxembourg, Spain and Portugal have been vocal against the inclusion of nuclear. Proposal for a Regulation of the European Parliament and the Council on the establishment of a framework to facilitate sustainable investment—Mandate for negotiations with the European Parliament 2018/0178(COD) 12360/2/19 REV 2 (Brussels, 24 September 2019), 4, 5.

[194] Denmark, Spain, Portugal and Austria. Mehreen Kahn and Daniel Domdey, 'Brussels faces threat of legal challenge over sustainable finance rules' (*Financial Times*, 23 January 2022).

### 3.4.5 International Law

International law compatibility is the third criterion for the externalization of environmental measures. As noted, it is not equally relevant to all case studies—much less so in cases of voluntary standards, like the Taxonomy—but it is all the more crucial to others. All three legislative institutions commit to international law compatibility at least rhetorically. However, the extent to which these commitments are consistently followed up on in the legal design of measures differs distinctly.

Amongst the legislative institutions, the Commission's approach is most clearly tailored to ensure international law compatibility. This can be attributed to its drafting initiative, its technical capacity and its representation of the Union in international trade matters (e.g. within the WTO), which makes it acutely aware of the trade effects of internal market externality. The Commission's cautious approach is most evidently demonstrated in the cases of the RED and CBAM.[195] Amongst others, WTO law prompted the EU to not subject imported biofuels to binding standards on biodiversity and soil degradation in the initial RED,[196] as well as social and agricultural norms.[197] Moreover, default values were initially more flexibly available for imports.[198] In the case of CBAM, the Commission explicitly avoided a cap on non-EU emissions 'in order to avoid restricting international trade.'[199] This is why the CBAM was designed as a so-called 'notional ETS': instead of simply including imports in the intra-EU ETS, it creates a separate, uncapped pool of allowances of which the price mirrors that of the ETS price of that week. A capped allowance pool for imports would constitute a quantitative restriction (Article XI GATT), which had to be avoided 'at all cost'.[200] Moreover, to avoid discrimination, carbon prices paid in third countries are exempted for and the EU can conclude agreements with third countries to account for this.[201]

In spite of this caution, the Commission does not always opt for the most international law-friendly variant. This choice can be motivated by environmental reasons. For example, the Commission opted against an exception from the CBAM based on levels of development, which some argue would be necessary to observe the principle of common but differentiated responsibilities (CBDR).[202] The Commission argued

[195] Recital 20 Directive 2018:2001. See also Tyson and Meganingtyas 2022; Mayr et al. 2020, p. 247; Scott 2011, p. 832.

[196] Barrett Lydgate, 1.

[197] As such, imports does not have to comply with the criteria of the EU's Common Agricultural Policy. Article 17(6) Directive 2009/28/EC; European Commission, 'Staff Working Document—Annex to Impact Assessment' SWD(2008), 85.

[198] Article 19(3) Directive 2009/28/EC.

[199] Explanatory Memorandum to COM(2021) 564 final, 32.

[200] Krenek 2020, p. 6.

[201] Article 9 and 2(12), COM(2021) 564 final.

[202] Scott and Rajamani have argued that the absence of such an exception was a main reason why the EU's Aviation Directive created such strong international opposition. Lowe 2021; Scott and Rajamani 2013.

that CBAM will have a moderate impact on developing economies and exceptions run counter to the environmental objectives of the measures.[203] Other Commission choices which can be considered controversial from the international law perspective, however, are made for non-environmental reasons. For example, the CBAM will apply alongside free allowances under the ETS,[204] whereas international partners would likely prefer them to be fully phased out from the point when CBAM certificates must be acquired. Another big question is that of the CBAM revenue. The option to allocate revenue to developmental finance is generally deemed to enhance positive reception by international partners, both with a view to the principle of CBDR and WTO law.[205] By contrast, whilst the proposal does envision technical assistance to LDCs,[206] the Commission proposes that 'most' revenue of the CBAM will go to the general EU budget.[207] The rationale behind this is part of a broader effort to expand the EU's own resources and help repay the €750 billion Recovery Instrument for the Covid-19 pandemic,[208] as per an Inter-institutional Agreement (IIA).[209]

The Parliament strongly commits to international law compliance rhetorically, as illustrated by titling its March 2021 Resolution "Towards a WTO-compatible' CBAM.[210] Often, this translates into tangible positions. For example, it sought to add that potential bilateral equivalency agreements 'shall not lead to undue preferential treatment' and 'take into account any carbon pricing mechanisms'.[211] Moreover, contrary to the Commission, it is set to spend an amount 'at least equal' to CBAM revenues to de-carbonization in developing countries.[212]

However, the Parliament's aforementioned outspoken environmental ambition often translates into positions which are at odds with international law. For example, it seeks to codify that 'the possible practices of circumvention in third countries' of

---

[203] COM(2021) 564 final, 19–21.

[204] Article 31(1), Regulation (EU) 2023/956; Article 31, COM(2021) 564 final.

[205] Marcu et al. 2021, p. 39; Mehling et al. 2019, p. 44; Lowe 2021.

[206] Preamble 55, COM(2021) 564 final

[207] COM(2021) 564 final, 10. Whilst there has been talk of an estimated revenue of 5–14 euros, the Commission in its Proposal speaks of 2.1 million euros by 2030. European Parliament resolution of 10 March 2021 towards a WTO-compatible EU carbon border adjustment mechanism, para 35; Opinion of the Committee on Budgets, para 9; COM(2021) 564 final, 47.

[208] Council Regulation (EU) 2020/2094 of 14 December 2020 establishing a European Union Recovery Instrument to support the recovery in the aftermath of the COVID-19 crisis.

[209] In the negotiations on the 2021–2027 Multiannual Financial Framework (MFF). Interinstitutional Agreement between the European Parliament, the Council of the European Union and the European Commission on budgetary discipline, on cooperation in budgetary matters and on sound financial management, as well as on new own resources, including a roadmap towards the introduction of new own resources LI 433/28 (22 December 2020).

[210] European Parliament resolution of 10 March 2021 towards a WTO-compatible EU carbon border adjustment mechanism.

[211] Amendment 53, TA/2022/0248.

[212] Amendments 40 and 130, TA/2022/0248.

the CBAM will be monitored by the Commission,[213] which is likely to be considered controversial from an international law perspective. Also controversial are its unsuccessful attempts to codify the environmental protection threshold for third-country legislation, such as requiring third-country forestry standards to be in line with EU standards,[214] and prohibiting obtaining agricultural biomass from countries not party to the Paris Agreement.[215] Most overtly contrary to international law, however, are its pursuit of quantitative restrictions such as a palm oil ban,[216] and a freeze of the use of 'primary woody biomass'.[217] Finally, similar to that of the Commission, friction between the Parliament's position and international law sometimes emerges for internal, non-environmental reasons. For example, it resisted designing the CBAM as a tax, which would have made sense from a WTO perspective, as this would have considerably decreased its influence over the legislative process.[218] Moreover, it was in favor of introducing export rebates for green EU companies which seek to export their goods, which would almost certainly conflict with WTO law.[219]

Like its co-legislators, the Council pledges that EU internal environmental measures, in spite of their externalization, shall contribute to the 'core value' of the EU's 'commitment to multilateralism'[220] and avoid trade barriers.[221] It is certainly sensitive to the international law controversy surrounding EU environmental externality—reportedly, some Council members were crucial in the EU's eventual decision to withhold the application of the Aviation Directive.[222] However, this sensitivity is not always reflected in its position on the design of measures. Like the Commission, it envisions CBAM revenue as an own resource,[223] and for the CBAM to apply alongside free allowances. On several occasions, its positions deviate from the Commission towards a less conscientious approach. For example, whilst the Commission proposes to 'facilitate' the adaptation of least developed countries to the CBAM, the Council only wants to 'contribute to ensuring' this adaptation.[224] More contentious are its

[213] Amendments 13 and 31, TA/2022/0248.

[214] Amendment 240, P9_TA(2022)0317.

[215] Article 1(1)(18)(da), P9_TA(2022)0317.

[216] Amendment 307, P8_TA (2018)0009.

[217] To be frozen at 2017–2022 average. When extracted in a sustainable manner and resulting from land use planning in an outermost region where forests cover at least 90% of the territory of that outermost region. Article 1(18)(gb), P9_TA(2022)0317.

[218] European Parliament resolution of 10 March 2021 towards a WTO-compatible EU carbon border adjustment mechanism, para 16.

[219] More specifically Article 3.1(a) of the WTO Agreement on Subsidies and Countervail Measures, which importantly could not be covered under an Article XX GATT exception.

[220] Gernot Blümel et al., 'To fight climate change, fight carbon leakage' (*Politico*, March 23, 2021).

[221] General Secretariat of the Council, 'Press Release: 2854th Council meeting Transport, Telecommunications and Energy' C/08/45 (Brussels, 28 February 2008), 11.

[222] Wurzel et al. 2017, p. 288.

[223] Council of the European Union, 'Special meeting of the European Council' EUCO 10/20 (Brussels, 21 July 2020).

[224] General Secretariat of the Council, 'CBAM: Commission proposal / Council general approach / Position of the EP' 13063/22 (Brussels, 3 October 2022), Recital 55.

positions on CBAM enforcement. It seeks to specify the practices of third-country circumvention much further, especially targeting mechanisms to absorb domestic CO2 prices (such as subsidies or taxes) or a domestic CO2 price selectively applied to EU exports.[225] Moreover, it seeks to impose CBAM penalties higher than those within the ETS, 'in order to be effective and dissuasive'.[226] This imbalance would be highly problematic from a non-discrimination perspective. Finally, by way of exception, the European Council publicly took a detailed position on a legislative instrument by indicating its support for a specified RED target to be achieved from biofuels alone,[227] which would have similarly raised WTO concerns and was hence not included in the Directive.

It follows that all EU legislative institutions rhetorically commit to international law compliance. This commitment is most tangibly translated into the legislative process by the Commission. The Parliament and the Council are less cautious. However, neither of the institutions consistently prioritizes the external implication of internal measures. In the Commission and the Parliament's positions, friction with international law is mainly created by the pursuit of high environmental ambitions. For the Council, its positions can be traced to a variety of non-environmental, internal incentives. These variances underline that the EU is a composite international actor. The role of internal measures in its capacity as a proponent of environmental integration in the international arena is subject to significant internal contention and division.

## 3.5 Conclusion

This chapter examined how the EU uses internal measures to integrate sustainability in its external relations law. Spurred by the environmental integration principle, the EU extensively externalizes environmental norms via internal market standards. The chapter observed that the EU has become increasingly coercive in articulating the legislative ambition to set green global standards, as well as in the legal mechanisms which extend the reach of internal measures over non-EU actors.

It can enhance its chances of achieving this ambition by paying attention to certain factors in the legislative process. These factors include the timing and innovation, stringency and international law compatibility of measures. The trade-offs between these factors illuminate that there is no obvious playbook to ensure the effectiveness of environmental externalization. The EU is treading a complex landscape of intertwined and, at times, conflicting considerations.

---

[225] General Secretariat of the Council, 13063/22, Article 27.

[226] General Secretariat of the Council, 13063/22, Article 26a, Recital 24.

[227] The European Council was in favour of a specified 10% biofuel target in transport, but target is 'to be achieved from renewable sources as a whole, and not from biofuels alone'. Recital 9 Directive 2009/28/EC, Council, "Presidency Conclusions" 7224/1/07 Rev 1 (Brussels, 2 May 2007), Annex I, para 7; Recital 18 Directive 2009/28/EC.

In this balancing act, it is pivotal to appreciate the legislative institutions in their capacity to construct the EU as an external actor. The chapter therefore assessed to what extent each of the legislative institutions prioritize the factors which contribute to effective externalization. It appears that the Commission contributes most to EU environmental leadership, as it is most ambitious and strategic on the external objectives of internal measures, as well as most cautious to ensure international law compliance. By contrast, the co-legislators are generally less cognizant of external objectives, at times slows down the legislative process which inhibits the EU's first-mover advantage, whilst their positions are less inhibited by external limitations following from international law. Nonetheless, the Parliament plays an important role by pushing the environmental ambition of measures, which helps establish the EU as a green global standard-setter via stringent standards. The Commission rather tempers its environmental ambition by balancing it with the implementability and international law compliance of measures, or the risk of being a preference outlier. The Council is more hesitant in its environmental ambition, sometimes subject to strong internal divisions, which has inhibited the potential for EU leadership on several counts. In fact, the Council appeared in this chapter to be least mindful of the effectiveness of EU externalization, since it is also least considerate of international law compliance.

Overall, this chapter demonstrated that the EU's endeavors to set global green standards are neither straightforward nor concerted. Rather, the external environmental ambitions of its internal measures are a result of a composite legal order with convoluted interests and compromises between different priorities. Overall, the legislative institutions are increasingly aware of and strategic on the external dimension of EU internal measures. Notwithstanding, the EU's effectiveness in this external dimension is not seldomly—and understandably—trumped by internal or non-environmental considerations. Consequently, the place of internal measures in EU external sustainability law will raise ever more legal questions in the years to come.

## References

Bäckstrand K and Elgström O (2013) The EU's Role in Climate Change Negotiations: From Leader to Leadiator. Journal of European Public Policy 20 10:1369

Barrett Lydgate E (2012) Biofuels, sustainability, and trade-related regulatory chill. Journal of International Economic Law 15

Bradford A (2013) The Brussels Effect. Nw. U. L. Rev. 107:1

Bradford A (2020) The Brussels effect: How the European Union rules the world. Oxford University Press

Biedenkopf K (2015) EU Chemicals Regulation: Extending Its Experimentalist REACH. In: Zeitlin J (ed) Extending Experimentalist Governance? The European Union and Transnational Regulation. Oxford University Press

Biedenkopf K (2018) Chemicals: Pioneering Ambitions with External Effects. In: Adelle C, Biedenkopf K and Torney D (eds) European Union External Environmental Policy: Rules, Regulation and Governance Beyond Borders. Palgrave Macmillan

Birger Skjærseth J (2018) Implementing EU climate and energy policies in Poland. Environmental Politics 27 3:498

Bogojević D (2012) Legalising Environmental Leadership: A Comment on the CJEU'S Ruling in C-366/10 on the Inclusion of Aviation in the EU Emissions Trading Scheme. Journal of Environmental Law 24:345

Borghesi S, Verde SF and Zhu T (2018) The International Dimension of the EU Emissions Trading System. European University Institute

Bretherton C and Vogler J (2006) The European Union as a Global Actor. Routledge

Bretherton C and Vogler J (2013) A Global Actor past Its Peak?. International Relations 27:3

Cremona M (2017) The internal market and external economic relations. In Koutrakos P and Snel J (eds) Research Handbook on the Law of the EU's Internal Market. Edward Elgar Publishing

Damro C (2012) Market Power Europe. Journal of European Public Policy 19 5:682

Dero-Bugny D and Motte-Baumvol J (2021) Shaping EU External Relations Beyond Treaty-Making: The Scope of Extraterritorial EU Legislation and Its Enforcement Challenges. In Douma W et al. (eds) The Evolving Nature of EU External Relations Law. T.M.C. Asser Press

De Ville F and Gunst S (2021) The Brussels Effect: How the GDPR Conquered Silicon Valley. European Foreign Affairs Review 26:3

Diez T and Pace M (2011) Normative Power Europe and Conflict Transformation. In Whitman R (ed) Normative Power Europe: Empirical and Theoretical Perspectives. Palgrave Macmillan

Drezner D (2005) Globalization, harmonization, and competition: the different pathways to policy convergence. Journal of European Public Policy 12 5:841

Drezner D (2007) All Politics is Global: Explaining International Regulatory Regimes. Princeton

Dobson NL (2021) Extraterritoriality and Climate Change Jurisdiction: Exploring EU Climate Protection under International Law. Bloomsbury Publishing

Dupont C, Oberthür S and Biedenkopf K (2018) Climate Change: Adapting to Evolving Internal and External Dynamics. In Adelle C, Biedenkopf K and Torney D (eds) European Union External Environmental Policy. Springer 2018

Franklin C (2011) The burgeoning principle of consistency in EU law. Yearbook of European Law 30 1:76

Fishman A and Obidzinski K (2014) European Union Timber Regulation: Is it legal?. Review of European, Comparative & International Environmental Law 13 2:258

Gstöhl S (2008) The Internal Market's External Dimension' in Pelkmans J, Hanf D and Chang M (eds) The EU Internal Market in Comparative Perspective: Economic, Political and Legal Analyses. Peter Lang

Gerber DJ (1984) Beyond balancing: International law restraints on the reach of national laws. Yale J. Int'l L. 10:185

Geraets D and Natens B (2014) The WTO Consistency of the European Union Timber Regulation. Journal of world trade 48:433

Groen L and Niemann A (2011) Challenges in EU External Climate Change Policy-Making in the Early Post-Lisbon Era: The UNFCCC Copenhagen Negotiations. In EU External Relations Law and Policy in the Post-Lisbon Era. TMC Asser Press

Goh G (2004) The World Trade Organization, Kyoto and Energy Tax Adjustments at the Border. Journal of World Trade 38:3

Hadjiyianni I (2019) The EU as a Global Regulator for Environmental Protection: A Legitimacy Perspective. Hart Publishing

Hillman JA (2013) Changing Climate for Carbon Taxes: Who's Afraid of the WTO?. Georgetown University Climate & Energy Policy Paper Series

Holzer K (2014) Carbon-related border adjustment and WTO law. Edward Elgar Publishing

Howse R (2012) Commentary: The Political and Legal Underpinnings of Including Aviation in the ETS' in Lorand Bartels, 'The Inclusion of Aviation in the EU ETS: WTO Law considerations. ICTSD Trade and Sustainable Energy Series 6

Intergovernmental Panel on Climate Change (2020) Climate Change and Land: An IPCC Special Report on climate change, desertification, land degradation, sustainable land management, food security, and greenhouse gas fluxes in terrestrial ecosystems, Summary for Policymakers
Jaremba U (2020) Non-economic values and objectives in EU trade policy: Different models of externalization and enforcement. In Global Politics and EU Trade Policy. Springer
Jordan A (ed) (2021) Environmental policy in the EU: Actors, institutions and processes, 2nd edn. Routledge
Krenek A (2020) How to implement a WTO-compatible full border carbon adjustment as an important part of the European Green Deal. ÖGfE Policy Brief 02
Kulovesi K (2012) Climate change in EU external relations: Please follow my example (or I might force you to). In Morgera E (ed) The External Environmental Policy of the European Union: EU and International Law Perspectives. Cambridge University Press
Kulovesi K and Muñoz M (2011) Environmental integration and multi-faceted international dimensions of EU law: Unpacking the EU's 2009 climate and energy package. Common Market Law Review 48:3
Lavenex S (2004) EU external governance in wider Europe. Journal of European Public Policy 11 4:680
Lavenex S (2022) Common market, normative power or super-state? Conflicting political identities in EU asylum and immigration policy. In Constructing the EU's Political Identity. Palgrave Macmillan
Lowe S (2021) The EU's carbon border adjustment mechanism: How to make it work for developing countries. Centre for European Reform Policy Brief
Marcu A, Mehling M and Cosbey C (2021) CBAM for the EU: a Policy Proposal. ERCST Roundtable on Climate Change and Sustainable Transition
Marín Durán G (2020) EU External Environmental Policy. In Larik J and Wessel RA (eds) EU External Relations Law - Text, Cases and Materials. Hart Publishing
Marín Durán G and Morgera E (2012) Environmental Integration in the EU's External Relations: Beyond multilateral dimensions. Bloomsbury Publishing
Mayr S, Hollaus B and Madner V (2020) Palm oil, the RED II and WTO law: EU sustainable biofuel policy tangled up in green?. Review of European, Comparative & International Environmental Law 30
Newman A and Posner E (2015) Putting the EU in its Place: Policy Strategies and the Global Regulatory Context. Journal of European Public Policy 22:9
Nicolaïdis K and Egan M (2001) Transnational market governance and regional policy externality: why recognize foreign standards?. Journal of European Public Policy 8:3
Öberg L (2020) Internal Market Acquis as a Tool in EU External Relations: From Integration to Disintegration. Legal Issues of Economic Integration 47:2
Oberthür S and Rabitz F (2014) On the EU's Performance and Leadership in Global Environmental Governance: The Case of the Nagoya Protocol. Journal of European Public Policy 21 1:39
Parker CF and Karlsson C (2015) Climate leadership. In Bäckstrand K and Lövbrand E (eds) Research handbook on climate governance. Edward Elgar Publishing
Poletti A and Sicurelli D (2018) The political economy of normative trade power Europe. Springer
Quaglia L (2014) The European Union, the USA and International Standard Setting by Regulatory Fora in Finance. New Political Economy 19:3
Renckens S, Skogstad G and Mondou M (2017) When normative and market power interact: The European Union and global biofuels governance. JCMS: Journal of Common Market Studies 55:6
Ryngaert C (2015a) Jurisdiction in International Law, 2nd edn. Oxford University Press
Ryngaert C (2015b) The Concept of Jurisdiction in International Law. In Orakhelashvili A (ed) Research handbook on jurisdiction and immunities in international law. Edward Elgar Publishing
Schimmelfennig F and Sedelmeier U (eds) (2005) The Europeanization of Central and Eastern Europe. Cornell University Press

Schunz S (2019) The European Union's environmental foreign policy: from planning to a strategy?' International Politics 56 3:339
Schunz S and Damro C (2020) Expanding actorness to explain EU external engagement in originally internal policy areas. Journal of European Public Policy 27:1
Schunz S, Gstöhl S and Damro C (2018) Analytical Framework: Understanding and Explaining EU External Engagement. In Damro C, Gstöhl S and Schunz S (eds) The EU's Evolving External Engagement – Towards New Sectoral Diplomacies?. Routledge
Scott J (2011) The Multi-Level Governance of Climate Change. In Paul Craig and Grianne de Burca (eds) The Evolution of EU Law. OUP
Scott J (2013) Extraterritoriality and Territorial Extension in EU law. American Journal of Comparative Law 62
Scott J and Rajamani L (2012) EU Climate Change Unilateralism. The European Journal of International Law 23:2
Scott J and Rajamani L (2013) Contingent Unilateralism: International Aviation in the European Emissions Trading Scheme. In Van Vooren B, Blockmans S and Wouters J (eds) The EU's Role in Global Governance The Legal Dimension. OUP
Simintzis A (2022) Externalizing the European Union's Climate Policies: Ensuring the Brussels Effect has an impact. In European Horizons (ed), Cornerstones for an Evolving Europe: New Policy Approaches to Economic Development, Environmental Policy, and Human Rights. Nomos
Sjåfjell B (2015) The Legal Significance of Article 11 TFEU for EU Institutions and Member States. In Sjåfjell B and Wiesbrock A (eds) (2015) The Greening of European Business under EU Law: Taking Article 11 TFEU Seriously. Routledge
Smith K (2003) European Union Foreign Policy in a Changing World. Cambridge Polity Press
Stattman SL and Gupta A (2015) Negotiating Authority in Global Biofuels Governance: Brazil and the EU in the WTO. Global Environmental Politics 15 1:41
Tene O and Wolf C (2013) Overextended: Jurisdiction and Applicable Law under the EU General Data Protection Regulation. White Paper (The Future of Privacy Forum)
Tyson A and Meganingtyas E (2022) The Status of Palm Oil Under the European Union's Renewable Energy Directive: Sustainability or Protectionism?. Bulletin of Indonesian Economic Studies 58 1:31
Unger C (2021) Linking the European Union Emissions Trading System Political Drivers and Barriers. Taylor & Francis
Van Alsenoy B (2017) Reconciling the (extra) territorial reach of the GDPR with public international law. In Vermeulen G and Lievens E (eds) Data Protection and Privacy under Pressure. Maklu
Vogel D (1997) Trading up and governing across: transnational governance and environmental protection. Journal of European public policy 4:556
Vogel D (2012) The Politics of Precaution: Regulating Health, Safety, and Environmental Risks in Europe and the United States. Princeton University Press
Voigt C (2012) Up in the Air – Aviation, the EU Emissions Trading Scheme and the Question of Jurisdiction. In Barnard C (ed) Cambridge Yearbook of European Legal Studies. Hart Publishing
Walker H and Biedenkopf K (2018) The historical evolution of EU climate leadership and four scenarios for its future. In Minas S and Ntousas V (eds) EU Climate Diplomacy: Politics, Law and Negotiations. Routledge
Wurzel RKW and Connelly J (2016) Conclusion: re-assessing European Union climate leadership' in The European Union in International Climate Change Politics. Routledge
Wurzel RKW, Connelly J and Liefferink D (eds) (2017) The European Union in international climate change politics. Routledge
Wurzel RKW, Liefferink D and Di Lullo M (2019) The European Council, the Council and the Member States: changing environmental leadership dynamics in the European Union. Environmental Politics 28 2:248
Wouters J, Hoffmeister D, De Baere G and Ramopoulos T (2021) The Law of EU External Relations. Oxford University Press

Wouters J, Marx A and Geraets D (2015) Global Governance through Trade: EU Policies and Approaches. Edward Elgar Publishing

Young AR (2014) Europe as a global regulator? The limits of EU influence in international food safety standards. Journal of European Public Policy 21 6:905

Young AR (2015) The European Union as a global regulator? Context and comparison. Journal of European Public Policy 22:9

Zeitlin J (2015) Extending Experimentalist Governance? The European Union and Transnational Regulation. Oxford University Press

**Eva Pander Maat** holds a PhD from the University of London and is an independent researcher.

# Chapter 4
# 'Green Conditionality' in the EU's Trade and Investment Policy: *Quo Vadis?*

**Belen Olmos Giupponi and Hannes Hofmeister**

**Contents**



**Abstract** With the entry into force of the Lisbon Treaty in 2009, the external dimension of the European Union's (EU) trade and investment policy has prominently emphasized sustainable development as a pivotal component of its external agenda. This emphasis has solidified over the years through the negotiation of various treaties and has received legal support from the Court of Justice of the European Union (CJEU). Nevertheless, significant variations and subtleties exist in how sustainability is conceptualized and incorporated into international agreements concluded by the EU. A pivotal moment in this context was Advisory Opinion 2/15 concerning the Free Trade Agreement between the European Union and the Republic of Singapore. This Opinion offered critical insights into the scope and character of sustainable development chapters within the framework of trade and investment policy, taking into consideration key EU constitutional law provisions such as Article 218(11) TFEU, Articles 3(5) and 21 TEU, and Articles 11 and 191 TFEU. Subsequent agreements

B. Olmos Giupponi (✉)
Middlesex University London, The Burroughs, Hendon, London NW4 4BT, United Kingdom
e-mail: belen_olmos_giupponi@biari.brown.edu

H. Hofmeister
University of Bozen, Piazza Universita', 1, 39100 Bolzano, BZ, Italy
e-mail: Hannes.hofmeister@unibz.it

 
R. A. Wessel et al. (eds.), *EU External Relations Law and Sustainability*, Global Europe: Legal and Policy Issues of the EU's External Action 4,
https://doi.org/10.1007/978-94-6265-655-0_4

have embraced distinct approaches to sustainability, particularly within the realm of environmental protection, where provisions range from general principles to comprehensive regulations governing specific aspects. This chapter aims to examine the concept of 'green conditionality' with the Sustainable Development Chapters of the EU's Trade and Investment Agreements, with a particular focus on the environmental provisions. It seeks to shed light on the extent to which these agreements prioritize and enforce environmental sustainability, thus contributing to a more nuanced understanding of the EU's evolving approach to green conditionality within its trade and investment policy framework.

**Keywords** Sustainable Development Chapters · EU External Commercial Policy · Trade and Investment Agreements · Multilateral Environmental Agreements (MEAs) · Sustainable Development Goals · Sustainability · Sustainability Impact Assessments

## 4.1 Introduction

Following the entry into force of the Lisbon Treaty in 2009, sustainable development has emerged as a predominant dimension within the European Union's (EU) trade and investment policy in its external sphere.[1] This trend has consolidated in recent years through proactive treaty negotiations, and it has received validation from the rulings of the Court of Justice of the European Union (CJEU). Nevertheless, notable distinctions and subtleties emerge concerning the conceptualization and incorporation of sustainability within the Free Trade and Investment Agreements (FTIAs) that the EU has concluded during this period.[2]

Within and outside its borders, the EU has intricately crafted a comprehensive policy framework pertaining to sustainable development and trade. This multifaceted strategy seeks to stimulate economic advancement while concurrently upholding environmental integrity, fostering societal welfare, and perpetuating sustainable development within its partner nations. The foundation of the EU's sustainable development and trade policy is firmly rooted in a tripartite structure, comprising economic, social, and environmental dimensions. The economic dimension is primed to propel sustainable economic growth, the social facet is resolutely committed to upholding principles of social justice and the safeguarding of human rights, while the environmental aspect is steadfastly dedicated to environmental preservation and the efficient management of finite natural resources.

Indicators of sustainability play a pivotal role, illustrating varying degrees of commitment within the EU, including leadership demonstrated by certain Member

[1] Damro 2012, pp. 55–75.

[2] European Commission. Negotiations and Agreements. Available at https://policy.trade.ec.europa.eu/eu-trade-relationships-country-and-region/negotiations-and-agreements_en (last accessed on 3 March 2024).

States. The manner in which these nations report their progress also serves as a significant aspect to consider in this context.

The Trade and Sustainable Development (TSD) chapters within the EU's trade agreements are specifically designed to harmonize trade activities with the imperatives of sustainable development. Their core objectives encompass the safeguarding of environmental integrity and the advancement of labor rights.

In this realm, the EU's priority is to meet Sustainable Development Goals and deliver sustainability globally. Against this backdrop, the chapter's main purpose is to explore the extent of the so-called "green conditionality" in the Sustainable Development Chapters (SDCs) of the Trade and Investment Agreements (TIAs) entered into by the EU with a focus on the enforcement of the provisions in cases of breach.

This Chapter comprehensively explores several crucial dimensions within the domain of Sustainable Development Clauses (SDCs). One significant focus is the diverse typologies of "sustainability" clauses, each encompassing specific content related to environmental law, intellectual property (IP) law, Corporate Social Responsibility (CSR), energy law, and more. The analysis extends beyond the mere identification of these clauses to an in-depth examination of their substantive provisions and implications.

A pivotal aspect of this Chapter involves an investigation into the applicability of EU constitutional law principles within the context of SDCs. This includes an examination of treaty provisions endowed with constitutional significance, accompanied by an exploration of prevailing theoretical underpinnings. Such scrutiny is crucial for a more nuanced understanding of the legal framework governing sustainable development.

Furthermore, this Chapter critically dissects the intricate interactions between EU law and both international and domestic realms of environmental and labor law. This exploration aims to unravel the complexities arising from the intersection of these legal domains, shedding light on potential synergies or conflicts.

In tandem, this Chapter will delve into the nuanced interplay between trade, investment, and sustainable development. This involves dissecting the ways in which trade and investment activities align with or challenge the principles of sustainable development. By doing so, the paper contributes to the discourse on the role of economic activities in fostering or hindering sustainable practices.

Lastly, the Chapter directs its attention to the mechanisms available for ensuring compliance and enforcement in the event of a breach of SDCs. This includes an analysis of the legal tools and frameworks designed to uphold the commitments laid out in sustainable development agreements. By examining these questions, we aim to shed light on the issue of sustainability in terms of "green conditionality" in the arena of trade and investment external relations pinpointing constraints and progress achieved so far.

## 4.2 Conceptualising "Sustainability" in the European Union

In the EU, sustainable development is a core principle of the Treaty of the EU, and also a priority objective for the EU's internal and external policies. On the international level, the EU is committed to achieving the 17 Sustainable Development Goals (SDGs) included in the United Nations 2030 Agenda. Under EU law, rather than a comprehensive framework obligating Member States to raise sustainability across diverse contexts, there are various provisions and obligations scattered in different texts. Moreover, a duty exists to perform environmental impact assessments.

Sustainability, as a concept, is shown through a multitude of norms, policies and actions in the EU. Within the EU, sustainable development is a pivotal objective that is intricately linked with other aims and it is embedded in different legal tools Member States have at their disposal. The synergy of these measures, coupled with due diligence facilitates cooperation within specific policy areas such as with regard to addressing deforestation and fosters opportunities for inter-state cooperation.

Collaboration in the realm of external relations is equally vital. Sustainable development clauses included in international agreements concluded by the EU indicate that they do not only deal with carving out sustainable development protection but that they are also about sanctioning the breach of specific sustainable development provisions. In each of the Sustainable Development chapters, the non-regression principle has been upheld within commercial aspects to prevent a race to the bottom. The legal framework that underpins the preservation and protection of sustainability within the EU comprises an array of various instruments and policies. Nestled beneath the expansive cover of the EU's sustainable development doctrine, various trade policy measures, initiatives, and agreements are incorporated within this framework, which includes a wide range of both soft law and hard law instruments, of multilateral and regional scope, such as:

- *Sustainable Development Goals (SDGs).* At the multilateral level, the EU is committed to attaining and cooperating to achieve the 17 SDGs set by the United Nations, which include ending poverty, reducing inequality, and protecting the environment.[3]
- *The EU's Generalized Scheme of Preferences (GSP).* In the international trade realm and within the WTO, the EU has committed to assist developing countries through preferential trade arrangements. The GSP scheme provides preferential access to the EU market for products from developing countries that meet certain social and environmental standards.[4]

---

[3] European Commission. Sustainable Development Goals. Available at https://commission.europa.eu/strategy-and-policy/sustainable-development-goals_en#:~:text=Sustainable%20development%20is%20a%20core,Sustainable%20Development%20Goals%20(SDGs) (last accessed on 3 March 2024).

[4] European Commission, EU's Generalized Scheme of Preferences (GSP), https://policy.trade.ec.europa.eu/development-and-sustainability/generalised-scheme-preferences_en (last accessed on 3 March 2024).

- *The EU's Circular Economy Action Plan.* This plan aims to promote a more sustainable and resource-efficient economy by reducing waste, promoting recycling, and increasing the use of renewable energy sources.[5]
- *The EU's Green Deal.* This initiative aims to make the EU climate-neutral by 2050 by reducing greenhouse gas emissions and promoting sustainable agriculture, forestry, and fisheries.[6] The Green Deal primarily focuses on domestic actions and the implementation of additional measures to confront the global environmental impact while fostering novel sustainability paradigms.

At the international level, the EU's sustainable development and trade policy encompasses a main aim to ensure that environmental and labor clauses are respected in partner countries.[7] For many years, the allocation of competences between the EU and Member States in external relations was controversial.[8] Opinion 2/15 on the Free Trade Agreement between the EU and the Republic of Singapore[9] clarified the legal basis, scope and nature of sustainable development chapters in the context of the new trade and investment policy, in light of Article 218(11) TFEU and other relevant EU constitutional law provisions, such as Articles 3(5) and 21 TEU, and Articles 11 and 191 TFEU.[10]

Different international agreements concluded after the adoption of this Opinion, embody "differential" approaches to sustainability, i.e. they present innovations in terms of the provisions included and mechanisms to ensure compliance. In the environmental protection realm, these agreements comprise a panoply of diverse provisions ranging from generic clauses referring to Multilateral Environmental Agreements (MEAs) to more detailed and specific clauses regulating specific aspects of international environmental protection. These chapters are in line with relevant international labor and environmental law frameworks.

[5] European Union, Circular Economy Action Plan, available at https://ec.europa.eu/environment/circular-economy/pdf/new_circular_economy_action_plan.pdf (last accessed on 3 March 2024).

[6] European Commission, The Green Deal Industrial Plan: putting Europe's net-zero industry in the lead (1 February 2023, available at https://ec.europa.eu/commission/presscorner/detail/en/ip_23_510 (last accessed on 3 March 2024).

[7] Pelkmans 2020, pp. 391–400.

[8] The allocation of competences between the EU and Member States in external relations was controversial. After the Lisbon Treaty, Opinion 2/15 on the Free Trade Agreement.

[9] Yotova 2018, pp. 29–32.

[10] G. Van der Loo, The Court's Opinion on the EU-Singapore FTA: Throwing off the shackles of mixity?, No 2017/17, May 2017, available at https://www.ceps.eu/wp-content/uploads/2017/05/PI2017-17 (last accessed on 3 March 2024).

## 4.3 Types of "Sustainability" Clauses/Provisions Included in EU Free Trade and Investment Treaties: Content and Commitments

Recent years witnessed a considerable expansion of trade and investment treaties and important changes in the ambition, scope, and density of sustainability provisions in the EU's internal and external policies. The EU has articulated a comprehensive approach to sustainability to encompass various elements as discussed below. Notwithstanding this progress, the broader picture of sustainability in international law looks quite scattered and fragmented, offering different levels of enforcement and compliance which is revealed in the relationships of the EU with third countries. Law-making in this specific area of EU External Relations is vested in the EU Commission with a considerable participation of the EU Parliament, under the monitoring of the CJEU. More recently, the EU Ombudsman has gained a more active role, especially regarding the sustainability impact assessments that are conducted before the signature and entry into force of the Trade and Investment Agreements.[11]

In relation to the content of the Sustainable Development (SD) Chapters of EU's Free Trade and Investment Agreements (FTIAs), they include references to MEAs and to the International Labor Organization's (ILO) core labor standards as regulated in the main conventions.[12] It is worth mentioning that sustainability provisions can be also placed in other chapters of the FTIAs and that are not necessarily contained in the environmental and labor law clauses. To illustrate, these can be provisions regarding Intellectual Property (IP) rights and law, Corporate Social Responsibility, and those concerning energy law as incorporated in other chapters.[13] The content and shape of these SD chapters vary greatly across EU free trade agreements. Hence, characterizing and assessing the nature of the provisions and the institutionalization that surrounds them is far from easy.

In an attempt to systematize them, SD provisions in international trade and investment agreements concluded outside the EU sphere, vary considerably. SD chapters constitute now a core component of the external dimension of the revised EU's trade and investment policy. Since the Lisbon Treaty entered into force in 2009, the EU's TSD chapter on environmental and labor matters combining both types of clauses have been shaped in different ways. Although these SD provisions are binding upon the parties, compliance with them has not been required in practice.[14] Clearly, the main motivation for seeking consensus around these environmental and labor issues relates to the EU's role in global climate change negotiations. Furthermore, the EU

[11] Sustainability impact assessments, https://environment.ec.europa.eu/law-and-governance/environmental-assessments/environmental-impact-assessment_en (last accessed on 3 March 2024).

[12] Environmental and labour clauses in Sustainable Development Chapters.

[13] See, for instance, the Agreement with New Zealand.

[14] The Lisbon Treaty introduced modifications to the Common Commercial Policy by adding investment and sustainable development under that framework.

holds a considerable responsibility in global value chains as it generally outsources economic, social and environmental impacts.[15]

Looking at the broader picture, the current legal framework encompasses the European Green Deal (EGD) launched in 2019, which increases the reach of green provisions in the Union's trade and investment policy, calling for mainstreaming of social and environmental sustainability goals into the EU's trade external trade policy.[16] The EGD was followed by the EU Trade Policy Review in 2021, placing sustainability at the center of the EU external policy.[17]

It is worth noting that not all sustainability issues can be effectively dealt with by trade agreements. Hence, the Commission has indicated that there are many other tools at the EU, Member State, and regional levels to tackle the negative effects of globalization. Notably, the 2017 European Consensus on Development, provides a common EU and Member State response to the 2030 Agenda from a development perspective. In the Commission's words, "sustainability issues feature prominently and stronger coherence between trade and development policies".[18]

While EU FTIAs have included environmental protection provisions and commitments on labor standards, there are still shortcomings concerning their application. This is observed in three particular areas: protection of the environment, mitigation of negative impacts of trade, and their use to enhance sustainability on a global scale.[19] To address these shortcomings, the EU institutions are seeking to overhaul the current policy as stated in the Communication "The power of trade partnerships: together for green and just economic growth".[20] In the document, the European Economic and Social Committee outlines strategies to further strengthen the implementation and enforcement of the Trade and Sustainable Development (TSD) chapters of the EU's trade agreements.

Considering this context, in the next sections this Chapter delves into the execution and enforcement of Sustainable Development (SD) Chapters in Free Trade and Investment Agreements (FTIAs) established by the EU. It explores the concept of

---

[15] Eurostat, EU and global value chains, available at https://ec.europa.eu/eurostat/web/economic-globalisation/globalisation-in-business-statistics/global-value-chains (last accessed on 3 March 2024).

[16] European Commission, Statistics for the European Green Deal, available at https://ec.europa.eu/eurostat/cache/egd-statistics/ (last accessed on 3 March 2024).

[17] Brussels, 18.2.2021 COM(2021) 66 final, Communication from the Commission to the European Parliament, The Council, the European Economic And Social Committee And The Committee of The Regions, Trade Policy Review—An Open, Sustainable and Assertive Trade Policy, available at https://trade.ec.europa.eu/doclib/docs/2021/february/tradoc_159438.pdf (last accessed on 3 March 2024).

[18] European Commission, *Trade and Sustainable Development (TSD) chapters in EU Free Trade Agreements (FTAs),* available at https://trade.ec.europa.eu/doclib/docs/2017/july/tradoc_155686.pdf (last accessed on 3 March 2024).

[19] Blot and Kettunen 2021.

[20] European Economic and Social Committee, *The power of trade partnerships: together for green and just economic growth*, available at https://www.eesc.europa.eu/en/our-work/opinions-information-reports/opinions/power-trade-partnerships-together-green-and-just-economic-growth (last accessed on 3 March 2024).

green conditionality and its integration into associated domains. While "green conditionality" lacks a universally agreed-upon definition, in this Chapter, it refers to the conditions mandated in international treaties, whether trade or cooperation agreements, which necessitate adherence to environmental regulations for trade and aid cooperation.[21]

## 4.4 Examining Sustainability Through the Lens of EU Law: EU Constitutional Law Principles and Prevailing Theoretical Underpinnings

From a constitutional EU law's perspective, TSD chapters raise the issue of their integration into the legal framework of the EU and the choice of the appropriate legal basis for secondary legislation.[22] Traditionally, the distribution of sustainable development competences in the context of the external commercial policy has been controversial due to the mismatch between internal and external allocation, i.e. environmental policy is a shared competence, whereas the external commercial policy is an EU's exclusive competence. Opinion 2/15 of the CJEU clarified that TSD provisions should be included under the umbrella of the EU's exclusive competence.[23] In the new scenario that emerged after the Lisbon Treaty, this would imply the possibility of carrying forward legal actions in cases of treaty breach by any of the parties to the EU FTIAs.

Though the AO shed light on competence allocation, the underlying problem continues to be compliance with the specific SD provisions of the agreements. Taking a critical viewpoint, Meunier and Nicolaidis contend that the efforts to align the EU's international investment agreements with sustainable development goals have reinforced the interests of individual States and heightened the politicization of the EU's Common Commercial Policy.[24]

Taking stock of the EU treaty-making experience, different cases brought before the EU Ombudsman illustrate controversial issues as observed in practice. In light of these cases, main obstacles to building a coherent and consistent sustainable development policy in line with SDGs become clear. These EU FTIAs are also

---

[21] The concept was coined in the 1990s to describe the greening of international aid, in other to promote environmental policies in developing countries donors seek to promote environmental policies in developing countries, according to their own objectives.

[22] From an EU constitutional law standpoint, sustainable development has been incorporated into different provisions of the treaties, see the Preamble to the TEU, Article 3(3) TEU and Article 11 TFEU, and Article 37 CFREU. Additionally, in terms of the EU's external action, see Article 3(5) TEU, Article 21(2)(d) TEU, and Article 21(2)(f) TEU See, also, Morgera and Marín Durán 2014, pp. 987–989, 992–994.

[23] Opinion 2/15 of the Court, of 16 May 2017, available at https://curia.europa.eu/juris/document/document.jsf?text=&docid=190727&doclang=EN (last accessed on 3 March 2024). For an analysis of the Opinion and its implications for EU constitutional law, see Cremona, 2018, pp. 231–259.

[24] Meunier and Nicolaidis 2019, pp. 103–113.

disclosing new perspectives and contributing recommendations to improving current EU policies, enhancing its significant role in preserving sustainability globally.

Consistent with this approach, Sustainability Impact Assessments (SIAs) and Human Rights Impact Assessments (HRIAs) carried out before finalizing FTIAs with other regions or third-party states are commonly deployed as mechanisms to identify specific areas of concern and action for both the EU and its external partners. SIAs' main goals are to identify, address and lessen the negative or detrimental impacts of EU FTIAs. SIAs are particularly relevant to monitor compliance with environmental standards and the protection of labor rights during the negotiations and before the entry into force of the agreements.

In some specific cases, civil society organizations have even complained about the timing for the HRIAs and the lack of an appropriate level of protection, with cases reaching the EU Ombudsman. In the EU Ombudsman's words "[t]he EU projects its values through its trade deals. Concluding a trade agreement before its potential impact has been fully assessed risks undermining those values and the public's ability to debate the merits of the deal. It also risks weakening European and national parliaments' ability to comprehensively debate the trade agreement."[25]

Amongst these controversial cases, in a complaint regarding the EU-Mercosur Agreement, as the negotiations were protracted, the claim was that the European Commission should have concluded an updated sustainability impact assessment (SIA) before the EU-Mercosur trade deal was agreed. The European Ombudsman urged that, in future trade negotiations, such assessments should be finished ahead of the final agreement.[26] As the European Commission failed to ensure the finalization of the SIA in good time (i.e. before the conclusion of the EU-Mercosur trade negotiations), there was maladministration. In a different case, the EUO examined the complaint submitted by a group of civil society organizations, which were concerned about the European Commission's failure to ensure human rights risks were assessed before providing support to African countries to develop surveillance capabilities, notably in the context of the EU Emergency Trust Fund for Africa (EUTFA).[27] The complainants contended that, before agreeing to support projects with potential surveillance implications, such as introducing biometric databases or mobile phone monitoring technologies, the Commission should have carried out prior risk and impact assessments to ensure that the projects do not result in human rights violations (such as the right to privacy).[28] In both cases, the Ombudsman found that the measures taken were not appropriate to guarantee a proper assessment of

[25] EU Ombudsman. EU Commission 'failed' on assessing Mercosur trade deal, available at https://euobserver.com/green-economy/151302 (last accessed on 3 March 2024).

[26] Decision in case 1026/2020/MAS concerning the failure by the European Commission to finalise an updated 'sustainability impact assessment' before concluding the EU-Mercosur trade negotiations, available at https://www.ombudsman.europa.eu/en/decision/en/139418 (last accessed on 3 March 2024).

[27] Decision on how the European Commission assessed the human rights impact before providing support to African countries to develop surveillance capabilities (case 1904/2021/MHZ). Decision. Case 1904/2021/MHZ—Opened on Tuesday | 30 November 2021—Decision of 28 November 2022.

[28] Ibid.

the human rights impact of EUTFA. The implication of this main outcome of the inquiry is that no adequate mitigation could have been put in place. To tackle this, the Ombudsman recommended that future EU Trust Fund projects would always include a prior human rights impact assessment.[29]

Another related area in which the EU has achieved significant progress is corporate due diligence. In its approach to dealing with sustainability, the EU has adopted three different pieces of legislation. The proposed directive on CSDD would require EU Member States to adopt legislation regulating due diligence aspects of adverse human rights and environmental effects. In a nutshell, the Directive aims at requiring large EU companies and foreign companies operating in the EU to meet due diligence obligations with respect to human rights and environmental standards, setting forth an enforcement mechanism with possible sanctions and civil liabilities for non-compliance.[30] Another piece of EU legislation which is relevant in this field is the Corporate Sustainability Reporting Directive (CSRD) in force since January 2023.[31]

Specific milestones to improve the negotiation and implementation of EUFTIAs depend upon different scenarios. Some other existing tools include post-assessments and the new EU taxonomy.[32] In addition, the adoption of a Rapid Response Mechanism (RRM) to be incorporated into the EUFTIAs has been suggested.[33] A similar RRM has been incorporated under the USMCA (formerly known as NAFTA), which can be triggered by some contracting parties "for expedited enforcement of workers' free association and collective bargaining rights at the facility level".[34] The RRM is understood as an innovative dispute settlement mechanism that can be relied upon to prompt enforcement of TSD provisions.

The effectiveness of Strategic Environmental Assessments (SIAs) varies according to the individual agreement, the timing of the assessment, the scope of its

---

[29] Decision on how the European Commission assessed the human rights impact before providing support to African countries to develop surveillance capabilities (case 1904/2021/MHZ), of 22 November 2021, available at https://www.ombudsman.europa.eu/en/decision/en/163491 (last accessed on 3 March 2024).

[30] EU Commission, Proposal for a Directive of The European Parliament and of the Council on Corporate Sustainability Due Diligence and amending Directive (EU) 2019/1937 (COM/2022/71 final).

[31] The Directive was published in the Official Journal of the European Union on 16 December 2022, after the European Parliament and the Council of the European Union formally adopted it. The CSRD entered into force on 5 January 2023. Available at https://finance.ec.europa.eu/capital-markets-union-and-financial-markets/company-reporting-and-auditing/company-reporting/corporate-sustainability-reporting_en (last accessed on 3 March 2024).

[32] European Commission, EU taxonomy for sustainable activities, available at https://finance.ec.europa.eu/sustainable-finance/tools-and-standards/eu-taxonomy-sustainable-activities_en (last accessed on 3 March 2024).

[33] See Blot et al. 2022.

[34] USMCA, Chapter 31 Annex A; Facility-Specific Rapid-Response Labor Mechanism, available at https://ustr.gov/sites/default/files/files/agreements/FTA/USMCA/Text/31-Dispute-Settlement.pdf (last accessed on 3 March 2024).

recommendations, and the degree of acceptance of those recommendations.[35] Recent assessments of the mechanism have found that the majority of the SIA processes reviewed can be considered good practice in terms of timely stakeholder engagement (i.e., engaging with stakeholders during the initial stages of the SIAs) and in terms of concluding negotiations after the delivery of the SIA, thus allowing the SIA recommendations—in principle—to feed into the negotiation process.

However, there are examples of failures to synchronize FTA negotiations with the SIA process. The EU-Mercosur agreement concluded negotiations before the final SIA report was published, and the EU-Singapore and EU-Vietnam agreements were negotiated using an obsolete SIA (EU-ASEAN). The involvement of trade partners takes place not only during the negotiations but also throughout the life of the agreement and during its revision. This ties in with the internal and external allocation of competences between the EU and Member States. Internally, the SIAs fall under the competence of different DGs (namely, trade and environment).

Regarding the evolution of different "generations" of FTIAs/TSD Chapters, specific wording and tailor-made clauses have varied over the years. Some of the early TSD clauses contained a reference to Multilateral Environmental Agreements (MEAs) as a way to reinforce the application of international environmental law norms, such as the climate change law. Hence, the baseline for compliance is determined by specific commitments as set out in the respective environmental treaty. More modern FTIAs include a binding framework aimed at evaluating compliance with different TSD commitments. In more recent agreements, for instance, the binding force of the Paris Agreement is reflected in the treaty provisions.

It is worth underlining that environmental and labor-related clauses are also found outside the specific TSD chapters, such as in chapters/provisions dealing with specific trade issues. The examination of sustainability within treaties extends beyond the confines of specific TSD chapters. It encompasses diverse dimensions woven into other treaty chapters, addressing aspects such as trade tariffs and various trade-related measures, including Technical Barriers to Trade (TBT) and Sanitary and Phytosanitary Measures (SPS). The broader canvas of sustainability considerations within these chapters forms a critical facet of EU Free Trade and Investment Agreements (EUFTIAs), incorporating various types of TSD clauses. This nuanced approach acknowledges the multifaceted integration of sustainability principles across diverse treaty sections, reflecting the intricate intersection of trade and sustainable development objectives. Sustainability standards vary across treaties and depend on the TSD provisions on environmental and labor issues included in the treaties.

An additional clause of a procedural nature pertains to the "trigger" clause, designed to instigate a dispute settlement process in the event of a violation of the sustainable development provisions outlined in the agreement. Dispute settlement

[35] IEEP, *Enhancing sustainability in EU Free Trade Agreements* (2022). The report's main findings indicate that the majority of SIA processes reviewed showed a timely stakeholder engagement. Sustainable development in the European Union—A statistical glance from the viewpoint of the UN Sustainable Development Goals, available at https://ec.europa.eu/eurostat/documents/15234730/15228899/KS-02-16-996-EN-N.pdf/e4f0a6bc-c051-1b0a-1d14-78f2f99c73de?t=1667248035644 (last accessed on 3 March 2024).

chapters serve a crucial function in upholding compliance with TSD provisions, facilitating the engagement of civil society organizations throughout the resolution process. Due to the imbalance often noted in agreements signed by the EU with developing or emerging contracting parties, in those agreements, it becomes imperative to ensure the assurance of technological transfer and financial support for the realization of sustainability commitments.

During the review of existing FTIAs, new proposals for enhancing sustainability commitments have been put forward. The first EU Free Trade Agreement to include the new generation of TSD clauses was the EU-South Korea FTA treaty and the 2023 EU-New Zealand treaty. The former establishes a structured foundation for collaborative engagement concerning sustainability and agricultural development. This encompasses areas of significance, notably sustainable food systems, which hold the potential for enhancing the resilience of farmers in mitigating the impacts of climate change and environmental degradation.[36]

The recently concluded Free Trade Agreement (FTA) between the EU and New Zealand signifies a substantial enhancement of sustainability commitments within EU FTAs. This agreement reflects, in part, a more progressive stance on sustainability matters. The agreement mandates both contracting parties to actively pursue the objectives outlined in the Paris Agreement and adhere to fundamental conventions established by the International Labor Organization (ILO).[37] Additionally, the agreement seeks to bolster collaboration in the realms of fossil fuel subsidy reform and the advancement of sustainable food systems.[38]

## 4.5 Synergies and Crossroads: EU Law's Interplay with International and Domestic Environmental and Labor Regulations

Another significant issue that these TSD Chapters pose concerns the implementation of treaty clauses in domestic environmental and labor law. One of the stumbling blocks in this respect is the absence of a level playing field in practice, with a considerable unevenness of levels of protection across the different trade partners. An additional hurdle is the exhortatory language used in the environmental and labor provisions of the TSD chapters. Taking the environmental realm as an example, the agreements usually compel the parties to comply with the commitments acquired under various Multilateral Environmental Agreements (MEAs). However, there is a

---

[36] The EU-New Zealand Agreement was signed on 9 July 2023. Following signature, both Parties must complete the required legal processes before the FTA can enter into force, which is expected to occur in the first half of 2024. https://policy.trade.ec.europa.eu/eu-trade-relationships-country-and-region/countries-and-regions/new-zealand/eu-new-zealand-agreement/agreement-explained_en (last accessed on 3 March 2024).

[37] EU-New Zealand Agreement, Article 19.6. Trade and climate change.

[38] Ibid., Article 19.7. Trade and fossil fuel subsidy reform.

weak follow-up of compliance and, often, no consequences are foreseen for the lack of observance of the treaty or related to the withdrawal from the MEA.[39]

In the environmental law realm, areas of particular concern are those relating to climate change provisions and deforestation. In terms of deforestation, most trade and investment treaties do not deal with these issues as demonstrated in specific cases. For instance, despite discussions about halting negotiations between the EU and MERCOSUR due to the forest fires in the Amazon region, no specific action was taken.[40] To address this concern, the EU has recently introduced legislation through a regulation to minimize EU-driven deforestation and forest degradation.[41]

Turning now to analyzing labor provisions in TSD Chapters, according to the International Spillover Index, EU Member States also create negative socio-economic spillovers outside the region.[42] To illustrate this, the 2021 report indicates that "imports of clothing, textiles, and leather products into the EU are related to 375 fatal workplace accidents and 21,000 non-fatal accidents every year".[43] In response to this trend, the EU has tried to progressively cover social aspects through its Free Trade Agreements. These labor provisions are modelled upon the clauses of the EU-Korea FTA. In practice, all EU FTIAs refer to ILO core labor standards as set out in the fundamental ILO Conventions, particularly acknowledging and protecting, freedom of association, the right to organize and collectively bargain, and the elimination of forced labor. Looking at the TSD Chapters from a comparative analysis perspective, a difference can be appreciated between the different TSD chapters of the FTAs between the EU and South Korea (EU-Korea), Canada (CETA) and Japan (JEFTA).[44]

The new generation of EU trade and investment treaties has introduced improved transparency in the TSD dispute settlement process compared to earlier agreements. This is evidenced by the strengthened language, such as the use of "shall", which imparts a sense of commitment in these agreements. Nevertheless, there remains inconsistency and variation across different treaties.

Civil society has played a crucial role in these FTIAs by creating networks of diverse NGOs that help with compliance with national legislation. These networks offer valuable feedback and foster cooperation on critical issues. However, the aspect of dispute resolution presents challenges, as the FTIAs lack sufficient safeguards to ensure compliance and the delivery of treaty outcomes. For instance, there is a gap

---

[39] *In arguendo*, this would mean in practice that if a trade partner would decide to leave the Paris Agreement no specific treaty mechanism would be set in motion.

[40] EU Observer, *Amazon deforestation and the EU-Mercosur trade deal*, 2 August 2021, available at https://euobserver.com/opinion/152572 (accessed on 3 March 2024).

[41] EU Commission, *Proposal for a regulation on deforestation-free products*, available at https://environment.ec.europa.eu/publications/proposal-regulation-deforestation-free-products_en (last accessed on 3 March 2024).

[42] Sustainable Development Report 2022, available at https://dashboards.sdgindex.org/map/spillovers (last accessed on 3 March 2024).

[43] Ibid.

[44] Van't Wout 2022, pp. 81–98.

in follow-up mechanisms when either party fails to adhere to the expert panel's decision.[45]

A controversial question revolves around the existence of a genuine "green conditionality" in the EU's trade and investment policy, particularly, in terms of a weak enforcement system and the absence of a robust sanctions scheme. Altogether, these factors suggest a relatively weak green conditionality, which has prompted the revision of the enforcement and compliance mechanisms.

A consistent and coherent approach to sustainable development in external policy would have a tangible impact on EU citizens. This impact can be particularly observed in specific supply chains, where contentious issues arise concerning the importation of "risk commodities". In this regard, there is emerging case law in the EU about sustainability and trade, for instance, the case of palm oil serves as an instructive example, notably highlighting concerns related to its elevated indirect land-use change risk (ILUC-risk). This particular instance has triggered the potential for climate change-related litigation within the EU, as evidenced by *Santiga Lipids v. Commission* (2019).[46] The legal matter in this case revolves around the exclusion of palm oil from the category of renewable sources. Such cases underscore the intricate legal and environmental implications associated with palm oil production and its broader consequences on climate change.

The ongoing revision of the Trade and Sustainable Development chapters is pivotal in addressing emerging concerns in the realm of trade agreements. Concurrently, the European Commission has taken proactive measures to enhance compliance with obligations by introducing a comprehensive 15-point Action Plan on Trade and Sustainable Development.[47] This multifaceted plan not only seeks to bolster participation from civil society but also aims to fortify regulatory oversight through the implementation of effective dispute-resolution mechanisms. The evolving landscape of international trade agreements, underpinned by these revisions and proposals, reflects the broader commitment to fostering sustainability and accountability. It remains to be seen whether the application of the new rules would extend to agreements already in force or in those under negotiation, such as those with South Korea and MERCOSUR, and on the possibility of expanding the rules to cover raw materials and energy goods.

---

[45] Paolo Mazzotti, "Stepping Up the Enforcement of Trade and Sustainable Development Chapters in the European Union's Free Trade Agreements: Reconsidering the Debate on Sanctions" (European Law Institute Paper 2021), available at https://www.europeanlawinstitute.eu/fileadmin/user_upload/p_eli/YLA_Award/Submission_ELI_Young_Lawyers_Award_Paolo_Mazzotti_2021.pdf (last accessed on 3 March 2024).

[46] Order of the General Court (Fourth Chamber) of 11 June 2020. *Lípidos Santiga, SA v European Commission.* Action for annulment—Energy—Directive (EU) 2018/2001 on the promotion of the use of energy from renewable sources, available at https://curia.europa.eu/juris/liste.jsf?num=C-402/20&language=EN (last accessed on 3 March 2024).

[47] European Commission, European Parliament resolution of 6 October 2022 on the outcome of the Commission's review of the 15-point action plan on trade and sustainable development (2022/2692(RSP), available at https://www.europarl.europa.eu/doceo/document/TA-9-2022-0354_EN.html (last accessed on 3 March 2024).

## 4.6 Navigating Trade, Investment, and Sustainable Development

Due to the substantial volume of trade encompassed by the EU-signed agreements, the EU possesses the capacity to take a global leadership role in establishing environmental and labor standards. As we have seen, the sustainable development elements in EU Trade and Investment Treaties encompass environmental and labor issues. In international law, international trade and investment systems take different approaches to mainstreaming sustainable development. Overall, sustainable development and climate change issues are part of broader initiatives at both the EU and the global level to promote decarbonization and articulate a just transition.

Yet, the SDGs offer a soft law framework by providing standards for assessing success in different areas. In the overall context of international trade and investment law, the SDGs emphasize certain areas that are considered a priority, but some significant transformations must take place in external relations to reinforce collaborative external relationships with other countries, such as capacity-building. There are major SDGs challenges for EU Member States in terms of the external policy, to offset the negative spillover effects that European Member States cause outside the region. This is measured by the Spillover Index, determining the transboundary impacts of one State's action on other States. These negative effects include CO2 emissions, biodiversity threats and profit shifting, biodiversity loss, poor labor standards and deforestation, which have a limiting effect on the achievement of the SDGs.[48] The index demonstrates that there are intrinsic relations between economic growth and environmental spill-overs from EU activities.

Regarding the manner in which the treaties integrate the Multilateral Environmental Agreements (MEAs), the core of conditionality is made up of the various commitments made through the Paris Agreement, other Multilateral Environmental Treaties, and the environmental and climate change clauses specifically incorporated into trade and investment treaties. In the context of multilateral environmental diplomacy, the EU has historically taken a leading role in negotiations. This leadership is exemplified, for instance, by its initiatives during COP27, where it advocated for the decarbonization of the economy and the widespread adoption of renewable energy sources.[49] To ensure coherence within the diverse domains of external policy, it is imperative to bolster adherence to environmental provisions within the sustainable development chapters. The recent convening of COP27 offered a valuable opportunity to contemplate the formulation of more robust objectives and strategies, assess regional advancements, devise recovery plans, and leverage the insights gained from the challenges posed by the pandemic.

---

[48] Spillover Rankings. The spillover performance of all 193 UN Member States https://dashboards.sdgindex.org/rankings/spillovers (last accessed on 3 March 2024).

[49] United Nations—Climate Change, COP27 Reaches Breakthrough Agreement on New "Loss and Damage" Fund for Vulnerable Countries, 20 November 2022, available at https://unfccc.int/news/cop27-reaches-breakthrough-agreement-on-new-loss-and-damage-fund-for-vulnerable-countries (last accessed on 3 March 2024).

In turn, the SDGs serve as a foundational "road map" delineating a comprehensive global framework for international cooperation in the realm of sustainable development, encompassing diverse dimensions including economic, social, environmental, and governance aspects. The far-reaching impact of the pandemic was compounded by deficiencies in social protection, fragmented healthcare systems, and the exacerbation of disparities among countries, notably intensifying poverty indicators. Consequently, imperative emphasis must be placed on prioritizing social inclusion and striving for heightened equality, aligning these goals with the fundamental tenets of the SDGs. Simultaneously, the implementation of the Paris Agreement exhibits variances in the pace at which objectives, such as achieving "net zero" emissions and decarbonization, are being realized, necessitating a rigorous legal analysis of the diverse trajectories in climate action.

Within the sustainable development domain, it is crucial to recognize that labor and social standards constitute a fundamental component of this multifaceted concept. This entails a holistic examination that extends to labor rights, thereby engendering a series of intricate inquiries. In the context of the EU's trade agreements, an integral feature is the inclusion of a ]TSD chapter, which amalgamates the stipulations of multilateral environmental agreements (MEAs) with the core labor standards established by the International Labor Organization (ILO). Under the aegis of the TSD chapters, trade, labor, and environmental considerations are intrinsically linked. However, despite the advancements in this realm, the interplay between these differing sets of priorities remains contentious, marked by a discernible diversity in the standards of protection across nations. It is noteworthy that the foundational framework of labor standards hinges on four primary ILO conventions, each carrying a significant weight in shaping the landscape of international labor rights.

This intricate interplay between trade, labor, and environmental concerns within TSD chapters underscores a critical area of scrutiny and debate in the realm of international law and policy. The amalgamation of environmental and labor standards in trade agreements represents a complex endeavor fraught with challenges, as it strives to strike a balance between fostering economic growth, ensuring the protection of labor rights, and promoting sustainable environmental practices. The variances in the levels of protection across countries further compound the complexities of harmonizing these priorities within the SD framework, necessitating rigorous legal analysis and policy deliberations to advance a more equitable and sustainable global trade landscape.

The objectives within EU Treaties serve as the foundational framework guiding the Union's external policy initiatives. This alignment is manifest in the numerous trade agreements, such as the 2011 EU-Korea Free Trade Agreement (FTA), which underscores the significance of these objectives. Specifically, the Trade and Sustainable Development (TSD) chapters within these agreements encompass a spectrum of standards. Notably, these chapters incorporate substantive standards, which necessitate the commitment of the involved parties to the effective execution of pertinent MEAs or ILO Conventions.[50] Secondly, within the framework of TSD chapters,

---

[50] Harrison et al. 2019.

there exists a set of procedural provisions that facilitate and regulate the discourse and collaboration between the involved parties. These provisions encompass critical aspects such as dialogues, cooperation mechanisms, and the ongoing assessment of the sustainability impacts resulting from the agreement's implementation. It is through these mechanisms that the parties engage in a continuous process of monitoring and revising the sustainability objectives and outcomes, thereby ensuring that the agreement remains aligned with its intended goals.

Thirdly, an essential facet of these TSD chapters is the rigorous adherence to the commitments outlined within the agreement. This aspect serves as a linchpin, embodying the fundamental principles upon which the agreement is founded. Typically, the enforcement of these commitments is entrusted to the dispute settlement section, which provides a structured mechanism for addressing any deviations or breaches that may arise during the agreement's tenure. This dispute settlement process plays a pivotal role in upholding the integrity of the TSD chapters by providing recourse for resolving disputes and reinforcing the commitment to the agreed-upon sustainability standards.

It is essential to note that the structural arrangement of TSD chapters and the interplay of these multifaceted provisions are contingent upon the specific treaty in question. The precise configuration and the level of integration of these elements may vary, reflecting the unique characteristics and objectives of each individual agreement. As such, the intricacies of the TSD chapters are intricately linked to the particularities of the treaty at hand, thereby underscoring the need for a case-by-case analysis when evaluating their design and effectiveness. To illustrate, the EU-UK Trade and Cooperation Agreement (TCA) includes the "Level playing field" chapter encompassing sustainable development provisions incorporating standards, but it does not comprise a robust dispute settlement mechanism.[51]

## 4.7 Ensuring Compliance: Mechanisms in Sustainable Development Chapters

Regarding the institutional mechanisms responsible for supervising the implementation of TSD chapters, they encompass a diverse array of options. These options encompass a panoply of options ranging from a panel of experts tasked with scrutinizing grievances and furnishing pertinent recommendations to more sophisticated mechanisms including quasi-judicial bodies. In cases of non-compliance, the dispute resolution apparatus embedded within the TSD chapters primarily entails government-to-government consultations. It also involves an impartial panel

---

[51] Briefing Paper Number 9190, 20 May 2021, *The UK-EU Trade and Cooperation Agreement: Level Playing Field*, available at https://researchbriefings.files.parliament.uk/documents/CBP-9190/CBP-9190.pdf (last accessed on 3 March 2024).

composed of trade, labor, and environmental experts, along with the ongoing process of implementation monitoring.[52]

When considering adherence to TSD chapters, it is imperative to underscore three specific aspects relating to enforcement mechanisms that merit consideration. These aspects encompass specific labor and environmental obligations articulated within the treaty, the carefully crafted mechanisms aimed at securing compliance with these commitments, and the various stages of the enforcement process.

Chapters on trade and sustainable development in the free trade agreements of the EU reflect international consensus around main issues. Most of the trade-related issues could also be dealt with in the framework of the WTO system, however, the current dispute system is stalled. There are some useful precedents in the WTO that could be relied upon for the future.[53] Precisely, in the WTO realm, the compatibility of EU measures with the multilateral system has long been debated. Along these lines, in the framework of the Committee on Trade and Environment (CTE), WTO members have addressed the trade-related dimensions of the European Green Deal. Additionally, various developing countries (EU partners) have outlined their respective national environmental efforts.[54]

With regard to the current situation and the characteristics of treaty provisions, there is no strict "green conditionality" in the foreign policy of the EU. In view of this, there are new proposals aimed at strengthening sustainability and compliance with environmental regulations within the framework of the common commercial policy. The inclusion of chapters pertaining to trade and sustainable development within free trade agreements occurs within the broader context of the legal framework provided by the WTO and the principles of international law governing foreign investments.

At this point, it is worth noting that TSD chapters in trade agreements embrace environmental and labor aspects. When viewed through an environmental law lens, these chapters show connections with significant international treaties, such as the Paris Agreement. Notably, these multilateral environmental agreements (MEAs) often incorporate precise mechanisms for resolving disputes. Nonetheless, despite the introduction of the SDGs, which provide a novel framework for action, these goals do not offer explicit strategies for putting them into practice.

While emphasizing the opportunity for countries to enhance their commitments in various areas, the Paris Agreement has shown limited progress in achieving emission

---

[52] Mazzotti, P., 'Stepping Up the Enforcement of Trade and Sustainable Development Chapters in the European Union's Free Trade Agreements: Reconsidering the Debate on Sanctions', available at https://www.europeanlawinstitute.eu/fileadmin/user_upload/p_eli/YLA_Award/Submission_ELI_Young_Lawyers_Award_Paolo_Mazzotti_2021.pdf (last accessed on 3 March 2024).

[53] Non-Paper of the Commission services (2018), Feedback and way forward on improving the implementation and enforcement of Trade and Sustainable Development chapters in EU Free Trade Agreements (26/02/2018), available at https://www.politico.eu/wp-content/uploads/2018/02/TSD-Non-Paper.pdf (last accessed on 3 March 2024).

[54] EU-Palm Oil (Indonesia), available at https://www.wto.org/english/tratop_e/dispu_e/cases_e/ds593_e.htm (last accessed on 3 March 2024).

reductions, notably in meeting the Nationally Determined Contributions (NDCs).[55] While it enshrines core provisions related to adaptation, sustainable development, participation, inclusion, and transparent progress monitoring systems, countries are struggling to effectively reduce emissions to limit global temperature increases to 1.5 °C. The absence of a clear link between non-compliance with the Paris Agreement and the Sustainable Development chapters indicates some shortcomings in these commitments.

To address these issues, the European Commission has initiated the examination of proposals for "green conditionality" while also recommending the reform of Sustainable Development chapters. In early 2021, the European Commission published a new communication on EU trade policy. [56] In light of increasing civil society activism against trade agreements, the EU Commission expedited the review of its 2018 15-point action plan on trade and sustainable development, originally slated for 2023.[57]

Yet, at present, the deepening of SD commitments is not guaranteed. Many efforts were made to include commitments in the treaty between the EU and South Korea, however, ILO commitments were breached.[58] The Republic of Korea was required to adjust its labor laws and practices to comply with the principle of freedom of association. Despite these recommendations, the Republic of Korea failed to take steps towards the ratification of fundamental ILO Conventions. Another controversial area concerns risk commodities is a sensitive field for external trade policy.

Another controversial area concerns the possibility of resorting to sanctions to reinforce compliance with TSD provisions. Proponents of this option submit that sanctions will reinforce compliance with TSD provisions.[59] Conversely, those who are opposed to the use of sanctions, allege that they are futile, and argue that soft measures based on litigation would be more effective to encourage compliance. The critics of the use of sanctions and the view that sanctions could lead to the opposite outcome by undermining the attainment of commitments in TSD Chapters. Table 4.1 presents a summary of the main features of SD in FTAs recently signed or negotiated by the EU.

---

[55] United Nations. (2018). 'Sustainable Development Goals'. Available at https://sustainabledevelopment.un.org/?menu=1300 (last accessed on 3 March 2024).

[56] United Nations. (2018). 'Sustainable Development Goals'. Available at https://sustainabledevelopment.un.org/?menu=1300 (last accessed on 3 March 2024).

[57] European Commission, Non-paper of the Commission services, Feedback and way forward on improving the implementation and enforcement of Trade and Sustainable Development chapters in EU Free Trade Agreements, available at https://trade.ec.europa.eu/doclib/docs/2018/february/tradoc_156618.pdf (last accessed on 3 March 2024).

[58] European Commission, Panel of experts confirms the Republic of Korea is in breach of labour commitments under our trade agreement, available at https://ec.europa.eu/commission/presscorner/detail/en/ip_21_203

[59] European Commission, Sustainable development in EU trade agreements (2022), https://policy.trade.ec.europa.eu/development-and-sustainability/sustainable-development/sustainable-development-eu-trade-agreements_en (last accessed on 3 March 2024).

**Table 4.1** EU FTAs—Sustainable Development Chapters

| EU Trade Agreement | Type of Agreement | Status | Sustainable Development Chapter |
|---|---|---|---|
| EU-Chile agreement | EU-Chile Advanced Framework Agreement | Being adopted or ratified | Chapter 26 includes reinforced cooperation regarding climate change, forests, biodiversity. In terms of labor law provisions, it encompasses specific provisions on core labor issues |
| EU-India agreement | EU-India Free Trade Agreement, Investment Protection Agreement and Geographical Indications Agreement | Being negotiated | TBD (To Be Determined)<br>Negotiations are still ongoing. The agreement may introduce a tailored approach. The Trade Sustainability Impact Assessment (SIA) in Support of Negotiations with India has already taken place[60] |
| EU-Kenya agreement | EU-Kenya Economic Partnership Agreement | Being adopted or ratified | The agreement includes strong and binding provisions on labor standards, climate change and biodiversity, and gender equality. In addition, it prevents both parties from lowering labor and environmental standards to attract trade or investment. These commitments are binding and enforceable |
| EU-Mercosur agreement | EU-Mercosur Trade Agreement | Being adopted or ratified | TBD<br>The Chapter has been hailed as putting forward an advanced SD approach. However, scholars like Harrison conclude that the SD Chapter does not significantly reinforce countries' climate change commitments as set out in the Paris Agreement. Nor does it provide a strong framework for addressing other important environmental and social issues. There is also a lack of a proper monitoring system of the treaty's effects, so that State parties can react accordingly[61] |
| EU-Mexico agreement | Trade part of new EU-Mexico Association Agreement | Being adopted or ratified | TBD<br>Modernization of the agreement, the EU and Mexico have agreed to a common sustainable development framework establishing a set of rules for both labor and the environment[62] |

(continued)

[60] https://policy.trade.ec.europa.eu/eu-trade-relationships-country-and-region/countries-and-regions/india/eu-india-agreement/documents_en

[61] Harrison and Paulini 2020.

[62] European Commission, EU-Mexico agreement explained (europa.eu) (last accessed 3 March 2024).

**Table 4.1** (continued)

| EU Trade Agreement | Type of Agreement | Status | Sustainable Development Chapter |
|---|---|---|---|
| EU-New Zealand agreement | EU-New Zealand Trade Agreement | Being adopted or ratified | TBD<br>The EU-New Zealand Free Trade Agreement was signed on 9 July 2023.[63] The Trade and Sustainable Development (TSD) Chapter of the NZ–EU FTA incorporates key provisions relating to labor standards, gender equality, climate change, emissions trading, environmental goods and services, forests, fisheries subsidies, and fossil fuel subsidy reform |
| EU-Singapore agreement | EU-Singapore Free Trade Agreement, Investment Protection Agreement, Digital Trade Agreement | In place | The EU and Singapore Free Trade Agreement and Investment Protection Agreement entered into force on 21st November 2019. From a legal perspective, the EU-Singapore FTA ought to be comprehended in tandem with two intertwined instruments |
| EU-South Korea agreement | EU-South Korea Free Trade Agreement, Digital Trade Agreement | In place | The EU and South Korea have agreed on high labor and environmental standards that protect workers and the environment. The agreement sets up mechanisms to make sure these commitments are met, including through involving civil society |

*Note*: Authors' own elaboration based on the information available on the European Commission's website[64]

A persistent point of contention revolves around the disparities in approaches between the EU and the United States regarding trade and sustainable development chapters. While the United States has chosen to adopt a robust system featuring specific punitive measures, the EU's model emphasizes a more cooperative approach that does not incorporate punitive actions.[65] The paradigmatic case at hand related to the dispute between the US and Guatemala in the textile sector which did not lead to an effective resolution as it did not solve the non-compliance issue.[66]

---

[63] Ballingall 2023.

[64] The EU is also negotiating other agreements, e.g. EU-Indonesia Free Trade Agreement.

[65] Hradilová and Svoboda 2018, pp. 1019–1042.

[66] In the Matter of Guatemala—Issues Relating to the Obligations Under Article 16.2.1(a) of the CAFTA-DR, available at https://ustr.gov/issue-areas/labor/bilateral-and-regional-trade-agreements/guatemala-submission-under-cafta-dr (last accessed on 3 March 2024).

## 4.8 Conclusion

The Trade and Sustainable Development chapters within the EU free trade and investment treaties have undergone a significant evolution. The EU's trade policy is now intrinsically linked to its environmental objectives, reflecting a growing emphasis on sustainability. However, amid these advancements, the issue of consistency remains a paramount challenge when it comes to compliance with the pertinent sustainable development provisions. A notable observation within the current framework is the existence of a relatively weak "green conditionality", which pertains to the assurance of compliance. One notable feature is the absence of a centralized mechanism for monitoring and ensuring compliance. Notably, some clauses integrated into Free Trade and Investment Treaties mirror obligations found within international environmental agreements.

A development on the horizon is the proposal to institute sanctions against EU trading partners found in violation of labor or environmental standards. This proposal is a subject of ongoing debate, and irrespective of its final outcome, it serves as a catalyst for generating innovative ideas that could potentially lead to the implementation of a more efficient compliance mechanism.

Within the global arena, the EU's ambition to establish a regional legal framework that aligns with ambitious goals and objectives is driving reforms that herald a new era for trade and investment treaties. These transformative changes signify a shift towards a more holistic and sustainable approach to international trade and investment, reflecting the EU's commitment to environmental and social responsibility.

## References

Blot, E. A. Oger and J. Harrison (2022), Enhancing sustainability in EU Free Trade Agreements: The case for a holistic approach, Policy report, Institute for European Environmental Policy (IEEP), available at https://www.greens-efa.eu/files/assets/docs/enhancing_sustainability_in_eu_free_trade_agreements._the_case_for_a_holistic_approach__ieep__2022_.pdf (last accessed on 3 March 2024).

Blot, E. and Kettunen, M. (2021). Environmental credentials of EU trade policy – A comparative analysis of EU free trade agreements. Institute for European Environmental Policy, Brussels and London.

Court of Justice of the European Union, Opinion 2/15 of the Court, of 16 May 2017, available at https://curia.europa.eu/juris/document/document.jsf?text=&docid=190727&doclang=EN (last accessed on 3 March 2024).

Court of Justice of the European Union. Order of the General Court (Fourth Chamber) of 11

Cremona, M., "Shaping EU Trade Policy post-Lisbon: Opinion 2/15 of 16 May 2017" (2018) 14 European Constitutional Law Review 23, pp. 231–259.

Damro, C. (2012), The post-Lisbon institutions and EU external environmental policy, in E. Morgera (Ed.), The External Environmental Policy of the European Union: EU and International Law Perspectives (pp. 55–75). Cambridge: Cambridge University Press.

EU Observer, Amazon deforestation and the EU-Mercosur trade deal, 2 August 2021, available at https://euobserver.com/opinion/152572 (accessed 3 March 2024).

EU Ombudsman. EU Commission 'failed' on assessing Mercosur trade deal, available at https://euobserver.com/green-economy/151302 (last accessed on 3 March 2024).

EU-New Zealand Agreement, Article 19.6. Trade and climate change.

European Commission, EU taxonomy for sustainable activities, available at https://finance.ec.europa.eu/sustainable-finance/tools-and-standards/eu-taxonomy-sustainable-activities_en (last accessed on 3 March 2024).

European Commission, EU's Generalized Scheme of Preferences (GSP), https://policy.trade.ec.europa.eu/development-and-sustainability/generalised-scheme-preferences_en (last accessed on 3 March 2024).

European Commission, EU-Mexico agreement explained (europa.eu) (last accessed 3 March 2024).

European Commission, European Parliament resolution of 6 October 2022 on the outcome of the Commission's review of the 15-point action plan on trade and sustainable development (2022/2692(RSP), available at https://www.europarl.europa.eu/doceo/document/TA-9-2022-0354_EN.html (last accessed on 3 March 2024).

European Commission, Statistics for the European Green Deal, available at https://ec.europa.eu/eurostat/cache/egd-statistics/ (last accessed on 3 March 2024).

European Commission, The Green Deal Industrial Plan: putting Europe's net-zero industry in the lead (1st February 2013), available at https://ec.europa.eu/commission/presscorner/detail/en/ip_23_510 (last accessed on 3 March 2024).

European Commission, Trade and Sustainable Development (TSD) chapters in EU Free Trade Agreements (FTAs), available at https://trade.ec.europa.eu/doclib/docs/2017/july/tradoc_155686.pdf (last accessed on 3 March 2024).

European Commission. Communication from the Commission to the European Parliament, The Council, the European Economic and Social Committee and The Committee of the Regions, Trade Policy Review - An Open, Sustainable and Assertive Trade Policy, available at https://trade.ec.europa.eu/doclib/docs/2021/february/tradoc_159438.pdf (last accessed on 3 March 2024).

European Commission. Decision in case 1026/2020/MAS concerning the failure by the European Commission to finalise an updated 'sustainability impact assessment' before concluding the EU-Mercosur trade negotiations, available at https://www.ombudsman.europa.eu/en/decision/en/139418 (last accessed on 3 March 2024).

European Commission. Decision on how the European Commission assessed the human rights impact before providing support to African countries to develop surveillance capabilities (case 1904/2021/MHZ). Decision. Case 1904/2021/MHZ - Opened on Tuesday | 30 November 2021 - Decision of 28 November 2022.

European Commission. Decision on how the European Commission assessed the human rights impact before providing support to African countries to develop surveillance capabilities (case 1904/2021/MHZ), of 22 November 2021, available at https://www.ombudsman.europa.eu/en/decision/en/163491 (last accessed 3 March 2024).

European Commission. Negotiations and Agreements. Available at https://policy.trade.ec.europa.eu/eu-trade-relationships-country-and-region/negotiations-and-agreements_en (last accessed on 3 March 2024).

European Commission. Proposal for a Directive of The European Parliament and of the Council on Corporate Sustainability Due Diligence and amending Directive (EU) 2019/1937 (COM/2022/71 final).

European Commission. Proposal for a regulation on deforestation-free products, available at https://environment.ec.europa.eu/publications/proposal-regulation-deforestation-free-products_en (last accessed 3 March 2023).

European Commission. Sustainable Development Goals. Available at https://commission.europa.eu/strategy-and-policy/sustainable-development-goals_en#:~:text=Sustainable%20development%20is%20a%20core,Sustainable%20Development%20Goals%20(SDGs (last accessed on 3 March 2024).

European Economic and Social Committee, The power of trade partnerships: together for green and just economic growth, available at https://www.eesc.europa.eu/en/our-work/opinions-information-reports/opinions/power-trade-partnerships-together-green-and-just-economic-growth (last accessed (last accessed on 3 March 2024).

European Union, Circular Economy Action Plan, available at https://ec.europa.eu/environment/circular-economy/pdf/new_circular_economy_action_plan.pdf (last accessed on 3 March 2024).

European Union. From an EU constitutional law standpoint, sustainable development has been incorporated into different provisions of the treaties, see the Preamble to the TEU, Art 3(3) TEU and Art 11 TFEU, and Art 37 CFREU. Additionally, in terms of the EU's external action, see Art 3(5) TEU, Art 21(2)(d) TEU, and Art 21(2)(f) TEU.

European Union. Sustainability impact assessments, https://environment.ec.europa.eu/law-and-governance/environmental-assessments/environmental-impact-assessment_en (last accessed on 3 March 2024).

Eurostat, EU and global value chains, available at https://ec.europa.eu/eurostat/web/economic-globalisation/globalisation-in-business-statistics/global-value-chains (last accessed on 3 March 2024).

Harrison, J. and Paulini, S., The Trade and Sustainable Development Chapter in the EU-Mercosur Association Agreement | ClientEarth 15 July 2020.

Hradilová, K. and Svoboda, O., 'Sustainable Development Chapters in the EU Free Trade Agreements: Searching for Effectiveness', Journal of World Trade, Volume 52, Issue 6 (2018) pp. 1019–1042.

IEEP, Enhancing sustainability in EU Free Trade Agreements (2022). The report's main findings indicate that the majority of SIA processes reviewed showed a timely stakeholder engagement. Sustainable development in the European Union - A statistical glance from the viewpoint of the UN Sustainable Development Goals, available at https://ec.europa.eu/eurostat/documents/15234730/15228899/KS-02-16-996-EN-N.pdf/e4f0a6bc-c051-1b0a-1d14-78f2f99c73de?t=1667248035644 (last accessed on 3 March 2024).

Mazzotti, P., 'Stepping Up the Enforcement of Trade and Sustainable Development Chapters in the European Union's Free Trade Agreements: Reconsidering the Debate on Sanctions', available at https://www.europeanlawinstitute.eu/fileadmin/user_upload/p_eli/YLA_Award/Submission_ELI_Young_Lawyers_Award_Paolo_Mazzotti_2021.pdf (last accessed on 3 March 2024).

Meunier, S., and Nicolaidis, K. (2019) The Geopoliticization of European Trade and Investment Policy. JCMS: Journal of Common Market Studies, 57, pp. 103–113. https://doi.org/10.1111/jcms.12932 (last accessed on 13 October 2023.

Morgera, E. and Marín Durán, G., "Article 37 - Environmental Protection", in Steve Peers and others (eds), The EU Charter of Fundamental Rights: A Commentary (Hart Publishing 2014), pp. 987–989 and pp. 992–994.

Pelkmans, J. Linking "values" to EU trade policy—a good idea? Eur Law J. 2020; 26(5-6), pp. 391–400.

Spillover Rankings. The spillover performance of all 193 UN Member States https://dashboards.sdgindex.org/rankings/spillovers (last accessed on 3 March 2024).

Sustainable Development Report 2022, available at https://dashboards.sdgindex.org/map/spillovers (last accessed on 3 March 2024).

UK Parliament. Briefing Paper Number 9190, 20 May 2021, The UK-EU Trade and Cooperation Agreement: Level Playing Field, available at https://researchbriefings.files.parliament.uk/documents/CBP-9190/CBP-9190.pdf (last accessed on 3 March 2024).

United Nations – Climate Change, COP27 Reaches Breakthrough Agreement on New "Loss and Damage" Fund for Vulnerable Countries, 20 November 2022, available at https://unfccc.int/news/cop27-reaches-breakthrough-agreement-on-new-loss-and-damage-fund-for-vulnerable-countries (last accessed 23 January 2023).

USMCA, Chapter 31 Annex A; Facility-Specific Rapid-Response Labor Mechanism, available at https://ustr.gov/sites/default/files/files/agreements/FTA/USMCA/Text/31-Dispute-Settlement.pdf (last accessed on 13 October 2023).

Van der Loo, G., The Court's Opinion on the EU-Singapore FTA: Throwing off the shackles of mixity?, No 2017/17, May 2017, available at https://www.ceps.eu/wp-content/uploads/2017/05/PI2017-17 (last accessed on 3 March 2024).
Van't Wout, D., "The enforceability of the trade and sustainable development chapters of the European Union's free trade agreements", (2022) Asia Europe Journal 20, pp. 81–98.
Yotova, R. (2018). Opinion 2/15 of the CJEU: Delineating the Scope of The New EU Competence in Foreign Direct Investment. The Cambridge Law Journal, 77(1), pp. 29–32.

**Belen Olmos Giupponi** is a Professor of International Environmental Law, Middlesex University London, The Burroughs, Hendon, London, NW4 4BT, United Kingdom.

**Hannes Hofmeister** is a Professor of European Law, University of Bozen, Piazza Universita', 1, 39100, Bolzano (BZ), Italy.

# Chapter 5
# The Decisive Link Between Sustainability and Security: Mandate, Scope and Purpose

**Stefania Rutigliano**

**Contents**



**Abstract** The interpretation of "security" has evolved and has enclosed the necessity for nations to take action against emerging new threats, such as climate change. Although there are scientific uncertainties about the potential capacity of environmental stress to become a cause of conflict, there is a broad consensus among institutions and literature that climate change is a "risk multiplier", since it can increase resource demands and environmental degradation and exacerbate existing fragility, challenging the stability of States and societies. Yet safeguarding States from the consequences of climate change or preventing these outcomes has not been explicitly included in the mandate of UN or EU missions. The UN has a broad mandate to maintain international peace and security and climate-related security risks could fall within this conflict prevention competence. Likewise, on the European front, the establishment of Common Security and Defense Policy (CSDP) missions specifically aimed at countering climate change's negative consequences seems necessary. In principle, the broad mandate on CSDP missions to prevent conflicts and strengthen international security could also include *lato sensu* climate-related security risks. The EU has a wide range of instruments to counter the negative effects of climate change: besides initiatives to reduce carbon footprint, it could invest in the creation of new missions in areas most at risk of conflict due to climatic conditions, with the specific aim of preventing conflicts from arising, expanding CSDP missions' mandate or,

S. Rutigliano (✉)
University of Cassino and Southern Lazio, Cassino, Italy
e-mail: stefania.rutigliano1@uniba.it



R. A. Wessel et al. (eds.), *EU External Relations Law and Sustainability*, Global Europe: Legal and Policy Issues of the EU's External Action 4,
https://doi.org/10.1007/978-94-6265-655-0_5

alternatively, and develop ad-hoc capabilities through the establishment of Permanent Structured Cooperation (PESCO) projects, which enjoy greater flexibility in the approval process.

**Keywords** Security · Sustainability · Climate-related security risks · Climate change · UN missions · CSDP missions · CSDP mandate

## 5.1 Evolution of the Concept Of "Security" through Three Conceptual Lenses

The concepts of security and sustainability have mostly been treated as separate and barely related topics by literature. But nowadays, climate change and climate disasters pose an existential threat to the entire human race and biodiversity, and, for this reason, sustainability has become a ubiquitous theme that overrides all spheres.

Indeed, climate change has serious implications for international stability and security, acting as a wake-up call for the international community. For this reason, this chapter explains why we consider security and sustainability as highly interconnected concepts.

First, it should be noted that the notion of "security policy" is not unambiguous: it has expanded considerably, and it has taken on different meanings depending on the evolving interpretations given by scholars. The older interpretation was "State-centered" and identified security with the aim of preserving society from any perceived threat from outside its territory that could endanger its survival, integrity, economy or independence, particularly if the State was in a condition of vulnerability.[1]

More recently, security policy has been interpreted through three conceptual lenses: globalization, human security and securitization.[2] Globalization has increased the interconnectedness of societies and States, led to a contraction of space and time, and diminished the ability of any State to manage security threats and risks on its own. Human security, on the other hand, has advocated a people-centered, universalist and non-military approach that takes due account of threats to human life.[3]

Moreover, through the securitization process, specific problems are constructed as security issues: when a concern is identified and declared as posing an existential

---

[1] Opposed to 'global security'—in the Copenhagen School's separation—there would be 'societal security' or 'inner security', understood as the collective security of the state through protection of its population and territory, which broadens the context of the threats to be considered by including economic, environmental, social and political factors. Rothe 2020; Sebesta 1991; Buzan 1991, as well as Sjursen 2001, Barbé 1995, Risi 2010.

[2] Le Gloannec et al. 2013.

[3] Tadjbakhsh and Chenoy 2006, Thakur and Newman, eds. 2004.

threat to a referent-object, it requires the adoption of extraordinary or even emergency measures.[4]

The idea of securitization has been widened, not without criticism:[5] as early as the 1970s the need to broaden the understanding of "national security" concepts beyond the realm of military security and other than the nation State emerged.[6]

Since the early 1980s, while military threats remained at the core of international security concerns, additional Non-Traditional Security Threats (NTST) arose and began to be identified. The NTS threats transcend national geographical boundaries, are non-State-centered, could lead to armed conflicts, and most importantly, they cannot be faced through traditional national security policies.[7]

Therefore, the concept of security has been broadened and deepened: it started to include in the security agenda non-military threats such as terrorism and security challenges, inter alia environmental scarcity and pandemics and to consider referent-objects other than the State, such as individuals and social groups.[8]

As a result of the combined application of the three conceptual lenses, the concept of national security has evolved from a "State-centric" interpretation to a "societal well-being" approach and has highlighted the necessity for nations to take action

---

[4] This concept was first presented in the Working Paper "Security the Speech Act: Analysing the politics of the world" (1989) by Ole Wæver, exponent of the Copenhagen School (CS). According to this faction of the doctrine, a security issue does not normally exist concretely but is constructed discursively through the securitisation process, i.e., when a "securitising actor uses a rhetoric of existential threat and thereby takes an issue of what under those conditions is normal politics". See Haacke and Williams 2008, p. 775; Buzan et al. 1998.

[5] The most important criticism of social security is that it is based on a conceptualisation of identity as fixed rather than constructed. See McSweeney 1999; Huysmans 1998.

Another critic concerns the inability of the concept of securitisation to identify 'the silent security dilemma': situations in which the potential subject of insecurity has no, or limited, possibility of speaking about its security problems. See Hansen 2000; Balzacq 2007.

Methodologically, there is a certain ambiguity in securitisation theory: what a securitisation might entail has not been further explored, and the majority of the theory leans for a more explicit verbal speech act methodology. Buzan et al. 1998.

[6] Buzan and Hansen 2009, p. 187.

[7] Non-traditional security threats are characterised by some common elements:

- Non-military for nature
- not stemming from states competition or power shifts
- Threats gather from impact of humans on the environment
- Consequences originated from threats difficult to reverse
- Solutions at national level is ineffective

Caballero-Anthony 2015; Craig 2007.

Some scholars, part of the widening debate during the 1970s and 1980s, were trying to address economic and especially environmental issues as part of international security (*inter alia* Buzan 1983; Ullman 1983; Mayall 1984). From the late 1980s, widening and deepening opened the way for a much more thorough debate about this concept than had occurred during the Cold War (Haftendorn 1991; Rothschild 1995; Wæver 1995, pp. 46–86).

[8] Paris 2001 pp. 87–102; Stern and Öjendal 2010; Hettne 2010; Schlag et al. 2016; Ligtvoet et al. 2017; Keskinen et al. 2019.

against emerging new threats to international stability and individuals' lives, such as terrorism and climate change, rather than focusing merely on military force.[9]

## 5.2 Climate Change Identified as a Security Threat

The global climate securitization has been presented over the years as an increasingly worrying crisis and an existential threat for all individuals. From 1988, when consequences of this phenomena were judged to be second "only to nuclear war",[10] to the early 2000s, when a proliferation of reports highlighted the security dimensions of climate change.[11]

In terms of scientific evidence, the link between climate change and conflicts remains hard to grasp, since there have been different approaches and studies leading to non-univocal results. Nonetheless, climate change might disrupt access to vital resources such as clean water, hence it becomes an increasingly relevant topic to assess.

Research on environmental security has first led to the denial that environmental variables alone would be sufficient to cause violent conflict, mass migration or other instabilities and that further specific factors that determine the vulnerability of societies to climate change were needed to be examined in tandem.[12] As a matter of fact, there are "notable uncertainties about climate-conflict links", and there is "limited understanding" of these connections: although climate change may be one of many factors leading to or exacerbating conflict, it can be arduous to pinpoint its exact role or the relative strength of its impact.[13]

Some others not only criticized the empirical basis of climate security discourses but explicitly warned against the potential and indirect negative implications of discussing climate change in terms of security and survival, since it might prompt the

---

[9] This broader perspective is consistent with the view that threats to "human security"—defined by General Assembly resolution 66/290 as an "approach to assist Member States in identifying and addressing widespread and cross-cutting challenges to the survival, livelihood and dignity of their people"—not only heighten the risk of conflict but also undermine efforts to sustain peace and stability in a long-term, durable way. UNGA resolution 66/290 2012, para 3.

[10] A conference about "The Changing Atmosphere: Implications for Global Security" was held in Toronto. See Davoudi 2014.

[11] Oels 2012, pp. 185–205.

[12] These scholars argued that there is no linear and deterministic relation between the variable "climate change" for the occurrence of violent conflicts and other security threats.

Similarly, also the relationships between desertification or floods and mass migration that could lead to violent reactions in recipient countries should be examined carefully. See Homer-Dixon 1991; Barnett and Adger 2007; Gleditsch et al. 2007; Theisen et al. 2012, pp. 585–602. In an article based on interviews with 12 leading social scientists, the authors conclude that low socioeconomic development, low state capacity, intergroup inequality, and a past history of conflict are most associated with conflict risk and that "climate variability is low on the ranked list of most influential conflict drivers across experience to date". Mach, Kraan et al. 2019, p. 4.

[13] See Mach and Kraan 2021, p. 169; Salehyan 2008; Sakaguchi et al. 2017, p. 641.

wrong policy measures and bear the risk of an increasing militarization of climate politics.[14]

However, there are persuasive arguments to support the connection between climate change and conflicts and, ipso facto, the link between climate change and security. Some studies find evidence that "climate change variables are associated with higher levels of conflict"—even if the role played by climate change in causal mechanisms related to conflict remains unclear[15]—and some other scholars see climate change as the source of "climate wars"[16] or even as a hyper-variable affecting global security in the twenty first century.[17]

Summarizing the vast academic debate, while some scholars consider implications of climate change as a national security concern,[18] other analysts see it as a clear case of human security.[19] A third view, conceptual combination of the first two, stresses the international and collective security dimension of climate change.[20]

Despite scientific uncertainties surrounding the potential capacity of environmental stress to become a cause for conflict constellations, experts clearly warn that further climate change will amplify conflict risk in the future: they estimate that a 2 °C increase in the global temperature will increase the risk of conflict by 13%.[21]

More specifically, scholars believe that a drastic change in temperatures and normal precipitation would systematically increase conflict risk and different forms of interpersonal violence, and that this positive relationship will continue, predicting that each degree Celsius increase in global temperature will increase homicide rates by 6%.[22]

---

[14] See Buckland 2007; Detraz and Betsill 2009, p. 314.

[15] See Sakaguchi et al. 2017, p. 640; UN Security Council 2021, The UN Security Council and Climate Change, Research Report 2021, n. 2.

[16] See Dyer 2009; Fetzek and Mazo 2014.

[17] See Smith 2007; Rothe 2015.

[18] See Busby 2008.

[19] Climate change is a case of human or societal security given its global scope and environmental, social and economic causes and impacts. See Tuchman 1989.

[20] See Kaldor et al. 2007.

[21] See Mach et al. 2019; Hsiang et al. 2013.

[22] See Mares and Moffetti 2016; Ranson 2014.

Altogether, there is a broad agreement among institutions[23] and literature[24] that climate change is a "risk multiplier", since it can increase resource demands, environmental degradation and exacerbate existing fragility and contextual factors of conflict, thereby challenging the stability of States and societies.

## 5.3 The UN and the EU Mandate to Counter Climate Change Consequences

The UN Security Council is conferred by Articles 1 and 24 of the UN Charter with the primary responsibility of prevention and removal of threats to international peace and security, in accordance with the Purposes and Principles of the United Nations.[25]

---

[23] Among others, see: in its Global Strategy (2016) the EU recognised that the international environment had become more hostile and competitive due to regional and great powers asserting themselves. European External Action Service 2016, Shared Vision, Common Action: A Stronger Europe A Global Strategy for the European Union's Foreign And Security Policy; European Commission 2019, The European Green Deal, COM (2019) 640 final. In 2019, the European Council labelled climate change as "a direct and existential threat, which will spare no country. Council of the European Union, Council Conclusions on Climate Diplomacy, February 18, 2019.

Moreover, the connection between climate change consequences and security has been identified by the European Commission and the High Representative since sudden shocks and chronic pressures—such as floods and droughts—can strain state capacities and heighten resource competition, sparking friction between countries or communities and that chronic environmental deterioration can sap social cohesion and create instability, conflict and fragility. European Commission and the High Representative 2017, A Strategic Approach to Resilience in the EU's External Action.

Furthermore, in the strategic document "Strategic Compass" there is a whole paragraph dedicated to climate change, environmental degradation and natural disasters to enhance resilience since these will "impact our security landscape over the next decades and are proven drivers for instability and conflict around the globe". Council of the European Union 2022, A Strategic Compass for Security and Defence For a European Union that protects its citizens, values and interests and contributes to international peace and security, n. 7371/22.

Recently, nearly 60 speakers in an open debate told the Security Council that people and countries most vulnerable to climate change also are most vulnerable to terrorist recruitment and violence and that the negative synergy between the two crises threatens to undermine States and international security itself. UN Security Council 2021, SC/14728 8923rd meeting (AM & PM).

[24] More specifically, climate change has the potential to:

• increase the likelihood of competition or even armed conflict over access to scarce resources such as water or oil or the control of waterways;

• affect revenues from exports, particularly of hydrocarbons, and with it the ability to fund armed forces relative to potential protagonists;

• render existing defence equipment less effective or unusable in harsher conditions, or require new capabilities;

• threaten critical infrastructure through extreme weather events, distract armed forces to domestic crises, and thus increase vulnerability to hostile actions by other states or non-state actors. In more vulnerable countries and regions, the impacts of climate change are more likely to have severe security implications.

See Meyer et al. 2021, p. 9; Detges et al. 2020, p. 11.

[25] Article 24 UN Charter:

Security implications of climate change appear not to fit into conventional notions of "international peace and security", although Article 24(1) of the UN Charter does not specify the exact nature of it. In the understanding of the aftermath of the Second World War, the reference in the Charter to international peace and security focused mostly on the absence of military violence, which had a very war-related restrictive interpretation, disregarding other aspects of security which were not manifest at the time.

As we have already noted in the previous paragraphs, the controversial boundary between climate change and conflicts is not unequivocally proven. Yet, climate change is considered a "risk multiplier" with direct or indirect links to conflicts[26] and it is within the Security Council's prerogative to decide which matter should be addressed and what "international peace and security" entails.[27]

In this regard, the Charter makes it clear that the responsibility of the United Nations is "to save succeeding generations from the scourge of war"[28] and that the UN Security Council may investigate any "dispute" or "situation" which might lead to international friction or give rise to a dispute.[29]

---

"1. In order to ensure prompt and effective action by the United Nations, its Members confer on the Security Council primary responsibility for the maintenance of international peace and security, and agree that in carrying out its duties under this responsibility the Security Council acts on their behalf.

2. In discharging these duties the Security Council shall act in accordance with the Purposes and Principles of the United Nations. The specific powers granted to the Security Council for the discharge of these duties are laid down in Chapters VI, VII, VIII, and XII.

3. The Security Council shall submit annual and, when necessary, special reports to the General Assembly for its consideration."

[26] As already assessed, there is no unanimity among scholars on the causal relationship of climate change and conflicts. Nevertheless, it is undisputed that consequences on climate change influence how violence arises and develops. See notes n. 16, 17, 18.

[27] According to Article 39 of the UN Charter: "The Security Council shall determine the existence of any threat to the peace, [...] to maintain or restore international peace and security."

An example of the discretionary power in assessing a direct link between conflicts and climate change effects can be found in the speech made by former UN Secretary-General Ban Ki-Moon regarding Darfur, where he said "Darfur is also an environmental crisis—a conflict that grew at least in part from desertification, ecological degradation and a scarcity of resources, foremost among them water." See UN Secretary General 2007, What I saw in Darfur.

See also Penny 2018, p. 34 "the Charter itself does not define 'threat to the peace, breach of the peace or act of aggression', instead, the meaning of this concept has developed through actual organisational practice."

[28] The Preamble of the Charter of the United Nations begins with the words "We the peoples of the United Nations determined to save succeeding generations from the scourge of war, which twice in our lifetime has brought untold sorrow to mankind."

[29] Article 34 UN Charter: "The Security Council may investigate any dispute, or any situation which might lead to international friction or give rise to a dispute, in order to determine whether the continuance of the dispute or situation is likely to endanger the maintenance of international peace and security."

Terms such as "situation" and "dispute" are wide enough to possibly include climate change.[30] Indeed, the Security Council has recently tried to approve a draft resolution that would have integrated climate-related security risk as a central component of UN conflict-prevention strategies, but some States rejected it, emphasizing that climate change is primarily a sustainable development issue, outside the Security Council's remit.[31]

Conversely, States with a less traditional view of peace and security have sought to address climate-security threats as an element of the UN conflict prevention work, and in more recent years, as an important part of the UN's peacekeeping, peacebuilding and sustaining peace efforts.[32]

Thus, the UN's charter-given mandate to maintain international peace and security does not expressly mention the protection of the environment or the safeguard of Member States from the consequences of climate change, but, according to a broad definition of "security", interpreted specifically through the conceptual lenses of globalization and securitization, these topics could fall within the UN's conflict prevention remit, since environmental issues can exacerbate the risk of conflict or deteriorate into violence.

Nevertheless, since there is no common agreement regarding the link between climate change consequences and conflict prevention on an institutional or doctrine level, the UN has tried to limit the negative consequences of the threats of climate change with other means than environmental security.

The first and most important achievement was the institution of the United Nations Environment Programme (UNEP),[33] the leading global environmental authority that works against climate change, in favor of environmental protection and sustainable use of natural resources with an autonomous mandate to provide guidance to its Member States on environmental issues and policy guidance.

---

[30] As Luck has argued: "Article 34's references to 'disputes' and 'situations' undoubtedly encompassed economic and trade issues, as well as social and human rights matters", Luck 2014, p. 129.

[31] An example of the rift that have characterised international climate change negotiations can be detected in UN Security Council discussion on climate change as a security threat held on 23 February 2021: representatives of China, India and Russia questioned the strength of the causal link between climate and security, then whether this is really a topic for the UNSC. UN Secretary General 2021, Secretary-General's remarks to the Security Council—on addressing climate-related security risks to international peace and security through mitigation and resilience building, p. 6. The same thing also happened on 13 December 2021, when India and the Russian Federation voted against a draft Resolution Integrating Climate-Related Security Risk into Conflict-Prevention Strategies. UN Security Council 2021, SC/14732, 8926th meeting (AM).

[32] Among the most favourable to the incorporation of climate change in the mandate of the UN's peacekeeping operations there are: Canada, Australia, New Zealand, Switzerland, Malta, Italy, Austria, Indonesia, Vietnam, Kazakhstan, and Latvia. See notes 61 et ss.

[33] Since its inception in 1972, the United Nations Environment Programme (UNEP) has been the global authority that sets the environmental agenda, promotes the coherent implementation of the environmental dimension of sustainable development within the UN system and serves as an authoritative advocate for the global environment.

Moreover, the UN has adopted numerous acts over the years: the first global Convention on climate change, UN Framework Convention on Climate Change, (UNFCCC),[34] which contained a commitment to investigate the problem and draw up plans to combat global warming and build international cooperation[35] and the Kyoto Protocol that committed the Organization for Economic Co-operation and Development (OECD) countries and the economies "in transition" to reduce greenhouse gas (GHG) emission.

In 2015, the international community was able to agree on another shared plan with a specific proposal towards climate change: the Paris Agreement which, together with the establishment of limits for the emission of GHG, gave a pattern for implementing the cooperation on climate change.[36]

In addition, all UN Member States adopted the Millennium Development Goals in 2000, then superseded by the Sustainable Development Goals in 2015, a universal call to action to end poverty, protect the planet, as part of the 2030 Agenda for Sustainable Development which set out a 15-year plan to achieve the Goals.[37] In particular, goal 13 calls for urgent action to combat climate change impacts and it is intrinsically linked to all 16 of the other Goals of the 2030 Agenda.

On the European front, the EU is attempting to provide a collective response to tackle the growing negative effects of climate change, and, in theory, this provision would have its legal basis in Articles 3(5)[38] and 21(2)(c)[39] of the Treaty on European Union (TEU), according to which the Union shall contribute to peace, international security and sustainable development of Earth and preserve and improve the quality

---

[34] United Nations Framework Convention on Climate Change (UNFCCC) adopted during the Rio Conference in 1992 and also known as "Rio Convention". FCCC/INFORMAL/84 GE.05-62220 (E) 200705.

[35] Sands 1992; Arts and Rüdig 1995, p. 481.

[36] World leaders at the UN Climate Change Conference (COP21) in Paris reached the historic Paris Agreement. United Nations Framework Convention on Climate Change (UNFCCC) 2016, FCCC/CP/2015/10/Add.1 See also Oberthür 2016.

[37] The agenda sets 17 objectives for sustainable development that cover numerous areas related to sustainability in its various dimensions and are in turn associated with 169 targets: in order to promote and support their realisation, the assembly also renewed the monitoring system and the global partnership, to which goal 17 is dedicated. United Nations 2015, The 2030 Agenda for sustainable development transforming our world: the 2030 agenda for sustainable development, A/RES/70/1. See also Ingravallo 2022.

[38] Article 3(5) TEU: "In its relations with the wider world, the Union shall uphold and promote its values and interests and contribute to the protection of its citizens. It shall contribute to peace, security, the sustainable development of the Earth, solidarity and mutual respect among peoples, free and fair trade, eradication of poverty and the protection of human rights, in particular the rights of the child, as well as to the strict observance and the development of international law, including respect for the principles of the United Nations Charter."

[39] Art. 21(2) TEU: "The Union shall define and pursue common policies and actions, and shall work for a high degree of cooperation in all fields of international relations, in order to: […] c) preserve peace, prevent conflicts and strengthen international security, in accordance with the purposes and principles of the United Nations Charter, with the principles of the Helsinki Final Act and with the aims of the Charter of Paris, including those relating to external borders".

of the environment and the sustainable management of global natural resources, to ensure sustainable development.

As we will see in the next paragraph, the climate-related security risks appear integrated and closely linked to the EU's Common Foreign and Security Policy (CFSP)[40] and climate change and environmental consideration could be considered as part of the Common Security and Defense Policy sector (CSDP).[41]

Building on this framework, the EU has elevated climate security risks to the center of EU defense, development, and peacebuilding policy priorities: in 2020, the Climate Change and Defense Roadmap[42] and in Parliament's resolution of 7 June 2022 thereon[43] proposed to fully incorporate climate factors into CSDP and concrete actions through capability development and diplomatic outreach.

The Roadmap clearly heralds closer links between climate change and security policy: climate change and environmental degradation could ultimately lead to increased violence and political instability, which could lead to a growing demand for CSDP military and civilian missions and operations.

In addition, the Resolution states that climate change worsen existing social, economic and environmental risks that can fuel unrest and potentially result in violent conflict and that, combined with other factors, aggravate pre-existing vulnerabilities, tensions and risks.

---

[40] At first, the EU's approach to climate security lacked a univocal and clear mandate and institutional competence since this topic 'falls into the gaps between ministerial portfolios and institutional mandates'. Youngs 2014, p. 49. Indeed, it did not appear clear where the concepts of 'environmental' or 'climate security' belong in the intricate institutional web of expertise of external relations: climate-related security risks are regularly mentioned in the Foreign Affairs Council conclusions, in documents from the Commission, the EEAS and the European Parliament, but few of them are also published by Commission's specific units as DG CLIMA, DG DEVCO, DG ECHO, DG DEFIS.

Bremberg 2019, Remling and Barnhood 2021. See also EUR-Lex—52018SC0461—EN—EUR-Lex (europa.eu) accessed on 28 December 2022.

[41] European Commission 2019, The European Green Deal, COM (2019) 640 final, para 3 recognises that climate policy implications should become an integral part of the EU's thinking and action on external issues, including in the context of the CSDP. And Strategic Compass considered "mainstream climate change and environmental considerations throughout our civilian and military CSDP missions and operations". Council of the European Union 2022, A Strategic Compass for Security and Defence For a European Union that protects its citizens, values and interests and contributes to international peace and security, n. 7371/22, p. 38.

[42] EEAS 2020, Climate Change and Defence Roadmap—EEAS(2020)1251. A number of CSDP missions and operations are already deployed in climate-challenged regions: EUTM / EUCAP Mali, EU NAVFOR Atalanta, EUCAP Somalia having as main objectives to fight off some of the threats listed above. However, the climate factors are neither the primary nor even the secondary circumstances or reasons for deployment, although, the aggravating climate change process is directly proportional with the hazards being fought off by the specified missions.

[43] European Parliament 2022, Resolution of 7 June 2022 on the EEAS's Climate Change and Defence Roadmap (2021/2102(INI)) *OJ C 493, 27.12.2022, pp. 19–31,* see paragraph on "strategy and concept": "Article 21 TEU provides an appropriate legal basis for making the Union's external action and the common security and defence policy (CSDP) fit to meet the 21st century's main challenges, of which climate change and climate-based factors are main drivers".

## 5.4 The EU's Possible Effective Actions through CSDP Missions

Today, it is undeniable that the EU is making a strong effort to put climate change and its prevention at the center of its foreign and security policy since it has acknowledged and then prompted environmental security in many of the above-mentioned documents and initiatives.

Nevertheless, in our opinion, what seems necessary at this point would be the establishment of CSDP missions specifically aimed at countering climate change's negative consequences. Although scholars and institutions, as we have assessed, are convinced that climate change is a "risk multiplier" and must be considered in future security scenarios, climate security considerations do not have a direct influence on either the establishment or management of EU's operations and missions.

As a matter of fact, climate-related conflict risks and responses have not been incorporated—even if there would be an identified need—into the mandate of any EU CSDP missions. Notably, the remit of these missions, according to Article 42(1) TEU[44] is to draw on civil and military means, also outside the EU, for peace-keeping, conflict prevention and to strengthen international security in accordance with the UN Charter.

In principle, the broad mandate of CSDP missions to prevent conflicts and strengthen international security could also include *lato sensu* climate-related security risks, as analogously noted in the previous section regarding UN missions' mandate.

Nonetheless, CSDP missions have not yet included explicit or direct operational elements related to climate change. The recent comprehensive evaluation of EU conflict prevention and peacebuilding activities did not assess climate factors as an issue, but, instead, it merely considered it as a "complementary" factor to the EU's core conflict strategies.[45]

Even if CSDP missions have been deployed in climate-stressed areas, environmental aspects have never been considered the reason for their deployment: out of the ongoing twenty CSDP civilian and military missions, seven[46] have been established

---

[44] Art. 42(1) TEU: "The common security and defence policy [...] shall provide the Union with an operational capacity drawing on civilian and military assets. The Union may use them on missions outside the Union for peace-keeping, conflict prevention and strengthening international security in accordance with the principles of the United Nations Charter. [...]"

[45] European Commission 2020, Final Report on External Evaluation of EU's Support to Conflict Prevention and Peacebuilding 2013–2018. Climate change was also mentioned as hybrid threat in the EDA's 2023 EU Capability Development Priorities regarding Critical Infrastructure Protection and Energy Security: "appropriate capabilities to address the protection of Critical Energy Infrastructure against hybrid threats, including climate change cascading effects and natural disasters, must be developed." EDA 2023, The 2023 EU Capability Development Priorities. Ducrotté 2012, p. 6.

[46] Civilian and military missions in Sahel: EUCAP SAHEL MALI (European Union Capacity Building Mission); EUCAP SAHEL NIGER; EU RACC SAHEL (The Eu Regional Advisory And Coordination Cell For The Sahel); EUTM MALI (European Union Training Mission In Mali—Military Mission).

in the Sahel and Horn of Africa, with training and counterterrorism capacity building, not climate factors, even if they operate in the world's most impacted countries by climate change.[47]

With the EU emphasizing its holistic approach towards security, it strikes that these factors haven't been even taken into account. Rather than focusing on this "direct security-oriented approach", the EU is targeting emissions to reduce risks and instability in the long term as part of an indirect "security policy by default"[48] and this approach could overshadow how the EU will really deal with climate security challenges.[49]

European institutions, while appreciating the EU's efforts to combat climate change in the field of CSDP through the attempt to "green the military" and make missions ecologically oriented,[50] are urging for the establishment and development of civilian and military capabilities to mainstream the potential effect of climate change on the security sector.[51]

---

Civilian and military missions in Horn of Africa: EUCAP SOMALIA (European Union Capacity Building Mission); EUNAVFOR SOMALIA (European Union Naval Force Somalia Military Operation Atalanta); EUTM SOMALIA (European Union Training Mission In Somalia—Military Mission). See also Rehrl 2021, p. 219.

[47] Meyer et al. 2021. European Commission and the High Representative, "Climate Change and International Security"; EEAS, "The European Union's Global Strategy: Three Years On, Looking Forward."; Lazards and Youngs, eds 2021 (Michael 2021). The EU's Strategic Framework for the Horn of Africa defines climate change as posing an "additional challenge to all countries in the region", requiring integration of climate change into development strategies. Council conclusions on the Horn of Africa. Council of the European Union, Press Release, Foreign Affairs Council 14 November 2011.

[48] In taking the above commitments forward, the EU has in practice developed what might be termed an indirect climate security policy. It has worked increasingly hard to shape preparatory principles and the contextual factors around climate geopolitics, while undertaking relatively little direct action of a security or geopolitical nature.

[49] European Commission Directorate-General for International Cooperation and Development (DG DEVCO) 2018, Environment and Climate Change Mainstreaming in EU Development Cooperation, Briefing note for the OECD DAC peer-learning visit; see also Youngs 2021.

[50] The EU pledged to improve the climate issue within its narrow mandate and to reduce its negative impact through some actions, aimed at "greening the military" and make missions ecologically oriented:

(a) Improving Operational Effectiveness & Reducing the Carbon Footprint;
(b) Helping to Build Resilience, Supporting Adaptation;
(c) Support to Humanitarian Assistance & Disaster Relief;
(d) Military Engagement in Climate Diplomacy.

For the complete analysis, see Rehrl 2021, p. 222.

[51] In October 2012 the European Parliament issued a resolution to prompt the High Representative and the Commission to analyse how the wider range of climate impacts requires far-reaching or systemic change to the EU's geoeconomic, military, development, migration, and other policies, paying special attention to the development of military capabilities and to the process of pooling and sharing'. European Parliament 2012, Resolution of 22 November 2012 on the role of the Common Security and Defence Policy in case of climate-driven crises and natural disasters (2012/2095(INI).

Likewise, European Council conclusions "underline ... the importance environmental issues and climate change have for security and defence," while simultaneously highlighting CSDP missions "as an essential part of the EU's integrated approach to conflicts and crises." Council of the European

A possible solution would be to expressly expand CSDP missions' mandate to help address the growing problems that intersect climate change with environmental degradation and to create a long-term plan to face potential resulting disaster.[52]

In this regard, military means per se are not necessarily the most effective, but military and civilian assets combined could be used to respond to climate related security risks and to build a more coherent approach to the underlying causes of conflicts and resource wars.

In particular, establishing CSDP military missions or operations with this specific mandate could complement civilian missions, obtain analytic assistance and early warning and implement relief operations to help populations cope with natural disasters.[53]

It is important to note how for CSDP military operations, an approval from the UN Security Council is a de facto condition for the launch of a deployment, unless they are established with the consent of the third state in question (intervention by invitation) or in the case of self-defense (individual or collective) under Article 51 of the UN Charter.[54]

However, we recognize that there are serious drawbacks and dilemmas about mainstreaming climate change into CSDP missions, especially military, since this could entail the risk to generate "further EU militarization"[55] and environmental military interventionism might aggravate those problems.

---

Union 2019, Council Conclusions on Security and Defence in the Context of the EU Global Strategy, June 17, 2019.

The Instrument for Stability, which came into effect in 2006, after being prompted by the Council and the Commission, had the overarching aim of supporting stability in third countries through crisis response and prevention. Moreover, the EU's 2020 Climate Change and Defence Roadmap announced more climate-related operations, although for the moment it is mainly about equipping EU militaries for extreme climates and finding ways to reduce the operations' dependency on local resources. EEAS 2020, Climate Change and Defence Roadmap—EEAS(2020)1251.

[52] See Gilbert 2012.

[53] The diagnostic challenge for anticipatory intelligence or foresight related to climate change lies in combining diverse data and forms of evidence from different disciplinary fields—forecasts of rising global temperatures and sea levels, acidification, droughts and extreme weather events—with a forward-looking assessment of state instability, mass atrocity risks or the threat posed by state and non-state actors. Meyer et al. 2021, p. 26.

[54] Although the EU's *status* has not been formalised, if the EU were to be considered as a 'regional organisation' falling under Chapter VIII of the UN Charter, it would be subject to the provisions contained therein, *inter alia* Article 54, which states that the Security Council shall at all times be kept 'fully informed of action taken or contemplated under regional agreements or by regional organisations for the maintenance of international peace and security'.

Nevertheless, we refer to a *de facto* practice: every military CSDP deployment since 2003 has been embedded in the legal framework of a UN authorisation, with the partial exception of EUTM RCA, which was undertaken at the host country's invitation before being welcomed by a UN Security Council (UNSC) resolution. See Klocker 2021.

[55] In 2012 a report was released by the European Parliament's Committee on Foreign Affairs on "The role of CSDP in case of climate-driven crises and natural disasters" that pointed the attention on the risk of "militarisation" of the EU's climate policy and the role of CSDP by saying that since the effects of climate change can destabilise states and the Union itself, any possible crisis should be prevented by "applying a comprehensive approach including the CSDP".

## 5.5 The EU's Different Means to Address Climate-Related Security Risks

The European Union has implemented a wide range of means to counteract climate change, but it is still searching for how it can most effectively address environmental-related security risks and how to relate with multiple actors within and outside the UN system working towards the same goal.

Embedding climate change into CSDP missions' mandate could be an opportunity for the EU to improve the overall effectiveness and impact of these missions. The EU would have several possibilities to do so: in the legal framework environmental security could fall within the broad mandate of CSDP missions and operations enshrined in Article 42 TEU, as already assessed in the previous paragraph, if it becomes widely accepted that the pace of climate change has created a state of 'emergency' and has a strong influence in increasing the risk of instabilities and conflicts.

With specific regard to the possible establishment of CSDP military operations to counteract the consequences of climate change, it would require a presentation of an EU's proposal to the UN Security Council. Despite the utter and growing importance of environmental security in the UN legal framework, it cannot be automatically assumed that the proposal would meet the favor of the Security Council.

An alternative possibility would be to act through a more flexible framework: the Permanent Structured Cooperation (PESCO), an EU defense initiative introduced by the Lisbon Treaty that allows Member States that meet higher criteria in terms of military capabilities and have made more binding commitments to strengthen collaboration in the area of armaments, harmonize security apparatus, share resources and coordinate logistics and training.[56]

Although PESCO falls into the category of sui generis enhanced cooperation, it is fully integrated into CSDP since it provides substantial support, in terms of means and capabilities, to CSDP operations and missions.[57]

(On the role of the Common Security and Defence Policy in case of climate-driven crises and natural disasters. European Parliament, Committee for Foreign Affairs, 2012/2095(INI).

Nevertheless, in our opinion, the likely means and instruments which would be used in this context are Chemical, biological, radiological and nuclear (CBRN) defence or engineering capabilities which are largely civilian or dual-use. Hence, if the armed forces deploy capabilities that can help people by providing water purification or disaster relief with engineering equipment, this would be warmly welcomed by the population.

[56] Such cooperation is regulated by Article 42, para. 6 TEU and Protocol No. 10, where it is ensured that PESCO acts in coherence with foreign policy objectives, in particular with regard to peacekeeping and conflict prevention. The introduction of the PESCO mechanism is to be considered the "most significant advance in the field of institutionalised flexibility applied to the security and defence sphere." Daviddi 2012, p. 125. More extensively, on the topic, see Dossi 2019; Béraud-Sudreau et al. 2019; Dorosh and Romanyk 2020; Biscop 2020; Blockmans and Crosson 2021.

[57] Notification On Permanent Structured Cooperation (Pesco) To The Council And To The High Representative Of The Union For Foreign Affairs And Security Policy. Commitments n. 12, subcommitment 12.4 of the ambitious and more binding common commitments in the five areas set out by Article 2 of Protocol No 10. COUNCIL DECISION (CFSP) 2017/2315 of 11 December 2017

The organizing principle of PESCO projects is "flexibility within flexibility": projects appear to be fully dynamic, with their governance left to the discretion of participating States and they should serve to deliver capabilities potentially available for CSDP missions and operations.

In essence, PESCO represents a "second coming" of CSDP and ensures greater coherence and strategic autonomy in the pursuit of EU's tasks and missions, with a much more streamlined approval process.[58]

This could possibly enable Member States participating in PESCO to propose projects that specifically address the effects of climate change by producing concrete capabilities, such as tools for early warning or establishing training assistance projects or rescue operations, particularly since the Council of the European Union has identified climate change as one of the areas where further attention is required from participating Member States.[59]

It is unclear whether these areas in which the Council of the EU calls for greater involvement, should be seen as legally but not explicitly contained in the mandate of CSDP missions conferred by Article 42 TEU, or as a specific and broader interpretation of that mandate, to be realized through PESCO.

## 5.6 Conclusion

It is undeniable that climate change and its consequences represent an urgent issue to be addressed. Since it is not possible to unambiguously identify a clear link between climate change and the outbreak of a conflict, expanding the mandate of UN peacekeeping operations and missions to include a focus on climate is consequently very challenging.

---

establishing permanent structured cooperation (PESCO) and determining the list of participating Member States.

[58] See Clancy 2018.

[59] Council of the European Union 2020, Recommendation of 15 June 2020 assessing the progress made by the participating Member States to fulfil commitments undertaken in the framework of permanent structured cooperation (PESCO), para. 20: "the Council stresses that it is necessary that pMS agree on the objectives and concrete deliverables […] also considering new areas of cooperation in light of the evolving security context, in particular, but not limited to, climate change […]"; Appendix 1 to the ANNEX Results of participating Member States' discussions related to the Commitments, the PESCO Projects and the Processes/Working methods Council Conclusions on the PESCO Strategic Review 2020; Council of the European Union 2020, Recommendation of 16 November 2021 concerning the sequencing of the fulfilment of the more binding commitments undertaken in the framework of permanent structured cooperation (PESCO) and specifying more precise objectives, and repealing the Recommendation of 15 October 2018, para. 20. More importantly, the more precise objectives referred to commitment n. 15 says that "[…] further attention is required from participating Member States on collaboration in the following areas: countering hybrid threats, climate change, energy security […]". Moreover, a PESCO project has already been established in this regard: the Deployable Military Disaster Relief Capability Package (DM-DRCP), whose objective is to face emergencies and exceptional events (such as public calamities and natural disasters) within and outside EU territory.

In particular, at the latest Security Council meetings on the topic,[60] many States urged the Council to incorporate climate change consideration in its conflict-prevention, conflict resolution and peacebuilding efforts with measures concerning early warning, risk assessment and information management.[61]

Indeed, the United Nations is investing in the capacity of its peace operations to anticipate and address the interlinkage between climate change and security.[62]

However, there are many differing opinions within the Security Council: several other States strictly refuse this interpretation and suggest that the topic should be discussed in more appropriate fora.[63]

Notwithstanding, the request that the Security Council includes climate risks as part of its peacekeeping and peacebuilding missions' mandate and practice, hence being "better attuned to a world facing a climate crisis"[64] appears to be an increasingly recurring issue coming from States and institutions,[65] pinpointing a possible path to be considered.

---

[60] Both the 9345th Meeting (AM & PM) SC/15318 of 13 June 2023 and 9547th Meeting (AM & PM) SC/15589 of 13 February 2024.

[61] Among the most favourable to the incorporation of climate change in the mandate of the UN's peacekeeping operations there are: Canada, Australia, New Zealand, Switzerland, Malta, Italy, Austria, Indonesia, Vietnam, Kazakhstan, and Latvia.

[62] At the 9345th meeting of the Security Council, Jean-Pierre Lacroix, Under-Secretary-General for Peace Operations, said most United Nations peace operations have faced a deteriorating security and political environment over the past several years. Alongside other cross-border challenges, environmental degradation and extreme weather events—amplified by climate change—have increasingly challenged missions' ability to carry out their mandates. The deployment of dedicated climate, peace and security capabilities in an increasing number of field missions, in particular, has been a game changer.

[63] At the 9345th meeting of the Security Council, the representative of Brazil said that the Conference of the Parties to the Framework Convention and the Paris Agreement are the most appropriate forums to discuss climate change. While the Council should support host countries' efforts—at their request—to increase local resilience, this does not mean that the organ has a mandate to address climate change.

Likewise, the representative of India, pointing to the little scientific evidence around the impact of climate change on peace and security, said that any attribution of a conflict's cause on climate change is an oversimplification. Choosing to place this issue in non-mandated forums will undermine the larger cause of securing climate justice.

Also, the representative of the Russian Federation stressed that despite the fact that climate elements can exacerbate conflict, discussions of climate change fall outside the Council's mandate, adding that the main causes of conflict are socio economic in nature. Finally, the representative of Bangladesh stated that in the absence of any clear and standard mandate, such issues should be dealt with on a case-by-case basis that accounts for on-the-ground realities.

[64] 9345th meeting (AM & PM) SC/15318 of 13 June 2023 "With Climate Crisis Generating Growing Threats to Global Peace, Security Council Must Ramp Up Efforts, Lessen Risk of Conflicts, Speakers Stress in Open Debate".

[65] The reference is to an interesting and innovative perspective suggested by the Director of Global Initiative and Head of Climate, Peace and Sustainable Development on the possibility of invoking Article 34 of the Charter of the United Nations.

According to this idea and starting from the thesis that climate magnifies existing realities beyond the countries on its agenda, the Council could investigate any situation which might lead to international friction or give rise to a dispute or situation likely to endanger the maintenance of

Besides the UN framework, where no clear and unambiguous willingness to incorporate climate risks into the mandate of missions can be discerned, it is likely that an understanding on the topic can be reached at the EU level, considering that most Member States have expressed themselves favorably in this regard within the Security Council.

The EU has been working towards a wide range of instruments at its disposal to counter the negative effects of climate change and to safeguard environmental security. In addition to initiatives to "green the military" and reduce carbon footprint,[66] the Union could also, in the future, invest in the creation of new autonomous operations and missions in the areas most at risk of conflict due to climatic conditions, with the specific aim of preventing conflicts from arising.

There are several ways to achieve this goal: not only through traditional CSDP missions and operations—recalling that for military ones it is a practice requiring the UN Security Council approval—but also through PESCO projects, which, as analyzed in the previous section, enjoy greater flexibility in the approval process.

Leaving aside, for the moment, the discussion on which possible and feasible way the UN and the EU will implement, it's worth noting that while the Climate Clock keeps ticking environmental security shapes as a multi-faceted issue that affects the whole world and which, if not properly addressed, could lead to an increasingly cross-cutting and worrying geopolitical situation.

## References

Arts B and Rüdig W (1995) Negotiating the 'Berlin mandate': Reflections on the first 'conference of the parties' to the UN framework convention on climate change, Environmental Politics, 4(3):481

---

peace and security. Members of the Council could set up a provisional investigative body to better understand the links between food security climate and conflict at current warming levels of 1.48°C.

According to our opinion—without possibility of further discussing the applicability of Art. 34 UN Charter in this context due to the sake of brevity—the interpretation provided, although disregards the practical applications of peaceful settlements over the years and beyond the scope, represents a clear indication of the widespread interest in the subject matter by institutions.

[66] Among other initiatives promoted at the EU level such as the Energy Efficiency Directive (EED) (Directive (EU) 2023/1791 of the European Parliament and of the Council of 13 September 2023 on energy efficiency and amending Regulation (EU) 2023/955 *OJ L 231, 20.9.2023, pp. 1–111*); Renewable Energy Directive (RED) (Directive (EU) 2023/2413 of the European Parliament and of the Council of 18 October 2023 amending Directive (EU) 2018/2001, Regulation (EU) 2018/1999 and Directive 98/70/EC as regards the promotion of energy from renewable sources, and repealing Council Directive (EU) 2015/652 *OJ L, 2023/2413, 31.10.2023*), see also Consultation Forum for Sustainable Energy in the Defence and Security Sector (CF SEDSS) managed by the European Defence Agency to assist the European Union Ministries of Defence to move towards green, resilient, and efficient energy models and Incubation Forum for Circular Economy in European Defence (IF CEED) to apply the circularity approach of the EU Green Deal to European defence by engaging a cooperative community, including EU defence ministries, industry.

Balzacq T (2007) Securitization Theory: How Security Problems emerge and dissolve. Routledge, London.

Barbé E (1995) La seguridad en la nueva Europa: Una aproximación institucional: Unión Europea, OTAN y UEO. Catarata, Madrid

Barnett J Adger WN (2007) Climate change, human security and violent conflict. Political Geography 26(6):639–655

Béraud-Sudreau L Efstathiou Y-S Hannigan C (2019) Keeping the momentum in European defence collaboration: an early assessment of PESCO implementation, The International Institute for Strategic Studies, 14 May.

Biscop S (2020) European Defence and PESCO: Don't Waste the Chance, in EU Integration and Differentiation for Effectiveness and Accountability Policy Paper 1

Blockmans S Crosson DM (2021) A Force for Positive Integration in EU Defence, in European Foreign Affairs Review, 26:87–110

Bremberg N (2019) EU foreign and security policy on climate-related security risks SIPRI Policy Brief, Brussels

Buckland B (2007) A climate of war? Stopping the securitization of climate change, Sage Publications, Newbury

Busby J (2008) Who Cares about the Weather? Climate Change and US National Security. Security Studies, 17:468–504

Buzan B (1983) People, States and Fear: The National Security Problem in International Relations. Harvester Wheatsheaf, London

Buzan B (1991) People, States and Fear: an agenda for international security studies in The post-cold War Era. ECPR, New York

Buzan B and Hansen L (2009) The Evolution of International Security Studies. Cambridge University Press, Cambridge.

Buzan B, Wæver O, de Wilde J (1998) Security: a new framework for analysis. Lynne Rienner Publisher, Boulder.

Caballero-Anthony M (2015) An introduction to non-traditional security studies: a transnational approach. Sage Publications, Newbury

Clancy P (2018) Arise, Sleeping Beauty: What PESCO means for Ireland, Irish Yearbook of International Law 13:79–98.

Craig S L (2007) Chinese Perceptions of traditional and nontraditional security threats. U.S. Army War College, Carlisle.

Daviddi G (2012) Verso un'eurozona della difesa. Sviluppo delle flessibilità istituzionali nelle politiche europee di sicurezza e difesa. Edizioni Nuova Cultura, Roma.

Davoudi S (2014) Climate change, securitisation of nature and resilient urbanism, in Environment and Planning C Government and Policy 32(2):360–375.

Detges A Klingenfeld et al (2020) 10 Insights on Climate Impacts and Peace: A Summary of What We Need to Know. Adelphi, Potsdam Institute for Climate Impact Research, Berlin

Detraz N and Betsill MM (2009) Climate change and environmental security: for whom the discourse shifts, International Studies Perspective 10:3

Dorosh L Romanyk V (2020) PESCO, CARD, EDF: Strategy, Analysis, and Financing in the European Union Security Guarantee, Political Sciences, 6(2)

Dossi A (2019) La coopération d'armement PESCO: potentiel et failles en Politique de sécurité, in analyses du Center for Security Studies (CSS), ETH Zürich, 241

Ducrotté F (2012) The Impact of Climate Change on International Security: Prospects for an Environmental Dimension in CSDP Missions, European Security Review, 6.

Dyer G (2009) Climate wars: how peak oil and the climate crisis will change Canada (and our lives). Vintage Canada, Toronto

Fetzek S and Mazo J (2014) Climate, scarcity and conflict, Survival 56(5):143–170

Gilbert E (2012) The Militarization of Climate Change, ACME: An international Journal for Critical Geographies 11(1)

Gleditsch NP, Nordås R, Salehyan I (2007) Climate change and conflict: The migration link. Coping with Crisis Working Paper Series. International Peace Academy, New York
Haacke J, Williams P (2008) Regional Arrangement, Securitisation and Transnational Security Challenges: The African Union and the Association of the Southeast Asian Nations Compared, Security Studies 17:775
Haftendorn H (1991) The Security Puzzle: Theory-building and Discipline Building in International Security, International Studies Quarterly, 35
Hansen L (2000) The little mermaid's silent security dilemma and the absence of gender in the Copenhagen school, Millenium: Journal of International Studies, 29:285–306
Hettne B (2010) Development and Security: Origins and Future, Secur. Dialogue 31:41
Homer-Dixon T (1991) Environmental Changes as Causes of Acute Conflict. International Security, 16(2):76–116
Hsiang SM, Burke M, Miguel E (2013) Quantifying the influence of climate on human conflict. Science 341:1235367
Huysmans J (1998) Dire et écrire la sécurité : le dilemme normatif des études de sécurité, Cultures & Conflits, 31:177–202
Ingravallo I (2022) Lo Sviluppo Sostenibile a Cinquant'anni dalla Conferenza di Stoccolma, Il diritto della comunità internazionale tra caratteristiche strutturali e tendenze innovative. Editoriale Scientifica, Napoli
Kaldor M Martin M and Selchow S (2007) Human Security: a new narrative for Europe, Royal Institute of International Affairs 83(2):273–288
Keskinen M, Sojamo S and Varis O (2019) Enhancing Security, Sustainability and Resilience in Energy, Food and Water, Sustainability 11(24):7244.
Klocker C (2021) The UN Security Council and EU CSDP operations : exploring EU military operations from an outside perspective, UCL Press. 5(1):3
Le Gloannec A-M et al (2013) New and Evolving Trends in International Security, Transworld 13:2–43
Ligtvoet W et al. (2017) Water, Climate and Conflict: Security Risks on the Increase? Planetary Security Initiative—Clingendael Briefing, The Hague
Luck EC (2014) Change and the United Nations Charter. In: Shapiro I and Lampert J (eds.) Charter of the United Nations Together with Scholarly Commentaries and Essential Historical Documents. Yale University Press, New Haven
Mach JK and Kraan CM (2021) Science–policy dimensions of research on climate change and conflict, Journal of Peace Research 58(1):168–176
Mach JK Kraan CM et al (2019) Climate as a risk factor for armed conflict, Nature 571:193–197
Mach KJ et al. (2019) Climate as a risk factor for armed conflict, Nature, 571.7764:193–197
Mares D, Moffetti KW (2016) Climate change and interpersonal violence: A "global" estimate and regional inequities. Clim. Change 135:297–310
Mayall J (1984) Reflections on the "New" Economic Nationalism, Review of International Studies, 10:313–21.
McSweeney B (1999) Security, Identity, and Interests: A Sociology of International Relations. Cambridge University Press, Cambridge
Meyer C, Vantaggiato F, Youngs R (2021) Preparing the CSDP for the new security environment created by climate change, European Parliament In-depth study, Brussels
Michael D (2021) Climate Security, Conflict Prevention, and Peacebuilding. In: Lazards O. and Youngs R. (eds) The EU and Climate Security: toward ecological diplomacy. Carnegie, the Hague.
Oberthür S (2016) Where to go from Paris? The European union in climate geopolitics, in Global Affairs 2:119–130
Oels A (2012) From 'Securitization' of Climate Change to 'Climatization' of the Security Field: Comparing Three Theoretical Perspectives, Climate Change, human security and violent conflict. Springer, Berlin.
Paris R (2001), Human Security. Paradigm Shift or Hot Air?, International Security, 26:87–102

Penny CK (2018) Climate Change as a 'Threat to International Peace and Security', in: Ku C and Shirley V (eds.) Climate Change and the UN Security Council. Edward Elgar Publishing, Northampton
Ranson M (2014) Crime, weather, and climate change. J. Environ. Econ. Management. 67:274–302
Rehrl J (2021), Handbook on CSDP The Common Security and Defence Policy of the European Union, Directorate for Security Policy of the Federal Ministry of Defence of the Republic of Austria
Remling E Barnhoorn A (2021) A Reassessment Of The European Union's Response To Climate related Security Risks, SIPRI Insights on Peace and Security (2).
Risi C (2010) L'azione esterna dell'Unione europea dopo Lisbona. Editoriale Scientifica, Napoli
Rothe D (2015) Securitising global warming: a climate of complexity. Routledge, London
Rothe D (2020) Securitizing Global Warming: A Climate of Complexity. Routledge, Abingdon and New York
Rothschild E (1995) What is Security?, Daedalus, 124:53–98.
Salehyan, I (2008) From climate change to conflict? No consensus yet, Journal of Peace Research, 45(3):315–326. Sakaguchi K Anil V and Graeme A (2017) Climate wars? A systematic review of empirical analyses on the links between climate change and violent conflict, International Studies Review 19(4): 622–645
Sands P (1992) The United Nations framework convention on climate change, in Review of European Community & International Environmental Law 1:270–277
Schlag G, Junk J and Daase C (2016) Transformations of Security Studies: Dialogues, Diversity and Discipline. Routledge, London
Sebesta L (1991) L'Europa indifesa. Sistema di sicurezza atlantico e caso italiano. Ponte alle grazie, Firenze
Sjursen H (2001) New forms of security policy in Europe, Arena working Papers
Smith P (2007) Climate change, mass migration and the military response, in Orbis 51:3
Stern M and Öjendal J (2010) Mapping the Security—Development Nexus: Conflict, Complexity, Cacophony, Convergence?, Secur. Dialogue 5:41
Tadjbakhsh S, Chenoy A (2006) Human Security. Concepts and Implications. Routledge, Abingdon and New York
Thakur R, Newman E (2004) Broadening Asia's Security Discourse and Agenda. Political, Social, and Environmental Perspectives. United Nations University Press, Tokyo
Theisen O et al (2012) Climate wars? Assessing the claim that drought breeds conflict, International Security 36:585–602
Tuchman M. J (1989) Redefining Security. Foreign Affairs, 68(2)
Ullman R (1983) Redefining Security, International Security, 8:129–53
Wæver O (1995) Securitization and Desecuritization, in: Lipschutz R. D. (ed.) On Security. Columbia University Press, New York
Youngs R (2021) The EU's indirect and defensive approach to climate security. In: Lazard O Youngs R (eds.) The EU and Climate Security: towards ecological diplomacy. Carnegie, the Hague.
Youngs R (2014) Climate Change and European Security. Taylor & Francis, London

**Stefania Rutigliano** University of Cassino and Southern Lazio, PhD in International and EU Law, Università degli Studi di Bari Aldo Moro.

# Part II
# Sustainable Trade Measures

# Chapter 6
# The Carbon Border Adjustment Mechanism: Reconciling the Principles of Sustainability and Free Trade in the EU's External Action?

**Patrick Abel**

**Contents**



**Abstract** The Carbon Border Adjustment Mechanism (CBAM) is key for the European Union's (EU) climate action and trade policy. Yet, many criticize it as a disguised protectionist measure that violates the EU's obligations under World Trade Organization (WTO) law. This Chapter proposes a different view. It claims that CBAM may be considered to conform with WTO law subject to certain conditions. CBAM should be understood as a regulation by which the EU contributes to reshaping WTO law that is currently undergoing a fundamental crisis. The Chapter argues that the European Treaties allow and even call upon the EU to engage in a balanced reshaping of WTO law. CBAM is a way by which the EU tests the waters how to reconcile

P. Abel (✉)
Assistant Professor (Akademischer Rat a.Z.), University of Passau, Passau, Germany
e-mail: patrick.abel@uni-passau.de



R. A. Wessel et al. (eds.), *EU External Relations Law and Sustainability*, Global Europe: Legal and Policy Issues of the EU's External Action 4,
https://doi.org/10.1007/978-94-6265-655-0_6

various constitutional goals that it is obliged to follow in its external actions: to combat climate change by internalizing the costs for polluting the atmosphere in a coherent manner, while at the same time promoting free trade and respecting the positions of developing countries.

**Keywords** Climate change · Sustainability · Carbon border adjustment mechanism · Developing countries

## 6.1 Introduction

The Carbon Border Adjustment Mechanism (CBAM) is a new EU unilateral and extraterritorial climate change law: Importers of certain goods into the EU will have to pay a price for greenhouse gasses (GHG) emitted abroad in the production process. CBAM is a crucial element in the EU's external relations, and developing countries in particular claim that CBAM does not comply with WTO law. This chapter analyzes this debate and proposes a dynamic, process-oriented perspective on the issue. After introducing climate change mitigation as an internal and external problem of sustainability for the EU (Sect. 6.2), it explains the main features of CBAM and how it aims at promoting coherence in the EU's sustainability policy, internally and externally (Sect. 6.3). Then, the analysis turns to how CBAM conforms with WTO law based on WTO jurisprudence (Sect. 6.4). Lastly, it explains that CBAM should also be understood as a dynamic contribution by the EU to reshape and recreate WTO law (Sect. 6.5).

## 6.2 Climate Change Mitigation as an Internal and External Problem of Sustainability

Following the famous Brundtland Report of 1989, we can define "sustainability" as a policy that meets the needs of the present generations without compromising the ability of future generations to do the same.[1] Broadly speaking, modern understandings of sustainability refer to a triangle of the economy, ecology and social policy, and call for a balance between these three dimensions in all governmental and societal actions.[2] Climate change challenges the ability of future generations to live healthy lives and to prosper in a stable economy.[3]

---

[1] UN (1987) Report of the World Commission on Environment and Development: Our Common Future ('Brundtland Report'). UN Doc A/42/427 Annex, para 27.

[2] Beyerlin 2013, para 11.

[3] See the most recent report by IPCC 2023, paras A.1 et seq.

Climate change mitigation constitutes a well-established part of the EU's environmental policy pursuant to Articles 191 and 192 TFEU.[4] Specifically, Article 191(1) TFEU stipulates that the Union's environmental policy "shall contribute to [...] promoting measures at international level to deal with regional or worldwide environmental problems, and in particular combating climate change." It is fair to read this passage as having a constitutional function, that is, to set a binding objective for the EU to address climate change in its external relations, similar to comparable provisions in the TEU[5] and TFEU.[6] Indeed, it seems that the EU must pursue this objective with a certain urgency, since it is supposed to "combat" (and not only, for example, merely "address") this global environmental problem. Given the GHG-emissions of third states, especially China and the USA, the EU alone cannot successfully halt global warming even if all the EU GHG-emissions were stopped immediately. The atmosphere is a global public good, a "common concern of mankind".[7] Therefore, any climate action must necessarily look outward, and indeed, EU secondary law on climate action is inextricably linked to international climate law, in particular the Paris Agreement.[8]

Conversely, it follows that at least some, if not all "internal" EU climate mitigation measures have an effect on third countries. Indeed, the Treaties do account for the external dimensions of internal action in Article 21(3)(1) TEU. It states that the Union shall respect the principles and pursue the objectives set out in Article 21(1) and (2) TEU in the development and implementation not only of the EU's external action, but also "of the external aspects of its other policies". More specifically, the EU must act coherently (Article 21(3)(2) TEU) by preventing contradictions in its internal and external policies, attempt to find cooperative, preferably regional and multilateral solutions (see Article 21(1)(2) TEU) and respect international law (Article 21(1) TEU). Furthermore, according to the integration clause in Article 11 TFEU, the EU must integrate environmental protection requirements into the definition and implementation of its policies and activities, internally and externally. While Articles 11 TFEU and 21 TEU impose legally binding restraints, the EU organs enjoy substantial discretion in their implementation. In particular, due to their general nature, the

---

[4] Treaty on the Functioning of the European Union (opened for signature 13 December 2007) OJ 2012 C 326/47 (entered into force 1 December 2009) TFEU.

[5] Treaty on the European Union (opened for signature 13 December 2007) OJ 2012 C 326/13 (entered into force 1 December 2009) TEU.

[6] On the constitutional and binding character of the EU's objectives in the European Foreign and Security Policy, see for example Kiss 2003, pp. 107 et seq. on the constitutional character of the environmental goal of the EC Treaty; Wessel 2016, pp. 444 et seq.; Thym 2022, § 18 para 33 on the objectives of EU primary law for the EU's external affairs as a form of "horizontal constitutionalization".

[7] The principle is mentioned for example in UN General Assembly, Res. 43/53, 6 December 1988, no 1; UN Framework Convention on Climate Change, preamble; Paris Agreement, preamble; for a comprehensive analysis of the principle, see Cottier 2021.

[8] See for example Scott 2011, pp. 25 et seq.

principles and objectives that the EU must observe often collide and require reconciliation by political decisions that allow and demand for setting priorities.[9] As we will see, the case of CBAM is an excellent example of a complex balancing of policy goals.

## 6.3 CBAM as a Means to Promote Coherence in the EU's Sustainability Policy

CBAM can be understood as a means to promote coherence in the EU's internal and external climate change mitigation strategy. It complements the market-based sustainability strategy of the EU emissions trading system (ETS) (Sect. 6.3.1) by finding a new way to prevent carbon leakage (Sect. 6.3.2) and mitigate climate change (Sect. 6.3.3).

### *6.3.1 The EU ETS as a Market-Based Sustainability Strategy*

The EU's main internal instrument for reducing GHG emissions is the ETS which was first introduced in 2005.[10] The ETS is a so-called "cap-and-trade"-mechanism.[11] It sets a maximum amount of GHG that can be emitted in a year. Above this "cap", the ETS regulation prohibits emitting any further GHG. The ETS divides this "cap" into allowances which function as permits to emit GHG, measured in $CO_2$-equivalents per ton. Companies that want to emit GHG, for example by generating electricity in coal-fired power plants, must acquire the respective number of allowances that corresponds to the amount of GHG emitted.[12] Most of the allowances must be purchased through auctions.[13] Allowance holders can also trade allowances instead of using them as permits for their own GHG emissions. Currently, the EU ETS covers the energy sector, large industrial installations, and aviation, which together account for around 45 per cent of total EU GHG emissions.[14] In this way, the EU ETS puts a price on GHG emissions and internalizes the cost to society of polluting the atmosphere.

---

[9] Cf Cremona 2012, pp. 39 et seq.

[10] Directive 2003/87/EC of 13 October 2003, OJ L 275/32 (hereafter: "EU ETS").

[11] See Hsu 2016, pp. 244 et seq. for a comparison of international market mechanisms.

[12] Hsu 2016, pp. 244 et seq.

[13] See in particular Article 10 EU ETS.

[14] Knodt 2023, p. 205. In 2024, the EU ETS will be extended to the maritime sector.

### 6.3.2 The External Dimension of the EU ETS: Carbon Leakage

The EU ETS presented above is an internal EU mechanism that requires consideration of how it operates with imports. The ETS has already proven to reduce GHG emissions.[15] However, it suffers from the structural problem of carbon leakage.[16] In a globalized economy, energy-intensive industries based in the EU may decide to relocate their production to third states that are not subject to an ETS or other types of carbon pricing in order to save costs. From there, they can then export their products into the EU, potentially to a more competitive price than EU-producers: They do not need to buy any ETS allowances and will thus have lower production costs. For example, the EU cement industry could build new plants in the US and export cement from the US to the EU. If the price of allowances required under the EU ETS to produce cement in the EU is sufficiently high, it may exceed the transport costs and any tariffs applied to cement imported from the US. Then, there would be a business case for the company to leave the EU. As a result, the EU would lose an important industry without contributing to climate change mitigation, as GHG would continue to be emitted in the US. This would run counter to the EU's obligation to pursue coherent environmental protection in all its internal and external policies and activities, as reflected in Articles 11 TFEU and 21(2) TEU.[17]

Currently, the EU ETS accounts for carbon leakage by granting "transitional free allocation" of allowances to certain energy-intensive industries.[18] It follows that the EU ETS does not internalize the costs that these corporations cause by polluting the atmosphere. While this prevents carbon leakage, it also exempts the very industries that emit particularly large amounts of GHG from the scope of the ETS. The market-based mechanism to incentivise decarbonisation does not apply to these corporations even though the environmental need for it remains urgent. For this reason, the EU considers free allocation as a temporary solution only.[19]

### 6.3.3 CBAM as a Tool for Internal and External Climate Change Mitigation

The Carbon Border Adjustment Mechanism (CBAM), a regulation adopted and in force since 2023, aims to provide an alternative solution to the problem of carbon

---

[15] The price for ETS allowances has increased sharply since 2018, leading more and more to the desired steering effect, see for example Knodt 2023, p. 205.

[16] For a general legal analysis of carbon leakage, see for example Shoyer et al. 2016, pp. 285 et seq.

[17] On coherence of the EU's environmental policy, see Cremona 2012, pp. 36 et seq.

[18] Articles 10a and 10b EU ETS; for further analysis of EU regulation on carbon leakage, see Carević 2015, pp. 47 et seq.

[19] See Regulation (EU) 2023/956 of 10 May 2023, OJ L 130/52 (hereafter: "CBAM") rec. 12.

leakage.[20] From 2026 onwards,[21] EU importers of certain energy-intensive products originating in a third country will have to purchase CBAM certificates from national authorities. To do so, they must declare the GHG emissions embedded in their products, including indirect emissions of some goods from the production of electricity consumed during the production process.[22] The price of the certificates corresponds to the price of the ETS allowances that importers would have had to pay if they had produced the imported goods in the EU (and thus been subject to the EU ETS).[23] The products covered are cement, iron and steel, aluminum, fertilizers (which receive free allocations under the EU ETS so far), hydrogen and electricity[24] including imported processed products from these aforementioned goods that underwent an inward processing procedure referred to in Article 256 of Regulation (EU) No. 952/2013.[25] CBAM thus prevents carbon leakage in these sectors, as there is no competitive advantage to be gained by relocating the EU production to third countries and exporting goods from there to the EU to avoid paying for ETS allowances. In turn, as CBAM is being phased in, free allocations under the EU ETS will gradually be phased out.[26] The idea is that then, also the energy-intensive industries concerned will have an incentive to decarbonize. This is highly desirable from the perspectives of climate change mitigation, the EU's international climate law obligations, and of sustainability. CBAM also promotes the coherence of the EU's climate action policies in line with Article 21(3)(2) TEU.

## 6.4 CBAM and the EU's Obligations to Find Cooperative Solutions and to Respect International Law in its External Actions

While CBAM serves to protect the integrity of the EU ETS and promotes the EU's policy coherence, at first glance, being a unilateral measure, it does not rest very well with the obligation to promote cooperative, pluri- and multilateral solutions to common problems enshrined in Article 21(1)(2) TEU. What is more, the EU must

---

[20] See n 19. Cf Trachtman 2017, pp. 469 et seq. on the potentials of a national carbon consumption tax as an alternative policy measure.

[21] Until then, CBAM defines a transitional period with reporting obligations only, see Article 32 CBAM.

[22] Articles 6 et seq. CBAM. Specifically on indirect emissions, see Article 23(22), Article 7(7) and Annex IV CBAM.

[23] Article 7 CBAM.

[24] Article 2(1) and Annex I CBAM. On the sectors defined as being at risk of carbon leakage, see Commission, Delegated Decision (EU) 2019/708 of 15 February 2019, OJ L 120/20.

[25] Article 2(1) CBAM and Regulation (EU) No 952/2013 of 9 October 2013, OJ L 269/1.

[26] Article 10a(1a) EU ETS. In line with that, Article 1(3) CBAM explicitly defines that CBAM aims to replace the mechanism for the free allocation of allowances established under the EU ETS to prevent the risk of carbon leakage. See Espa et al. 2022, pp. 16 et seq.

respect the rule of law and comply with WTO law as a branch of international law (Article 21(1) TEU).[27] This leads to the question if CBAM violates WTO law.

The WTO Agreement contains multilateral disciplines on trade in goods, the most relevant for the present purpose to be found in the General Agreement on Tariffs and Trade of 1994 (GATT 1994) which will be analyzed in this section. The GATT adopts a two-tiered system. First, there are numerous obligations that the WTO members have to comply with, such as most-favored nation treatment (MFN) and national treatment, and indeed, CBAM does violate some of these disciplines (Sect. 6.4.1). Second, even if WTO members violate a GATT obligation, this violation may be justified under Article XX GATT (Sects. 6.4.2 and 6.4.3). As a starting point, the analysis will look at CBAM based on the WTO's jurisprudence.

### *6.4.1 Breach of GATT 1994 Obligations*

Most observers agree that CBAM violates at least some GATT obligations.[28] The exact legal analysis on national treatment depends on how one qualifies the type of trade barrier that CBAM creates.[29] For the sake of brevity, at the very least, CBAM violates the MFN obligation (Article I(1) GATT) which prohibits discrimination between like foreign goods with origins in different WTO members. The very purpose of CBAM is to discriminate in pursuit of climate protection. WTO members have varying domestic environmental and climate protection standards. EU importers producing in countries with weaker standards will have to buy more CBAM certificates, making their products more expensive and thus creating a competitive disadvantage. An even more open discrimination is the fact that CBAM explicitly exempts the members of the European Economic Area and Switzerland from the system because they have linked their ETS with the EU's.[30]

For a long time, both the literature and the WTO jurisprudence have discussed whether one could argue that there is no discrimination (and thus no GATT 1994

[27] Because the EU itself is a member of the WTO, the WTO Agreement itself has the status of EU law of superior hierarchy to secondary EU law (Article 216(2) TFEU).

[28] See for example Leal-Arcas et al. 2022, pp. 230 et seq.; Dias et al. 2020, pp. 16 et seq.

[29] Broadly speaking, the GATT distinguishes between trade-restrictive measures at the border and internal measures. There is no scholarly consensus on how to qualify CBAM. This may lead to additional breaches of GATT disciplines, for example the obligation to respect "bound tariffs" subject to Article II(1) GATT. For more details on this question, see for example Pauwelyn 2020, 8 et seq.; Quick 2020, pp. 571 et seq.; Venzke and Vidigal 2022b, pp. 10 et seq. On the breach of national treatment, see for example Leonelli 2022, p. 625. Cf with the comparable question of how to qualify the integration of aviation into the EU ETS, as analysed by Bartels 2012, pp. 438 et seq.

[30] Article 2(4) and Annex III point 1 CBAM. On the linkages, see Agreement on the European Economic Area of 3 January 1994, OJ L 1/3, Article 74 and Annex XX no 21al; Agreement between the European Union and the Swiss Confederation on the Linking of their Greenhouse Gas Emissions Trading Systems of 7 December 2017, OJ L 322/3.

violation) because the products compared here are not "like". The idea was to distinguish "green" products with low embedded GHG emissions from other products—to say that, for example, steel produced with renewable electricity was not "like" steel produced with electricity generated by coal plants, and that WTO members would, hence, not discriminate if they treated "green" steel better. Yet, it is well-established WTO jurisprudence that this line of reasoning has no merit. This is not the place to go into the details of this case-law. In essence, the WTO Appellate Body rejected to distinguish products according to so-called non-product-related process and production methods; to determine "likeness", only consumer preferences and habits, end-uses tariff classification, physical characteristics of the products may be considered which usually do not change because products have low embedded GHG emissions.[31]

### *6.4.2 Provisional Justification Under Article XX(b) and (g) GATT 1994*

The main and decisive arena for determining whether CBAM complies with WTO law is Article XX GATT. This provision covers the "general exceptions" for justifying violations of GATT obligations. Under certain conditions, these exceptions allow unilateral trade-restrictive measures. WTO jurisprudence has upheld them even if they exert an extraterritorial effect,[32] a case-law that has remained controversial until the present day.[33] Article XX GATT provides two relevant justification grounds for the present context. CBAM could be a measure "necessary to protect human, animal or plant life or health"[34] or a measure "relating to the conservation of exhaustible natural resources if such measures are made effective in conjunction with restrictions on domestic production or consumption".[35] Indeed, CBAM pursues both policy objectives. Climate change mitigation reduces the negative impacts of global warming on human, animal and plant life and health. And the atmosphere qualifies as a global public good and an exhaustible natural resource.

Under Article XX(b) and (g) GATT, the EU would have to show that CBAM is "necessary" or that it "relates to" the respective policy goals. These requirements

[31] See for example Dobson 2021, pp. 64 et seq. On the four criteria that are allowed to determine discrimination, see WTO, Appellate Body Report, *EC—Asbestos*, 12 March 2001, WT/DS135/AB/R, paras 87 et seq.

[32] See for example WTO, Appellate Body Report, *US—Shrimp*, 12 October 1998, WT/DS58/AB/R, para 121; *EC—Seals*, 22 May 2014, WT/DS400, 401/AB/R, para 5.173.

[33] See for example Dobson 2021, pp. 86 et seq. with further references.

[34] Article XX(b) GATT 1994.

[35] Article XX(g) GATT 1994.

enshrine a proportionality test,[36] but with different thresholds. The standard of "relating" is more lenient than that of "necessity",[37] which is why this paper will focus on the former for the sake of brevity. According to the Appellate Body, the test requires "a close and genuine relationship of ends and means" between the measure under scrutiny and the conservation objective.[38]

As a *general* concept, CBAM should meet the test of "relating to" the conservation of the atmosphere.[39] It seems plausible that CBAM may hinder the energy-heavy EU industry from realizing competitive advantages by moving their production to states with poorer climate protection standards. Notwithstanding that empirical proof is generally not necessary under Article XX(g) GATT, the EU is well-advised to prepare substantiated data on carbon leakage and CBAM's positive effect on climate change mitigation,[40] in particular in light of some interventions that carbon border tax measures—if designed poorly—may be ineffective to combat climate change.[41] More importantly, the EU will have to make it plausible that preventing carbon leakage overall contributes to climate change mitigation on a global scale—and not only to protect the EU's own industries.[42]

### *6.4.3 Chapeau of Article XX GATT 1994*

Even if there are solid grounds to find for a provisional justification under Article XX (g) GATT—and perhaps even (b)—, CBAM must also conform with the *chapeau* of Article XX GATT.[43] It prohibits applying measures in a manner amounting to arbitrary or unjustifiable discrimination between countries where the same conditions prevail or amounting to a disguised restriction on international trade—three alternatives which the Appellate Body has treated as one uniform standard so far.[44] The

---

[36] Cottier et al. 2017, pp. 644 et seq.

[37] WTO, Appellate Body Report, *China—Rare Earths*, 7 August 2014, WT/DS431-433/AB/R, paras 5.87 et seq.

[38] WTO, Appellate Body Report, *US—Shrimp*, 12 October 1998, WT/DS58/AB/R, para 136; *China—Raw Materials*, para 355; *China—Rare Earths*, 7 August 2014, WT/DS431-433/AB/R, para 5.90.

[39] See for example Pauwelyn 2020, 11; Venzke and Vidigal 2022a, pp. 203 et seq.

[40] But see the critical analysis of Pirlot 2022, pp. 25 et seq. that CBAM contributes less to climate change mitigation because it does not cover EU exports.

[41] See for example Colares and Rode 2015, pp. 423 ff.

[42] Cf Dias et al. 2020, p. 22; Sato 2022, pp. 390 et seq. on the problematic rationale of leveling the playing field for European industries.

[43] See WTO, Appellate Body Report, *US—Gasoline*, 29 April 1996, WT/DS2/AB/R, p. 22; *Korea—Meat*, 11 December 2000, WT/DS161,169/AB/R para 157; *Indonesia—Importation of Horticultural Products, Animals and Animal Products*, 9 November 2017, WT/DS477/AB/R, paras 5.95 et seq.

[44] See already WTO Appellate Body Report, *US—Gasoline*, 20 May 1996, WT/DS2/AB/R, p. 25; cf Bartels 2015, p. 97, who considers that panels and the Appellate Body have "ignored" the requirement of "disguised restriction on international trade" so far.

*chapeau* is thus about how the relevant WTO member applies the measure in practice.[45] In WTO dispute settlement, the *chapeau* has proven to be the greatest hurdle to justify trade restrictive measures under Article XX GATT, which is why its exact requirements remain controversial until the present day.[46] Looking back how Panels and the Appellate Body have interpreted and applied the *chapeau* in other cases, one may identify three main concerns against CBAM that will be discussed in the following subsections: It would have to maintain a rational relationship between how the financial burden is imposed on EU importers and CBAM's actual climate change mitigation effect (Sect. 6.4.3.1). It must refrain from coercing other WTO members to use carbon pricing as the only viable climate protection policy (Sect. 6.4.3.2). And the EU would have to negotiate bi- and multilateral agreements for climate change protection in a serious and even-handed manner (Sect. 6.4.3.3).

#### 6.4.3.1 Rational Relationship Between CBAM and Climate Change Mitigation Effect?

In *Brazil–Retreaded Tyres*, the Appellate Body held that discrimination is "arbitrary or unjustified" if there is no rational relationship of the discriminatory state action to the policy objective pursued and provisionally justified under Article XX GATT.[47] This means that the EU must always interpret and apply the conditions and exceptions of CBAM in a manner that actually contributes to the goal of climate change mitigation. This resonates well with the EU's obligation to act coherently in its external actions (Article XXI(3)(2) TEU). Furthermore, the measure must be applied even-handedly between countries in which the same conditions prevail. In doing so, one must take account of context-specific conditions related to the policy objective under the applicable subparagraph of Article XX GATT[48]—thus, here, all conditions that relate to climate action policies.

In the first place, the EU would have to make sure that CBAM internalizes the costs of damaging the atmosphere, reflecting as closely the real harmful effect as possible. This means that the system must operate in a manner that accurately takes account of the GHG of goods imported into the EU. It is for CBAM declarants to present verified calculations of the emissions to the authorities.[49] Thus, much depends on the methodology that CBAM provides for the calculation—for direct emissions from the production processes of goods, including emissions from the production of heating and cooling that is consumed during the production processes and (for some goods)

---

[45] WTO Appellate Body Report, *US—Gasoline*, 20 May 1996, WT/DS2/AB/R, p. 22; for a critique, see Bartels 2015, pp. 98–101.

[46] On the lack of clarity, see e.g. Bartels 2015, p. 96; Durán 2016, p. 474; Riffel 2018, p. 143: "The chapeau is rife with ambiguity […]".

[47] WTO, Appellate Body Report, *Brazil—Retreaded Tyres*, 3 December 2007, WT/DS/332/AB/R, paras 227, 246.

[48] See for example WTO, Appellate Body Report, *EC—Seals*, 22 May 2014, WT/DS400, 401/AB/R, paras 5.298 et seq.

[49] See Article 8(1) CBAM.

for the indirect emissions from the production of electricity that is consumed during the production processes.[50]

CBAM sets a methodology for calculation of emissions that generally seems capable of reflecting the actual harm caused to the atmosphere by producing a good. To that end, Annex IV contains a framework with general rules for calculation. It distinguishes between "simple" goods, produced only out of raw materials, "complex" goods (defined as all goods other than simple goods)[51] and electricity. Direct emissions of simple and complex goods must be calculated based on the actual values or, if not available, from the literature that may be country- or region-specific for the exporting state.[52] If such values are not available or not reliable, default values apply.[53] In the case of indirect emissions, default values apply as the standard method but the declarant can present actual values if adequate.[54] How to determine default values differs depending on whether they are related to assessing the emissions of the electricity used. In the case of default values for production processes, as the last resort, the EU can use the average emissions intensity of the X% worst performing EU ETS installations for that type of goods. It is for the Commission to set the value of "X" in an implementing act.[55]

This example shows that default values can be problematic in case they do not adequately reflect the real emissions in the exporting country. For example, the X% rule could inadequately benefit goods from an exporting country in which production processes cause much higher GHG emissions than the worst EU installations. On the other hand, default values as a residual rule seem inevitable—and thus not arbitrary or unjustified—to make CBAM practically applicable for scenarios with a lack of available reliable data.[56] Indeed, also the ETS provides for default values as a calculation method of last resort for EU companies.[57] Therefore, much will depend on how the Commission concretizes CBAM calculation methodology through implementing acts.[58] To provide for even-handedness with domestic cases, it should mirror as closely as possible the EU ETS emissions calculation method.

That CBAM *completely exempts* goods from the regulation if they have their origin in countries that have an ETS linked with the EU ETS appears less problematic in most cases. Generally, such linkages will prevent any carbon leakage effect between the respective partners. A rational relationship to climate change mitigation would

[50] Article 3(21) and (22) CBAM. See Espa et al. 2022, pp. 21 et seq. who generally welcome the orientation towards actual emissions as favouring GATT-compliance.

[51] Annex IV, No 1(b) CBAM.

[52] Article 7 and Annex IV CBAM.

[53] Article 7(2 s. 2) CBAM.

[54] Article 7(2 s. 2) and (4) CBAM.

[55] Annex IV, no 4.2 CBAM.

[56] See Dias et al. 2020, p. 23 on questions of practicability; but see Sato 2022, pp. 389 et seq. criticizing scenarios in which emissions cannot be accurately determined.

[57] See especially Article 31 of the Commission Implementing Regulation (EU) 2018/2066 of 19 December 2018, OJ L 334/1.

[58] Article 7(6) CBAM.

thus be present. However, this may only be true in some cases. The EU would have to make sure that the newly linked ETS overall provides for the same or even lower maximum number of emissions allowed than the two ETS foresaw before the linkage.

It is particularly contentious in the public and in the legal literature how to take account of the conditions prevailing in developing countries, the development of which the EU is obliged to foster pursuant to Article XXI(2)(d) TEU. They often do not have the economic capacity to employ costly environmental measures without compromising their competitiveness and development goals.[59] While the EU may counter that it is allowed to make a unilateral call for a high climate action standard under Article XX GATT notwithstanding, there is another argument that is harder to rebut: One may say that applying the *chapeau*, the conditions for climate policy vary between developing and developed countries because the former have historically contributed much less to global warming. The principle of "common but differentiated responsibilities and respective capabilities", enshrined for example in Article 2(2) Paris Agreement, thus calls on developed countries such as the EU to take the lead and carry a greater share of the transitory burden.[60] Reading the principle into Article XX GATT based on Article 31(3)(c) Vienna Convention on the Law of Treaties (VCLT), it seems possible to argue that CBAM arbitrarily discriminates products from developing countries by expecting the same climate action ambition in their production processes as from developed countries such as the US.[61] Nevertheless, one may also take a perspective that looks into the future: Then, any GHG emissions inevitably contribute to further warming the planet and CBAM's comprehensive coverage of products from developing and developed countries alike could be seen as rational. All in all, because the *chapeau* requires a holistic analysis of the application of the measure, much will depend on other problematic aspects of how CBAM addresses developing countries to which the analysis will turn now.

#### 6.4.3.2 Coercion to Adopt Carbon Pricing as the Only Viable Climate Protection Policy?

The *chapeau* requires that measures based on Article XX GATT should not coerce other WTO members to apply specific policy instruments (chosen by the trade restricting state) to overcome the trade restriction. Rather, in restricting trade and construing possible exceptions, states must give due regard to alternative policy options of other states having a similar effect on the policy goal that the trade-restricting measure pursues. The Appellate Body elaborated on this in *US–Shrimp*. The US had enacted an import ban on shrimp if they were not caught in line with

---

[59] See for example Leal-Arcas et al. 2022, p. 233.

[60] On the principle, see for example Stone 2004, pp. 276 et seq.; Rajamani 2018, pp. 291 et seq. For a critical analysis of CBAM in light of the EU's obligations under the Paris Agreement, see Quick and Das 2023, pp. 76 et seq.

[61] Hertel 2011, pp. 660 et seq.; Ladly 2012, p. 79; Pauwelyn 2020, p. 11; Espa et al. 2022, pp. 20 et seq. See Quick 2020, pp. 584 et seq.; Durán 2023, pp. 80 et seq. who consider this argument to lead to the incompatibility of CBAM with the GATT.

US guidelines that assured that shrimp trawl vessels do not accidently catch and kill sea turtles. The Appellate Body held that the US had unjustifiably discriminated by exerting "intended and actual coercive effect" on other countries to "adopt *essentially the same* policies and enforcement practices".[62] It found that in the way the US drafted implementing guidelines and applied the rules on the ground, they exclusively checked if the country of origin applied a standard that is essentially the same as the US one.[63] In the words of the Appellate Body, the import ban thus did "not allow for any inquiry into the appropriateness of the regulatory programme for the conditions prevailing in those exporting countries".[64]

Nevertheless, WTO members may condition market access by requiring a programme "comparable in effectiveness" (instead of: "essentially the same") operable in the country of origin.[65] In the enforcement stage (Article 21(5) DSU) of *US—Shrimp*, the Panel and the Appellate Body considered a modified US import ban to be WTO-conformant. The changed US ban explicitly allowed importing shrimps if "the government of a harvesting nation demonstrates that it has implemented and is enforcing a comparably effective regulatory program to protect sea turtles in the course of shrimp trawl fishing".[66] The Appellate Body considered this ban to comply with the *chapeau*. It ruled that trade-restrictive measures must only be sufficiently flexible to take into account the specific conditions "prevailing in *any* exporting Member" but do not need to provide for "specific provisions in the measure aimed at addressing specifically the particular conditions prevailing in *every individual* exporting Member".[67]

At first glance, CBAM appears to force a "market-based approach" on third countries because only linkages of ETS and carbon pricing mechanisms in third states can lead to an exemption or reduction of CBAM levy—and only by doing so can third countries receive the revenues rather than leaving them for the EU.[68] Yet, CBAM would in many cases also give due regard to alternative policies such as command-and-control legislation (e.g. technical standards for the production of cement, for example on energy efficiency) and incentives (e.g. subsidies for the use of green hydrogen in the production of cement). To the extent these alternative instruments actually reduce the GHG emissions in the production of the respective goods,

---

[62] WTO Appellate Body Report, *US—Shrimp*, 12 October 1998, WT/DS58/AB/R, para 161 (emphasis in the original).

[63] WTO Appellate Body Report, *US—Shrimp*, 12 October 1998, WT/DS58/AB/R, paras 162–163.

[64] WTO Appellate Body Report, *US—Shrimp*, 12 October 1998, WT/DS58/AB/R, para 165.

[65] Appellate Body Report, *US—Shrimp (Article 21(5) DSU)*, 22 October 2001, WT/DS58/AB/RW, para 144.

[66] Appellate Body Report, *US—Shrimp (Article 21(5) DSU)*, 22 October 2001, WT/DS58/AB/RW, para 146.

[67] Appellate Body Report, *US—Shrimp (Article XXI(5) DSU)*, 22 October 2001, WT/DS58/AB/RW, para 149.

[68] Leonelli 2022, p. 627; Sato 2022, pp. 402 et seq. For that reason, Quick 2020, p. 584 considers CBAM to be incompatible with the GATT. For a general critique on the use of CBAM revenues, see Espa et al. 2022, pp. 26 et seq.

these products will also receive better treatment under CBAM because the calculated embedded GHG emissions will be lower accordingly.[69]

However, CBAM could put countries at a disadvantage that adopt non-market-based climate change mitigation measures which distribute GHG emissions unevenly between different national sectors. These might be third countries which aim at developing certain "national champion" industries with high GHG emissions while offsetting these in other industry sectors (without also applying a market-based mechanism). Similarly, negative emissions techniques, such as growing forests, reduce the net emissions of a country without an effect on the embedded emissions in a product, and thus do not appear to be adequately grasped by CBAM. To comply with the *chapeau* of Article XX GATT, it thus seems that the EU product based-methodology for calculating embedded GHG emissions would have to account for such non-market based, cross-sectoral distribution of GHG reduction measures and for negative emissions technologies, for example in the implementing regulation of the Commission.[70]

#### 6.4.3.3 Serious and Even-Handed Attempt to Negotiate a Bi- or Multilateral Agreement?

Furthermore, in *US—Shrimp*, the Appellate Body required "serious, across-the-board negotiations with the objective of concluding bilateral or multilateral agreements for the protection and conservation of sea turtles" to conform with the *chapeau*.[71] The Panel and the Appellate Body pointed out that the US had not actually attempted to negotiate an international agreement on sea turtle conservation.[72] This had shown that there was a real alternative to a unilateral import ban that the US should at least have attempted to pursue, and that the US had only seriously negotiated with some, but not with other WTO members that export shrimp.[73]

However, the Appellate Body Report in the dispute settlement enforcement stage (Article XXI(5) DSU) relativized this inclination for bi- and multilateral cooperation to some extent. As seen, the US had kept the import ban in place with some modifications. The Appellate Body held that the US were now complying with Article XX GATT as long as "in particular the ongoing serious good faith efforts

[69] In the same vein Leonelli 2022, p. 626; Galiffa and Bercero 2022, pp. 198 et seq.

[70] Otherwise, charging imports would amount to a discriminatory "double counting" of GHG emissions, see Bartels 2012, p. 457.

[71] WTO, Appellate Body Report, *US—Shrimp*, 12 October 1998, WT/DS58/AB/R, para 166.

[72] The Appellate Body pointed out that this omission ran counter to the facts that US Congress had directed the Secretary of State to enter into negotiations, that the Rio Declaration as well as the Conventions on Biological Diversity and on Migratory Species of Wild Animals underlined the importance of bilateral and multilateral cooperation in this area, and that the US had already concluded the Inter-American Convention that provided for protection of sea turtles, see Appellate Body Report, *US—Shrimp*, 12 October 1998, WT/DS58/AB/R, paras 166–171 with further references.

[73] WTO, Appellate Body Report, *US—Shrimp*, 12 October 1998, WT/DS58/AB/R, para 172.

to reach a multilateral agreement" remained satisfied; it was not necessary to actually conclude bi- or multilateral agreements but to attempt comparable negotiations with all affected WTO members.[74] This rests well with the EU's obligation to find cooperative, preferably pluri- and multilateral solutions (Article XXI(1)(2) TEU).

Therefore, the mere fact that the EU plans to exclude goods with origins in Switzerland and the European Economic Area from CBAM alone does not run counter to the *chapeau*. Rather, it is decisive if the Commission will follow equal attempts to negotiate with all interested third countries. For a fair negotiation attempt, the Commission will at least have to start negotiations with all countries from which substantial imports of the goods to be covered by CBAM originate. Arguably, the EU has already attempted to negotiate a multilateral solution for a long time under the auspices of the UNFCCC, in particular by the EU's involvement in the negotiation of the Paris Agreement and subsequent conferences on detailing its rules.[75]

## 6.5 CBAM as WTO-Norm (Re-)Creation by the EU in the Spirit of Sustainability

Having analyzed how CBAM stands in light of the WTO's jurisprudence, it is submitted that the analysis must take one step further. The WTO (and indeed, multilateralism in general) has now been in a severe crisis for some years. As will be shown in the following subsections, there are strong calls by the Members and by scholars that WTO law must change (Sect. 6.5.1). In this light, CBAM can be understood as a regulation by which the EU contributes to reshaping WTO law (Sect. 6.5.2). It is submitted that the European Treaties allow and even call upon the EU to walk the fine line between complying with international law and creatively changing and developing it to promote sustainability (Sect. 6.5.3).

### *6.5.1 The WTO in Crisis and Reform Discussions*

The analysis of WTO law presented in the previous section builds on the jurisprudence of the Appellate Body before its paralysis in 2019 which by and large led the dispute settlement system to a deadlock.[76] Some of the Appellate Body's core findings relevant for the present analysis go back to reports from the 1990s. Then,

[74] WTO, Appellate Body Report, *US—Shrimp (Article 21(5) DSU)*, 22 October 2001, WT/DS58/AB/RW, paras 115 et seq.; direct quote in para 153.

[75] For a comprehensive analysis of EU climate diplomacy, see Minas and Ntousas (eds) 2018. For a different perspective, see for example Mbengue and Cima 2022, pp. 120–121 who criticize that the EU moves forward unilaterally.

[76] The US has been blocking the reappointment of new Members of the Appellate Body. Since December 2019, Appellate Body proceedings cannot be conducted anymore, causing any appeals to lead "into the void", see for example Pauwelyn 2019, pp. 297 et seq.

the global political landscape (or at least the widespread perception of it) was very different than it is today. There was trust in a growing multilateralism, reflected in the move to international trade law as a "rule-based system" with the WTO dispute settlement as one of its elementary pillars.[77] Today, multilateralism is in a crisis, pushed back by a renaissance of geo-economic conflicts that call for strengthening national security and the reduction of dependencies on trade with (potentially) problematic states.[78] At the same time, international climate law has taken various important steps, leading to the elaborate Paris Agreement.

In light of these challenges, there is a widespread opinion amongst WTO members that the law and the institutional setting of the WTO must change. Members heatedly discuss possible reforms of the WTO dispute settlement system to overcome its deadlock.[79] The EU has been a pioneer in defending the adjudicatory, rule-based trade system by supporting the creation of the Multi-Party Interim Appeal Arbitration Agreement.[80]

Necessarily, this development has an effect on the interpretation of WTO law and the provisions of the GATT analyzed above. WTO law is at a crossroad. It is not very likely that the jurisprudence of the 1990s and even the 2000s and 2010s contain the answers to the contemporary problems of free trade and climate change.[81] A "static" legal analysis as conducted in the previous section is thus prone to disregard that most probably, WTO law will in some form adapt to the challenges presented in the 2020s. There are different avenues for that change. Of course, the WTO agreement provides for the possibility for treaty changes or amendments and authoritative interpretations by the WTO members.[82] But a multilateral coalition for a comprehensive "green reform" of the WTO does not appear realistic in the current political climate.

### *6.5.2 CBAM as an EU Contribution to Reshape WTO Law*

Apart from formal treaty changes, WTO law may also be reshaped on the level of the interpretation of the WTO agreement and the application of its norms to the facts. Importantly, the rules of treaty interpretation under Articles 31 and 32 VCLT generally allow for an "evolutionary" interpretation of international law if that conforms with the original intention of the parties. As the ICJ held—among other decisions—in the *Navigational and Related Rights*-case, especially treaties that have been concluded to apply over a long time usually allow for an evolutive meaning of treaty terms to adapt to changing circumstances in the real world and

---

[77] Pauwelyn 2019, p. 319.

[78] On the broader backlash against international law, see for example Krieger 2019, pp. 971 et seq.

[79] From the ample literature on this matter, see for example Gao 2021, pp. 534 et seq.

[80] WTO, Multi-Party Interim Appeal Arbitration Agreement Pursuant to Article 25 of the DSU, 30 April 2020, JOB/DSB/1/Add.12.

[81] In the same vein Pauwelyn 2020, p. 5.

[82] See in particular Article 10 WTO Agreement.

developments in international law[83]—in our case, an escalating climate crisis, a crisis of multilateralism and the developments in international climate law. Indeed, and famously, the Appellate Body has already engaged in evolutive interpretation of Article XX(g) GATT and the term of "exhaustible natural resources" in *US—Shrimp*.[84] Furthermore, interpretation of treaties must take account of the subsequent practice in the application of the treaty which establishes the agreement of the parties (Art 31(3)(b) VCLT).[85]

It is true that any evolutive reinterpretation of multilateral treaties is difficult[86] and leads to complex legal questions on how the various paragraphs of Articles 31 and 32 VCLT are to be interpreted and applied to WTO law that cannot be discussed here.[87] Indeed, the Appellate Body has been reluctant to identify a subsequent interpretive practice of the WTO Members so far.[88] Notwithstanding, CBAM can well be understood as an attempted contribution by the EU to a dynamic (re-)interpretation of WTO law. In essence, CBAM is the EU's try to craft a new and fairer way of understanding international competition in trade in goods—one that internalizes the costs of damaging the global atmosphere by emitting GHG. By drafting CBAM and concretizing its content in its upcoming administrative practice, the EU contributes to the global discussions on how we should understand and apply WTO law, or how we should change its text in a possible treaty reform. Naturally, the EU is only one voice among many other WTO Members, even if an important one, and much will depend on the EU's capacity to persuade others to join in. An interpretive evolution of WTO law would require other WTO members to agree with the EU position. With CBAM, the EU sets a political precedent and steers the discussion by proposing a carbon border mechanism that in its view conforms with WTO law. In a potential future dispute settlement procedure on the WTO-compliance of CBAM, memorials and pleadings by the EU thus not only serve to fend off any claims by the complainant but also speak to the right (re-)interpretation of WTO law for the future.

The potential for such reinterpretation is significant, appreciating that some key questions of WTO law, relevant for examining the legality of CBAM, are considered

---

[83] ICJ, *Dispute Regarding Navigational and Related Rights (Costa Rica v Nicaragua)*, Judgment of 13 July 2009, ICJ Rep 2009, p. 213, paras 63 et seq.

[84] WTO, Appellate Body Report, *US—Shrimp*, 12 October 1998, WT/DS58/AB/R, paras 128 et seq. For a comprehensive analysis of other Panel and Appellate Body decisions that applied (or refused to apply) evolutive treaty interpretation, see van den Bossche 2019, pp. 224 et seq. For further examples of how WTO dispute settlement has engaged in evolutionary interpretation related to renewable energies, see Grigorova 2019, pp. 241 et seq.

[85] See ILC, Draft Conclusions on Subsequent Agreements and Subsequent Practice in Relation to the Interpretation of Treaties (2018) UN Doc A/73/10, Chapter IV. On the possibility of treaty modification by subsequent practice, see Buga 2018.

[86] See for example van den Bossche 2019, pp. 221 et seq.

[87] On the many complex questions on the interpretation of Articles 31 and 32, see Dörr (2018a) and Dörr (2018b) with further references.

[88] See for example WTO, Appellate Body Report, *Japan—Alcoholic Beverages*, 4 October 1996, WT/DS8,10,11/AB/R, pp. 12 et seq.; discussed by Nolte 2011, pp. 138 et seq. with further references to WTO case law in comparison to the jurisprudence of the ICJ; Cook et al. 2019, pp. 184 et seq. with further critical analysis.

to remain unresolved even under the WTO's jurisprudence until 2019. This is, for example, the case for the exact requirements of the *chapeau* of Article XX GATT, as shown above. But also the application of the law to the facts requires the interpreter to concretize the legal norm to the specific factual circumstances.[89] This leaves room for a context-sensitive application of WTO law. Even more so, the *chapeau* of Article XX GATT operates as a legal test that takes account of all circumstances and details of the trade-restrictive measure at stake. Being an expression of the good faith principle,[90] it provides for substantial judicial leeway in ascertaining and evaluating CBAM in the relevant global context that the EU faces.

### 6.5.3 Complying with and Shaping International Law in the EU's External Action for Sustainability

From a bird's-eye perspective, the presented dynamic perspective on CBAM and WTO law may serve as a case study for a more fundamental observation on the EU's external actions for sustainability. Article 3(5) TEU states that the EU "shall contribute to the strict observance and the development of international law". It is submitted that this calls upon the EU to walk the fine line between complying with international law as understood at the relevant point in time, and creatively changing and developing it, for example to combat climate change and further sustainability.[91]

To be clear, this is not to say that the EU may disregard international law, which would be contrary to Article XXI(1) TEU.[92] However, in contrast to domestic law, it is a defining feature of international law that it lacks a central legislator. Treaty law, customary law and general principles are thus always potentially subject to constant and successive interpretive change. For the EU, this means that promoting international law not only entails conserving the (perceived) current state of the art,

---

[89] Kelsen 1934, pp. 94 et seq. describes this as the "constitutive function" of the judicial decision. Cook 2019, pp. 182 et seq. describes this as "evolutionary application" of the law to the facts in distinction from "evolutionary interpretation" of the law itself, drawing on traditions of domestic constitutional law in various jurisdictions.

[90] WTO Appellate Body Report, *US—Gasoline*, 20 May 1996, WT/DS2/AB/R, p. 22; *US—Shrimp*, 12 October 1998, WT/DS58/AB/R, para 151, 158; *Brazil—Retreaded Tyres*, 3 December 2007, WT/DS/332/AB/R, paras 224.

[91] Generally on the EU's contribution to the development of international law, see Hoffmeister 2008, pp. 38 et seq.; Klabbers 2015, pp. 53–54; for a critical assessment of the EU as an exporter of European values, see Herlin-Karnell 2015, pp. 1227 et seq.; for a complex analysis of the EU as a "regulatory actor" that shapes international law through the bidirectional interaction of EU and international law, see Cremona 2019, pp. 100 et seq. based on the example of money laundering. For a progressive position that criticises an overemphasis on the "strict observation of international law" in the interpretation of Article 3(5) TEU, see Dunbar 2021, pp. 479 et seq.

[92] Indeed, with Cremona 2019, p. 70, "the EU derives its power from and through law" and is thus "profoundly committed to law and to the legality of its actions, both internally (constitutional legality) and externally (with respect to international law)."

but also to support its development in line with the EU's values, for example, by engaging in treaty (re-)interpretation.

Bringing the presented static and dynamic stances to international law together, the EU must on the one hand respect and comply with the established understanding of international law at a given point in time, for example, with uncontroversial aspects of WTO law and with interpretative questions that the Appellate Body has clarified. On the other hand, the EU can also engage with unclear, precarious or contested international norms. The EU may promote its understanding of these rules, and act accordingly and unilaterally to the extent it does not undermine multilateral institutions.[93] If and to the extent multilateral solutions could not be found, the EU can push the development of international law as a "norm entrepreneur".[94] It seems that this also serves to comply with the EU's obligation to act coherently in its external affairs (Article 21(3)(2) TEU) because such action contributes to consolidating and clarifying international law.

Reconciling a static and dynamic stance to applying international law also means allowing a different perspective on the various, often-colliding constitutional goals that the European Treaties call upon the EU to pursue in its external relations. Following this line of argument, we may say that CBAM does not cause such a severe conflict between the goals to promote international law, free trade (Article 21(1)(1), (2)(e) TEU) and combating climate change in the EU's external relations (Article 191(1) TFEU) as sometimes brought forward.[95] Rather, in the WTO's currently precarious status, the EU may be seen as a potential (re-)creator of WTO law, hopefully aligning climate action and free trade better than WTO law used to do until 2019.

As a result, future WTO law may then better mirror the EU's internal values (such as combating climate change) in line with the EU's obligations under the European Treaties to promote its internal values in its external relations (Article 21(1)(1) TEU). To be sure, this can only prevent an overly Eurocentric position if the EU engages sufficiently with developing countries in line with the legal analysis presented above. The point made here is that the EU Treaties, correctly interpreted, require the EU to take account of the interests of lesser developed countries and find a balance with the objective to combat climate change. Importantly, as seen, this includes giving due regard to the interests of developing countries in the "Global South".

---

93 In the same vein Odermatt 2020, pp. 49 et seq who submits that unilateral and extraterritorial action of the EU may contribute to developing international law but may also risk undermining multilateralism.

94 Scott and Rajamani 2012, p. 472.

95 See for example Dobson (2023), pp. 375–379 with further references.

## 6.6 Conclusion

CBAM will be at the center of political and legal debates on the EU's external policies in the coming years, particularly in relation to developing countries. Not least because of its complex nature and its connection to the EU ETS, it is not easy to assess whether it is compatible with WTO law. Based on the WTO jurisprudence until 2019, CBAM would have to meet several conditions in the way it is interpreted and applied in practice in order to be consistent with WTO law. However, this paper argues that the legal analysis should be taken one step further. Given the crisis of the WTO and the calls for reform, it is necessary to take a dynamic perspective on WTO law. In this light, CBAM can be understood as a regulation by which the EU does not only test the boundaries of WTO law, but also contributes to the reshaping of WTO law in a discursive interaction with other concerned WTO members. It is submitted that the European Treaties allow and even require the EU to walk the fine line between adhering to WTO law and creatively changing and developing it to promote sustainability.

## References

Bartels L (2012) The WTO Legality of the Application of the EU's Emission Trading System to Aviation. European Journal of International Law 23:429–467

Bartels L (2015) The Chapeau of the General Exceptions in the WTO GATT and GATS Agreements: A Reconstruction, AJIL 109:95–125

Beyerlin U (2013) Sustainable Development. In: Peters A (ed), Max Planck Encyclopedia of Public International Law, Oxford University Press, Oxford

Buga I (2018) Modification of Treaties by Subsequent Practice, Oxford University Press, Oxford.

Carević M (2015) Carbon Leakage in the EU in the Light of the Paris Climate Agreement, Croatian Yearbook of European Law & Policy 11:47–72

Colares JF and Rode A (2017) The Opportunities and Limitations of Neutral Border Tariffs. American Law and Economics Review 19:423–463

Cook G (2019) The Illusion of "Evolutionary Interpretation" in WTO Dispute Settlement. In: Abi-Saab G et al (eds) Evolutionary Interpretation and International Law, Hart Publishing, Oxford, pp 181–194

Cottier T (ed) (2021) The Prospects of Common Concern of Humankind in International Law, Cambridge University Press, Cambridge

Cottier T et al (2017) The Principle of Proportionality in International Law. The Journal of World Investment & Trade 18:628–672

Cremona M (2012) Coherence and EU External Environmental Policy. In: Morgera E (ed), The External Environmental Policy of the European Union: EU and International Law Perspectives, Cambridge University Press, Cambridge, pp 33–54

Cremona M (2019) Extending the Reach of EU Law: The EU as an International Legal Actor. In: Scott J and Cremona M (eds) EU Law Beyond EU Borders: The Extraterritorial Reach of EU Law, Oxford University Press, Oxford, pp 64–111

Dias A et al (2020) EU Border Carbon Adjustment and the WTO: Hand in Hand Towards Tackling Climate Change, Global Trade and Customs Journal 15:15–23

Dobson NL (2023) Climate Protection versus Trade: Dilemmas for the EU. In: Rayner T et al (eds) Handbook on European Union Climate Change Policy and Politics, Edward Elgar, Cheltenham, pp 367–382

Dobson NL (2021) Extraterritoriality and Climate Change Jurisdiction: Exploring EU Climate Protection under International Law. Hart Publishing, Oxford

Dörr O (2018a) Article 31. In: Dörr O/Schmalenbach K (eds) Vienna Convention on the Law of Treaties, A Commentary, 2nd ed, Springer, Berlin, pp 559–616.

Dörr O (2018b) Article 32. In: Dörr O/Schmalenbach K (eds) Vienna Convention on the Law of Treaties, A Commentary, 2nd ed, Springer, Berlin, pp 559–616.

Dunbar R (2021) Article 3(5) TEU a Decade On: Revisiting 'Strict Observance of International Law' in the Text and Context of other EU Values, Maastricht Journal of European and Comparative Law 28:479–497

Durán GM (2016) Measures with Multiple Competing Purposes after EC – Seal Products: Avoiding a Conflict between GATT Article XX-Chapeau and Article 2.1 TBT Agreement, JIEL 19:467–495

Durán GM (2023) Securing Compatibility of Carbon Border Adjustments with the Multilateral Climate and Trade Regimes, International and Comparative Law Quarterly 72:73–103

Espa I et al (2022) The EU Proposal for a Carbon Border Adjustment Mechanism (CBAM): An Analysis under WTO and Climate Change Law, WTI Working Paper No 06/2022

Galiffa C and Bercero IG (2022) How WTO-Consistent Tools Can Ensure the Decarbonization of Emission-Intensive Industrial Sectors, American Journal of International Law Unbound 116:196–201

Gao H (2021) Finding a Rule-Based Solution to the Appellate Body Crisis: Looking Beyond the Multiparty Interim Appeal Arbitration Arrangement, Journal of International Economic Law 24:534–550

Grigorova J (2019) Energy Trade in the WTO, Yesterday, Today and Tomorrow. In: Abi-Saab G et al (eds) Evolutionary Interpretation and International Law, Hart Publishing, Oxford, pp 241–250

Herlin-Karnell E (2015) The EU as a Promoter of Values and the European Global Project, German Law Journal 13:1225–1246

Hertel M (2011) Climate-Change-Related Trade Measures and Article XX: Defining Discrimination in Light of the Principle of Common but Differentiated Responsibilities, Journal of World Trade 45:653–678

Hoffmeister F (2008) The Contribution of EU Practice to International Law. In: Cremona M (ed) Developments in EU External Relations Law, Oxford University Press, Oxford, pp 37–127

Hsu S-L (2016) International Market Mechanisms. In: Gray KR (ed) The Oxford Handbook of International Climate Change Law, Oxford University Press, Oxford, pp 239-256

IPCC (2023) Climate Change 2023, Synthesis Report, Summary for Policymakers.

Kelsen H (1934) Reine Rechtslehre [Pure Theory of Law], Franz Deuticke, Vienna

Kiss A (2003) European Environmental Law and the Constitution, Pace Environmental Law Review 21:103–117

Klabbers J (2015) Straddling the Fence: The EU and International Law. In: Chalmers D and Arnull A (eds) The Oxford Handbook of European Union Law, Oxford University Press, Oxford, pp 52–72

Krieger H (2019) Populist Governments and International Law, European Journal of International Law 30:971–996

Knodt M (2023) Instruments and Modes of Governance in EU Climate and Energy Policy: From Energy Union to the European Green Deal. In: Rayner T and Szulecki K (eds) Handbook on European Union Climate Change Policy and Politics, pp 202–215

Ladly SD (2012) Border Carbon Adjustments, WTO-Law and the Principle of Common But Differentiated Responsibilities, International Environmental Agreements 12:63–84

Leal-Arcas R et al (2022) A Legal Exploration of the European Union's Carbon Border Adjustment Mechanism, European Energy and Environmental Law Review 31:223–240

Leonelli GC (2022) Carbon Border Measures, Environmental Effectiveness and WTO Law Compatibility: Is There a Way Forward for the Steel and Aluminium Climate Club?, World Trade Review 21:619–632

Mbengue MM and Cima E (2022) "Clubbing in the Club": Could Climate-Related Trade Arrangements Set the Pace for Future Climate Cooperation?, American Journal of International Law Unbound 116:219–224

Minas and Ntousas (eds) (2018) EU Climate Diplomacy, Politics, Law and Negotiations, Routledge, Abingdon

Nolte G (2011) Subsequent Practice as a Means of Interpretation in the Jurisprudence of the WTO Appellate Body. In: Cannizzaro E (ed) The Law of Treaties Beyond the Vienna Convention, Oxford University Press, Oxford, pp 138–144

Odermatt J (2020) Convergence through EU Unilateralism. In: Fahey E (ed) Framing Convergence with the Global Legal Order: The EU and the World, Hart Publishing, Oxford, pp 49–64

Pauwelyn J (2019) WTO Dispute Settlement Post 2019: What to Expect?, Journal of International Economic Law 22:297–321

Pauwelyn J (2020) Trade Related Aspects of a Carbon Border Adjustment Mechanism, A Legal Assessment, Briefing Requested by the INTA Committee of the European Parliament, PE603502

Pirlot A (2022) Carbon Border Adjustment Measures: A Straightforward Multi-Purpose Climate Change Instrument?, Journal of Environmental Law 24:25–52

Quick R (2020) Carbon Border Adjustment, A Dissenting View on its Alleged GATT-Compatibility, Zeitschrift für Europarechtliche Studien 23:549–597

Quick R and Das I (2023) Can Paris Strike Back? On the Paris Agreement's Inability to Cope with Unilateral Trade-Related Carbon Measures Such as the European Commission's CBAM-Proposal. In: Baetens F and Van den Bogaert S (eds) Liber Amicorum Marco Bronckers, Wolters Kluwer, Alphen aan den Rijn, pp 75–102

Rajamani L (2018) Common But Differentiated Responsibilities. In: Faure M, Krämer L and Orlando E (eds) Elgar Encyclopedia of Environmental Law, Volume VI: Principles of Environmental Law, Edward Elgar Publishing, Cheltenham, pp 291–302

Riffel C (2018) The Chapeau: Stringent Threshold or Good Faith Requirement, Legal Issues of Economic Integration 45:141–176

Sato SY (2022) EU's Carbon Border Adjustment Mechanism: Will It Achieve Its Objective(s)?, Journal of World Trade 56:383–404

Scott J (2011) The Multi-Level Governance of Climate Change, Carbon & Climate Law Review 5:25–33

Scott J and Rajamani L (2012) EU Climate Change Unilateralism, European Journal of International Law 23:469–494

Shoyer A et al (2016) Carbon Leakage and the Migration of Private CO2 Emitters to other Jurisdictions. In: Gray KR (ed) The Oxford Handbook of International Climate Change Law, Oxford University Press, Oxford, pp 285–312

Stone CD (2004) Common But Differentiated Responsibilities in International Law, American Journal of International Law 98:276–301

Thym D (2022) In: von Arnauld A Bungenberg M (2022) Europäische Außenbeziehungen, Enzyklopädie Europarecht Band 12 [European External Affairs, EU Law Encyclopedia Volume 12], 2nd edition, Nomos, Baden-Baden, pp 1129–1264

Trachtman JP (2017) WTO Law Constraints on Border Tax Adjustments and Tax Credit Mechanisms to Reduce the Competitive Effects of Carbon Taxes, National Tax Journal 70:469–494

Van den Bossche P (2019) Is There Evolution in the Evolutionary Interpretation of WTO Law?. In: Abi-Saab G et al (eds) Evolutionary Interpretation and International Law, Hart Publishing, Oxford, pp 221–228

Venzke I and Vidigal G (2022a) Of False Conflicts and Real Challenges: Trade Agreements, Climate Clubs and Border Adjustments?, American Journal of International Law Unbound 116:202-207

Venzke I and Vidigal G (2022b) Are Unilateral Trade Measures in the Climate Crisis the End of Differentiated Responsibilities? The Case of the EU Carbon Border Adjustment Mechanism (CBAM), Amsterdam Law School Legal Studies Research Paper No. 2022-02

Wessel RA (2016) Lex Imperfecta: Law and Integration in European Foreign And Security Policy, European Papers 1:439–468

**Patrick Abel** is an Assistant Professor (Akademischer Rat a.Z.) at the University of Passau, Passau, Germany.

# Chapter 7
# What is There to Enforce? Multilateral Agreements and Standards in the Trade and Sustainable Development Chapters of the EU Free Trade Agreements

**Saide Esra Akdogan**

**Contents**



**Abstract** For years, the discussions on the enforceability of the Trade and Sustainable Development (TSD) chapters of the European Union (EU) Free Trade Agreements (FTAs) focused on the shortcomings of the institutional mechanism. Stakeholders have criticized the collaborative approach where no sanctions were available for breaches, combined with the exclusion from the general Dispute Settlement Mechanism (DSM) of the FTA. In 2022, the European Commission announced a more assertive approach to remedy these shortcomings. Long-awaited developments in the procedural area raise questions about the substantive content of the TSD chapters. This Chapter aims to provide the needed reality check on to what extent this assertive approach in the procedure is backed up by the substance of the obligations.

---

This chapter is based on the Chapter 5—Article bringing obligations regarding multilateral agreements and standards in Akdogan, S. E. (2024). Decrypting Sustainability in EU Free Trade Agreements: Substantive content and the enforceability of the Trade and Sustainable Development chapters. [Thesis fully internal (DIV), University of Groningen]. University of Groningen. In detail analysis of the substantive content of other TSD articles can be found in the other chapters of the said dissertation.

---

S. E. Akdogan (✉)
Law Group, Wageningen University and Research (WUR), Wageningen, The Netherlands
e-mail: esra.akdogan@wur.nl

 
R. A. Wessel et al. (eds.), *EU External Relations Law and Sustainability*, Global Europe: Legal and Policy Issues of the EU's External Action 4,
https://doi.org/10.1007/978-94-6265-655-0_7

Through textual analysis, the Chapter scrutinizes multilateral labor and environmental standards articles under all the FTAs with a TSD chapter to date and assesses the commitments to explore the trends in the EU's approach. It looks at the clarity of the language used; the substance, nature, and the enforceability of the obligations.

**Keywords** European Union · Free trade agreements · Trade and sustainable development chapters · European Commission · Enforceability · Substantive obligations · MEAs · ILO conventions

## 7.1 Introduction

Currently, TSD chapters are excluded from the general Dispute Settlement Mechanism (DSM) of EU Free Trade Agreements (FTAs).[1] Instead, the parties have recourse to a Panel of Experts (PoE) which is the main body of a TSD DSM if government consultations fail.[2] The PoE can provide suggestions[3] but there is no follow-up mechanism or sanction for non-compliance. This *status quo* led to intense criticisms towards the enforcement of TSD chapters, and the '*compliance gap*' between the commitments undertaken and their realization has become the center of the discussion.[4]

The Commission engaged in dialogue with stakeholders several times over the years to improve the TSD chapters which resulted in non-papers and communications.[5] A need for a more assertive approach has been a recurring theme. Several stakeholders, including the European Parliament, had called for expanding the standard DSM of the FTAs to the TSD chapters and recourse to trade sanctions in the event of breaches.[6] Up until now, the Commission has rejected these suggestions.

The communication of the Commission of June 2022 changed this situation.[7] The new assertive approach will extend the compliance stage of the DSM of the FTA to monitor the implementation of the PoE report.[8] Additionally, for the first time, the use of trade sanctions has been named as a possible last-resort option for breaches of the Paris Agreement and fundamental ILO conventions.[9] However, for all EU FTAs

[1] See for example Article 13.16, Free trade Agreement between the European Union and its Member States, of the one part, and the Republic of Korea, of the other part, Official Journal of the European Union, L 127, 14 May 2011, available at https://eur-lex.europa.eu/legal-content/EN/ALL/?uri=OJ%3AL%3A2011%3A127%3ATOC, last access in 05.11.2022 [South Korea FTA].

[2] See for example Article 13.15 (1), South Korea FTA.

[3] See for example Article 13.15 (2), South Korea FTA.

[4] See for example Prévost and Alexovičová 2009; European Economic and Social Committee 2018; Gruni and Bronckers 2019; Harrison et al. 2019; Krämer 2021.

[5] European Commission 2017; 2018; 2022.

[6] European Parliament 2016.

[7] European Commission 2022.

[8] European Commission 2022.

[9] European Commission 2022.

that are currently in force, the procedure remains the same until they are updated. This is except for the New Zealand FTA, which came into force on May 1, 2024, fleshing out the new approach in an FTA for the first time.[10]

These procedural improvements to the TSD enforcement system are certainly hopeful. These changes, however, can only have a meaningful effect on the enforcement of the TSD chapters if they are backed up by legally binding obligations in the substantive articles. The question, therefore, remains whether the TSD content also needs a stepping up to match the new procedural approach. Only then can we talk about the integrated assertiveness of the TSD chapters.

Against this background, this Chapter addresses the following question: *"What is the level of enforceability of TSD chapters substantive obligations regarding multilateral labor and environmental standards?"* It looks at multilateral labor standards and agreements and the multilateral environmental agreements articles which are one of the main sources of substantive obligations regarding labor and environment in the TSD chapters with a comparative perspective based on doctrinal constructivism.[11]

This necessitates an interpretive exercise.[12] Articles 31 and 32 of the Vienna Convention on the Law of Treaties (VCLT) are accepted to reflect customary international law on treaty interpretation and applied as such by international courts.[13] Although the scope of VCLT in principle only covers the international agreements of states,[14] The Court of Justice of the European Union (CJEU) has consistently applied VCLT rules of interpretation to the international agreements of the EU.[15] International agreements of the EU also include the EU FTAs. Accordingly, they

---

[10] EU-New Zealand FTA, available at https://policy.trade.ec.europa.eu/eu-trade-relationships-country-and-region/countries-and-regions/new-zealand/eu-new-zealand-agreement_en, last access on 13.05.2024 [New Zealand FTA].

[11] Von Bogdandy 2009; van Gestel and Micklitz 2011.

[12] *"Whenever a subject of international law invokes, applies or goes about implementing a treaty, it can only do so on the basis of a certain understanding of its terms, ergo on the basis of an interpretation."* Dörr and Schmalenbach 2018, p. 567.

[13] For an example of ICJ's case law see: Territorial Dispute (Libyan Arab Jamahiriya/Chad), 1994, para 41; Oil Platforms (Islamic Republic of Iran v. United States of America), 1996, para 23). A Kasikili/Sedudu Island (Botswana v. Namibia) *"Customary international law found expression in Article 31 of the Vienna Convention"*, para 18. For an example of WTO Appellate Body's case law see I.3.1.1 US—Gasoline, p. 17, DSR 1996:I, para 16.

[14] Article 1, VCLT, United Nations, Vienna Convention on the Law of Treaties, 23 May 1969, United Nations, Treaty Series, vol. 1155, p. 331. [VCLT] is titled *Scope of the present convention* and states that *"The present Convention applies to treaties between States"* and Article 2 (1) (a), VCLT defines a treaty as *"an international agreement concluded between States [...]."* For the international agreements of international organization, a separate agreement was adopted by ILC in 1986: Vienna Convention on the Law of Treaties between States and International Organizations or between International Organizations (VCLT-IO) which is yet to come in force. Some authors have noted the gap in CJEU's reasoning as to why it is the Article 31 VCLT and not the identical Article 31 of VCLT-IO that applies, see for example Odermatt 2021. In practice in makes little difference to which Convention the CJEU refers to. The CJEU applies the Article 31 VCLT in its capacity to reflect customary international law and Article 31 VCLT-IO, as it is identical to Article 31 VCLT, necessarily shares this status.

[15] Agreement on the European Economic Area 1992, OJ 1994 L 1/3 (Case T-115/94, Opel Austria.

should be interpreted *"in good faith in accordance with the ordinary meaning to be given to the terms of the treaty in their context and in the light of its object and purpose."*[16]

The novelty of the study lies in its focus on the substantive obligations under the TSD chapters and the background of the recent assertive approach the EC has announced for the procedural aspects that the study tests this substantive content against. The doctrine[17] almost exclusively focused on the procedural shortcomings surrounding the TSD chapters, and policy suggestions on improvements. Studies of this nature often focus on either labor standards or environmental protection. Two studies have been identified that also cover some of the substantive content.[18] In both studies, the substantive content is a short part of the articles before moving on to the procedural aspects.

The present Chapter aims to add to these studies through an extensive coverage of all EU FTAs with a TSD chapter to this date, both environmental and labor-related obligations, and through its exclusive focus on the substantive content of these articles and their clarity and enforceability. Additional value is provided with an outlook on an integrated assertiveness of the TSD, both procedurally and substantively, following the recent approach of the European Commission.

Before the analysis of the substantive obligations, the role of the PoE within the TSD system should be looked at. The PoE is the second tier of the TSD DSM after the first tier of government consultations.[19] However, being the only body of the TSD DSM, the PoE naturally gains the function of being an interpretive body. Where the articles of the TSD chapter fail to provide clarity, the interpretation made by the PoE is key in understanding the obligations under the articles and subsequently increasing their enforceability. Hence, the only example of the TSD DSM being triggered to date, the EU-South Korea dispute[20] regarding the multilateral labor standards and the report delivered thereunder will be referred to throughout the analysis below.

Although preferable to no guidance or interpretation at all, clarity through the DSM has disadvantages. Firstly, the interpretations of PoE are dependent on a dispute being brought before the PoE on a specific provision. Considering the scarce number of disputes so far with only a single example and the hesitance of the parties to

---

GmbH v. Council of the European Union [EU:T:1997:3]), Brita v Hauptzollamt Hamburg Hafen, C-386/08, EU:C:2010:91, para 7.4; IATA and ELFA, C-344/04, EU:C:2006:10.

[16] Article 31, VCLT.

[17] See Prévost and Alexovičová 2019; Harrison et al. 2019; Van den Putte 2015; Nissen 2022.

[18] Van't Wout 2022; Bronckers and Gruni 2021; Füller 2018.

[19] For example, see Article 13.16, South Korea FTA: *"For any matter arising under this Chapter, the Parties shall only have recourse to the procedures provided for in Articles 13.14 and 13.15."* For more information on the dispute settlement mechanism of the TSD chapters see Velut et al. 2022.

[20] For a timeline and all documents of the dispute see Bilateral Disputes> Korea Labour Commitments on the website of the European Commission, available at https://policy.trade.ec.europa.eu/enforcement-and-protection/dispute-settlement/bilateral-disputes/korea-labour-commitments_en, last access on 22.11.2022.

trigger the process, this is likely to be a long wait before a set of PoE reports on TSD accumulate.

Secondly, PoEs are established per dispute on an *ad hoc* basis.[21] There is no permanent institution. This is a challenge for the consistency of interpretations if different PoEs interpret the same words or articles differently. Investment treaty arbitration, for example, has faced this problem chronically.[22] Decentralized DSM, broad legal terms, and the investment law being a relatively new area of law have been identified as contributing factors to the inconsistency of arbitral awards.[23] The parallels in these characteristics make it very likely the same problems to be experienced by the TSD system as well in a future where more PoE have issued reports.

For these reasons, the ideal situation is for the primary law, the articles themselves, to provide clarity itself. It is therefore crucial that the EC does not only step up the procedural aspects of the TSD but also its substantive content and opt for making clearer and better enforceable versions of the articles in the future. With this in mind, we will now look at the substantive content of the TSD obligations.

## 7.2 Multilateral Labor Standards and Agreements

The first paragraph of the *Multilateral Labour Standards and Agreements* article typically reaffirms the importance of international cooperation in labor rights matters. The *2006 Ministerial Declaration on Full Employment and Decent Work,*[24] which appears in some FTAs seems to be abandoned and replaced with "*full and productive employment and decent work for all*" instead. Such a first paragraph shows preamble characteristics in operative articles. Following this reaffirmation, the *Multilateral Labour Standards and Agreements* article typically goes on and regulates two main categories of obligations:

- Ratification obligation: Ratifying fundamental and non-fundamental ILO conventions.[25]

---

[21] For example see Article 13.15 (3), South Korea FTA.

[22] See Franck 2004; Kaufmann-Kohler 2008, p. 541; Cate 2012, p. 418.

[23] International Bar Association 2018.

[24] 2006 Ministerial Declaration of the UN Economic and Social Council on Full Employment and Decent Work, available at https://digitallibrary.un.org/record/593242, last access on 13.07.2023.

[25] Initially, eight ILO conventions were classified as fundamental ILO conventions by the Governing Body of the ILO. With the amendments over the years, there are currently eleven of them: (1) Freedom of Association and Protection of the Right to Organise Convention, 1948 (No. 87), (2) Right to Organise and Collective Bargaining Convention, 1949 (No. 98), (3) Forced Labour Convention, 1930 (No. 29) (and its 2014 Protocol), (4) Abolition of Forced Labour Convention, 1957 (No. 105), (5) Minimum Age Convention, 1973 (No. 138), (6) Worst Forms of Child Labour Convention, 1999 (No. 182), (7) Equal Remuneration Convention, 1951 (No. 100), (8) Discrimination (Employment and Occupation) Convention, 1958 (No. 111), (9) Occupational Safety and Health Convention, 1981 (No. 155), (10) Promotional Framework for Occupational Safety and Health Convention, 2006 (No.

- Implementation obligation: Respecting, promoting, and realizing certain principles in parties' laws and practices.[26]

### 7.2.1 *Implementation Obligation*

To illustrate, the related paragraph in the EU—South Korea FTA[27] reads as follows:

"The Parties, in accordance with the obligations deriving from membership of the ILO and the ILO Declaration on Fundamental Principles and Rights at Work[28] and its Follow-up, adopted by the International Labour Conference at its 86th Session in 1998, *commit to respecting, promoting, and realizing,* in their laws and practices, the principles concerning fundamental rights, namely:

(a) freedom of association and the effective recognition of the right to collective bargaining;
(b) the elimination of all forms of forced or compulsory labor;
(c) the effective abolition of child labor; and
(d) the elimination of discrimination in respect of employment and occupation.

The Parties reaffirm the commitment to effectively implementing the ILO Conventions that Korea and the Member States of the European Union have ratified respectively."[29]

Commitment to respect, promote, and realize these fundamental rights are repeated in each TSD. While the wording differs slightly under some agreements, the essence of the obligation stays the same. The PoE of the EU-South Korea dispute dealt with what is meant by this article which is a good place to start in understanding the obligation.

187). More information available at https://www.ilo.org/global/standards/introduction-to-international-labour-standards/conventions-and-recommendations/, last access on 22.11.2022.

[26] According to the VCLT *"ratification [...] means the international act so named whereby a State establishes on the international plane its consent to be bound by a treaty."* (Article 2 (1) (b), also see Article 14) Implementation however regards to the timeframe after the ratification of a treaty where the parties are to perform obligations they undertook by ratifying. While ratification is a single procedural act, implementation regards to the functioning of the treaty and a process where the parties make necessary changes to their domestic law and practices to adhere to the articles of the treaty.

[27] Article 13.4 (3), South Korea FTA.

[28] The ILO Declaration on Fundamental Principles and Rights at Work which was adopted in 1998 and amended in 2022 affirms the obligations and commitments that are inherent in membership of the ILO which are listed below under the 13.4 of the South Korea FTA. See ILO Declaration on Fundamental Principles and Rights at Work, available at https://www.ilo.org/declaration/lang-en/index.htm, last access on, 05.11.2022.

[29] These rights are enshrined under the initial eight fundamental ILO conventions. See *supra note 30.* A safe and healthy working environment was added to the ILO Declaration on Fundamental Principles and Rights at Work through the amendment of June 2022 as a fifth fundamental principle. See *supra note 38.*

The EU claimed that South Korea failed to "*respect, promote and realize*" certain aspects of freedom of association as understood in the ILO system: the definition of workers under South Korea labor law, the eligibility for trade union leader candidacy, and workers' protection due to union membership.[30] South Korea claimed that the word *to commit* only expressed an aspiration to "*engage in good faith behavior toward the overall objective of respecting, promoting and realizing the principles,*"[31] thus suggesting an effort obligation. South Korea stated that "'*commit to respecting, promoting, and realizing' means something less than 'shall respect, promote and realize'.*"[32]

The PoE rejected this interpretation of the word *to commit* and declared that the ordinary meaning of the word is *"to bind oneself to a course of action"* and that the article represents a legally binding obligation.[33] The choice to use the word *"commit"* instead of, for example the word *"will"* was appropriate as the Parties are referring to an already existing external obligation arising from ILO membership. This referral however does not prevent the obligation of respecting, promoting, and realizing from being a stand-alone obligation under the FTA. The PoE found that the parties "*reaffirm the existing obligations […] as a separate and independent obligation*" under the TSD chapter.[34]

Therefore, the PoE affirmed the word "*commit*" does not only show an aspiration or require a showing of efforts towards the general aim of the principles, but creates a separate, independent, and binding legal obligation arising from the TSD. Thus, the legal obligation on this matter arises not from the 1998 ILO Declaration but from Article 13.4.3 of the FTA, i.e., the multilateral labor standards and agreements article.[35]

The PoE then went on to analyze the meaning of the words respecting, promoting, and realizing and found the following:

- respecting: a negative obligation to *"show respect for, refrain from injuring, harming, insulting, interfering with, or interrupting"*[36]
- promoting: a positive obligation to *"further the development, progress, or establishment of (a thing), encourage, help forward, or support activity"*[37]
- realizing: *"commitment to attain or make real"*[38]

---

[30] PoE of Experts Proceeding Constituted Under Article 13.15 of The EU-Korea Free Trade Agreement Report of The PoE of Experts, 20 January 2021, available at https://trade.ec.europa.eu/doclib/docs/2021/january/tradoc_159358.pdf, last access on 05.11.2022, pp. 53–70 [Report of the PoE of Experts].

[31] Report of the PoE of Experts, para 126, Korea Written Submissions para 41.

[32] Ibid.

[33] Report of the PoE of Experts, paras 124–126.

[34] Report of the PoE of Experts, para 107.

[35] Ibid.

[36] Report of the PoE of Experts, para 131.

[37] Report of the PoE of Experts, para 132.

[38] Report of the PoE of Experts, para 133.

In this context, an obligation to respect, promote, and realize involves both a negative obligation to refrain from interfering or harming as well as a positive obligation to further develop. progress and make necessary changes. Thus, any law or common practice that intervenes with these fundamental rights should be rectified by the state as well as being on course for further improvement.

It is important to point out how the PoE depends its analysis only on the ordinary meaning of the words used in the article. VCLT Article 31 has other elements of interpretation as well: the context, object and purpose of the treaty at hand. While interpreting a treaty article, all these elements must be taken into account equally, with the same status and authority.[39] None is superior or inferior to the others.[40] They are not a list of steps but rather constitute as a single exercise where they are to be applied cumulatively.[41]

The PoE is not alone in giving more weight to the ordinary meaning and regards others *"as discretionary or supplementary."*[42] Even the International Court of Justice (ICJ), in some cases, stated "*interpretation must be based above all upon the text of the treaty.*"[43] Some find that VCLT Article 31 indeed follows a *"textual primacy"*[44] and that the text of a treaty *"obviously counts more"*.[45] Such a deduction can only be possible by an overly textual interpretation of Article 31 itself[46] which is an "*excessive molecularization*" of Article 31.[47] Without diving further into public international law, this discussion suffices to reasonably conclude that the PoE exhibited an overreliance on the ordinary meaning of the terms used.

Alongside the preambles and articles on the objectives of the FTA in general, each TSD chapter starts with an article titled *Context and Objectives.*[48] These should, as per the regime of Article 31 VCLT, be considered while interpreting a TSD. PoE could have provided a stronger ground for its arguably liberal interpretation of implementation obligation through the importance given to sustainable development by the context of the FTA and its objectives regarding sustainable development.

Having noted this, on the EU's claim regarding South Korea's failure to respect, promote, and realize related principles, the PoE ruled that South Korea indeed was in breach of this obligation under the TSD chapter and advised South Korea to make the necessary changes to its domestic law. These changes included the removal of certain provisions and amending the definitions of certain terms. Thus, the remedy of a breach of this article appears to change the domestic law and bring the measure in question in conformity with the ILO Conventions.

---

[39] Regan 2017, p. 1050.

[40] Jennings and Watts 1992, p. 114.

[41] Gardiner, p. 172.

[42] For possible reasons for such behavior see Villiger, pp. 115–117.

[43] ICJ, Territorial Dispute, para 41; Legality of Use of Force, para 100.

[44] Orakhelashvili, p. 309.

[45] Ratner p. 82.

[46] Villiger p. 122.

[47] Ibid.

[48] For example Article 13.1, South Korea FTA.

In addition to these fundamental rights, under some FTAs[49] the parties reaffirm their commitment to effectively implement other ILO conventions and recommendations that they are party to. In later FTAs, the conventions covered by the effective implementation obligation are listed as *"the fundamental, the priority and other ILO conventions"*.[50] Formulation of this obligation under the Agreement in Principle (AiP) with Mexico[51] stands out with its binding language, which uses the word *shall*, and states that the parties "*shall respect, promote and realize*"[52] these other ILO documents. Effective implementation obligation and what it entails is dealt with below under the Multilateral Environmental Agreements (MEAs) article where it has a more central role.

For now, it suffices to underline that in the light of the interpretation of the PoE of the word *commit*, both *shall* and *commit* are binding and create obligations for the parties. Bearing in mind the aforementioned disadvantages of providing clarity through PoE reports, another PoE could adopt a less/non-binding interpretation of the word to *commit*, similar to what South Korea claimed. The word *shall* doubtlessly create a binding obligation, is therefore less likely or prone to be interpreted fundamentally differently by another PoE, and as such should be preferred over the word commit in future FTAs.

### *7.2.2 Ratification Obligation*

The second category of obligations regarding the multilateral labor standards is the ratification obligation. One of the four claims raised by the EU in the EU-South Korea dispute regards this obligation. The degree of clarity and the language used vary per agreement and category of ILO conventions, i.e., fundamental, and non-fundamental.[53] Three versions of the article are observed under different FTAs which can be seen in Table 7.1. An analysis of each version of the article follows.

---

[49] See Table 7.1 for the exact distribution. FTAs with Georgia, Central America, Singapore, Moldova, Vietnam.

[50] Priority conventions are a set of conventions that The ILO Governing Body has designated as governance (priority) instruments and encourages the member States to ratify them. Currently there are four priority conventions: Labour Inspection Convention, 1947 (No. 81), Employment Policy Convention, 1964 (No. 122).

Labour Inspection (Agriculture) Convention, 1969 (No. 129), Tripartite Consultation (International Labour Standards) Convention, 1976 (No. 144). See https://www.ilo.org/global/standards/introduction-to-international-labour-standards/conventions-and-recommendations/lang--en/index.htm, last access on 28.11.2022.

[51] EU-Mexico The agreement in principle as announced on 21.04.2018, available at https://policy.trade.ec.europa.eu/eu-trade-relationships-country-and-region/countries-and-regions/mexico/eu-mexico-agreement/agreement-principle_en, last access on 05.11.2022 [Mexico AiP].

[52] Article 3(3) under Chapter XX, Mexico AiP.

[53] See supra note 26.

**Table 7.1** Different levels of commitment to ratify ILO conventions chronologically. *Source* Author

| Year of the FTA | FTA | Fundamental ILO conventions | | Non-fundamental ILO conventions | | |
|---|---|---|---|---|---|---|
| | | Continued and sustained efforts | Exchange of information | Continued and sustained efforts | Consideration | Exchange of information |
| 2011 | South Korea | X | | X | | |
| 2013 | Central America | | | | | X |
| 2013 | Andean Community | | | | | X |
| 2016 | Moldova | | | | X | X |
| 2016 | Georgia | | | | X | X |
| 2017 | Ukraine | | | | X | |
| 2018 | Singapore | X | X | | X | X |
| 2018 | Armenia | | | | X | X |
| 2019 | Vietnam | X | X | | X | X |
| 2019 | Japan | X | X | | | X |
| 2020 | Canada | X | X | | | X |
| 2023 | New Zealand | X | X | | | X |
| 2023 | Kenya | X | X | | | X |
| – | Mexico | X | X | | | X |
| – | Australia | X | X | | | X |
| – | Mercosur | X | X | X | | X |
| – | India | X | X | | | X |
| – | Chile | | X | | | X |
| – | Indonesia | | X | | | X |

In the context of this study 'non-fundamental ILO conventions' are used to cover any ILO convention that is not fundamental, thus both priority and other conventions. The consideration obligation in the Singapore FTA is weaker than the others. "*Parties will also consider the ratification [...], taking into account domestic circumstances.*" Domestic circumstances can be used as an argument as to why a party has not considered ratification. Similarly, the continued and sustained efforts obligation in the Japan EPA is weaker than the rest. The commitment is left to the parties' initiative for the agreements they consider appropriate. As even the South Korea version of the article, which has a clear obligation not left to parties' appreciation was not enough to convince the PoE of a violation amid a specific schedule, the author believes that successfully enforcing this article is not possible.

#### 7.2.2.1 Continued and Sustained Efforts

Different categories of ILO conventions are subject to different terms regarding their ratification by the parties. The strongest language among the versions of the article reads as follows:

> "The parties will make continued and sustained efforts towards ratifying the fundamental ILO conventions."[54]

As seen in Table 7.1, except for FTAs concluded between 2013–2017, this version is dominant for the Fundamental Conventions. The highest priority seems to be given to these conventions by the parties. What does this highest priority entail? "*Continued and sustained efforts*" is a formulation that requires the party to show efforts. Also called a "*best endeavor obligation*",[55] it demands no result. Therefore, a party not ratifying the relevant conventions will not constitute as a breach as long as it can show that proper efforts were made. In other words, a party is obliged to merely show efforts towards ratification but the ratification itself is not guaranteed or promised. Thus, in case of a breach, the required remedy would be for the breaching party to start showing efforts. The question that naturally arises from here is: when are continued and sustained efforts demonstrated or, more importantly, when are they not? When has a party failed to show such efforts? What is the legal threshold for a breach?

The EU-South Korea dispute settlement proceedings are once again a useful tool to answer these questions. Firstly, South Korea contested that the word *will* is less strong than the word *shall* and indicated more of an intention. However, the PoE found that considering the absence of any evidence to suggest otherwise and the general rules of treaty interpretation, these two words are used interchangeably,[56] and *will* establish a binding legal obligation.[57]

Regarding the interpretation of the phrase *continued, and sustained efforts,* the threshold applicable for the required efforts is one of the most central parts of the PoE report. South Korea asserted a passive meaning of not sliding back from the *status quo* or not taking measures jeopardizing future ratification.[58] The EU submitted a more active interpretation of taking all appropriate measures uninterruptedly.

The PoE opted for an interpretation somewhere in the middle of what the parties claimed. Accordingly, while the Article creates an ongoing legally binding obligation,[59] it requires '*best endeavors*' from the parties to ratify the fundamental ILO conventions. Without specific targets or a schedule foreseen, the obligations are aimed at the efforts and not at results. According to the PoE, this is a standard higher

---

[54] Article 13.4 (4) South Korea FTA.

[55] Bronckers and Gruni 2021, p. 26.

[56] Report of the PoE of Experts, para 263.

[57] Report of the PoE of Experts, paras 268–269.

[58] Report of the PoE of Experts, pp. 73–74.

[59] Report of the PoE of Experts, p. 75.

than merely protecting the *status quo* or taking minimal steps but lower than taking all measures available.[60]

The PoE interpreted this phrase as requiring '*best endeavors*' from the parties. Without specific targets or a schedule, there being no ratification of any of the four conventions in question within the eight years of the FTA being in force did not amount to a breach of the article, although it was *"less than optimal."*[61] The PoE pointed to efforts made by South Korea *"at least since 2017"* which were also acknowledged by the EU.[62] Among these efforts are the foundation of an expert group to analyze the differences between the ILO conventions and the national law and putting the ratification of the conventions into the agenda of the parliament.[63] The PoE found that while slow, these satisfied the legal threshold,[64] without mentioning what happened between 2011 and 2017.

There are several aspects of the decision of the PoE that will have a positive effect on the EU's TSD enforcement. Being the first of its kind and, therefore a precedent, the establishment of jurisdiction by the PoE carries a lot of importance for similar future proceedings. South Korea claimed that the issues brought forward by the EU were not covered by the TSD chapter because they had no direct effect on trade.[65] It based this argument on Article 13.2.1 of the FTA stating that *'Except as otherwise provided in this Chapter, this Chapter applies to measures adopted or maintained by the Parties affecting trade-related aspects of labor and environmental issues.'*

The PoE rejected this argument and assumed jurisdiction. According to the PoE, Article 13.4.3 regarding the ratification of ILO conventions is one of these exceptions for issues having to relate to trade.[66] A decision otherwise would have ruled out the option to invoke DSM foreseen for TSD chapters unless a direct and provable link with trade is present. Such a link is a hard, if not impossible, bar to reach, as seen in the example of the US-Guatemala labor dispute.[67]

Similarly, the establishment of breaches of South Korea regarding freedom of association is an important win and will enable various actors, from the EU to local NGOs, to increase the pressure on South Korea for improvement. From the EU perspective, it is also meaningful that the EU finally showed some teeth regarding TSD enforcement. This is an important message to other countries that the EU can and will initiate proceedings for TSD breaches.

---

[60] Report of the PoE of Experts, para 277.

[61] Report of the PoE of Experts, para 292.

[62] First Written Submission by the European Union, available at https://circabc.europa.eu/ui/group/09242a36-a438-40fd-a7af-fe32e36cbd0e/library/487dcb44-facd-4634-b3bf-1c1f29a156f7/details, last access on 13.07.2023, paras 79–80 and 93–94.

[63] Report of the PoE of Experts, paras 80–82.

[64] Report of the PoE of Experts, para 293.

[65] Report of the PoE of Experts, para 56.

[66] Report of the PoE of Experts, para 63.

[67] In the very first labour dispute arising from a trade agreement in history the arbitral PoE interpreted the phrase *"in a manner affecting trade"* strictly. The decision of the PoE has been commented on as a *"failure by design"* in Cross 2017 and as making *"it difficult or unworkable to prove"* any breach had such a connection to trade in Banks 2020.

Seen these aspects, a prima facie reading can lead some to think of the decision as a victory for the EU. Surely the EU itself sees it that way.[68] A closer look into the decision reveals, if not the opposite, quite a different picture.

The most important negative outcome of the decision is the PoE's finding no breach of the commitment to show best endeavors for ratification. Some commented that this sends a strong and unfortunate message to other trading partners of the EU.[69] The message is that despite the calls from the EU and civil society, stalling and showing no effort to ratify the ILO conventions for eight full years is excusable if the trading partner shows *'less than optimal'* efforts in the last few years leading to the proceedings. If the absence of efforts to ratify did not amount to a breach under the oldest new-generation FTA, it is highly unlikely that a more recent one will for a long time to come.

One cannot help but wonder what would have happened had the EU brought the claims of a breach before 2017, before South Korea started showing efforts. Would the PoE then have found a breach of the ratification obligation faced with eight years of inaction and no recent change in this behavior? Considering the central role the efforts since 2017 played in the reasoning, this could have very well been the case. The takeaway for the Commission should be not to be reluctant to trigger the TSD DSM in the future.

The importance of this commitment should also be underlined within the big picture of the TSD chapter. Ratifying the ILO conventions is one of the clearest obligations under TSD. Vagueness and unclear formulation of the articles under the TSD have been a central point in the debate regarding their enforcement. What is vital to see here is that if even one of the most clearly worded articles of these chapters is not enough to establish a breach, the rest surely is not. In this context, the decision has been commented on as showing the need for a stronger language,[70] as opposed to the EU's narrative of it showing the effectiveness of the cooperation-based approach.[71] A way to remedy these shortcomings and overcome similar future decisions is for the EU to consider including a concrete schedule for the ratification of the ILO conventions and other obligations into the future FTAs.

Some have identified ratifications that took place within a short time after the conclusion of the FTA negotiations as ratifications that *"can be reasonably linked*

---

[68] The press release of the EU on the PoE report reads as follows: *"The experts also agreed that the commitment to take steps towards the ratification of fundamental ILO Conventions requires ongoing and substantial efforts. Finally, the PoE confirmed the [obligation] is legally binding."* It fails to communicate that, despite indeed accepting the legal bindingness of the ratification obligation, the PoE did not find a breach of it, available at https://ec.europa.eu/commission/presscorner/detail/en/ip_21_203, last access on 13.05.2024.

[69] Gruni 2017; Peers 2021.

[70] Ruling confirms strong language is needed in TSD chapters, available at https://www.eurogroupforanimals.org/news/ruling-confirms-strong-language-needed-tsd-chapters, last access on 15.07.2023.

[71] The EU has portraited the report of the PoE as a victory. Executive Vice-President and Commissioner for Trade, Valdis Dombrovskis, stated that the PoE report *"shows the effectiveness of our cooperation-based approach to trade and sustainable development."* Available at https://ec.europa.eu/commission/presscorner/detail/en/ip_21_203, last access on 14.07.2023.

*to the FTA"*.[72] South Korea ratified three of the four fundamental ILO conventions that were the subject of the dispute with the EU within the same year of the PoE report.[73] These ratifications of South Korea ratifications can be reasonably linked to the report.[74] Despite the current lack of a compliance mechanism, a PoE report finding a breach can still have the so-called "naming and shaming"[75] effect and induce change. However, this effect seems to remain incidental. Seven *ratification gaps* remain present for the following countries, each of which has an FTA with the EU, besides the one remaining of South Korea: Singapore, Japan, and Vietnam[76] are also certainly hopeful for the actual impact of the TSD chapters in improving the labor standards in trading partners' countries. This can be shown as a success of the soft approach to the TSD enforcement the EU has pursued so far. When all of the ratification gaps disappear remain to be seen.

#### 7.2.2.2 Consideration or Exchange of Information

There are two other versions of the ratification obligation available: obligation to consider the ratification of non-fundamental ILO agreements and exchange of information obligation on the ratification status.

Consideration obligation appears as a standard under the FTAs in the period from 2016 to 2019, starting with the Moldavian FTA and ending with the Vietnamese FTA. The only example outside of this period is the Armenian FTA of 2021. This obligation usually takes the following form: "*The Parties will also consider the ratification of the remaining priority and other conventions [...].*"[77]

Concerns regarding the enforceability of the *"continued and sustained efforts"* version of the article have already been expressed. Keeping in mind the already low level of enforceability of requiring efforts, the consideration version goes even lower than this level. It requires not even effort but a mere consideration. What naturally comes to mind is how can a state show that it has considered ratifying the related conventions. Considering the interpretation of the PoE for the effort obligation and the standard to meet it, requiring only a consideration does not predict a very bright future for its enforcement. Proving a party failed to consider ratification does not seem possible as this would require proving the absence of consideration altogether. It therefore overlaps with the "*continued and sustained effort*" obligation as to the

---

[72] Blot et al. 2022, p. 10.

[73] ILO, Ratifications for Republic of Korea, available at https://www.ilo.org/dyn/normlex/en/f?p=NORMLEXPUB:11200:0::NO::P11200_COUNTRY_ID:103123, last access on 05.11.2022.

[74] Blot et al. 2022, p. 10.

[75] García 2022, pp. 58–67.

[76] Blot et al. 2022, p. 10.

[77] Article 365.4, Association Agreement between the European Union and the European Atomic Energy Community and their Member States, of the one part, and the Republic of Moldova, of the other part OJ L 260, 30.8.2014, p. 4-738, available at https://eur-lex.europa.eu/legal-content/EN/TXT/?uri=uriserv%3AOJ.L_.2014.260.01.0004.01.ENG, last access on 05.11.2022 [Moldavian AA].

need to show complete inaction, although the absence of effort should be easier to show than the absence of consideration. The remedy to the breach of consideration is as vague as the obligation itself. If a breach is established, the failing party should start "*considering*" ratification. Exhibition of any consideration should satisfy. This can be the reason why the EU has abandoned the use of the consideration articles.

The trend seems to be going back to the effort obligation for the ratification of fundamental ILO conventions if there is a ratification gap and an exchange of information obligation for the ratification of both fundamental and non-fundamental ILO conventions. The first example of an exchange of information obligation is found under the Central American Association Agreement (AA): "*The Parties will exchange information on their respective situation and advancement as regards the ratification of the other ILO Conventions.*"[78]

This additional obligation can be beneficial to improve the transparency for TSD chapters should this information be made publicly available. However, its enforceability depends on the type of ratification obligation it regards. When combined with consideration, it is unlikely that the exchange of information obligation will create any additional incentive for parties as a simple notification of "*Ratification is considered but no ratification has taken place.*" seems to satisfy this obligation. Thus, even though the obligation to exchange information itself is defined clearly, this does not automatically mean high enforceability. The only possible scenario of claiming and successfully enforcing a breach would be if one of the parties completely disregards the obligation and does not provide any information on the ratification status.

Exchange of information, in combination with continued and sustained efforts towards ratification, however, is clearer in terms of the content of the communication and thus easier to enforce. The parties need to exchange information about the ratification status where they need to show continued and sustained efforts. Therefore, the information should necessarily include what efforts are being shown toward ratification.

The Vietnam and Singapore FTAs seem to be the overlapping era of two versions of the article, regarding ratification as they foresee an effort obligation for the fundamental ILO conventions and a consideration obligation for the other conventions. The proposed text for an AiP with Mercosur on the other hand stands out as the strongest structure for multilateral labor standard obligations so far.[79] It has a continued and sustained efforts-obligation for all ILO conventions, fundamental and non-fundamental alike, for the first time since the South Korea FTA. In addition, it brings an exchange of information obligation for the ratification of all of the ILO conventions.

---

[78] Article 286.3, Agreement establishing an Association between the European Union and its Member States, on the one hand, and Central America on the other, OJ L 346, 15.12.2012, p. 3-2621, available at https://eur-lex.europa.eu/legal-content/EN/ALL/?uri=CELEX%3A22012A1215%2801%29, last access on 05.11.2022 [Central American AA].

[79] Agreement in Principle announced by the European Commission, 28.06.2019, available at https://policy.trade.ec.europa.eu/eu-trade-relationships-country-and-region/countries-and-regions/mercosur/eu-mercosur-agreement/text-agreement_en, last access on 05.02.2023 [Mercosur AiP].

This strong version is, however, not repeated in the FTAs that followed. The drafts with India and Kenya go back to the trend of effort obligation for fundamental ILO conventions and exchange of information for other ILO conventions. Chile and Indonesia stand out by only having exchange of information obligation for all ILO conventions, despite both countries having unratified fundamental ILO conventions.[80] This illustrates the fluctuating and non-uniform treaty practice of the EU. Such weak obligations are quite unlikely to incentivize ratification.

## 7.3 Multilateral Environmental Agreements

Each TSD chapter in EU FTAs has an article that regulates parties' obligations regarding MEAs. The obligations under this article can be categorized into three whose overview can be seen in Table 7.2.

1. Effective implementation obligation for (certain) MEAs,
2. Ratification obligation and exchange of information on ratification status for certain MEAs,
3. Commitment to the ultimate objective of the United Nations Framework Convention on Climate Change (UNFCCC) and other certain MEAs.

It is important to note that there is no more enforceable-less enforceable relationship between most of them, unlike the obligations regarding multilateral labor agreements. Only the effective implementation obligation has more than one version. For other obligations, there is almost always a single version and the difference between the FTAs is whether or not this version appears.

### *7.3.1 Effective Implementation Obligation*

All the TSD chapters bring an obligation for the parties to effectively implement the MEAs, protocols, and amendments that they have ratified. An example of it can be found under the South Korea FTA:

> "The Parties reaffirm their commitments to the effective implementation in their laws and practices of the multilateral environmental agreements to which they are party."[81]

---

[80] Chile has not ratified C155—Occupational Safety and Health Convention 1981; Indonesia has not ratified C155—Occupational Safety and Health Convention, 1981 (No. 155); P029—Protocol of 2014 to the Forced Labour Convention, 1930. See https://www.ilo.org/dyn/normlex/en/f?p=1000:11210:0::NO:11210:P11210_COUNTRY_ID:102588 and https://www.ilo.org/dyn/normlex/en/f?p=1000:11210:0::NO:11210:P11210_COUNTRY_ID:102938.

[81] Article 13.5 (2), South Korea FTA.

**Table 7.2** Categories of obligations regarding MEAs. *Source* Author

| Year of the FTA | FTA | Effective implementation | | | Ratification | Exchange of information on ratification | The ultimate objective of UNFCC and others |
|---|---|---|---|---|---|---|---|
| | | Reaffirm commitment | Shall | UNFCC and Paris Agreement* | | | |
| 2011 | South Korea | X | | | | | X |
| 2013 | Central America | X | | | X | | |
| 2013 | Andean Community | X | | | | | |
| 2016 | Moldova | X | | | | X | X |
| 2016 | Georgia | X | | | | X | X |
| 2017 | Ukraine | X | | | | | |
| 2018 | Singapore | | X | X | | | X |
| 2018 | Armenia | X | | X | | X | |
| 2019 | Vietnam | X | | X | | X | |
| 2019 | Japan | X | | X | | X | X |
| 2020 | Canada | X | | | | X | |
| 2023 | New Zealand | | X | X | | X | |
| 2023 | Kenya | | X | X | | X | |
| – | Mexico | | X | X | | X | |
| – | Mercosur | X | | X | | X | |
| – | Australia | | X | X | | X | |
| – | India | | X | X | | X | |
| - | Chile | | X | X | | X | |
| – | Indonesia | | X | X | | X | |

*Effective implementation of the UNFCCC and the Paris Agreement is always regulated under the separate *"Trade and Climate Change"* article

However, the varied language changes the nature of the obligation per FTA. Most of the agreements follow the above-given example of the South Korea FTA. What this phrase entails is dealt with below.

The second and stronger version of the article appears consistently in the recent FTAs and only the Singapore FTA among those already in force. This version states that the parties "*shall effectively implement*" the MEAs that they are parties to which defines the legal obligation more clearly and imperatively and necessitates the result of effective implementation. Parties are not only under a negative obligation to refrain from contradicting the aims of the MEAs but also an active obligation to carry out their effective implementation and enact changes to their national laws if necessary.

The trend of the EU choosing this version in her new FTAs is, therefore, a welcomed one.

Most of the TSDs formulate the obligation to effectively enforce MEAs in an open list where all the MEAs the parties have ratified are covered. All the MEAs the parties ratify in the future automatically fall under the scope, making the article future-proof. On the other hand, this might induce reluctance to agree to this version of the article.[82]

One exception to this is the Andean FTA which adopted a closed list of MEAs that the parties reaffirm their commitment to effectively enforce.[83] This list can be changed by the Trade Committee founded under the FTA.[84] No future MEA is automatically covered. As the Trade Committee decisions require a positive vote from both parties,[85] one of the parties could keep future MEAs out of the scope of the effective implementation obligation. Thus, this version is not future-proof. Where possible the treaty makers should opt for the open list approach.

It must be underlined that the recent FTAs of the EU all have a new article titled "Trade and Climate Change" which states that each party *"shall effectively implement the UNFCCC and the Paris Agreement established thereunder."*[86] In addition, newer FTAs all have the "*shall*" version of the effective implementation obligation for the MEAs that they are party to. The possible consequences of this will be discussed further below under Section 2.2.1.4.

The Paris Agreement was adopted in December 2015[87] and the FTAs dated before that naturally do not mention the Paris Agreement but the Kyoto Protocol in the

---

[82] This effect relates closely to complementary jurisdiction, i.e., an act of a state constituting a breach not only under the relevant MEA but also the FTA through this article. The notion of complementary jurisdiction is explained further below.

[83] Article 270 (2), Trade Agreement between the European Union and its Member States, of the one part, and Colombia, Peru and Ecuador, of the other part, OJ L 354, 21.12.2012, p. 3-4065, available at https://eur-lex.europa.eu/legal-content/EN/TXT/?uri=CELEX%3A02012A1221%2801%29-20181213, last access on 05.02.2023 [Andean FTA], reads as follows: *"The Parties reaffirm their commitment to effectively implement in their laws and practices the following multilateral environmental agreements: the Montreal Protocol on Substances that Deplete the Ozone Layer adopted on 16 September of 1987, the Basel Convention on the Control of Transboundary Movements of Hazardous Wastes and their Disposal adopted on 22 March 1989, the Stockholm Convention on Persistent Organic Pollutants adopted on 22 May 2001, the Convention on International Trade in Endangered Species of Wild Fauna and Flora signed on 3 March 1973 (hereinafter referred to as "CITES"), the CBD, the Cartagena Protocol on Biosafety to the CBD adopted on 29 January 2000, the Kyoto Protocol to the United Nations Framework Convention on Climate Change adopted on 11 December 1997 (hereinafter referred to as "Kyoto Protocol") and the Rotterdam Convention on the Prior Informed Consent Procedure for Certain Hazardous Chemicals and Pesticides in International Trade adopted on 10 September 1998."*

[84] Article 270 (3), Andean FTA.

[85] Article X, South Korea FTA.

[86] Article 6 (2) (a), Mercosur AiP.

[87] More information on the Paris Agreement can be found at https://unfccc.int/process-and-meetings/the-paris-agreement, last access on 13.07.2023.

effective implementation obligation. CETA stands out as the only FTA that does not mention the Paris Agreement, even though it is dated 2020.[88]

Some of the newer FTAs also include a cooperation obligation, which will not be covered in this study. With these initial considerations, words used in this article should be analyzed to understand what obligation the parties undertake.

#### 7.3.1.1 "Effectively Implement"

To start with, what '*effective implementation*' entails, i.e., when implementation is effective should be defined. This notion is not defined in the TSD itself or has been the subject of a PoE report. Thus, we have to look elsewhere for a possible answer. EU law, where the effective implementation notion also exists, can be an inspiration and show how such a notion is understood at least by one of the FTA partners.

The EU Member States have to implement the directives into national law in an effective way[89] but are free to choose the form and methods.[90] Under international law, states undertake to ensure that their domestic laws conform with the international agreements they are party to[91] ensure the applicability and enforceability of the rights arising from these agreements.[92] This means that the content of the obligation can only be known once put into the context of the obligations arising from a specific MEA.

Thus, parties to the FTAs should ensure that their domestic law conforms with the MEAs that they are parties to, while they are free to choose the means to do so. In case of a failure, the remedy would be removing or amending the domestic law that conflicts with the MEA or putting an end to the domestic practice that is breaching the MEA.

#### 7.3.1.2 "Commitment"

Under the second version of the effective implementation obligation where the parties "*reaffirm their commitment to effectively implement*" the MEAs, we need to look at the meaning of the words '*reaffirm*' and '*commitment*'. There is no report of a PoE to guide the interpretation.

---

[88] Comprehensive Economic and Trade Agreement (CETA) between Canada, of the one part, and the European Union and its Member States, of the other part, OJ L 11, 14.1.2017, p. 23-1079, available at https://eur-lex.europa.eu/legal-content/EN/TXT/?uri=celex:22017A0114(01), last access on 06.02.2023 [CETA].

[89] Article 197, Consolidated version of the Treaty on the Functioning of the European Union. OJ C 326, 26.10.2012, pp. 47-390, available at https://eur-lex.europa.eu/legal-content/EN/TXT/?uri=celex%3A12012E%2FTXT, last access on 05.11.2022. [TFEU] More on this see: Dimitrakopoulos 2001.

[90] Article 288, TFEU.

[91] Villiger 2009; Buergenthal and Murphy 2007.

[92] Neudorfer and Wernig 2010, p. 415.

As parts of the same chapter of the same agreement, it is reasonable to interpret the same words under the MEA article the same way as the labor rights article. This stems from the integrity of the TSD chapter and context as understood in VCLT. It is reasonable to assume that for similar articles in the same chapter of the FTA, the treaty makers understood the same thing from the same or similar words. When dealing with identical or similar phrases as in the EU-South Korea dispute, the interpretations of the PoE should be followed, so long as there is no different interpretation warranted by the nature of the covered area. In this regard, the word "*commitment*" should be interpreted as not showing only an aspiration but as creating a legally binding obligation.[93]

This interpretation of the PoE is also in line with the words used in the versions of the FTAs in other official EU languages[94] where the "*binding oneself to a course of action*" meaning is clearer than the English word *commitment*. For example, in the Dutch version, the word "*verbintenis*" is used[95] which is interpreted by several Dutch courts as requiring "*prestatie*"[96] which translates as performance, achievement[97]: therefore, a result. Thus, even if the interpretation of the PoE on the labor standards obligation was not applied to environmental obligations, the word "*commitment*" should be understood as creating a binding obligation of result.

---

93 For detailed discussion of the PoE of Experts, see Section 2.1.1.

94 Article 33 of the VCLT is titled *Interpretation of treaties authenticated in two or more languages*, third paragraph of which states that. *"The terms of the treaty are presumed to have the same meaning in each authentic text."* All the agreements of the EU are authenticated in all of its 24 official languages of the EU, see Šarčević 2013. Multilingual law-making and legal (un) certainty in the European Union. International journal of law, language & discourse, 3(1), 1–29. Shelton states that "*[...]a comparison of different texts may help to resolve an ambiguity inherent in a term or phrase used in one language, making clearer the intention of the drafters.*" in Shelton 1996. Thus, when the meaning of an article in one language version of an FTA is unclear, other languages versions can and should be consulted. For an in-depth analysis of multilingualism in the EU see Otero Fernández 2020.

95 Article 13.5 (3), South Korea FTA in its Dutch version reads as follows: *"De partijen herbevestigen hun verbintenis ten aanzien van het bereiken van de uiteindelijke doelstelling van het Raamverdrag van de Verenigde Naties inzake klimaatverandering en het bijbehorende Protocol van Kyoto."* Similarly, Article 13.5 (3), South Korea FTA in its German version uses the word "Verpflichtung" and reads as follows: *"Die Vertragsparteien bekräftigen ihre Verpflichtung auf das oberste Ziel des Rahmenübereinkommens der Vereinten Nationen über Klimaänderungen und des beigefügten Kyoto Protokolls."*

96 "Dit maakt dan de vraag relevant tot welke prestatie de verbintenis verplicht.", Arrest Hoge Raad: ECLI:NL:HR:2020:141, available at https://uitspraken.rechtspraak.nl/inziendocument?id=ECLI:NL:PHR:2019:962, last access on 23/11/2022; "Heeft de prestatie niet aan de verbintenis beantwoord, dan wordt deze vergoeding [...], Arrest Rechtbank Noord-Holland: ECLI:NL:RBNHO:2021:8937, available at https://uitspraken.rechtspraak.nl/inziendocument?id=ECLI:NL:RBNHO:2021:8937, last access on 23/11/2022.

97 Van Dale Woordenboek, available at https://www.vandale.nl/gratis-woordenboek/nederlands-engels/vertaling/prestatie, last access on 23.11.2022.

#### 7.3.1.3 "Reaffirm"

The interpretation of the only verb in the sentence "*Parties reaffirm their commitment to effectively implement the […] MEAs*" is central. Looking at the word *reaffirm* in a similar way the PoE looked at the words that needed further interpretation under the labor standards article, dictionaries are the first place to look at.[98]

The ordinary meaning of the word reaffirm appears as "*to affirm (something) again especially to strengthen or confirm*"[99] or "*to state something as true again*".[100] In international treaty-making, the word *reaffirm* is usually found in the preamble sections of the treaties which do not create obligation but "*add color, texture and shading*" to the treaty interpretation and show the intention of the treaty makers.[101]

In the case at hand, the article is located in the operational part of the agreement, where it should have "*the capacity to create rights and obligations for parties.*"[102] The choice of the treaty makers in doing so should have a meaning. In several legal systems, the legislature is assumed to be a rational body[103] that acts reasonably, consciously, or purposely, having regard to its purpose in making a law.[104] The same should be assumed of the legislators of the treaty as well.

Read in conjunction with the place of the article and the ordinary meaning of the word to reaffirm,[105] it should be concluded that if the treaty makers placed such an article into the operative clauses of the treaty, they made the conscious choice to strengthen and confirm their commitments regarding the MEAs under this article of the TSD. By doing so, they incorporated these MEAs and created a separate legal obligation to effectively implement them in the FTA. This can be also seen as an effort of the treaty makers to ensure a level playing field for the matters covered by the other chapters of the FTA: it should be ensured that environmental (and labor) standards are respected by both parties so that the trade can be "*free.*"

---

[98] Notwithstanding, the ordinary meaning of words should be constructed within the general context and the object and purpose of the treaty. See above for a discussion on the overreliance of the EU-South Korea dispute PoE on ordinary meaning.

[99] Merriam Webster Dictionary, available at https://www.merriam-webster.com/dictionary/reaffirm, last access on 05.02.2023.

[100] Cambridge Dictionary, available at https://dictionary.cambridge.org/dictionary/english/rea ffirm, last access on 05.02.2023.

[101] WTO-AB Shrimp case.

[102] Rajamani 2016a, p. 343.

[103] Principle of "*Kanun koyucu abesle iştigal etmez.*" in Turkish legal system; everything the legislature does has a purpose. In the US legal system: "*Courts must presume that a legislature says in a statute what it means and means in a statute what it says there.*" Connecticut Nat'l Bank v. Germain, 112 S. Ct. 1146, 1149 (1992). "Congress is presumed to act intentionally and purposely when it includes language in one section but omits it in another." Estate of Bell v. Commissioner, 928 F.2d 901, 904 (9th Cir. 1991).

[104] Middleton 2016, p. 626.

[105] The German version of the Treaty uses the word "*bekräftigung*", which means "*affirming*", not reaffirming.

Despite some advising the trade negotiators that using the word *reaffirming* a commitment elsewhere cannot give rise to a binding commitment itself,[106] the only TSD PoE to date also confirms the conclusion of the opposite. For a similarly formulated obligation in the labor standards article, the PoE in the EU-South Korea dispute stated that the FTA "*reaffirms the existing obligations of the Parties under the ILO Constitution, and has incorporated these obligations, as they are defined within the ILO system, as separate and independent obligations* under [...]"[107] the TSD chapter.

Keeping this in mind, this referral to existing obligations of the parties undertaken under MEAs creates separate and independent obligations regarding the MEAs, arising from the FTAs themselves. However, considering the lengthy interpretative exercise needed to reach this conclusion and the possible opposite interpretation, it is a positive development that the EU opts for the word *shall* instead in its recent FTAs.

#### 7.3.1.4 Complementary Jurisdiction for MEAs through Reference in the TSD

A separate and independent obligation regarding MEAs in the TSD has an interesting consequence: complementary jurisdiction.

Complementary jurisdiction refers to the incorporation of MEAs into the TSD chapters, creating an additional layer of enforcement for a breach of an MEA. For the same breach, a party can be held responsible not only for a breach of the MEA through the mechanisms present under the MEA in question but also for a breach of the FTA, through the mechanism available to the TSD chapters. The complementary jurisdiction can bear an even more interesting interplay between the two mechanisms especially when the more assertive approach of the Commission takes its place in the future TSDs. Then the general compliance mechanism of the FTA is also applicable to TSD and there is the possibility of sanctions for certain breaches. Especially when no binding DSM is available under the MEA itself this would be particularly valuable.

Even if the complementary jurisdiction for a less bindingly formulated version that *reaffirms* parties' commitments to MEA can be questioned or an interpretation otherwise can be argued, complementary jurisdiction is certainly the case for the UNFCCC and the Paris Agreement, at the very least. Remembering that the Parties "*shall effectively implement the UNFCCC and the Paris Agreement*",[108] we can use this straightforward obligation for illustration. A breach of one of the two agreements shall thus constitute a breach of the FTA as well.

---

[106] International Institute for Sustainable Development, A Sustainability Toolkit for Trade Negotiators: Trade and investment as vehicles for achieving the 2030 Sustainable Development Agenda, available at http://www.iisd.org/toolkits/sustainability-toolkit-for-trade-negotiators/3-environmental-provisions/3-1-binding-or-non-binding-commitments/#jump, last access on 29.11.2022.

[107] Report of the PoE of Experts, para 107.

[108] Article 6 (2) (a), Mercosur AiP.

Imagine that an FTA partner of the EU breaches the Paris Agreement and the action steps in the recent communication of the Commission are part of the TSD chapter. While no sanctions are foreseen under the Paris Agreement itself, the breaching party can face trade sanctions under the FTA due to the breach of related TSD commitments as the Paris Agreement is named explicitly as a breach for which the failing party shall face trade sanctions. This breach can be a country withdrawing from the Paris Agreement, just as the Trump administration did for the US[109] as withdrawing certainly is *not* implementing. The types of breaches depend on the obligations of the UNFCCC and the Paris Agreement.[110] For example, under Article 4.2, the parties are to *"prepare, communicate and maintain successive nationally determined contributions* [NDCs]." If a party fails to do so, this will not only be a breach of the Paris Agreement but also the FTA.

Under the same article of the Paris Agreement, the parties also undertake to *"pursue domestic mitigation measures, with the aim of achieving the objectives of such NDCs." T*he legal character of this obligation is discussed in the doctrine as to whether it constitutes an effort obligation.[111] Bronckers and Gruni stated that the effective implementation obligation under the TSD chapter cleared all the doubts about this obligation regarding the NDC.[112] Firmed up by the FTA, the parties shall at the very least need to show best endeavors for domestic mitigation measures and a failure to work towards this goal can be pursued under the FTA DSM.[113]

Thus, the incorporation of MEAs into the TSD chapters can lead to an increased compliance of the MEAs. In addition to complementary jurisdiction, it can also clarify the legal character of an obligation under the MEA. This provides a new perspective to the decades-old discussions on trade versus environmental values: trade agreements and MEAs do not only co-exist but also reinforce each other. Considering the different incentives trade agreements give to parties for compliance and the explained disparities of two enforcement systems, this is an interesting avenue that is yet to be explored.

### *7.3.2 Ratification and Exchange of Information Obligation on Ratification Status*

There are two more categories of obligations regarding the ratification of MEAs, quite similar to the ones for ILO conventions: ratification and exchange of information on

---

[109] Climate change: US formally withdraws from Paris agreement, available at https://www.bbc.com/news/science-environment-54797743, last access on 14.07.2023.

[110] For an extensive overview of the obligations under the Paris Agreement and their legal character see Bodansky 2016; Rajamani 2016a, b.

[111] Bronckers and Gruni 2021, p. 21; Bodansky 2016.

[112] Bronckers and Gruni 2021, p. 29.

[113] Ibid.

ratification obligations. We will look at these obligations together due to their close relationship.

Only the Central American AA has obligations for the parties to ratify certain MEAs under two separate paragraphs:

> "The Parties *undertake to ensure* that they have ratified by the date of entry into force of this Agreement, the Amendment to Article XXI of CITES...." and "The Parties also undertake, to the extent they have not yet done so, to ratify and effectively implement, at the latest by the date of entry into force of this Agreement, the Rotterdam Convention on the Prior Informed Consent Procedure for Certain Hazardous Chemicals and Pesticides in International Trade."[114]

This is one of the most bindingly formulated obligations under the TSD chapters with a clear deadline. Considering the cruciality of schedule in EU-South Korea proceedings,[115] such a ratification obligation is more desired and has a better chance of successful enforcement in front of a PoE. The EU could use this version as a sample for stepping up the obligation to show "*continued and sustained efforts*" towards the ratification of multilateral agreements.

All the recent FTAS of the EU since 2018 without an exception have an exchange of information obligation regarding the ratification status of MEAs. This requires a regular dialogue between the parties. Due to the identicality of the provisions, the discussions made above in Section 2.1.2.2. for the exchange of information obligation relating to ratification of ILO conventions also apply here and will not be repeated. Thus, a simple notification of "*no ratification has taken place*" would satisfy this obligation.

There are differences between the formulation of two exchange of information obligations. Firstly, the MEA one uses the word "*shall*" while the ILO one uses "*will*". However, the PoE found that in the absence of specific evidence, there is no difference between the use of these two words in treaty-making and the obligations they bring are equally strong.[116]

The second and more interesting difference is the word "*regularly*". This provides some structure to the frequency of the exchanges, compared to the version there is no qualification at all. Once again considering the importance given to there being specified timelines by the PoE, this adds a temporal dimension to the exchange of information obligation. Consequently, a party can claim a breach of this temporal aspect if the other party does not share information regularly. It is unclear what regular entails and how long not sharing information constitutes a breach of the exchange of information. Thus, the treaty-makers should opt for defining the temporal aspects of obligations in a clearer way to increase enforceability, for instance by using *"twice a year"* or *"once every two years"*.

The related paragraph of CETA sheds light onto what an exchange of information under the TSD chapter can entail: "*a) implementation of multilateral environmental agreements, to which a Party is party; b) on-going negotiations of new multilateral*

---

[114] Article 287 (3) and (4), Central American AA.

[115] See Section 2.1.2.1. for PoE's reasoning and the discussion regarding schedule.

[116] Report of the PoE of Experts, paras 268–269.

*environmental agreements; and c) each Party's respective views on becoming a party to additional multilateral environmental agreements.*"[117] While not binding for other FTAs, it can be used as a guideline.

### *7.3.3 UNFCCC*

Several of the MEA articles under the TSD chapters have a paragraph stating that the parties *"reaffirm their commitment to reaching the ultimate objective"* of the UNFCCC and its Kyoto Protocol[118] and the development of a future international framework targeting climate change. The discussion above regarding the meaning of the word *"reaffirm"* and the legal character of the obligation it creates apply here as well.

The content of this article can only be understood by reading it in conjunction with the UNFCCC. The ultimate objective of the Convention is to *"achieve … stabilization of greenhouse gas concentrations in the atmosphere at a level that would prevent dangerous anthropogenic interference with the climate system […].*"[119]

It is unclear what marginal value this article brings. All the agreements that have this obligation also have the effective implementation obligation regarding all the MEAs the parties have ratified. and a separate effective implementation obligation for the UNFCCC and the Paris Agreement.[120] Considering this, a specific article re-committing parties to their aims is redundant which is possibly why it is abandoned. Explicitly singling out these agreements can show the political importance adhered to.

In general, the commitments regarding MEAs seem to have stabilized into two obligations in recent years: effective implementation and exchange of information. The wording used still varies for effective implementation. The new EU FTAs typically do not bring obligations for ratifying new MEAs other than exchange of information on ratification status. The Andean FTA and Ukrainian DCFTA stand out as the ones with the weakest protection with only effective implementation obligation and no mention of ratifying other MEAs.

[117] Article 24.4 (3), CETA.

[118] Under the Japan EPA, it is the Paris Agreement. Agreement between the European Union and Japan for an Economic Partnership, OJ L 330, 27.12.2018, p. 3-899, available at https://eur-lex.europa.eu/legal-content/EN/TXT/?uri=CELEX%3A22018A1227%2801%29, last access on 06.02.2023 [Japan EPA].

[119] Article 2, UN General Assembly, United Nations Framework Convention on Climate Change: resolution/adopted by the General Assembly, 20 January 1994, A/RES/48/189.

[120] This is except for the Ukrainian DCFTA which does not name any specific MEA.

## 7.4 Conclusion

While the new assertive approach announced by the Commission is certainly a step in the right direction procedurally, there is still a long way to go to achieve an integrated assertiveness of the TSD chapters where they are clear and enforceable.

This Chapter showed that several obligations regarding the multilateral agreements and standards under the TSD are bindingly formulated. While an interpretation exercise is needed to see what the obligation entails and what the required remedy would be for a breach, an FTA party can bring a claim of such breach against the other party in front of the PoE. This is particularly important in case of complementary jurisdiction, for example for the Paris Agreement, where a breach of the Paris Agreement will also constitute as a breach of the FTA for which the possibility of trade sanctions were announced under the new assertive approach of the EC. This shows how trade can not only co-exist with environmental protection but can also be used to reinforce it.

Where the FTAs, as the primary law, fail to provide clarity, it is the DSM that has to assume this role. Thankfully, the report of the PoE in the EU-South Korea dispute provides some guidance as to how certain phrases of TSD articles are to be understood. This increases the clarity and enforceability of the obligations, not only for the specific article interpreted by the PoE but also for other articles of the TSD chapter that use similar or identical phrases.

The clarity of the obligations being dependent on DSM has several problems. Firstly, for a PoE to interpret and clarify provisions, there should be cases being brought before it. In over a decade since the TSD chapters existed, there has been only one PoE. Considering the reluctance of the parties to trigger the mechanism the needed influx of disputes is unlikely in the short future.

Secondly, it is currently voluntary whether the parties will follow the recommendations of the PoE. Hopefully, the more assertive approach will solve this by expanding the compliance mechanism of the general DSM to TSD. Still, the possible inconsistency of decisions by *ad hoc* PoEs remains a problem. Of the interpretations possible and claimed by the Parties on certain matters, the PoE has sided with the more liberal and extensive interpretation that gives the TSD the strongest effect possible. This is certainly the case where the PoE established a separate and independent obligation in the FTA through the reference made to external obligations under the ILO system. The fact remains, however, that a different PoE could have easily adopted a more strict interpretation. Just as experienced by investment treaty arbitration, conflicting interpretations will further jeopardize the clarity of TSD articles.

Therefore, the PoE as a source of clarity is far ideal. It should be the primary law that defines the content of the obligation and the legal standard thereof clearly without a lengthy interpretation process. The following are suggested to the treaty-makers to ensure this:

(a) Abandon preamble-like language such as *reaffirming*. Use *shall* or *will* instead.
(b) Use language that defines the legal standard clearly so that the parties know what is expected of them and what constitutes a breach.

(c) Opt for an open-list approach while referring to external agreements where future agreements the parties ratify are covered automatically.
(d) Where applicable, use clear timelines or schedules when an obligation should be honored.

The content of the TSD should be stepped up and updated accordingly, to include more binding and clearer language with a clear temporal dimension, where appropriate, to increase enforceability. Only then the content of the TSD chapters will match the new assertive procedural approach and produce meaningful change for international sustainability. Until then, the changes made to the procedural mechanism will not have much value.

## References

2006 Ministerial Declaration of the UN Economic and Social Council on Full Employment and Decent Work, available at https://digitallibrary.un.org/record/593242, last access on 13.07.2023.

Agreement between the European Union and Japan for an Economic Partnership, OJ L 330, 27.12.2018, p. 3–899, available at https://eur-lex.europa.eu/legal-content/EN/TXT/?uri=CELEX%3A22018A1227%2801%29, last access on 06.02.2023. [Japan EPA]

Agreement establishing an Association between the European Union and its Member States, on the one hand, and Central America on the other, OJ L 346, 15.12.2012, p 3–2621, available at https://eur-lex.europa.eu/legal-content/EN/ALL/?uri=CELEX%3A22012A1215%2801%29, last access on 05.11.2022. [Central American AA]

Agreement in Principle announced by the European Commission, 28,06.2019, available at https://policy.trade.ec.europa.eu/eu-trade-relationships-country-and-region/countries-and-regions/mercosur/eu-mercosur-agreement/text-agreement_en, last access on 05.02.2023. [Mercosur AiP]

Alexvičová I & Prévost D (2019) Mind the compliance gap: managing trustworthy partnerships for sustainable development in the European Union's free trade agreements, International Journal of Public Law and Policy 6:3

Association Agreement between the European Union and its Member States, of the one part, and Ukraine, of the other part, OJ L 161, 29.5.2014, p 3–2137, available at https://eur-lex.europa.eu/legal-content/EN/TXT/PDF/?uri=CELEX:22014A0529(01)&from=EN, last access on 05.11.2022. [Ukrainian DCFTA]

Association Agreement between the European Union and the European Atomic Energy Community and their Member States, of the one part, and the Republic of Moldova, of the other part, OJ L 260, 30.8.2014, p 4–738, available at https://eur-lex.europa.eu/legal-content/EN/TXT/?uri=uriserv%3AOJ.L_.2014.260.01.0004.01.ENG, last access on 05.11.2022. [Moldavian AA].

Banks, K. (2020). Fit for Purpose? The Extent and Enforcement of International Trade Agreement Labor Obligations after the Guatemala-Labor Obligations Decision. Geo. J. Int'l L., 52, 639. https://worldtradelaw.typepad.com/files/caftaguatemalalabor.pdf

Blot A, Harrison J & Oger A (2022) Enhancing sustainability in EU Free Trade Agreements: The case for a holistic approach. Policy report, for European Environmental Policy (IEEP) https://ieep.eu/publications/enhancing-sustainability-in-eu-free-trade-agreements-the-case-for-a-holistic-approach Accessed 29.11.2022.

Bodansky D (2016) The legal character of the Paris Agreement, Review of European, Comparative & International Environmental Law 25.2 142–150.

Bronckers M & Gruni G (2019) Taking the enforcement of labour standards in the EU's free trade agreements seriously, CMLRev 56:6.

Bronckers M & Gruni G (2021) Retooling the sustainability standards in EU Free Trade Agreements, Journal of International Economic Law 24.1: 25–51.
Buergenthal T & Murphy S.D (2018) Public International Law in a Nutshell, WestAcademic.
Cate I (2012) The costs of consistency: precedent in investment treaty arbitration, Colum J Transnat'l L 51:418, https://heinonline.org/HOL/Page?public=true&handle=hein.journals/cjtl51&div=17&start_page=418&collection=journals&set_as_cursor=0&men_tab=srchresults Accessed 29 November 2022
Comprehensive Economic and Trade Agreement (CETA) between Canada, of the one part, and the European Union and its Member States, of the other part, OJ L 11, 14.1.2017, p. 23–1079, available at https://eur-lex.europa.eu/legal-content/EN/TXT/?uri=celex:22017A0114(01), last access on 06.02.2023. [CETA]
Consolidated version of the Treaty on the Functioning of the European Union, OJ C 326, 26.10.2012, p 47–390, available at https://eur-lex.europa.eu/legal-content/EN/TXT/?uri=celex%3A12012E%2FTXT, last access on 05.11.2022. [TFEU]
Cross, C. (2017). Focus: Failure by design: Did the US choose to lose the Guatemala labour dispute? International Union Rights, 24(3).
Dimitrakopoulos, D. G. (2001). The transposition of EU law: 'post-decisional politics' and institutional autonomy. European Law Journal, 7(4), 442–458.
Dörr O & Schmalenbach K (2018) Vienna Convention on the Law of Treaties - A Commentary. Springer Nature, Berlin.
Ellis J (2012) Shades of grey: soft law and the validity of public international law, Leiden Journal of International Law 25.2: 317.
EU-New Zealand (draft) FTA, Text is available at https://policy.trade.ec.europa.eu/eu-trade-relationships-country-and-region/countries-and-regions/new-zealand/eu-new-zealand-agreement/text-agreement_en, last access on 06.02.2023.
European Commission, Communication from The Commission to The European Parliament, The Council, The European Economic and Social Committee and the Committee of The Regions: The Power of Trade Partnerships: Together for Green and Just Economic Growth, Brussels, 22.6.2022 COM (2022) 409, available at https://eur-lex.europa.eu/legal-content/EN/TXT/PDF/?uri=CELEX:52022DC0409, last access on 05.11.2022. [Communication]
European Commission, Non-paper of the Commission services, Feedback and way forward on improving the implementation and enforcement of Trade and Sustainable Development chapters in EU Free Trade Agreements, 26.02.2018, available at https://trade.ec.europa.eu/doclib/docs/2018/february/tradoc_156618.pdf, last access on 21.11.2022.
European Commission, Non-paper of the Commission services, Trade and Sustainable Development (TSD) chapters in EU Free Trade Agreements (FTAs), 11.07.2017, available at https://trade.ec.europa.eu/doclib/docs/2017/july/tradoc_155686.pdf, last access on 21.11.2022.
European Parliament resolution of 5 July 2016 on implementation of the 2010 recommendations of Parliament on social and environmental standards, human rights and corporate responsibility, available at https://www.europarl.europa.eu/doceo/document/TA-8-2016-0298_EN.pdf, last access on 05.11.2022.
Franck S (2004) The Legitimacy Crisis in Investment Treaty Arbitration: Privatising Public International Law through Inconsistent Decisions, Fordham L. Rev, 73:1521 https://ir.lawnet.fordham.edu/cgi/viewcontent.cgi?article=4062&context=flr&httpsredir=1&referer= Accessed 29 November 2022.
Füller V (2018) How to make TSD chapters more effective, Trans European Policy Studies Association TEPSA Briefs.
García, M. J. (2022). Sanctioning capacity in trade and sustainability chapters in EU trade agreements: The EU–Korea case. Politics and Governance, 10(1).
Giupponi B.O (2021) International environmental law compliance in context: Mechanisms and Case Studies. Routledge, Oxforshire.
Gruchalla-Wesierski T (1984) A Framework for Understanding Soft Law, McGill L.J. 30:37.

Gruni G (2017) Labour Standards in the EU-South Korea Free Trade Agreement: Pushing Labor Standards into Global Trade Law? Korean J. Int'l & Comp. L. 6: 100,114.
Harrison J et al (2019) Governing Labour Standards Through Free Trade Agreements: Limits of the European Union's Trade and Sustainable Development Chapters, Journal of Common Market Studies 57:2, p 260–277.
Hulme M.H (2016) "Preambles in Treaty Interpretation" University of Pennsylvania Law Review 164.5 https://scholarship.law.upenn.edu/cgi/viewcontent.cgi?article=9527&context=penn_law_review&httpsredir=1&referer= Accessed on 23 January 2023.
International Bar Association. "Consistency, Efficiency and Transparency in Investment Treaty Arbitration." *IBA Arbitration Subcommittee on Investment Treaty Arbitration* (2018), available at https://uncitral.un.org/sites/uncitral.un.org/files/investment_treaty_report_2018_full.pdf, last access on 29.11.2022.
Jennings R.Y. & Watts A (eds.), Oppenheim's International Law, 1992, n. 632, as referred to in Villiger.
Kaufmann-Kohler G (2008) Is consistency a myth?" Precedent in International Arbitration https://lk-k.com/wp-content/uploads/Is-Consistency-a-Myth.pdf Accessed 29 November 2022.
Klabbers J (2018) "Treaties and Their Preambles", Conceptual and Contextual Perspectives on the Modern Law of Treaties: 172–200.
Krämer L (2021) A Lost Opportunity? The Environment and the EU–Mercosur Trade Agreement, Journal for European Environmental & Planning Law 18.1: 143–163.
Middleton J (2016) "Statutory interpretation: Mostly common sense?" Melb. UL Rev. 40 626
Neudorfer S & Wernig (2010) Implementation of International treaties into national legal orders: The protection of the rights of the child within the Austrian legal system, Max Planck Yearbook of United Nations Law, 14.
Nissen A (2022) Not That Assertive: The EU's Take on Enforcement of Labour Obligations in Its Free Trade Agreement with South Korea, European Journal of International Law 33.2: 607–630.
Odermatt J (2021) International Law and the European Union, Cambridge University Press.
Otero Fernández, I. (2020). Multilingualism and the Meaning of EU Law (Doctoral dissertation, European University Institute).
Peers S (2021) Free trade v freedom of association? The EU/South Korea free trade agreement and the PoE report on the EU challenge to South Korean labour law, http://eulawanalysis.blogspot.com/2021/01/free-trade-v-freedom-of-association.html Accessed 5 November 2022.
Rajamani L (2016a) "Ambition and differentiation in the 2015 Paris Agreement: Interpretative possibilities and underlying politics." International & Comparative Law Quarterly 65.2, 493–514 (b).
Rajamani L (2016b) The 2015 Paris Agreement: Interplay between hard, soft and non-obligations." Journal of Environmental Law 28.2: 337–358 (a).
Šarčević, S. (2013). Multilingual law-making and legal (un) certainty in the European Union. International journal of law, language & discourse, 3(1), 1–29.
Shelton, D. (1996). Reconcilable Differences--The Interpretation of Multilingual Treaties. Hastings Int'l & Comp. L. Rev., 20, 611.
Trade Agreement between the European Union and its Member States, of the one part, and Colombia, Peru and Ecuador, of the other part, OJ L 354, 21.12.2012, p. 3-4065, available at https://eur-lex.europa.eu/legal-content/EN/TXT/?uri=CELEX%3A02012A1221%2801%29-20181213, last access on 05.02.2023 [Andean FTA].
UN General Assembly, United Nations Framework Convention on Climate Change: resolution / adopted by the General Assembly, 20 January 1994, A/RES/48/189.
van den Putte L (2015) EU bilateral trade agreements and the surprising rise of labour provisions, International Journal of Comparative Labour Law and Industrial Relations 31.3.
van Gestel R and Micklitz H.W (2011) Revitalising Doctrinal Legal Research in Europe: What About Methodology? EUI Working Paper LAW 2011/05, https://cadmus.eui.eu/handle/1814/16825 Accessed 29 November 2022.

van't Wout D (2022) The enforceability of the trade and sustainable development chapters of the European Union's free trade agreements, Asia Europe Journal 20.2: 81–98.
Velut J.B et al. (2022) "Comparative Analysis of Trade and Sustainable Development Provisions in Free Trade Agreements." London School of Economics, Report commissioned by the European Commission 62 https://www.lse.ac.uk/business/consulting/assets/documents/TSD-Final-Report-Feb-2022.pdf Accessed 29 November 2022.
Villiger M.E (2009) Commentary on the 1969 Vienna Convention on the Law of Treaties, Brill.
von Bogdandy A (2009) The past and promise of doctrinal constructivism: A strategy for responding to the challenges facing constitutional scholarship in Europe, *International Journal of Constitutional Law* 7.3:364–400.

**Saide Esra Akdogan** is a Lecturer and Researcher at the Law Group of Wageningen University and Research (WUR), The Netherlands.

# Chapter 8
# Environmental Protection or Domestic Protectionism? The EU Deforestation-Free Regulation and its Shift from Exporter to Importer of Foreign Capital

**Thiago Ferreira Almeida**

## Contents



**Abstract** The principle of environmental integration is part of the European Union's (EU) Sustainable Development Strategy, which defines the bloc's long-term vision for environmental protection. The development of comprehensive sustainability and environmental protection policies in the EU involves bilateral and regional agreements on investment and trade. Therefore, the negotiations on foreign investments within (intra-EU) and outside the EU (extra-EU) are defined by the evolving jurisprudence of the Court of Justice of the European Union (CJEU), in particular the *Achmea*, *Micula* and *Komstroy* cases, Opinions 1/17 and 1/20. The EU investment and trade landscape is gradually becoming a repository of foreign investments, especially from emerging countries, such as Brazil, China and India. As a consequence, the EU adopts investment regulations instead of bilateral agreements of foreign investment protection, distancing itself from the traditional model of investor-State arbitration dispute settlements. The EU's adoption of mega agreements, particularly, the Canadian-EU agreement (CETA), establishes an elevated protection for the EU market. Although

T. F. Almeida (✉)
Federal University of Minas Gerais (UFMG), Brazil, Development Bank of Minas Gerais (BDMG), Belo Horizonte, Minas Gerais, Brazil
e-mail: thiago.almeida@bdmg.mg.gov.br

 
R. A. Wessel et al. (eds.), *EU External Relations Law and Sustainability*, Global Europe: Legal and Policy Issues of the EU's External Action 4,
https://doi.org/10.1007/978-94-6265-655-0_8

the European trend seems to aim at centralizing environmental and investment protection together, the treatment for investment reflects its repositioning as a recipient of foreign capital, changing its historical characteristic of capital exporter. When analyzing the EU's provisions on investment, there are new measures to block trade and investment liberalization using the strategy of environmental protection, as seen in the EU 2023/1115 Deforestation-Free Regulation, as well as the continuous opposition to the Mercosur-EU agreement by some EU Members, and the regulation 2024/1252 for critical minerals for energy transition. This Chapter concludes that the EU uses the principle of integration and environmental protection to oppose bilateral and regional agreements, securing trade protectionism. This behavior is justified on the grounds that the Union becomes a capital importer and the sustainable measures are effective in hampering foreign capital and goods.

**Keywords** European Union · Environmental integration · Foreign investments · Bilateral investment agreements · Mercosur · Deforestation-free regulation

## 8.1 Introduction

Since the Treaty of Lisbon—which entered into force in 2009—the European Union's (EU) Common Commercial Policy (CCP) has covered foreign direct investments (FDIs) and environmental and sustainable development policies, as EU competences. In this sense, the EU has progressively concentrated the subjects of investment and environmental protection in the EU bodies.

Such competences are subject to the concept of the EU environmental integration principle. The principle of integration recognizes the various environmental effects from a holistic understanding. Article 11 of the Treaty on the Functioning of the EU (TFEU), which states that the principle of environmental integration requires EU policies and activities to be directed towards sustainable development, encompassing issues on agriculture, energy, fisheries and transport, in line with environmental and climate commitments.[1]

The principle of integration is part of the EU's Sustainable Development Strategy, which sets out the bloc's long-term vision by combining economic activity, social cohesion and the commitment to environmental standards.[2] Such cohesion aligns the Union's exclusive competences with the actions of Member States on their territories. In addition, the principle of integration is also part of the Cohesion Policy, the main strategic framework related to investment and sustainable development.[3]

Regarding the treatment of foreign investments within the EU, there has been an important movement towards the centralization of that subject from two perspectives:

---

[1] European Union, 'Treaty on the Functioning of the European Union', Article 11.

[2] European Union, 'Communication from the Commission A Sustainable Europe for a Better World: A European Union Strategy for Sustainable Development'.

[3] European Union, 'New Cohesion Policy'.

(i) foreign investments between the Member States of the bloc (intra-EU); and (ii) European Union with third countries (extra-EU).

Considering intra-EU investments, the EU moved out from the traditional model of foreign investor protection in Bilateral Investment Treaties (BITs), where substantive rules of non-discrimination were established, such as National Treatment (NT) and Most-Favorable-Nation Treatment (MFN), and protection, Minimum Standard of Treatment (MST), Fair and Equitable Treatment (FET), Full Protection and Security (FPS), and compensation for direct and indirect expropriation. In addition, BITs have a specific dispute settlement mechanism, essentially applying the *ad hoc* investor-State arbitration model.[4]

The EU's orientation to terminate intra-EU agreements on investments could be connected with its gradual shift from a capital exporter to a capital importer, incorporating measures to secure its domestic market. In this sense, the traditional defense of capital (and trade) liberalization is no longer desirable for the EU Members. Therefore, over the last few years, new instruments have been developed indicating the EU's shift to a more protectionist bloc.

As for foreign investments outside the bloc (extra-EU), a different conduct can be witnessed. The EU, where needed together with the Member States, has the competence to conclude agreements which may include investment chapters, such as the groundbreaking agreement with Canada: the European Union and Canada Comprehensive Economic and Trade Agreement (CETA).[5] It has led to a new category of free trade agreements that involve trade, investment, environmental and social protection, intellectual property, services and establishes specific mechanisms for dispute settlements.[6]

[4] The list of the main clauses applied in bilateral investment treaties (BITs) and free trade agreements (FTAs) is the result of the analysis of specialized doctrine. See Bjorklund 2013, pp. 182–185; Dolzer and Schreuer 2008, pp.178–179, 182–183, 186–187; Dolzer et al. 2022, pp. 67, 69, 85, 413, 459–460; Alvarez 2011, p. 177; Hsu 2015, p. 139; Schill and Djanic 2018, pp. 2–3; Yannaca-Small 2006, pp. 31–32; Schill 2016, pp. 30–32; Newcombe and Paradell 2009, pp. 448, 453–454, 467–468.

[5] Eu-Canada Comprehensive Economic and Trade Agreement—CETA'; Tietje and Crow 2017.

[6] Although EU-Korea Free-Trade Agreement (FTA) was signed before CETA, only with the latter the Court of Justice of the European Union (CJEU) questioned the investment provisions, like seen in Achmea. Particularly, CETA included Investor-State dispute settlement mechanism, which are not seen in EU-Korean FTA. The EU dispositions for *ad hoc* arbitration for investor-state disputes after the Lisbon Treaty started only with CETA and it is permitted only with extra-EU agreements, as the present will be demonstrated.

Mega agreements have been a recent modality since 2015, as shown by for instance the CPTPP,[7] RCEP,[8] AfCFTA[9] and USMCA,[10] with their own arbitration rules for dispute settlement. In short, these new models of regional agreements comprise large geographic or continental groups and gather rules on trade and investment in a single instrument.

At the European level, examples include agreements with Japan,[11] Korea,[12] Vietnam,[13] Singapore,[14] and negotiations with Mercosur[15] and China.[16] The shift towards comprehensive free trade and investment models was triggered by the unsuccessful Doha Round of the World Trade Organization (WTO), and the deadlock of the Appellate Body of the WTO Dispute Settlement System, whose selection of judges has been hampered by the United States since December 2019.[17] The EU principle of integration is a mechanism for fostering sustainable development, which had an impact on intra and extra-EU investments.

Aligned with this understanding, it is proposed to present the evolution of the treatment of foreign investments within (intra-EU) and outside the EU (extra-EU), from the founding treaties and judgments of the Court of Justice of the European Union (CJEU), with emphasis on the *Achmea* case and Opinion 1/17. It is considered that the EU, as it gradually becomes a repository of foreign investments, even from emerging countries such as Brazil, China and India, turns to the concentration of Union legislation on the subject, replacing the traditional BIT model of universal protection for foreign investors, which was the paradigm largely sponsored by the Western European economies, such as Germany and France. As will

[7] 'Comprehensive and Progressive Agreement for Trans-Pacific Partnership—CPTPP', Available at https://www.mfat.govt.nz/en/trade/free-trade-agreements/free-trade-agreements-in-force/cptpp/comprehensive-and-progressive-agreement-for-trans-pacific-partnership-text-and-resources/#bookmark0.

[8] 'Regional Comprehensive Economic Partnership Agreement—RCEP', Available at https://rcepsec.org/legal-text/.

[9] 'Agreement Establishing the African Continental Free Trade Area—AfCFTA', Available at https://au.int/sites/default/files/treaties/36437-treaty-consolidated_text_on_cfta_-_en.pdf.

[10] The USMCA replaced NAFTA in 2020. See: 'United States-Mexico-Cananda Agreement', Available at https://ustr.gov/trade-agreements/free-trade-agreements/united-states-mexico-canada-agreement/agreement-between.

[11] European Commission, 'Agreement between the European Union and Japan for an Economic Partnership' 2019.

[12] European Commission, 'European Union and Republic of Korean Free Trade Agreement' 2011.

[13] European Commission, 'European Union and Vietnam Trade Agreement and Investment Protection Agreement' 2019.

[14] European Commission, 'European Union and Singapore Free Trade Agreement and Investment Protection Agreement' 2018.

[15] This agreement had its text approved in 2019, however it has not yet been signed. See: European Commission, 'EU-Mercosur' 2019.

[16] On December 30, 2020, the general principles for the agreement was finalized. It was called EU-China Comprehensive Agreement on Investment (CAI). See European Commission, 'EU-China Comprehensive Agreement on Investment' 2020.

[17] Almeida 2023.

be analyzed here, the EU-China Comprehensive Agreement on Investment (CAI) and the Mercosur-EU Agreement are examples of the advancement of negotiations for a comprehensive agreement on investments between the EU and China and with Mercosur, respectively.

The concentration of EU legislation is also observed in environmental protection, using the integration principle as a way of adopting domestic protection against the historical position for trade and capital liberalization. The EU creates measures of environmental protection and sustainable development as a way to justify its trade and investment protectionism. Therefore, the discourse of sustainable development and the integration principle are mechanisms to block the entry of foreign capital inside the EU.

The EU's movement to centralize these issues is a consequence of its re-positioning as a capital importer, changing its historical *status* as a capital exporter. Thus, the BIT model, usually used to exclusively secure foreign investors, is rejected, given the need of the home States to ensure the protection of their domestic public interests. The EU structure can be understood as a demonstration of the contemporary trend of challenging the system of international investment protection, opposing sustainable development and environmental protection to reduce competition inside the EU market.

In this sense, this Chapter shows environmental protection as a unilateral mechanism of the EU, against the adoption of trade and capital consensus in multilateral organizations, like the World Trade Organization, or even by bilateral or regional agreements. Later, it demonstrates how the Treaty of Lisbon and the *Achmea* case terminated intra-EU BITs. Following that, the CETA and Opinion 1/17 established the first understanding of EU regional agreements with third countries. This Chapter concludes that the Union's environmental protection and sustainable development legislation, along with the principle of integration, are mechanisms to secure EU protectionism, distorting international trade and capital flows. The shift of the international economic paradigm, after the rise of emerging economies like Brazil, India and China, turned the ancient sponsors of liberalization into new protectionists, hampering the development of emerging countries. It permits us to infer that the liberalization model for trade and investment, largely applied in the second part of the XX century, is not universal, only being maneuvered from the perspective of capital exporters.

## 8.2 The Environmental Protection in the International and European Context

The understanding of the healthy environment as an expression of a human right is observed previously in regional legal systems, as in the Protocol of San Salvador of 1988, embedded in the Inter-American Human Rights System.[18] The interdependent relationship between environmental protection, sustainable development and human rights was recognized by Advisory Opinion OC-23/17 of the Inter-American Court of Human Rights in 2017.[19]

Other international and regional systems have acknowledged the right to a healthy environment, as noted in the preamble to the 1998 Aarhus Convention,[20] in Article 24 of the 1991 African Charter on Human and Peoples' Rights,[21] and in Article 38 of the 2004 Arab Charter on Human Rights.[22]

Unlike the Inter-American Court of Human Rights, which already in 2017 opined to recognize the environment as a human right, the European Convention on Human

[18] "Article 11. Right to a Healthy Environment. (1) Everyone has the right to live in a healthy environment and to have access to basic public services. (2) States Parties shall promote the protection, preservation and improvement of the environment.". See: Brazilian Federal Decree nº. 3.321, of December 30th, 1999. The Inter-American Court of Human Rights, in the case of *Saramaka v. Suriname*, ruled that the logging concessions granted by the State of Suriname in the region traditionally inhabited by the Saramaka people damaged the environment, recognizing to this indigenous people the right to communal property. See: National Council of Justice, *Povo Saramaka v. Suriname.* Inter-American Court of Human Rights. Judgment on November 28th, 2007, 48, para 154.

[19] According to the Advisory Opinion: '55. As a consequence of the close connection between environmental protection, sustainable development and human rights (supra paras. 47 to 55), currently (i) multiple human rights protection systems recognize the right to a healthy environment as a right in itself, particularly the Inter-American human rights system, while there is no doubt that (ii) multiple other human rights are vulnerable to environmental degradation, all of which entails a series of States' environmental obligations to comply with their obligations to respect and guarantee these rights. Precisely, another consequence of the interdependence and indivisibility between human rights and environmental protection is that, in determining these State obligations, the Court may make use of the principles, rights and obligations of international environmental law, which as part of the international *corpus iuris* contribute decisively to establishing the extent of the obligations derived from the American Convention in this regard (supra paras. 43 to 45)." See: Brazilian Federal Prosecutor, 'Opinião Consultiva OC-23/17, de 15 de novembro de 2017, solicitada pela Colômbia', Inter-American Court of Human Rights 2017, paras 52, 55.

[20] "The Parties to this Convention (...) Recognizing that adequate protection of the environment is essential to human well-being and the enjoyment of basic human rights, including the right to life itself." See: United Nations Economic Commission for Europe, 'Convention on access to information, public participation in decision-making and access to justice in environmental matters', Aarhus Convention 1998.

[21] "All peoples shall have the right to a general satisfactory environment favourable to their development". See: African Union, 'African Charter on Human and Peoples' Rights' 1986.

[22] "Article 38. Every person has the right to an adequate standard of living for himself and his family, which ensures their well-being and a decent life, including food, clothing, housing, services and the right to a healthy environment. The States parties shall take the necessary measures commensurate with their resources to guarantee these rights." See: United Nations, 'Arab Charter on Human Rights' 2004.

Rights and the Charter of Fundamental Rights of the European Union do not yet contain any provision on the recognition of the right of the individual to a healthy environment. Later, at the international level, the 2015 Paris Agreement was the most prominent universal instrument to explicitly include the status of the environment as a human right.[23]

It is recently recognized at the international level that a healthy environment is a fundamental human right, based on the acknowledgement that every individual has the right to a healthy environment as an expression of human dignity. The United Nations Commission on Human Rights, by Resolution 48/13 of October 8, 2021, declared access to a healthy environment as a human right.[24] This decision was subsequently adopted by the UN General Assembly on July 28, 2022. At that session, 161 UN Member States voted in favor of recognizing a healthy environment to ensure a decent life and well-being for all human beings.[25]

Such acknowledgment by a significant number of subjects of international law is an important milestone for the effective implementation of measures to adapt and mitigate Greenhouse Gas (GHG) emissions and to promote climate resilience. In this sense, the UN General Assembly, based on the initiative of the State of Vanuatu, submitted a request for an advisory opinion to the International Court of Justice (ICJ) to analyze the obligations of States in international law regarding climate protection from anthropogenic emissions of GHG. It requires the ICJ to address the legal consequences of States' obligations, whether their actions or omissions have caused significant harm to the climate system, particularly to small island countries, considering the harm to individuals of present and future generations.[26]

Although the systematic evolution of the importance of protecting the environment and the concerns about climate change, along with the recent requests in international courts to make environmental protection enforceable, positioning it at the same level as a fundamental right, the bilateral and regional agreements on investment still locate environmental and climate protection as non-binding structure.

---

[23] In the preamble, "Acknowledging that climate change is a common concern of humankind, Parties should, when taking action to address climate change, respect, promote and consider their respective obligations on human rights, the right to health, the rights of indigenous peoples, local communities, migrants, children, persons with disabilities and people in vulnerable situations and the right to development, as well as gender equality, empowerment of women and intergenerational equity." See: United Nations Framework Convention on Climate Change, 'Paris Agreement' 2015.

[24] United Nations, 'Access to a healthy environment, declared a human right by UN rights council' 2021.

[25] United Nations, 'UN General Assembly declares access to clean and healthy environment a universal human right' 2022.

[26] United Nations, 'General Assembly Adopts Resolution Requesting International Court of Justice Provide Advisory Opinion on States' Obligations Concerning Climate Change' 2023. Moreover, the request to recognize States' obligations regarding climate change is seen at the Interamerican Court of Human Right and at the Tribunal of the Sea. See: Interamerican Court of Human Rights, 'Request for an advisory opinion on the Climate Emergency and Human Rights submitted to the Inter-American Court of Human Rights by the Republic of Colombia and the Republic of Chile' 2023; International Tribunal for the Law of the Sea, 'Requesto for an Advisory Opinion submitted by the Commission of Small Island States on Climate Change and International Law' 2023.

In the European Union, the environmental policy is linked with the principle of environmental integration, in Article 11 of the Treaty on the Functioning of the European Union (TFEU),[27] and with the objectives of the EU's external action, in Article 21 of the Treaty on the European Union (TEU).[28] On the other hand, environmental protection is argued to be balanced with other EU trade and social policies.[29] In particular, Article 3 TEU states that the bloc shall establish an internal market, taking into account the sustainable development of Europe, economic growth, price stability, market competition, social progress and employment, as well as providing a high level of protection and development of the quality of the environment.[30]

The large increase in cases involving environmental and foreign investment issues has become a key element to understand the paradigm shift in international investment law, where arbitral awards are commonly inconsistent and fail to take into account the domestic public interest, such as environmental protection, health, social security and national security. Such issues have been neglected in favor of defending the interest of foreign investors, as in *Tecmed v. Mexico,*[31] *Metalclad v. Mexico,*[32] and *Santa Elena v. Costa Rica.*[33]

This situation has also directly affected traditional capital exporters in Western Europe, such as Germany and Belgium, in *Vattenfall v. Germany*[34] and *Ping An v. Belgium,*[35] respectively, which recently have also become host countries for foreign capital from emerging economies, like India, China, Turkey and Brazil. In this context, the dispositions about environmental and climate protections are justifications for protecting the European market and investors.

The traditional Western countries, now in the position of host states, are moving away from the BIT model and are advocating for a more comprehensive framework to protect the host state's interest. They are defending more space to secure domestic public interest *vis-à-vis* foreign investment protection. This shift is the opposite of the traditional neoliberal agenda, implemented during the 1980s and 1990s. At that time, BITs were strongly supported by Western countries, among which the U.S., Germany, United Kingdom and France favored their BIT models to be dispersed mostly amongst developing countries. Two regions were highly influenced: Latin

[27] Article 11 do TFEU, *op. cit.* n. 2.

[28] "Article 21. 1. 1. The Union's action on the international scene shall be guided by the principles which have inspired its own creation, development and enlargement, and which it seeks to advance in the wider world: democracy, the rule of law, the universality and indivisibility of human rights and fundamental freedoms, respect for human dignity, the principles of equality and solidarity, and respect for the principles of the United Nations Charter and international law." See European Union, 'Treaty on European Union', Article 21.

[29] Wessel and Lark 2020.

[30] Kuijper et al. 2015, p. 591.

[31] *Técnicas Medioambientales Tecmed, S.A. v. The United Mexican States* 2003.

[32] *Metalclad Corporation v. The United Mexican States* 2000.

[33] *Compañia del Desarrollo de Santa Elena S.A. v. Republic of Costa Rica* 2000.

[34] *Vattenfall AB and others v. Federal Republic of Germany* 2020.

[35] *Ping An Life Insurance Company, Limited and Ping An Insurance (Group) Company, Limited v. The Government of Belgium,* 2015.

America and the former Soviet republics. Countries from these regions were characterized by sensible economic changes from the State to private orientation. At that time, there were few competitive companies and the Western bloc could fulfil these new markets occupying the absence of the State companies. Therefore, the United States and the Western European countries promoted the liberalization of trade and investments in order to reach markets from the Second and Third Worlds.[36]

During the beginning of the XXI century, even after the 2008 crisis, another group of countries, characterized by recent industrialization and relevant domestic markets, increased their economic performance and started not only to import capital but export as well. These are the emerging economies, commonly present in BRICS[37] and G20.[38]

Therefore, considering that investment protection still relegates environmental and climate protections as a second category, the European market changes in recent years push the European legislation to implement protectionism and use environmental protection as an excuse to bar foreign capital into the European market. The next sections will describe this context among EU States and, later, with third countries and the Union.

## 8.3 *Achmea* and the End of Intra-EU BITs

In the context of the EU, in the 2009 Lisbon Treaty, the competence of the European Commission was defined to deal with foreign direct investment (FDI). Therefore, there was a reorientation about investment protection and the use of BITs inside and outside the EU market. Article 207 of the TFEU determines the concentration of the FDI issue within the European common commercial policy, making it an exclusive competence of the EU.[39] In this sense, the Lisbon Treaty significantly changed the system of legal protection of investments between member countries (intra-EU) and with countries outside the European bloc (extra-EU). This EU concentration brought

---

[36] Sornarajah 2017, p. 6; Almeida 2024, pp. 43, 149, 167.

[37] O'Neil 2001; Jones 2012; Stuenkel 2015; Cooper 2016; Almeida and Silva 2018.

[38] As for its members, the G20 is composed of 8 advanced or ancient industrialized economies (Australia, Canada, Italy, Japan, France, Germany, the United Kingdom and the United States) and 11 developing countries (Argentina, Brazil, China, India, Indonesia, Mexico, South Korea, South Africa, Turkey and Russia). Completing its number, the European Union is the 20th member. Plus, African Union was invited to join in 2023. See: Brazilian Ministry of Foreign Affairs, 'O Brasil no G20'.

[39] "Article 207 1. The common commercial policy shall be based on uniform principles, particularly with regard to changes in tariff rates, the conclusion of tariff and trade agreements relating to trade in goods and services, and the commercial aspects of intellectual property, foreign direct investment, the achievement of uniformity in measures of liberalisation, export policy and measures to protect trade such as those to be taken in the event of dumping or subsidies. The common commercial policy shall be conducted in the context of the principles and objectives of the Union's external action." See: Article 207 do TFEU, *op. cit.* n. 2.

to an end the use of the intra-EU Bilateral Investment Treaty (BIT), which exclusively provided substantive and procedural protection rules for foreign investors, and defined the use of *ad hoc* investor-state arbitration as a dispute settlement mechanism (Investor-State Dispute Settlement—ISDS). Before 2009, the protection of foreign investments was commonly found in BITs, in which Germany and Switzerland were the first European countries to adopt specific instruments of investment protection. France, United Kingdom, Italy and Spain adopted it later. In the 1980s and 1990s, EU countries expanded the BITs with the new capitalist countries in the East, turning BIT into the sole model of investment protection. At that time, Western European companies exported capital to the former soviet countries and the so-called Third World. In this sense, international protection of investments in poor countries was fundamental to the expansion of transnational enterprises, particularly oil companies.

Accordingly, Regulation EU no 1219 of 2012,[40] it defined the conditions for the transition from the bilateral model (BIT) to the exclusive EU competence. In particular, as regards to the liability in ISDS, the EU Regulation nº. 912 of 2014[41] established the conditions on existing arbitrations with foreign investors.[42]

The decision of the Court of Justice of the European Union in *Achmea*,[43] ruled that investment agreements between EU countries (intra-EU) should be terminated. The case dealt with the BITs concluded between the Netherlands and Czechoslovakia in 1992, which stipulated an investor-state arbitration mechanism for dispute settlement. In 1993, Slovakia succeeded Czechoslovakia in the agreement and in 2004 it opened its market to private health insurance companies.

The Dutch company Achmea therefore started operating in the country as a foreign investor. In 2006 Slovakia reversed the liberalization measures in the health insurance sector and, according to the 2007 national legislation, the distribution of profits generated by these private companies was prohibited. In 2011, the Constitutional Court of Slovakia ruled the 2007 domestic legislation was unconstitutional, restoring the possibility of profit distribution. Finding that Slovakia's 2007 legislative change caused harm to Achmea, the company requested the establishment of an *ad hoc* arbitral tribunal against the State of Slovakia in 2008, in accordance with the 1992 BIT.

The forum of arbitration was Frankfurt, Germany, and the applicable law was German law. During the arbitral proceedings, the State of Slovakia declared that the arbitral tribunal lacked jurisdiction to arbitrate, as it had been a member of the EU since 2004 and therefore the arbitral mechanism would be incompatible with the European Union legal order. The arbitral tribunal dismissed the Slovakian objection,

---

[40] European Union, 'Regulation (EU) nº. 1219/2012 of the European Parliament and of the Council of 12 December 2012'.

[41] European Union, 'Regulation (EU) nº. 912/2014 of the European Parliament and of the Council of 23 July 2014'.

[42] Kuijper, Wouters et al., *op. cit.* n. 31, pp. 419–420.

[43] Court of Justice of the European Union (2018).

which was based on the incompatibility of the investor-State arbitration solution in the 1992 BIT and Articles 267[44] and 344[45] of TFEU.

Article 344 TFEU provides that the Court of Justice of the EU has jurisdiction in the event of divergence of interpretation and application of the treaties of the bloc. The judgment in the *Achmea* case held that Article 344 determines the autonomy of the EU legal system, in which each Member State must cooperate with the Court in accordance with Article 4(3) TEU. Furthermore, Article 344 TFEU ensures the exclusive jurisdiction of the Court of Justice of the EU, holding that intra-EU BITs do not contain any provision on the application of the European legal system. The European Court added that the dispute settlement model in an intra-EU BIT is incapable of ensuring the uniform application of European law, as provided for in Article 267 TFEU. In conclusion, the European Court ruled for the preservation of the principle of the European legal system's autonomy.

Furthermore, the CJEU found that European investors of a nationality other than the Netherlands or Slovakia would be in a less favorable position due to the absence of a bilateral investment agreement to allow the investor-State arbitration solution, remaining only the solution by the domestic jurisdiction of the host State. This situation, according to the Court, would constitute discriminatory treatment in relation to the other Member States, according to Article 18 of the TFEU.[46]

The advocacy of the autonomy of the EU legal system is justified, according to the *Achmea* case, by the common values shared among the Member States, which constitute the principles of mutual trust and sincere cooperation, expressed in Article 2 and § 3 of Article 4, both of the TEU.[47] The uniformity of the EU legal system is ensured by Article 267 of the TFEU, based on the preliminary ruling procedure, in

[44] "Article 267. The Court of Justice of the European Union shall have jurisdiction to give preliminary rulings concerning: (a) the interpretation of the Treaties; (b) the validity and interpretation of acts of the institutions, bodies, offices or agencies of the Union; Where such a question is raised before any court or tribunal of a Member State, that court or tribunal may, if it considers that a decision on the question is necessary to enable it to give judgment, request the Court to give a ruling thereon. Where any such question is raised in a case pending before a court or tribunal of a Member State against whose decisions there is no judicial remedy under national law, that court or tribunal shall bring the matter before the Court. If such a question is raised in a case pending before a court or tribunal of a Member State with regard to a person in custody, the Court of Justice of the European Union shall act with the minimum of delay." See: Article 267 do TFEU, *op. cit.* n. 2.

[45] "Article 344. Member States undertake not to submit a dispute concerning the interpretation or application of the Treaties to any method of settlement other than those provided for therein." See: Article 334 do TFEU, *op. cit.* n. 2.

[46] "Article 18. Within the scope of application of the Treaties, and without prejudice to any special provisions contained therein, any discrimination on grounds of nationality shall be prohibited. The European Parliament and the Council, acting in accordance with the ordinary legislative procedure, may adopt rules designed to prohibit such discrimination." See: Article 18 do TFEU, *op. cit.* n. 2.

[47] "Art. 2. The Union is founded on the values of respect for human dignity, freedom, democracy, equality, the rule of law and respect for human rights, including the rights of persons belonging to minorities. These values are common to the Member States in a society in which pluralism, non-discrimination, tolerance, justice, solidarity and equality between women and men prevail." See: European Union, 'Treaty on European Union', Article 2. Plus, see para 3º of Article 4º: "Pursuant to the principle of sincere cooperation, the Union and the Member States shall, in full mutual respect,

which a dialogue is established between the European Court and the courts of the Member States, ensuring the consistency, full effect and autonomy of EU law.

In this sense, the EU Court of Justice in that case held that the arbitral tribunal was not a part of the legal system of Slovakia or the Netherlands, being an exceptional jurisdiction, outside of the Article 267 of the TFEU. This decision significantly changed the relationship between investments within the bloc, removing international protection rules, as well as recognizing the exclusive European jurisdiction over investments (FDI).

The *Achmea* judgement demanded the termination of the intra-EU BITs, as noted in the 2019 political declarations of Member States and the Mutual Agreement for the Finalization of intra-EU BITs, signed in May 2020, in which the parties confirm that arbitration clauses in BITs would be contrary to EU treaties (Article 4).[48]

It is noted that the issue of bilateral treatment of investments by intra-EU BITs was not fully addressed in the process of expansion of Member States in 2004 and 2007 in the EU accession agreements, and the conflict between the arbitral solution and the preservation of the European legal system's uniformity was comprehensively analyzed in 2018 in the *Achmea* case.[49] On the other hand, BITs concluded between the United States and Eastern European countries in the process of integration into the EU were given special attention, with the aim of reducing incompatibilities between bilateral conventions with the United States and the EU rules.[50]

Furthermore, the cases involving the European Commission and Austria, Sweden and Finland have already dealt with the need to eliminate incompatibilities in BITs, according to the European legislation in which they were joining at the time, following the incorporation of the *acquis communautaire* and the impossibility of the arbitral tribunal to interpret EU's rules.[51]

Finally, the *Micula* case was analyzed by the European Court which understood the validity of the arbitral award of the Swedish company against Romania, since it was a violation of the BIT before Romania becoming an EU Member State. Therefore, the Court of Justice of the EU confirmed the legality of the indemnity claim set out in the

assist each other in carrying out tasks which flow from the Treaties." See: European Union, 'Treaty on European Union', Article 4.

[48] European Commission, 'Declaration of the Member States of 15 January 2019 on the legal consequences of the *Achmea* judgment and on investment protection' 2019; European Commission, 'Agreement for the termination of Bilateral Investment Treaties between the Member States of the European Union' 2020.

[49] Berger 2021, p. 28.

[50] United States of America, 'Understanding Concerning Certain U.S. Bilateral Investment Treaties, signed by the U.S., the European Commission, and acceding and candidate countries for accession to the European Union' 2003.

[51] Wu 2012, 377. Regarding the aforementioned cases, the Court of Justice of the European Union ruled that these countries had failed to promote the necessary measures to adapt the BITs signed prior to joining the European community. See: Court of Justice of the European Union , *Commission v. Austria* 2009; Court of Justice of the European Union, *Commission v. Sweden* 2009; Court of Justice of the European Union, *Commission v. Finland* 2009.

investor-State arbitration award.[52] In this sense, the *Achmea* case and the dispositions at the Lisbon Treaty demonstrate that intra-EU BITs are against the Union principles and those agreements needed to be terminated. The decision to terminate BITs shows a question about the international system of investment protection. If a domestic regulation of investments, as defended by the Union, is considered a better legal system than the liberalization of investments and its international protection, why do other countries, especially developing and the least developing ones still need to support BITs? In this sense, the EU demonstrates that domestic regulation will bring more benefits to its European companies. In consequence, it blocks foreign capital from entering into the EU market. The next section analyzes the different behavior of the ECJ regarding agreements between the EU and third countries.

Although there are decisions that show the acceptance of international investment protection, the EU judgments are non-conclusive, in which some absence of rule is recognizable, like the Energy Charter Treaty (ECT).

## 8.4 Opinion 1/17 and the EU Mega Agreements

Regarding the relationship between the EU and third parties outside the bloc, the European Commission leads the negotiations of trade and investment agreements. Article 207 of the TFEU, thus, establishes the exclusive competence to regulate FDI to the European Commission, as part of the Common Commercial Policy (CCP).

The EU's modifications have promoted a departure from the traditional BIT and greater control of the rules applied to the protection of foreign investors. Such behavior, therefore, reinforces the centralization and control of the investment issue by the EU, moving away from the position of previous decades of deregulation via BIT.

The concern to determine exactly the extent of investor rights comes from the need to ensure the protection of the bloc's interests. As the EU moves from the position of capital exporter to capital importer, including receiving investments from emerging economies such as China and India, it assumes the need to protect its domestic market. In this sense, the advocacy for the adoption of BITs models of universal investor protection, which was widely proclaimed in the 1980s–1990s, is dismissed by these countries in the 21st century.

The predominance of European competence demonstrates a centralization of the investment theme to a domestic sphere. This structure can be explained by the pluralist theory, which recognizes a parallelism of legal systems that coexist in a globalized and contemporary reality, which goes beyond the dualistic understanding centered on the division between the international and state spheres.[53]

---

[52] Court of Justice of the European Union, *Judgment of the Court (Grand Chamber) of 25 January 2022 European Commission v European Food SA and Others—Appeal* 2022.

[53] The Kadi cases against the European Council and Commission before the Court of Justice of the European Union addresses the debate on the EU's constitutional pluralism. See: Court of Justice

Concerning the agreements signed with third countries, the European Court, in Opinion 1/17, confirmed the compatibility of the ISDS system in the CETA agreement[54] in terms of the possibility of resolving investment disputes by arbitration (extra-EU).[55] With respect to dispute settlement, CETA provides for investor-State arbitration, in addition to defining rules for consultation and mediation. In particular, the parties may set up a specific tribunal to deal with matters arising under CETA, as well as their own court of appeal. Finally, CETA also allows for joining a multilateral investment court and its respective appellate body, in case they are created in future.[56]

On the establishment of trade and investment agreements between the EU and its Member States with third countries (extra-EU), Opinion 1/17 is an important milestone which ensures the possibility of establishing agreements with a dispute settlement system unrelated to the European legal system. In this sense, the CJEU ruled on Belgium's request by indicating that the European Court should have exclusive jurisdiction in the interpretation of EU law, as well as ensuring the autonomy of the bloc, in the face of the provisions of CETA. The decision of the European Court found that the principle of mutual trust, required of Member States and the EU, does not apply in the context of non-Member States and the EU. In particular, the dispute settlement mechanisms of the extra-EU agreement focus only on the rules of the treaty, not involving the interpretation of European regulations.[57] However, the Belgian government later requested a specific opinion of the CJEU regarding the effects of regional agreements with non-EU countries and its dispositions regarding ad hoc arbitration, as seen at the ECT.

The Opinion 1/20 of the CJEU of 2022 was based on a request from Belgium on whether the investor-State arbitration system provided for in the Energy Treaty Charter (ECT) would be contrary to the exclusivity of the European legal system between Member States (intra-EU). The European Court shied away from ruling, stating that there was insufficient information about the matter presented.[58]

The ECT was established in 1994 and came into force in 1998. The text was based on the European Energy Charter of 1991. In addition to its 54 members, the

---

of the European Union, *Kadi e Al Barakaat International Foundation/Council and Commission* 2008; Court of Justice of the European Union, *Al Barakaat International Foundation v. Council and Commission*; Court of Justice of the European Union, *Kadi v. Commission* 2010. See also De Búrca 2012, p. 130; De Witte 2012, p. 37; Halberstam 2012, p. 152; Krisch 2012, p. 220; Celeste 2020, pp. 99–100; Muñoz 2016, pp. 136–137.

[54] CETA, *op. cit.* n. 6.

[55] Court of Justice of the European Union, *Opinion 1/17* 2019.

[56] Schill 2017.

[57] Mersch et al. 2019, p. 22.

[58] Court of Justice of the European Union, Opinion 1/20 2022.

European Union[59] and EURATOM[60] are also part of the agreement.[61] The ECT is a regional agreement with the provision of the ISDS clause in the protection of foreign investors in the energy sector. In addition, ECT and European Energy Charter contracting members signed the International Energy Charter (IEC) in 2015 at the Ministerial Conference in The Hague, the Netherlands.[62]

The ECT was created at the historical period of the end of the Cold War when the economies of the former Soviet bloc began the transition to capitalism. These economies of Eastern Europe and Central Asia adopted significant changes in their rules of exploitation of energy resources. In particular, the entry into the capitalist model occurred through an accelerated process of privatization of state-owned enterprises that controlled this economic sector. In this sense, the ECT was based on the traditional BIT model that ensured the protection of the foreign investor from the host state, foreseeing exclusive solutions through investor-state arbitration.

In recent decades, Western European countries have been strongly critical of the ECT because of the huge financial losses, resulting from arbitral awards between investors and these countries as host states. Italy lost the arbitration case against Rockhopper Italia S.p.A., Rockhopper Mediterranean Ltd, and Rockhopper Exploration Plc, having to pay in total of 190 million euros for the application of discriminatory measures, violation of Fair and Equitable Treatment (FET) and illegal expropriation.[63] The United Kingdom, France, Luxembourg, Poland, Spain and the Netherlands, as well as the EU, have expressed their wish to withdraw from the ECT. Recently, Denmark, Ireland and Portugal announced their decision to leave unilaterally that agreement.[64]

In particular, the European Union on July 7, 2023, proposed a coordinated EU withdrawal from the ECT, maintaining the excuse that the agreement is not aligned

---

[59] The European Union has its origins in the 1951 Treaty of Paris, which created the European Coal and Steel Community (ECSC) in order to bring the coal and steel industries under a supranational regime. The founding countries of the ECSC were West Germany, France, Italy, the Netherlands, Belgium and Luxembourg. In 1957, the Treaty of Rome created two new institutions: the European Economic Community (EEC) and the European Atomic Energy Community (EURATOM). In 1992, by the Treaty of Maastricht, the term European Union is officially introduced. Currently, the Treaty of Lisbon of 2008 defines the EU's exclusive competences and those shared with its current 27 Member States. See European Union, 'EU History'; European Parliament, 'Treaty of Lisbon'.

[60] The 1957 Treaty of Rome established the European Economic Community (EEC) and the European Atomic Energy Community (EURATOM), with the aim of using nuclear energy for non-military purposes inside the European community. See: European Parliament, 'EURATOM Treaty'.

[61] Energy Charter, 'A little bit of history'; European Commission, 'Energy Charter'.

[62] Energy Charter, 'International Energy Charter' 2015.

[63] International Institute for Sustainable Development, 'ICSID tribunal finds that Italy committed an unlawful expropriation under ECT's Art. 13(1)'; Neslen 2022.

[64] European Parliament, 'EU withdrawl from the Energy Charter Treaty'; United Kingdom, 'UK departs Energy Charter Treaty' 2024; Euractiv, 'Netherlands follows Spain in quitting Energy Charter Treaty' 2022.

with the objectives of tackling climate change or under the European Green Deal.[65] However, the bloc evades in line with the Western European economies by withdrawing from the agreement based on the losing arbitration cases as a host state, ignoring that this group of countries was the main sponsor of the signing of the agreement in the period immediately after the fall of the Soviet regime.[66]

In *Moldova v. Komstroy*, the European Court analyzed the question of the jurisdiction of an investor-State arbitral tribunal held in Paris, France, in 2013. The dispute was between a Ukrainian energy company and the State of Moldova, involving a breach of ECT provisions. The forum chosen was Paris for the arbitral tribunal which sentenced Moldova to pay compensation to the Ukrainian company. Moldova appealed to the Paris Court of Appeal invoking the protection of public interest. The Paris Court of Appeal referred the matter to the European Court of Justice in order to assess whether the issue involved an EU rule. It is interesting to note that, as the chosen forum was in France, an EU member state, the European Court understood that it had jurisdiction over the matter, even if it was a host state outside the bloc and a company of non-EU state nationality. The understanding in the *Achmea* case is applied to the ECT when involving EU Member States, ruling out the application of the arbitration model in defense of the principle of autonomy of the EU legal system. Finally, the Court held on the merits that it was not an investment, as stipulated in the ECT and, therefore, dismissed the arbitral award.[67]

It is noted that the European Court upheld its jurisdiction because the arbitral forum chosen was an EU member country, even though the legal fact occurred in a state outside the bloc. Another interesting point, is the conclusion that the Court of Justice of the EU stated that, in principle, the ECT arbitration solution should be excluded in the event that both parties are EU members. Furthermore, the Court assessed the merits by interpreting the concept of investment in the ECT and whether it had repercussions on its legislation. Note that the ECJ analyzes the case in accordance with the EU legislation, even though none of the parties—the Ukrainian company and the state of Moldova—are parties of the Union.

Based on the judgments and opinions of the EU Court of Justice, there is a recognition of principles such as sincere cooperation and mutual trust. For intra-EU agreements, it is a solid position to guarantee the EU's exclusive jurisdiction. Regarding

[65] European Commission, 'Proposal for a Council Decision on the withdrawal of the Union from the Energy Charter Treaty' 2023.

[66] Carrington 2023; Stilton and Ainger 2023.

[67] "65. It follows that, although the ECT may require Member States to comply with the arbitral mechanisms for which it provides in their relations with investors from third States who are also Contracting Parties to that treaty as regards investments made by the latter in those Member States, preservation of the autonomy and of the particular nature of EU law precludes the same obligations under the ECT from being imposed on Member States as between themselves. 66. In the light of the foregoing, it must be concluded that Article 26(2)(c) ECT must be interpreted as not being applicable to disputes between a Member State and an investor of another Member State concerning an investment made by the latter in the first Member State." See: Court of Justice of the European Union, *Republic of Moldova v. Komstroy LLC* 2021.

extra-EU agreements, the European court understands the possibility of interpretation aimed to secure its autonomy, notwithstanding the identification of extensive rulings on the limits of the EU jurisdiction, as seen in *Moldova v. Komstroy*.

This approach is confirmed in the criticism of the ECT in the face of the gradual repositioning of Western European countries as capital importers,[68] supporting measures of greater domestic and regional protection, as against the advocacy for the liberalization of investment rules in the 1990s. The recognition of Europe's position as a capital importer, especially the importation of capital from emerging countries such as China and India, demonstrates how the liberalization of the capital was against the interest of the EU.

## 8.5 The EU Environmental and Climate Agenda as Protectionist Measures

However, the alignment of environmental and climate issues with the economic interests of the bloc involves strong resistance, compromising the institutional objectives of the EU.[69] The externalization of the environmental integration principle could be effectively applied in mega agreements on investment and trade that are under negotiation between the EU and countries outside the bloc, such as the EU-Mercosur agreement and the CAI.

The CAI had its final text announced on December 30, 2020, and is in the process of being signed and subsequently ratified by the parties.[70] The CAI has acknowledged political importance by signaling the *rapprochement* of China and the European Union in the field of foreign investments, as a counterweight to the traditional trade relations between the EU and the United States. It should be noted that in 2021, China overtook the United States as the largest trading partner.[71] However, in 2022, China remained the second largest partner. Among EU members, the largest trading partner with China is Germany.[72]

The Trade Agreement between Mercosur and the EU does not involve a separate chapter on investment. Its content is set out in the chapter on trade in services and rules on the establishment of companies. It should be noted that the agreement, whose negotiations began in 2000, reached the definition of the political text on June 28, 2019, and whose signature and subsequent ratification by the parties has not yet begun.[73]

---

[68] M. Sornarajah, *op. cit.* n. 37. pp. 131, 242, 410–413.

[69] Karageorgou 2023.

[70] European Commission, 'EU-China agreement in principle'.

[71] BBC News, 'China overtakes US as EU's biggest trading partner' 2023.

[72] Huld 2023.

[73] European Commission, 'EU-Mercosur Trade Agreement' 2019.

Mercosur was created by the Treaty of Asunción in 1991, and negotiations with the European Union began at the same time as discussions on the Free Trade Area of the Americas (FTAA) in the 2000s.[74]

In 2023, the agreement was again put up for discussion, and two contradictory positions within the EU were presented: (i) France is against the signature by upholding agricultural interests in the face of competition from Mercosur, indicating that there would be obstacles regarding the deforestation of the Amazon rainforest; and (ii) Germany is in favor due to the availability of essential minerals for the electric car industry, such as lithium, given that Argentina and Brazil possess considerable reserves. In addition, the conflict in Ukraine and the post-COVID-19 inflationary effects on the continent require the EU to seek short-term trade alternatives.[75]

The joint communiqué between the EU and Mercosur, resulting from the meeting held on 7 and 8 March 2023, indicated the intensification of dialogue aiming to finalize an agreement that is beneficial to both sides, alleging the existence of a new political and economic scenario in 2023.[76] What is observed, therefore, is the EU's reticence to force the application of more conditions in the agreement, since Mercosur had already accepted the terms in 2019.

The Mercosur-EU Agreement does not adopt the Investor-State dispute settlement, only the State-State model, which reinforces the contemporary position of moving away from investor-State arbitral solutions, as observed in the defense of the EU autonomy principle. The CAI proposal also rejects the investor-State arbitration model.

Later, the EU Deforestation-Free Regulation 2023/1115 targeted mainly developing and emerging economies, being a strong mechanism to halt competition between EU and third countries.[77] This Regulation uses environmental protection to hamper the importation of crucial commodities products to many developing economies, like Brazil and other Latin-American countries. The legislation aims to block the importation of relevant commodities into the EU market, namely cattle, cocoa, coffee, oil palm, rubber, soya and wood, which the EU customs authority has the competence to hinder the importation, alleging lack of proof of deforestation-free, not in line with EU legislation, and non-conformity with the EU due diligence statement, which encompasses risk assessment and risk mitigation.

---

[74] All South American countries are linked in some way to Mercosur. In addition to the four active and founding members of the bloc (Brazil, Argentina, Paraguay and Uruguay), Venezuela has been an official member since 2006 and has been suspended since 2016, according to the second paragraph of Article 5 of the Ushuaia Protocol on Democratic Commitment in Mercosur, while Bolivia, although approved as a member in 2015, is awaiting ratification by the other members. The other South American countries, meanwhile, are observer members of the bloc. See: Brazilian Federal Decree 4.210 of 2002, Ushuaia Protocol on Democratic Commitment with Mercosur, Chile and Bolivia.

[75] Aarup 2023; Silva 2023.

[76] The European position is prevalent when considering that the text approved in 2019 does not have a version in Portuguese and Spanish, official languages of Mercosur See: European Commission, 'Meeting of the European Union-Mercosur chief negotiators' 2023.

[77] European Commission, Regulation (EU) 2023/1115.

However, there are severe consequences on trade and development which are not mentioned in that legislation. In addition, the EU legislation admits product confiscation and application of fines, as the charge of 4% of the operator's or trader's total annual Union-wide turnover in the financial year preceding the fining decision (Article 25).[78]

Considering the implications of loss and damages to third countries, particularly emerging and developing countries, due to the fact that the EU legislation targets only commodities, a unilateral measure adopted by the Union confronts the main principles of the World Trade Organization, which is the legitimate forum to discuss multilateral measures related to trade.

## 8.6 Climate Action and Environmental Protection Policies of the Global North to Conceal Trade Protectionism: The perspective of the Global South on minerals for energy transition

The defense of a healthy environment and sustainable development based on the policies of the European Union contrasts with recent geopolitical and international trade manifestations.

The field of energy transition towards low-carbon production includes the use of renewable energy sources and energy efficiency. The use of electric vehicle batteries (EVs), for instance, is a global priority, raising questions about technological development, national industrialization policies, environmental protection and access to new minerals for such batteries.

In this sense, certain issues have become fundamental in the use of renewable energy sources and energy efficiency. Firstly, electric batteries qualify as new technologies and therefore require substantial investment to develop. In addition, trade protection measures have been adopted by economies concentrated in the Global North in order to resist the pioneering pace of Chinese technology in the electric battery development sector.[79]

[78] Reuters, 'Brazil says EU deforestation rules hamper Mercosur trade deal negotiation' 2023; Euronews, 'The EU's deforestation law was cheered here. Brazilian experts and farmers are skeptical' 2023.

[79] In the 21st century, the distinction between North and South can be difficult to be established clearly. Different attempts to divide countries into groups according to political and economic criteria (Three Worlds, Developed and Developing) were debated throughout the 20th century. From an emerging perspective, it is possible to understand the identification of developed countries as the Global North and emerging and developing countries as the Global South. Aware of the contemporary criticism of the attempt to place different countries in the same bloc, the Global South must take into account the historical past of colonialism, imperialism, slavery, racism and apartheid, along with the different economic and political processes in pursuit of the socio-economic development of these nations. See Almeida 2024.

One example is the return of intense trade protectionism by both European countries and the United States, on the subject of electric batteries and the trade and production of chips (semiconductors). In particular, during the Trump administration, the US government pushed for the creation of subsidiaries of Taiwanese chip companies in the United States, as seen in the state of Arizona.[80] Furthermore, it is estimated that, in the medium term, the United States could become autonomous in the production of chips and, consequently, would no longer require Taiwanese companies to purchase the product. The commercial relationship between Taiwan and the United States is also linked to national and military defense. Once the United States is no longer dependent on Taiwanese production, it will reduce its foreign policy of military protection of Taiwan, which would mean a change in its foreign policy in East Asia. The creation of subsidiaries on US territory was accompanied by strong pressure from the United States to prevent the sale of Taiwanese chips to the Chinese market, which was part of the US-China trade war.[81]

The enactment of two pieces of legislation by the Biden administration, the Inflation Reduction Act (IRA)[82] and the Chips and Science Act (CHIPS),[83] both of 2022, are aimed at injecting subsidies into the domestic industry, especially in the chip and electric car sectors, with the requirement that production takes place in the United States.[84] This means that such legislation reduces the competitiveness of other countries in the clean technology sector, generating criticism from European allies and other trading partners, such as Brazil.

At first, the EU opposed the North American measures by rejecting trade protection, indicating that this was contrary to the World Trade Organization (WTO). Subsequently, the EU inevitably adopted a similar protectionist policy and, on September 21, 2023, the European Commission announced the European Chips Act.[85] This is clear evidence of a repeat of the US protectionist strategy in the semiconductor technology sector.[86]

The adoption of subsidy policies has confronted the Global North's previous commitment to economic liberalism or neoliberalism since the end of the Cold War in 1991. The creation of the WTO in 1994, by the Treaty of Marrakech, is the main example of a reversal in the policies of the main economies at the time, in favor of international liberalism and the reduction of tariff and non-tariff barriers to goods and services, technology and investments. The recent shift towards protectionism by the developed economies is justified by the presence of new global players in the

---

[80] Nellis and Shepardson 2020.

[81] Lyengar 2024; Maizland 2024.

[82] United States of America, 'Statement by President Biden on Senate Passage of the Inflation Reduction Act' 2022.

[83] United States of America, 'Fact Sheet: CHIPS and Science Act Will Lower Costs, Create Jobs, Strengthen Supply Chains, and Counter China', 2022.

[84] Chu 2023.

[85] European Commission, 'European Chips Act'; European Parliament, 'EU's response to the US Inflation Reduction Act (IRA)'.

[86] Sutter et al. 2023.

international economic system, characterized by the emerging countries, to a greater extent China and India, and to a lesser extent Brazil, Turkey, Indonesia, South Korea, Colombia, Mexico and South Africa.

The paradigm shift from liberalism to protectionism in the Global North, therefore, stems from the economic and financial changes that have taken place since the 21st century, especially after the financial crisis of 2008.

The first decades of the 21st century were marked by strong growth in the annual GDP rates of emerging economies. China's strong economic growth, based on industrial production for export, increased demand for raw materials, driving up the price of commodities in the decade, such as oil, agricultural goods and minerals.[87] As a result, countries whose economies depend on few primary products experienced strong growth in the decade. This period is mostly known as the 'BRICS Decade', due to the economic performance of China, accompanied by India, Brazil, Russia and, to a lesser extent, South Africa. The BRICS, a term coined by Jim O'Neil in 2001, marks the group of countries with significant economic performance and the possibility of overtaking the world's largest Western economies in the next 30 years.[88] Since 2009, the heads of state of these five economies have been meeting formally and discussing issues of cooperation, development, trade and technology.[89]

As a result, emerging economies began to demand more representation in international financial institutions, with clear agendas for redistributing voting quotas in the IMF, which effectively happened in 2015, the World Bank and other forums. The change from the G7 to the G20 from 2008 with the financial crisis demonstrates the presence of emerging economies, in which 11 of the 19 nations come from the Global South (South Africa, Saudi Arabia, Argentina, Brazil, China, South Korea, India, Indonesia, Mexico, Russia and Turkey).

In particular, the economic shift observed by the presence of emerging economies is seen in the entry of foreign capital from companies from the Global South into the markets of the Global North. As a result, the EU, UK and US, traditional capital exporters, have recently become importers of Chinese and Indian capital and technology, observing the competition of products and technology from the Global North with new goods from emerging countries. The competition has not been in primary or low-tech products, but Chinese, Taiwanese, South Korean and Indian products are highly competitive in new technologies such as chips, electric batteries, cars, cell phones and electronic equipment.

The new competitiveness of emerging countries on the global scene was supposed to take place within the same paradigm of commercial liberalism, which advocated for freedom of trade and cross-border capital. However, the entry of new products and technologies has led to a loss of European and North American competitiveness.

---

[87] United Nations Conference on Trade And Development, 'The changing face of commodities in the twenty-first century', 2008; Black 2013.

[88] O'Neil 2001; Jones 2012; Stuenkel 2015; Cooper 2016; Almeida and Roberto 2018.

[89] Over the first decade of the 21st century, the BRIC countries accounted for 36.3% of the world's gross domestic product in purchasing power parity terms, as well as representing a quarter of global economic output. See Wilson et al. 2010.

As a result, the Global North began to support protectionist policies. In addition to the use of subsidies in green technologies, the United States has also been blocking the Appellate Body of the WTO Dispute Settlement System since December 2019. Palliative attempts at the WTO to create the MPIA and unilateral sanctions for non-compliance with international trade rules are confronted with the gradual discrediting of the Global North towards the WTO, an organization directly envisaged by developed countries to allow greater trade liberalism in the 1990s. With only 30 years of existence, the WTO has become irrelevant as developed countries resort to national measures to regulate their trade, in blatant disregard for international institutions.[90]

What we are witnessing, is an attempt by developed countries to evade the very rules they have created, on the understanding that international law should always favor them and when the economic situation is unfavorable to them, they believe they have the right to neglect international obligations and favor domestic rules, under baseless allegations of protecting national security and the domestic public interest.

The same can be seen in the international investment system, where developed countries have recently adopted a position of moving away from the liberalizing model of universal protection for foreign investors. The defense of the environment and domestic public interest has been claimed to support the blocking of foreign capital from entering the markets of the Global North. An example of this system is the application of investment screening, where Chinese and Indian capital is the most affected, using the subjective justification of protecting national security for the entry of capital from certain nationalities.[91]

In this respect, the traditional bilateral investment agreements with the modality of investor-state arbitration for conflict resolution, which were widely endorsed in the 1980s and 1990s in Latin American countries and former Soviet republics, based on models from the United States, Germany, the United Kingdom and France, are now being rejected by the countries of the Global North themselves. Some countries use models that allow exceptions to foreign investor protection, based on the protection of the public interest, while other countries have decided to move away from the dispute settlement model, such as Canada in the 2020 USMCA (United States-Mexico-Canada Agreement), in its Annex 14-D,[92] and the EU, following the 2018 *Achmea* case and the 2020 Intra-EU BIT Termination Agreement, in which EU member states confirmed that Investor-State Dispute Settlement (ISDS) clauses were contrary to EU treaties and therefore not applicable (Article 4).[93]

It is worth noting that the position of developing countries in rejecting the BIT model, which is harmful to the interests of their countries and has been claimed throughout the 20th century and, remotely, originated from the Latin American doctrines of Bello, Calvo and Drago in the 19th century, was disregarded as contrary

[90] Almeida 2023.

[91] Kovacs 2023; United Nations Conference on Trade And Development, 'World Investment Report 2023'.

[92] United States of America, 'United States-Mexico-Canada Agreement', Chapter 14, 2020.

[93] European Union, 'Agreement for the termination of Bilateral Investment Treaties between the Member States of the European Union', 2020.

to the paradigm of trade and capital liberalization at the end of the 20th century, following the movement towards the internationalization of global production, the expansion of multinationals and the processes of globalization.[94]

With the reverse flow of capital from emerging countries to the Global North, there has been a change in the direction of trade and investment policies. Protectionism, which was repudiated and seen in the 1980s and 1990s as contrary to development, is now defended as a measure to protect the national interests of countries. In turn, the liberalization of goods, services and capital, defended at the time, has been somewhat relativized.

Justifications such as environmental protection, the labor market, food and citizen security, and interests in protecting public policies such as health and education, were discredited when claimed by the countries of the Global South. As soon as the Global North began to demand this conduct, it became legitimate and possible to introduce this discussion for the reform of multilateral trade protection systems and the reform of foreign investor protection institutes.

Thus, it is clear that the grounds for the significant change in the global economic system do not result from a plea for security, environmental protection or public interest, but rather from competition between products, services and technology between emerging and developed countries.

The evident loss of market share and technological development by the countries of the North has led them to adopt unilateral national protection measures, contrary to international economic norms. This behavior is exemplified in the contemporary situation where the WTO dispute system is blocked, in the application of economic sanctions and the raising of tariff and non-tariff barriers, in the indiscriminate use of extraterritorial rules with a significant impact on the balance of trade of developing countries, in the impediment of reforms of international financial institutions to adapt to the new global reality, in the requirement to adopt environmental measures for developing countries disproportionately, and in the maintenance of subsidies for fossil companies.

From the aforementioned understanding, a significant change is also evident in the national and, above all, international processes for creating and negotiating international standards in the global economic system. Whereas in the post-Cold War period, when the United States was the only military, political and economic power, it was argued that international measures should favor trade and the free movement of factors of production (capital and labor). According to this analysis, it was the economy that set the international rules, as seen with the WTO and the BITs, which provide exclusive and universal protection for foreign investors. Since the financial crisis of 2008 and the recognized growth of emerging countries in the first decades of the 2000s, competition from emerging capital has caused fractures in the globalized system of trade and investment, leading to the adoption of protectionist measures and subsidies from the North to prevent the entry of technology and capital from emerging economies. In short, politics became the guideline for the production of international standards, altering even economic matters.

---

[94] Almeida 2021, 2022.

Therefore, having understood the first fundamental issue based on the context of foreign economic policies regarding the use of renewable energy sources and energy efficiency, it is necessary to look at a second analysis, involving the carbon generation chain of these new energy sources for the manufacture of green products.

The state of Minas Gerais, one of the 27 states of the Federative Republic of Brazil, characterized by a territory the size of metropolitan France, a population similar to Chile's, 20 million inhabitants, and with the 3rd largest Brazilian GDP, which is similar to Hungary's GDP, is historically known for its mineral production. With the discovery of gold and diamonds at the beginning of the 18th century, this region became responsible for producing wealth for its metropolis, the Portuguese Crown, for more than 100 years, guaranteeing Portugal's balance of trade deficit, as it purchased much more English products than it exported. Baroque cities such as Ouro Preto, Diamantina, Sabará, Congonhas do Campo, Mariana, Tiradentes, and São João Del Rey were created in the middle of the mountains. Even the capital of Brazil Colony, previously Salvador, in Bahia, was transplanted to Rio de Janeiro in 1753, due to its proximity to the gold mines, as sugar production was already in decline and had reached its peak in the 17th century.

At the end of the Brazilian gold rush, the region lost its splendor and Brazilian export production shifted to the coffee cycle in the 19th and early 20th centuries. Minas Gerais went into decline. In the 20th century, with the development of new technologies, another mining cycle began, now centered on the exploitation of iron, starting with the creation of the state-owned Vale do Rio Doce, in the municipality of Itabira, under the Getúlio Vargas government in 1952. Privatized in 1997, Vale is one of the main mining multinationals, with operations in Angola, Canada and Mozambique.[95] Since then, several foreign multinational companies have entered the state, starting a new mining cycle, now for iron. The historic region of the 18th century mines even became known as the Iron Quadrangle in the 20th century.[96]

However, mining confronts the need for environmental protection and sustainable development. Mining areas are often characterized by a situation of quasi-dependence on this production, and are responsible for the jobs, income and taxes of the mining municipalities. The consequences of mining are dire when you consider the end of exploitation. On the environmental side, there will be degraded areas and environmental destruction. On the social side, the end of mining will mean the end of income and jobs for a population dependent on ore.

In particular, mining in Minas Gerais has been marked by two episodes of disasters caused by mineral exploitation: in 2015, a dam broke in Mariana, and in 2018, another dam broke in Brumadinho. In both places, there were several deaths, contamination of the soil and watersheds, an increase in poverty and irreversible destruction of the environment.

Since then, mining in the state has been under debate, given that its production is necessary in order to ensure income and jobs for a population dependent on this economy, there are few discussions about a just transition, a constant instability of the

---

[95] Vale, Available at https://vale.com/pt/espaco-memoria.

[96] Peters Coelho 2012; Silva 2007.

risk areas because they are downstream of the dams, and a lack of clear public policies for environmental, social and climate protection in the face of mineral exploitation.[97]

In this local context, additional international considerations have to be taken into account. The concentration of production of essential minerals for electric batteries, electronic products and semiconductors, such as lithium and rare earths, has led to a reinterpretation of mineral exploitation. As we have seen, the European Union and the United States, determined to produce electric batteries and chips to compete with Chinese production, have started to relativize environmental and social concerns in mining.

The European Commission and multilateral banks such as the World Bank, the Inter-American Development Bank and the European Investment Bank, which used to prohibit any financing for mining companies, have adopted policies and regulations to allow the exploitation of minerals considered essential for the energy transition, i.e. for the production of electric batteries.

Therefore, the advocacy of mining would allow the European Union and the United States to produce electric batteries and avoid dependence on the supply of such products from China. The Chinese government, in the context of the trade war, has already blocked the export of rare earths for chips and batteries to the United States, in retaliation for the overcharging of other Chinese products.[98]

The defense of mining for the energy transition is based on the need to secure the market (electric batteries or chips) and reduce competition. Aspirations to promote climate action or environmental protection took second place. This fact is interesting as it demonstrates the Global North's recent interest in minerals for the energy transition in Brazil, with no mention of the situation in the state of Minas Gerais in the wake of recent natural disasters, such as Brumadinho and Mariana.

In particular, the EU applies the term "critical minerals for the energy transition", while the Brazilian government considers them to be "strategic minerals". For Europeans, these minerals are critical due to the absence of these minerals on their territory, which they must exploit in the Global South. From a Brazilian perspective, these minerals are strategic, allowing for a new market, technology transfer and the production of part of their output on Brazilian territory. Simply exporting raw minerals does not generate added value and financial return for the Brazilian citizens. In turn, Brazil believes that critical minerals are those used as fertilizers in agriculture, such as nitrogen, phosphorus and potassium, which are highly dependent on Russian imports and essential for Brazilian agricultural production. In fact, the issue of fertilizers for Brazil is critical and is one of the main factors behind Brazil's refusal to apply economic sanctions to Russia in the face of the conflict with Ukraine.[99]

In this context, minerals for energy transition are critical for the Global North due to its dependence, while such minerals would be strategic for Brazil and other emerging and developing nations.

---

[97] Harris 2014; Fabrício et al. 2021.

[98] Reuters, 'China gallium, germanium export curbs kick in; wait for permits starts', 2023.

[99] Nicas and Spigariol 2022; Tocci 2023.

From this, it can be inferred that the policies of states, and especially the European Union, which appear to legitimately promote environmental protection and climate action, have been adopted to meet essentially economic demands and ensure competition for their products in the global high-tech market. As noted, EU Regulation 2024/1252 establishes the framework for ensuring the sustainable supply of critical materials.[100]

Aware of this context, the conduct of the EU and other countries of the Global North are *deux poids, deux mesures*. The process of negotiating international standards for environmental protection and climate action is directly influenced by historical issues of colonialism, Eurocentrism and disputes over economic power. Thus, the principle of common but differentiated responsibilities must be strengthened, to understand that developing countries should not have the same level of exigencies as developed countries.

## 8.7 Conclusion

The evolution of green treatment at the international level has taken a center stage in the main economic forums. The environment has ceased to be a marginal issue and it is gradually being integrated as an essential aspect of individual protection, also constituting as a fundamental human right. Thus, the traditional division between economic development and environmental protection becomes anachronistic in light of rising global temperatures and frequent natural disasters resulting from anthropogenic actions and GHG emissions. It is essential to address environmental protection and sustainable development, considering actions that promote low-carbon manufacturing activities, renewable energy, energy efficiency and resilience to climate disasters. Plus, sustainable development cannot disregard social, regional and gender vulnerabilities, where populations in emerging and developing countries are the most affected by climate change.

On the other hand, there is an intrinsic relationship between environmental protection and economic development. The adoption of (unilateral) measures unilaterally to protect the environment does not constitute a reasonable solution. It hampers the critical economic and social development of many countries, particularly those less dynamic or non-economic diversified. The EU Deforestation-Free Regulation is a drastic example of how the EU considers environmental protection as a way to block trade competition and damage third-countries development.

[100] European Commission, 'Regulation (EU) 2024/1252; European Commission, 'Overview of the Critical Raw Materials Act; European Investment Bank, 'Rwanda and EIB agree new Critical Raw Materials investment partnership', 2023; European Investment Bank, 'Europe Must Get Serious About Critical Minerals, Werner Hoyer, President of the EIB, 2023; Luca 2024; European Investment Bank, 'EIB welcomes launch of European Raw Materials Alliance', 2020; World Bank Group, 'Minerals for Climate Action: The mineral intensity of the clean energy transition', 2020; Inter-american Development Bank, 'Leveraging the Growth in Demand for Minerals and Metals in the Transition to a Low Carbon Economy', 2022.

Furthermore, in the legal system of the European Union, there is a clear separation between the issues of environmental protection and investment. Although the Treaty of Lisbon defines the principle of environmental integration, the subject of investment is considered separate from matters of sustainable development. Moreover, the EU's treatment of foreign investment, since the judgment of the *Achmea* case in 2018, has deepened the European position of centralization of the theme and recognition of the principle of the bloc's autonomy. This centralization began with intra-EU agreements, in which any jurisdiction over investments other than those of the Member States and the EU was prohibited, ensuring the uniform interpretation and application of European rules. As for extra-EU agreements, the application of jurisdictions outside the EU is recognized, as in Opinion 1/17 on CETA, but the arbitration solution provided for in the ECT is still a matter of debate in EU bodies and EU Member States.

In order for the EU to truly apply environmental and climate-friendly measures, the economic implications ought to be considered in the international financial and trade systems. There are the WTO rules on trade and investment to reduce tariff and non-tariff barriers. The despise of the multilateral consensus regarding trade and investment in the EU market reaffirms the orientation of the EU legislation towards the adoption of substantial protectionism.

Therefore, the claim for the bloc's environmental integration principle, the EU Green Pact and the EU Deforestation-Free Regulation only provide EU profits and the reduction of the already drastic conditions for the developing and emerging countries to achieve better economic and social standards. In the end, the environmental protection measures of the EU are not driving countries to more sustainable development, but to trade and investment protectionism and enlargement of economic and social disparities between countries.

## References


Aarup S A (2023) EU-Mercosur deal faces moment of truth, Politico, February 2nd, 2023, <https://www.politico.eu/article/eu-mercosur-deal-truth-amazon-deforestation-trade-agreement-france-emmanuel-macron/>.

African Union (1986) African Charter on Human and Peoples' Rights', October 21st, 1986, <https://au.int/sites/default/files/treaties/36390-treaty-0011_-_african_charter_on_human_and_peoples_rights_e.pdf>.

'Agreement Establishing the African Continental Free Trade Area—AfCFTA', <https://au.int/sites/default/files/treaties/36437-treaty-consolidated_text_on_cfta_-_en.pdf>.

Almeida T F (2024) Investimentos Internacionais e Desenvolvimento em Países Emergentes: O Direito Internacional do Investimento nos níveis bilateral, regional e multilateral, na perspectiva dos países emergentes como importadores e recente exportadores de capital, Expert, <https://experteditora.com.br/wp-content/uploads/2024/03/Investimentos-Internacionais-e-Desenvolvimento-em-Paises-Emergentes.pdf>.

Almeida T F (2023) The Legitimacy of the MPIA's Decisions in the WTO Dispute Settlement System, Transnational Dispute Management, May, <https://www.transnational-dispute-management.com/journal-advance-publication-article.asp?key=1982>.

Almeida T F (2022) Is There a Latin American Approach to the International Investment Law? From the Legal Doctrines and Pan American Conferences in the XIX and XX Centuries to a Contemporary Approach in the XXI Century', Transnational Dispute Management. TDM Special Issue, v. 19, n. 4, <https://www.transnational-dispute-management.com/about-author-a-z-profile.asp?key=3672>.

Almeida T F (2021) A Antropofagia das Economias Emergentes: de Importadores a Exportadores de Capital e de Tomadores a Formadores do Direito Internacional do Investimento, in Rodrigues N C Accioly E, De Moura A B, Relações Comerciais e de Investimento entre União Europeia e América Latina, Anais do I Seminário Jean Monnet Network BRIDGE.

Almeida T F & Silva R L (2018) The Development Bank of BRICS', BRICS Law Jornal. v. 5, n. 4. University of Tyumen, Russian Federation.

Alvarez J E (2011) The Public International Law Regime Governing International Investment, The Hague Academy of International Law, AIL-Pocket.

BBC News China overtakes US as EU's biggest trading partner, February 17, 2023. <https://www.bbc.com/news/business-56093378>.

Bjorklund A (2013) Practical and legal avenues to make the substantive rules and disciplines of international investment agreements converge, in Echandi R & Sauté P (eds.), Prospects in International Investment Law and Policy. World Trade Forum, Cambridge University Press.

Black C (2013) Eventos relacionados ao superciclo de preços das commodities no século XXI, Revista de Indicadores Econômicos, Porto Alegre, v. 40, n. 2, <https://revistas.planejamento.rs.gov.br/index.php/indicadores/article/view/2811>.

Brazilian Federal Decree 4.210 of 2002, Ushuaia Protocol on Democratic Commitment with Mercosur, Chile and Bolivia, <http://www.planalto.gov.br/ccivil_03/decreto/2002/d4210.htm>.

Brazilian Federal Decree nº. 3.321, of December 30th, 1999, <https://www.planalto.gov.br/ccivil_03/decreto/d3321.htm>.

Brazilian Federal Prosecutor, Opinião Consultiva OC-23/17, solicitada pela Colômbia, Inter-American Court of Human Rights, <https://www.mpf.mp.br/atuacao-tematica/sci/dados-da-atuacao/corte-idh/OpiniaoConsultiva23versofinal.pdf>.

Brazilian Ministry of Foreign Affairs, O Brasil no G20, <http://antigo.itamaraty.gov.br/pt-BR/politica-externa/diplomacia-economica-comercial-e-financeira/15586-brasil-g20>.

Carrington D (2023) UK must quit climate-harming energy charter treaty, experts say, The Guardian, February 10th, 2023, <https://www.theguardian.com/environment/2023/feb/10/uk-must-quit-climate-harming-energy-charter-treaty-experts-say>.

Celeste E (2020) Digital Constitutionalism: The Role of Internet Bills of Rights', Thesis: supervisor TJ McIntyre (Ph.D.), University College Dublin.

Chu A (2023) Critics warn US Inflation Reduction Act could keep prices high, Financial Times, April 24th, 2023, <https://www.ft.com/content/3f8cdb59-587b-4809-80a9-1f950d0f5bce>.

'Comprehensive and Progressive Agreement for Trans-Pacific Partnership—CPTPP', <https://www.mfat.govt.nz/en/trade/free-trade-agreements/free-trade-agreements-in-force/cptpp/comprehensive-and-progressive-agreement-for-trans-pacific-partnership-text-and-resources/#bookmark0>.

Cooper A F (2016) The BRICS: A Very Short Introduction, Oxford University Press.

Court of Justice of the European Union, *Achmea BV v. Slowakische Republik*, C-284/16, Judgement on April 20th, 2018, <https://curia.europa.eu/juris/documents.jsf?num=C-284/16>.

Court of Justice of the European Union, *Commission v. Austria*. C-205/06, Judgment on March 3rd, 2009, <https://curia.europa.eu/juris/documents.jsf?num=C-205/06>.

Court of Justice of the European Union, *Commission v. Sweden*. C-249/06, Judgment on March 3rd, 2009, <https://curia.europa.eu/juris/documents.jsf?num=C-249/06>.

Court of Justice of the European Union, *Commission v. Finland*. C-118/07, Judgment on November 19th, 2009, <https://curia.europa.eu/juris/documents.jsf?num=C-118/07>.

Court of Justice of the European Union, *Judgment of the Court (Grand Chamber) of 25 January 2022 European Commission v European Food SA and Others—Appeal*, C-638/19 P, Judgment on January 25th, 2022, <https://curia.europa.eu/juris/document/document.jsf?docid=252641&doclang=en>.

Court of Justice of the European Union, *Kadi e Al Barakaat International Foundation/Council and Commission*, C-402/05 P, Judgement on September 3rd, 2008, <https://curia.europa.eu/juris/liste.jsf?num=C-402/05>.

Court of Justice of the European Union, *Al Barakaat International Foundation v. Council and Commission*, C-415/05 P, <https://curia.europa.eu/juris/liste.jsf?language=en&num=C-415/05%20P>.

Court of Justice of the European Union, *Kadi v. Commission*, T-85/09, Judgement on September 30th, 2010, <https://curia.europa.eu/juris/liste.jsf?language=en&num=T-85/09>.

Court of Justice of the European Union, *Opinion 1/17*, Judgment of April 30th, 2019, <https://curia.europa.eu/juris/document/document.jsf?docid=213502&doclang=EN>.

Court of Justice of the European Union, *Opinion 1/20*, Judgement on June 16th, 2022, <https://curia.europa.eu/juris/document/document.jsf?text=&docid=260993&pageIndex=0&doclang=en&mode=lst&dir=&occ=first&part=1&cid=3184590>.

Court of Justice of the European Union, *Republic of Moldova v. Komstroy LLC*, Judgment of the Court (Grand Chamber) of September 2nd, 2021, <https://curia.europa.eu/juris/document/document.jsf?text=&docid=245528&pageIndex=0&doclang=EN&mode=lst&dir=&occ=first&part=1&cid=1216548>.

Da Silva I M (2023) Negotiations between the two blocs—Mercosur includes Brazil, Argentina, Uruguay and Paraguay—started in 2000 and were concluded in June 2019, Euronews. February 1st, 2023, <https://www.euronews.com/my-europe/2023/02/01/explained-why-the-eu-mercosur-trade-deal-could-finally-be-ratified-this-year>.

De Búrca G (2012) The ECJ and the international legal order: a re-evaluation, in De Búrca G & Weiler J H H (eds.) The Worlds of European Constitutionalism, Cambridge University of Press.

De Witte B (2012) The European Union as an international legal experiment, in De Búrca G & Weiler J H H (eds.) The Worlds of European Constitutionalism, Cambridge University Press.

Dolzer R Kriebaum U Schreuer C (2022) Principles of International Investment Law (3) Oxford University Press.

Dolzer R & Schreuer C (2008) Principles of International Investment Law, Oxford University Press.

Energy Charter, A little bit of history', <https://www.energycharter.org/iec/history/>.

Energy Charter, 'International Energy Charter', May 20th, 2015, <https://www.energycharter.org/fileadmin/DocumentsMedia/Legal/IEC_EN.pdf>.

'EU-Canada Comprehensive Economic and Trade Agreement—CETA', <https://www.international.gc.ca/trade-commerce/trade-agreements-accords-commerciaux/agr-acc/ceta-aecg/text-texte/toc-tdm.aspx?lang=eng>.

Euractiv, 'Netherlands follows Spain in quitting Energy Charter Treaty', October 19th, 2022, <https://www.euractiv.com/section/energy/news/netherlands-follows-spain-in-quitting-energy-charter-treaty/>.

Euronews, 'The EU's deforestation law was cheered here. Brazilian experts and farmers are skeptical', August 8th, 2023, <https://www.euronews.com/my-europe/2023/08/31/the-eus-deforestation-law-was-cheered-here-brazilian-experts-and-farmers-are-skeptical>.

European Commission, 'Energy Charter', <https://energy.ec.europa.eu/topics/international-cooperation/international-organisations-and-initiatives/energy-charter_en#:~:text=>.

European Commission, 'Agreement between the European Union and Japan for an Economic Partnership', entered in force on February 1st, 2019, <https://policy.trade.ec.europa.eu/eu-trade-relationships-country-and-region/countries-and-regions/japan/eu-japan-agreement/eu-japan-agreement-chapter-chapter_en>.

European Commission, 'European Union and Republic of Korean Free Trade Agreement', provisionally applied since 2011 and formally ratified in December 2015, <https://eur-lex.europa.eu/legal-content/EN/ALL/?uri=OJ:L:2011:127:TOC>.

European Commission, 'European Union and Vietnam Trade Agreement and Investment Protection Agreement', signed on June 30th, 2019, <https://policy.trade.ec.europa.eu/eu-trade-relations

hips-country-and-region/countries-and-regions/vietnam/eu-vietnam-agreement/texts-agreements_en>.

European Commission, 'European Union and Singapore Free Trade Agreement and Investment Protection Agreement', signed on October 19th, 2018, <https://policy.trade.ec.europa.eu/eu-trade-relationships-country-and-region/countries-and-regions/singapore/eu-singapore-agreement/texts-agreements_en>.

European Commission, 'EU-Mercosur', <https://policy.trade.ec.europa.eu/eu-trade-relationships-country-and-region/countries-and-regions/mercosur/eu-mercosur-agreement/text-agreement_en>.

European Commission, 'EU-China Comprehensive Agreement on Investment', <https://policy.trade.ec.europa.eu/eu-trade-relationships-country-and-region/countries-and-regions/china/eu-china-agreement_en>.

European Commission, 'Declaration of the Member States of 15 January 2019 on the legal consequences of the *Achmea* judgment and on investment protection', January 17th, 2019, <https://finance.ec.europa.eu/publications/declaration-member-states-15-january-2019-legal-consequences-achmea-judgment-and-investment_en>.

European Commission, 'Agreement for the termination of Bilateral Investment Treaties between the Member States of the European Union', May 29th, 2020, <https://eur-lex.europa.eu/legal-content/EN/TXT/?uri=CELEX:22020A0529(01)>.

European Commission, 'EU-China agreement in principle', <https://trade.ec.europa.eu/doclib/press/index.cfm?id=2237>.

European Commission, 'Proposal for a Council Decision on the withdrawal of the Union from the Energy Charter Treaty', July7th, 2023, <https://eur-lex.europa.eu/legal-content/EN/TXT/PDF/?uri=CELEX:52023PC0447>.

European Commission, 'EU-Mercosur Trade Agreement', <https://policy.trade.ec.europa.eu/eu-trade-relationships-country-and-region/countries-and-regions/mercosur/eu-mercosur-agreement_en>.

European Commission, 'Meeting of the European Union-Mercosur chief negotiators', March 8th, 2023, <https://policy.trade.ec.europa.eu/news/meeting-european-union-mercosur-chief-negotiators-2023-03-08_en>.

European Commission, Regulation (EU) 2023/1115, <https://eur-lex.europa.eu/legal-content/EN/TXT/PDF/?uri=CELEX:32023R1115>.

European Commission, 'European Chips Act', <https://commission.europa.eu/strategy-and-policy/priorities-2019-2024/europe-fit-digital-age/european-chips-act_en>.

European Commission, 'Regulation (EU) 2024/1252 of the European Parliament and of the Council of 11 April 2024 establishing a framework for ensuring a secure and sustainable supply of critical raw materials and amending Regulations (EU) No 168/2013, (EU) 2018/858, (EU) 2018/1724 and (EU) 2019/1020, <https://eur-lex.europa.eu/legal-content/EN/TXT/?uri=OJ:L_202401252>.

European Commission, Overview of the Critical Raw Materials Act, <https://single-market-economy.ec.europa.eu/sectors/raw-materials/areas-specific-interest/critical-raw-materials/critical-raw-materials-act_en>.

European Investment Bank, 'Rwanda and EIB agree new Critical Raw Materials investment partnership', December 19th, 2023, <https://www.eib.org/en/press/all/2023-530-rwanda-and-eib-agree-new-critical-raw-materials-investment-partnership>.

European Investment Bank, 'Europe Must Get Serious About Critical Minerals, Werner Hoyer, President of the EIB, September 15th, 2023, <https://www.eib.org/en/stories/critical-raw-materials-europe>.

European Investment Bank, 'EIB welcomes launch of European Raw Materials Alliance', September 20th, 2020, <https://www.eib.org/en/press/news/eib-welcomes-launch-of-european-raw-materials-alliance>;.

European Parliament, 'Treaty of Lisbon', <https://www.europarl.europa.eu/about-parliament/pt/in-the-past/the-parliament-and-the-treaties/treaty-of-lisbon>.

European Parliament, 'EURATOM Treaty', <https://www.europarl.europa.eu/about-parliament/pt/in-the-past/the-parliament-and-the-treaties/euratom-treaty>.

European Parliament, 'EU withdrawl from the Energy Charter Treaty', December 2023, <https://www.europarl.europa.eu/RegData/etudes/BRIE/2023/754632/EPRS_BRI(2023)754632_EN.pdf>.

European Parliament, 'EU's response to the US Inflation Reduction Act (IRA)', <https://www.europarl.europa.eu/RegData/etudes/IDAN/2023/740087/IPOL_IDA(2023)740087_EN.pdf>.

European Union, 'EU History', <https://european-union.europa.eu/principles-countries-history/history-eu_pt>.

European Union, 'Treaty on the Functioning of the European Union'.

European Union, 'Communication from the Commission A Sustainable Europe for a Better World: A European Union Strategy for Sustainable Development (Commission's proposal to the Gothenburg European Council)', COM/2001/0264, <https://eur-lex.europa.eu/legal-content/EN/TXT/?uri=CELEX:52001DC0264>.

European Union, 'New Cohesion Policy', <https://ec.europa.eu/regional_policy/2021-2027_en>.

European Union, 'Treaty on European Union', Article 21, <https://eur-lex.europa.eu/eli/treaty/teu_2008/art_21/oj>.

European Union, 'Regulation (EU) n. 1219/2012 of the European Parliament and of the Council of 12 December 2012', establishing transitional arrangements for bilateral investment agreements between Member States and third countries, <https://eur-lex.europa.eu/LexUriServ/LexUriServ.do?uri=OJ:L:2012:351:0040:0046:En:PDF>.

European Union, 'Regulation (EU) n. 912/2014 of the European Parliament and of the Council of 23 July 2014', establishing a framework for managing financial responsibility linked to investor-to-state dispute settlement tribunals established by international agreements to which the European Union is party, <https://eur-lex.europa.eu/legal-content/EN/TXT/PDF/?uri=CELEX:32014R0912&rid=2>.

European Union, 'Treaty on European Union', <https://eur-lex.europa.eu/legal-content/EN/TXT/?uri=CELEX:12012M004>.

European Union, 'Agreement for the termination of Bilateral Investment Treaties between the Member States of the European Union', May 5th, 2020, <https://eur-lex.europa.eu/legal-content/EN/TXT/?uri=CELEX:22020A0529(01)>.

Fabrício S A Ferreira D D M & Borba J A (2021) A panorama of Mariana and Brumadinho: what do we know so far? Revista Eletrônica de Administração—READ. v. 27, n. 1. Janeiro/Abril 2021, <https://www.scielo.br/j/read/a/CjQDqqkvQvLqXDXnVVfTs9x/?format=pdf&lang=en>.

Halberstam D (2012) Local, global and plural constitutionalism: Europe meets the world, in De Búrca G & Weiler J H H (eds.) The Worlds of European Constitutionalism, Cambridge University Press.

Harris B (2014) As Minas goes, so goes Brazil', Financial Times, May 15th, 2024, <https://www.ft.com/content/0d0f1369-80d3-490f-91d5-afd5263d7614>.

Herdegen M (2016) Principles of International Economic Law' (2) Oxford University Press.

Hsu L (2015) Rule of Law and Foreign Investment, Rule of Law Symposium 2014: The importance of the rule of law in promoting development, Research Collection School of Law, Singapore Management University.

Huld A (2023) EU-China Relations—Trade, Investment, and Recent Developments, China Briefing, April 4th, 2023, <https://www.china-briefing.com/news/eu-china-relations-trade-investment-and-recent-developments/>.

Interamerican Court of Human Rights, 'Request for an advisory opinion on the Climate Emergency and Human Rights submitted to the Inter-American Court of Human Rights by the Republic of Colombia and the Republic of Chile', January 9th, 2023, <https://www.corteidh.or.cr/docs/opiniones/soc_1_2023_en.pdf>.

Interamerican Development Bank, 'Leveraging the Growth in Demand for Minerals and Metals in the Transition to a Low Carbon Economy', January 2022, <https://publications.iadb.org/en/leveraging-growth-demand-minerals-and-metals-transition-low-carbon-economy>.

International Institute for Sustainable Development, 'ICSID tribunal finds that Italy committed an unlawful expropriation under ECT's Art. 13(1)', <https://www.iisd.org/itn/en/2023/04/02/icsid-tribunal-finds-that-italy-committed-an-unlawful-expropriation-under-ects-art-131/>.

International Tribunal for the Law of the Sea, 'Requesto for an Advisory Opinion submitted by the Commission of Small Island States on Climate Change and International Law', June 15th, 2023, <https://www.itlos.org/fileadmin/itlos/documents/cases/31/written_statements/4/C31-WS-4-5-CIEL___GPI.pdf>.

Italaw, *Técnicas Medioambientales Tecmed, S.A. v. The United Mexican States*, ICSID Case No. ARB (AF)/00/2. Award, May 29th, 2003. <https://www.italaw.com/cases/1087>.

Italaw. *Metalclad Corporation v. The United Mexican States*, ICSID Case No. ARB(AF)/97/1. Award, August 30th, 2000, <https://www.italaw.com/cases/114>.

Italaw, *Compañia del Desarrollo de Santa Elena S.A. v. Republic of Costa Rica*, ICSID Case No. ARB/96/1. Award, February 17th, 2000, <https://www.italaw.com/cases/3413>.

Italaw, *Vattenfall AB and others v. Federal Republic of Germany*. ICSID Case No. ARB/12/12, <https://www.italaw.com/cases/1654>.

Italaw, *Ping An Life Insurance Company, Limited and Ping An Insurance (Group) Company, Limited v. The Government of Belgium,* ICSID Case No. ARB/12/29, <https://www.italaw.com/cases/3088>.

Jones S (2012) BRICS and Beyond. Executive Lessons on Emerging Markers, John Wiley & Sons.

Julien B (2021) International Investment Protection within Europe. The EU's Assertion of Control', Rutledge Research in International Economic Law, Rutledge.

Kovacs B (2023) Facilitating and Scrutinizing National Security Measures: Analysis of an Investment Screening Case in the European Union, <https://papers.ssrn.com/sol3/papers.cfm?abstract_id=4475471>.

Krisch N (2012) The case for pluralism in postnational law, in De Búrca G & Weiler J H H (eds.) The Worlds of European Constitutionalism, Cambridge University Press.

Kuijper P J Wouters J Hoffmeister F De Baere G Ramopoulos T (2015) The Law of EU External Relations. Cases, Materials, and Commentary on the EU as an International Legal Actor (2) Oxford University Press.

Lyengar R (2024) The Chipmaking World Hedges Its Taiwan Bets, Foreign Policy, April 11th, 2024, <https://foreignpolicy.com/2024/04/11/semiconductor-chips-taiwan-earthquake-tsmc-choke-point/>;

Luca F (2024) Brussels counts on internal EIB task force for critical commodities, February 8th, 2024, <https://www.eunews.it/en/2024/02/08/brussels-counts-on-internal-eib-task-force-for-critical-commodities/>.

Maizland L (2024) Why China-Taiwan Relations Are So Tense. Council of Foreign Relations' February 8th, 2024, <https://www.cfr.org/backgrounder/china-taiwan-relations-tension-us-policy-biden>.

Mercosur, <https://www.mercosur.int/pt-br/quem-somos/paises-do-mercosul/>.

Mersch Y Achtouk-Spivak L Affaki G, Cristina Contaterse C & Puig R V (2019) The new challenges raised by investment arbitration for the EU legal order', Legal Working Paper Series. n. 19/ October 2019, European Central Bank.

Muñoz J G P (2016) International Investment Disputes in South America: Rethinking Legitimacy in the Context of Global Pluralism, in Tanzi A Austerity A Lazo, R P & Turrini P (eds.) International Investment Law in Latin America / Derecho Internacional de las Inversiones en América Latina, Nijhoff International Investment Law Series, Brill Nijhoff.

National Council of Justice, *Povo Saramaka v. Suriname*. Inter-American Court of Human Rights. Jugdment on November 28th, 2007, <https://www.cnj.jus.br/wp-content/uploads/2016/04/cc1a1e511769096f84fb5effe768fe8c.pdf>.

Nellis S & Shepardson D (2020) Taiwan's TSMC to build Arizona chip plant as U.S.-China tech rivalry escalates, Reuters, May 15th, 2020, <https://www.reuters.com/article/idUSKBN22Q38T/>.

Neslen A (2022) Oil firm Rockhopper wins £210m payout after being banned from drilling, The Guardian, August 24th, 2022, <https://www.theguardian.com/business/2022/aug/24/oil-firm-rockhopper-wins-210m-payout-after-being-banned-from-drilling>.

Newcombe A & Paradell L (2009) Law and Practice of Investment Treaties, Kluwer Law International.

Nicas J & Spigariol A (2022) Good News for Food, Bad News for War: Brazil Buys Russian Fertilizer, The New York Times. May 8th, 2022, <https://www.nytimes.com/2022/05/08/world/americas/brazil-russian-fertilizer-sanctions.html>.

O'Neil J (2001) Building Better Global Economic BRICs, n. 66, Goldman Sachs.

Peters Coelho T (2012) Mineração e dependência no quadrilátero ferrífero, Universidade Federal de Juiz de Fora. Intra Textos. n. 3, <https://www2.ufjf.br/poemas//files/2014/07/Coelho-2012-Mineração-e-Dependência-no-Quadrilátero-Ferrífero.pdf>.

Schill S W & Djanic V (2018) International Investment Law and Community Interests, in Benvenisti E & Schill G (eds.) Community Interests Across International Law, University of Amsterdam.

Schill S W (2016) Derecho internacional de inversiones y derecho público comparado en una perspectiva latino-americana, in Tanzi A Asteriti A Lazo R P &Turrini P (eds.) International Investment Law in Latin America / Derecho Internacional de las Inversiones en América Latina, Nijhoff International Investment Law Series, Brill Nijhoff.

Schill S W (2017) Authority, Legitimacy, and Fragmentation in the (Envisaged) Dispute Settlement Disciplines in Mega-Regionals, in Griller S Obwexer W & Vranes E Mega-Regional Trade Agreements: CETA, TTIP, and TiSA; New orientations for EU external economic relations, Oxford University Press.

'Regional Comprehensive Economic Partnership Agreement—RCEP', < https://rcepsec.org/legal-text/>.

Reuters, Brazil says EU deforestation rules hamper Mercosur trade deal negotiation, November 8th, 2023, <https://www.reuters.com/world/americas/brazil-says-eu-deforestation-rules-hamper-mercosur-trade-deal-negotiation-2023-11-08/>.

Reuters, China gallium, germanium export curbs kick in; wait for permits starts, August 1st, 2023, <https://www.reuters.com/markets/commodities/chinas-controls-take-effect-wait-gallium-germanium-export-permits-begins-2023-08-01/>.

Silva F R. (2007) A Paisagem do Quadrilátero Ferrífero, MG: Potencial para o uso turístico da sua geologia e geomorfologia, Universidade Federal de Minas Gerais. Departamento de Geologia. Dissertação de Mestrado de Geografia.

Sornarajah M (2017) The International Law on Foreign Investment' (4) Cambridge University Press.

Stilton S & Ainger J (2023) EU Set to Exit From Controversial Energy Charter Treaty, BNN Bloomberg. July 4th, 2023, <https://www.bnnbloomberg.ca/eu-set-to-exit-from-controversial-energy-charter-treaty-1.1941365>.

Stuenkel O (2015) The BRICS and the future of global order, Lanham: Lexington Books.

Sutter K M Sargent Jr J F Singh M (2023) Semiconductors and the CHIPS Act: The Global Context, Congressional Research Service of the United States, September 28th, 2023, <https://crsreports.congress.gov/product/pdf/R/R47558>.

Tietje C & Crow K (2017) The Reform of Investment Protection Rules in CETA, TTIP, and Other Recent EU FTAs: Convincing? in Griller S Obwexer W Vranes E Mega-Regional Trade Agreements: CETA, TTIP, and TiSA; New orientations for EU external economic relations Oxford University Press.

Tocci N (2023) War in Ukraine is revealing a new global order—and the 'power south' is the winner' The Guardian. September 20th, 2023, <https://www.theguardian.com/commentisfree/2023/sep/20/war-ukraine-new-global-order-power-south-india-china>.

United Nations, 'Access to a healthy environment, declared a human right by UN rights council', October 8th, 2021, <https://news.un.org/en/story/2021/10/1102582>.

United Nations, 'UN General Assembly declares access to clean and healthy environment a universal human right', July 28th, 2022, <https://news.un.org/en/story/2021/10/1102582>.

United Nations, 'General Assembly Adopts Resolution Requesting International Court of Justice Provide Advisory Opinion on States' Obligations Concerning Climate Change', March 29th, 2023, <https://press.un.org/en/2023/ga12497.doc.htm>.

United Nations, 'Arab Charter on Human Rights', 2004, <https://digitallibrary.un.org/record/551368?ln=en>.

United Nations Conference on Trade And Development, 'The changing face of commodities in the twenty-first century', Note prepared by the UNCTAD secretariat, April 20th–25th, 2008, <https://digitallibrary.un.org/record/623254?ln=en>.

United Nations Conference on Trade And Development, 'World Investment Report 2023', <https://unctad.org/publication/world-investment-report-2023>.

United Nations Economic Commission for Europe, 'Convention on access to information, public participation in decision-making and access to justice in environmental matters', Aarhus Convention, June 25th, 1998, <https://unece.org/DAM/env/pp/documents/cep43e.pdf>.

United Nations Framework Convention on Climate Change, 'Paris Agreement', 2015, <https://unfccc.int/sites/default/files/english_paris_agreement.pdf>.

United Kingdom, 'UK departs Energy Charter Treaty, February 22nd, 2024, <https://www.gov.uk/government/news/uk-departs-energy-charter-treaty>.

'United States-Mexico-Cananda Agreement', <https://ustr.gov/trade-agreements/free-trade-agreements/united-states-mexico-canada-agreement/agreement-between>.

United States of America, 'Understanding Concerning Certain U.S. Bilateral Investment Treaties, signed by the U.S., the European Commission, and acceding and candidate countries for accession to the European Union', September 22nd, 2003. U.S. Department of State, <https://2009-2017.state.gov/s/l/2003/44366.htm>.

United States of America, 'Statement by President Biden on Senate Passage of the Inflation Reduction Act', August 7th, 2022, <https://www.whitehouse.gov/briefing-room/statements-releases/2022/08/07/statement-by-president-biden-on-senate-passage-of-the-inflation-reduction-act/>.

United States of America, 'Fact Sheet: CHIPS and Science Act Will Lower Costs, Create Jobs, Strengthen Supply Chains, and Counter China', August 9th, 2022, <https://www.whitehouse.gov/briefing-room/statements-releases/2022/08/09/fact-sheet-chips-and-science-act-will-lower-costs-create-jobs-strengthen-supply-chains-and-counter-china/>.

Vale, <https://vale.com/pt/espaco-memoria>.

Vasiliki (Vicky) Karageorgou, 'The environmental integration principle in EU law: normative content and functions also in light of new developments, such as the European Green Deal', European Papers v. 8, n. 1, 2023, <https://www.europeanpapers.eu/en/system/files/pdf_version/EP_eJ_2023_1_SS1_1_Vasiliki_Karageorgou_00645.pdf>.

Wessel R A & Lark J (eds.) (2020) EU External Relations Law: Text, Cases and Materials (2) Hart Publishing.

Wilson D Kelston A & Ahmed S (2010) Is this the 'BRICs Decade'?, BRICs Monthly, Goldman Sachs Global Economics, Commodities and Strategy Research. Issue n. 10/03. May 20, 2010, <https://www.goldmansachs.com/insights/archive/archive-pdfs/brics-decade-pdf.pdf>.

World Bank Group, 'Minerals for Climate Action: The mineral intensity of the clean energy transition', Climate Smart Mining. 2020, <https://pubdocs.worldbank.org/en/961711588875536384/Minerals-for-Climate-Action-The-Mineral-Intensity-of-the-Clean-Energy-Transition.pdf>.

Wu C (2012) Foreign Direct Investment as Common Commercial Policy: EU External Economic Competence After Lisbon, in Cardwell P J (ed.) EU External Relations Law and Policy in the Post-Lisbon Era, TMC Asser Press.

Yannaca-Small K (2006) Improving the System of Investor-State Dispute Settlement, OECD Working Papers on International Investment, 2006/01, OECD Publishing, <https://www.oecd.org/china/WP-2006_1.pdf>.

**Thiago Ferreira Almeida** is a Postdoctoral researcher of Climate Finance at Federal University of Minas Gerais (UFMG), Brazil, Doctor of international investment law UFMG, Brazil, Lawyer and Specialist in Public Policy at the Development Bank of Minas Gerais (BDMG), Brazil.

# Chapter 9
# Expanding European Climate Policy to International Shipping

**Manolis Kotzampasakis**

## Contents



**Abstract** The European Union (EU) has extended its climate policy to greenhouse gas emissions from international maritime voyages that start or end at European ports. This chapter explores certain international law dimensions of the upcoming EU climate policy expansion to international shipping, with a focus on international climate law, the law of the sea, and customary international law of state jurisdiction. Although there are sufficient jurisdictional grounds for implementing climate regulations in the form of conditions for entering European ports, the EU's climate policy for international shipping is not fully aligned with certain applicable legal limitations. Such limitations arise, among others, from the principle of common but differentiated responsibilities and respective capabilities (CBDR-RC), which in this context, requires the EU to more carefully consider the potential economic impacts of its extended carbon pricing regulation on vulnerable countries. Suggestions are formulated to increase alignment with the CBDR-RC principle, while complying with other applicable legal limitations, such as the obligation of non-discrimination against foreign ships.

**Keywords** EU ETS · FuelEU maritime · Shipping · Greenhouse gas emissions · International law · Jurisdiction

---

M. Kotzampasakis (✉)
University of Groningen, Groningen, The Netherlands
e-mail: m.kotzampasakis@rug.nl



R. A. Wessel et al. (eds.), *EU External Relations Law and Sustainability*, Global Europe: Legal and Policy Issues of the EU's External Action 4,
https://doi.org/10.1007/978-94-6265-655-0_9

## 9.1 Introduction

Shipping is the main mode of international commerce, as over 80% of all goods traded globally are transported by sea.[1] For centuries, ships were traversing the oceans relying on renewable energy (wind), but since the Industrial Revolution, they have been powered almost entirely by fossil fuels. Today's colossal merchant vessels provide the most carbon-efficient way to transport cargo over long distances.[2] Despite that, in absolute numbers international shipping is currently responsible for about 3% of all anthropogenic greenhouse gas (GHG) emissions.[3] In addition to their contribution to climate change, exhaust fumes from marine diesel engines also cause local air pollution, representing a public health risk, particularly in coastal areas.[4] In contrast to decarbonization efforts in other sectors, under a business-as-usual scenario, shipping emissions are projected to rise by mid-century, in tandem with an expected growth in global trade.[5]

Until recently, international maritime GHG emissions escaped both domestic and international regulation. Shipping is a globally operating sector with multiple international actors. The same vessel may fly Panama's flag, be owned by Greeks, be chartered by a multinational corporation based in Switzerland, and perform voyages between Brazilian and Dutch ports. In light of these complex features, in 1997 countries agreed to work on emissions reduction measures for international shipping collectively, under the auspices of the International Maritime Organization (IMO).[6] However, large divisions prevented the negotiating parties of the IMO from agreeing on any climate-related regulations until 2011. Since then, and until today, a bundle of measures adopted by the IMO has been inadequate for reducing maritime emissions in a trajectory aligned with international climate goals.[7]

In this context, the European Union (EU) decided to take matters into its own hands by including international maritime transport in its 'flagship' climate policy instrument, the EU Emissions Trading System (EU ETS).[8] From 2024, the EU ETS gradually imposes an emissions cap and charge a market-based carbon price on (a

---

[1] United Nations Conference on Trade and Development 2022, p. xv.

[2] European Environment Agency 2021.

[3] International Maritime Organization (IMO) 2021, p. 1.

[4] Contini and Merico 2021.

[5] IMO 2021, p. 3.

[6] Kyoto Protocol to the United Nations Framework Convention on Climate Change (adopted 11 December 1997, entered into force 16 February 2005) 2303 UNTS 148 (Kyoto Protocol) Article 2(2).

[7] International Energy Agency (IEA) 2022.

[8] Directive (EU) 2023/959 of the European Parliament and of the Council of 10 May 2023 amending Directive 2003/87/EC establishing a system for greenhouse gas emission allowance trading within the Union and Decision (EU) 2015/1814 concerning the establishment and operation of a market stability reserve for the Union greenhouse gas emission trading system [2023] OJ L130/134.

share of) the GHG emissions from voyages that start or end at European harbors.[9] In addition, the scheme is complemented by the FuelEU Maritime Regulation, which aims to progressively decrease the GHG intensity of maritime fuels combusted in those voyages and sets out requirements for on-shore power supply or zero-emission technology while ships are at berth.[10] All large commercial vessels that call (stop) at European ports will be obliged to comply with the EU's climate regulations, regardless of their flag. These EU initiatives triggered negative reactions from certain stakeholders in the international maritime sector,[11] bringing up memories from the early 2010s when the EU attempted to extend the EU ETS to flights between European and non-European airports. Its jurisdiction was contested by several countries, including the United States and China, and eventually the scope of the scheme was temporarily limited to intra-European flights only.[12]

Surprisingly, despite the possible jurisdictional tensions, the European Commission did not justify its maritime EU ETS and FuelEU Maritime proposals from the perspective of international law. This chapter explores certain international law dimensions of the EU climate policy expansion to maritime transport, with a focus on the EU ETS. The chapter is outlined as follows. Section 9.2 describes the scope of the new EU climate regulations for ocean shipping and analyses their consistency with international climate law. Section 9.3 then focuses on certain jurisdictional boundaries imposed by the law of the sea and general international law. Section 9.4 explores how the maritime ETS could comply with the common but differentiated responsibilities and respective capabilities (CBDR-RC) principle, without breaching the obligation of non-discrimination against foreign ships. Section 9.5 concludes.

[9] In the present chapter, the term 'European' narrowly refers to the European Economic Area (EEA), which consists of the 27 EU Member States, Norway, Iceland and Liechtenstein. The term 'intercontinental' refers to voyages between European and non-European ports.

[10] Regulation (EU) 2023/1805 of the European Parliament and of the Council of 13 September 2023 on the use of renewable and low-carbon fuels in maritime transport, and amending Directive 2009/16/EC [2023] OJ L234/48 (FuelEU Maritime Regulation).

[11] See, e.g., Maritimes [2021] ICS reaction to EU ETS shipping extension <https://maritimes.gr/en/maritimes-news/seagoing/41293-ics-reaction-to-eu-ets-shipping-extension> [Accessed 28 December 2022].

[12] Martinez Romera and van Asselt 2015, p. 275.

## 9.2 EU Maritime Climate Policy and International Climate Law

### *9.2.1 Expanding EU Climate Regulation to International Shipping*

The EU ETS is a market-based instrument that obliges companies to cover their annual GHG emissions with a corresponding number of tradable permits (allowances). By steadily reducing the number of new allowances issued each year, the EU ETS imposes an absolute declining cap on the total emissions under its scope. In 2021, the scheme was on course to achieve a 51% reduction of GHG emissions by 2030 from its covered sectors, namely electricity and heat producers, energy-intensive industries, and airlines performing flights within the European Economic Area (EEA).[13]

A necessary first step before the inclusion of shipping in the EU ETS was the collection of reliable data on GHG emissions from maritime transport. This was achieved with the introduction of the EU Monitoring, Reporting and Verification (MRV) Shipping Regulation, which obliges shipping companies to annually report the carbon dioxide ($CO_2$) emissions—and, from 2024, also the methane (CH4) and nitrous oxide ($N_2O$) emissions—of their ships to the Commission.[14] The MRV Regulation applies to all large merchant ships that stop at EEA ports 'to load or unload cargo or to embark or disembark passengers'.[15] Smaller vessels and certain ship types, such as warships and fishing vessels, are currently excluded.[16] Since 2018, shipping companies have been reporting the full range (100%) of their $CO_2$ emissions for intra-European voyages (between and within EEA ports), as well as for extra-European voyages (between EEA ports and non-EEA ports).[17]

The maritime EU ETS generally mirrors the MRV Shipping Regulation in terms of covered greenhouse gases, ship types, and inclusion thresholds,[18] but the scope of the ETS is narrower than the MRV Regulations with regard to intercontinental voyages. Shipping companies will be required to pay for all of their intra-European GHG

[13] Commission (EU) 'Proposal for a Directive of the European Parliament and of the Council Amending Directive 2003/87/EC Establishing a System for Greenhouse Gas Emission Allowance Trading within the Union, Decision (EU) 2015/1814 and Regulation (EU) 2015/757' (Communication) COM(2021) 551 final, 14 July 2021, Explanatory Memorandum 1.

[14] Regulation (EU) 2015/757 of the European Parliament and of the Council of 29 April 2015 on the monitoring, reporting and verification of carbon dioxide emissions from maritime transport, and amending Directive 2009/16/EC [2015] OJ L123/55 (MRV Shipping Regulation) Article 11(1).

[15] Ibid., Articles 2 and 3(b).

[16] Ibid., Article 2.

[17] Ibid.

[18] Directive 2003/87/EC of the European Parliament and of the Council of 13 October 2003 establishing a scheme for greenhouse gas emission allowance trading within the Community and amending Council Directive 96/61/EC [2003] OJ L275/32 (EU ETS Directive) Annex I.

emissions, but only for half of their GHG emissions from extra-European voyages.[19] This means that a company operating a ship that transports goods from São Paulo to Rotterdam (or vice versa) would have to surrender only 50 EU ETS allowances for every 100 tons of $CO_2$ or equivalent GHGs emitted in that voyage. Similarly, the FuelEU Maritime Regulation covers only 'a half of the energy used' in intercontinental voyages.[20] The European Commission and the co-legislators characterized this 'fifty-fifty' approach as 'a practical way to solve the issue of Common but Differentiated Responsibilities and Capabilities', which would allow non-European coastal states 'to decide on appropriate action in respect of the other share of emissions'.[21]

While this statement was not elaborated any further, the 50% scope for extra-European maritime voyages appears as an effort to avert international political backlash, similar to what happened with aviation. This political risk was highlighted in the public consultation for the EU ETS revision,[22] while the Commission acknowledged in the Impact Assessment for FuelEU Maritime that this narrower scope 'could be easier to accept by third-country operators, but would have a considerably lower impact'.[23] Yet, within this sensitive political context, an explicit justification of the scope based on an analysis of its compatibility with international law could have increased legal certainty, by placing (the rule of) international law at the heart of the political discussions between the EU and its global partners. The legal analysis in the present chapter contributes to this objective. More importantly, the EU is committed to 'the strict observance and the development of international law',[24] while its institutions are obliged to exercise their regulatory powers within the limitations that international law imposes.[25] The above necessitates a more careful examination of the relevant possibilities and limitations from the perspective of international law. This chapter focuses on international climate law, the law of the sea, and customary

19 Ibid., Article 3ga(1).

20 FuelEU Maritime Regulation Article 2(1)(d).

21 EU ETS Revision Proposal, recital 17; Directive (EU) 2023/959, recital 20.

22 Commission (EU) 'Impact Assessment Report Accompanying the Proposal for a Directive of the European Parliament and of the Council Amending Directive 2003/87/EC Establishing a System for Greenhouse Gas Emission Allowance Trading within the Union, Decision (EU) 2015/1814 and Regulation (EU) 2015/757' (Staff Working Document) SWD(2021) 601 final, 14 July 2021 (EU ETS Revision Impact Assessment), Part 1, pp. 150, 108.

23 Commission (EU) 'Impact Assessment Report Accompanying the Proposal for a Regulation of the European Parliament and of the Council on the use of renewable and low-carbon fuels in maritime transport and amending Directive 2009/16/EC' (Staff Working Document) SWD(2021) 635 final, 14 July 2021, 44. See also Finska and Ringbom 2022, p. 47.

24 Consolidated Version of the Treaty on European Union [2016] OJ C202/13 (TEU) Article 3(5).

25 Consolidated Version of the Treaty on the Functioning of the European Union [2016] OJ C202/47 (TFEU) Article 216(2); Case C-286/90, *Poulsen and Diva Navigation*, ECLI:EU:C:1992:453, para 9. For an overview of the interaction between EU law and international law, see Larik and Wessel 2020.

international law of state jurisdiction. While international trade law is also relevant, it is excluded from the present analysis, as existing literature suggests that the maritime EU ETS is unlikely to be in conflict with the World Trade Organization (WTO) agreements.[26]

## *9.2.2 Regulating International Shipping Under International Climate Law*

### 9.2.2.1 Possibilities

The 1992 United Nations Framework Convention on Climate Change (UNFCCC) declared climate change as a common concern of humankind and set a general obligation for all of its 197 parties to adopt climate mitigation measures, in accordance with their unequal historical responsibilities for the causes of climate change and their differing economic capabilities to combat it.[27] While the UNFCCC does not refer explicitly to maritime transport, it stipulates that climate mitigation policies and measures should 'cover all relevant sources' and 'comprise all economic sectors'.[28]

Subsequently, the 1997 Kyoto Protocol obliged countries with developed or transitional economies ('Annex I Parties') to implement climate policies, to individually or collectively achieve binding emissions reduction targets.[29] A specific reference to maritime transport was included in Article 2(2) of the Kyoto Protocol:

> The Parties included in Annex I shall pursue limitation or reduction of emissions of greenhouse gases not controlled by the Montreal Protocol from aviation and marine bunker fuels, working through the International Civil Aviation Organization and the International Maritime Organization, respectively.[30]

This provision can be interpreted as a requirement to cooperate towards a regulatory solution for maritime GHG emissions under the auspices of the IMO, but not as a clause that provides the organization with *exclusive* competence for their regulation, as it imposes an obligation of conduct ('shall pursue') on a limited number of states (Annex I Parties).[31]

---

[26] Even if the scheme were found to be inconsistent with a core provision of the General Agreement on Tariffs and Trade 1994 (adopted 15 April 1994, entered into force 1 January 1995) 1867 UNTS 187 (GATT), such as Articles I, III, V or XI, it could likely be justified under Article XX(b) or (g). See, e.g., Bäuerle et al. 2010, p. 93; Ringbom 2011, pp. 636–637; Rodriguez 2012, pp. 36–37; Dobson and Ryngaert 2017, pp. 315–325.

[27] United Nations Framework Convention on Climate Change (adopted 9 May 1992, entered into force 21 March 1994) 1771 UNTS 107 (UNFCCC), preamble, Articles 4(1) and 3(1).

[28] UNFCCC Article 3(3).

[29] Kyoto Protocol Articles 2 and 3. As stated in Article 10, the Kyoto Protocol did not introduce any new commitments for (developing) countries not included in Annex I.

[30] Ibid., Article 2(2).

[31] Ringbom 2011, p. 639; Heine et al. 2015, pp. 28–30.

Nonetheless, Article 2(2) of the Kyoto Protocol may have contributed to the fact that GHG emissions from international shipping have fallen into a prolonged regulatory limbo. As discussions at the IMO have been stalling, international maritime GHG emissions have remained unregulated, as they have been excluded from national or regional climate inventories and targets.[32] For example, the EU's climate target encompasses only intra-national shipping, as even emissions from voyages between EU Member States are not accounted for. Moreover, bunker fuels for international shipping have generally been exempted from taxation.[33] Following certain energy efficiency measures first adopted by the IMO in 2011, it was not until 2018 that the organization formulated an 'Initial GHG Strategy', in which it expressed an ambition 'to reduce the total annual GHG emissions [from international shipping] by at least 50% by 2050 compared to 2008'.[34] This ambition was strengthened in the revised IMO GHG Strategy of 2023: 'to reach net-zero GHG emissions by or around, i.e. close to 2050'.[35]

The Paris Agreement has set the collective goal to keep the rising global average temperature 'to well below 2 °C above pre-industrial levels', while pursuing efforts to restrict the temperature increase to 1.5 °C.[36] It obliges all parties to define their own GHG reduction targets, referred to as nationally determined contributions (NDCs), and pursue domestic measures to achieve them.[37] The treaty does not specify what sectors should be included in or excluded from the NDCs, nor does it contain a special reference to maritime transport (as an earlier draft in support of a sectoral climate target for international shipping did not make it to the final treaty text).[38] However, the Paris Agreement explicitly states that developed countries should lead climate mitigation efforts, by undertaking 'economy-wide absolute emission reduction targets'.[39]

Pursuant to the CBDR-RC principle, the EU must shoulder a large share of global mitigation efforts, due to its significant share of *responsibility* for climate change, but also due to its considerable economic, technological, and institutional *capabilities* to combat it.[40] Some legal scholars have even argued that the EU might have a due

---

[32] See United Nations Environment Programme 2021, pp. 25 and 27.

[33] International Transport Forum 2022, p. 20.

[34] IMO, 'Resolution MEPC.304(72), Initial IMO Strategy on Reduction of GHG Emissions from Ships' MEPC 72/17/Add.1 (13 April 2018) (Initial IMO GHG Strategy) Annex 11, para 3.1.3.

[35] IMO [2023] Revised GHG reduction strategy for global shipping adopted <https://www.imo.org/en/MediaCentre/PressBriefings/pages/Revised-GHG-reduction-strategy-for-global-shipping-adopted-.aspx> [Accessed 8 July 2023].

[36] Paris Agreement (adopted 12 December 2015, entered into force 4 November 2016) 55 ILM 743, Article 2(1).

[37] Ibid., Article 4(2).

[38] UNFCCC Ad Hoc Working Group on the Durban Platform for Enhanced Action 'Negotiating Text' UN Doc FCCC/ADP/2015/1 (25 February 2015), paras 40 and 116.5.

[39] Paris Agreement Article 4(4).

[40] See Holz et al. 2018.

diligence obligation to exercise its regulatory power in order to put international shipping on a Paris-aligned trajectory.[41] This argument is backed by the preventive and precautionary principles embedded in the UNFCCC framework.[42] These principles permit (or, according to some, necessitate) action to prevent a risk of 'serious or irreversible damage', even when such risk is not scientifically certain.[43] The current state of scientific knowledge indicates with high confidence that human-induced climate change is the main driver of increasingly frequent and intense extreme weather events globally.[44] As international maritime emissions contribute a significant share to the concentration of GHGs in the atmosphere, the above suggests that international climate law allows (if not obliges) the EU to adopt regulations for their mitigation.

#### 9.2.2.2 Impacts on Vulnerable States

In light of the CBDR-RC principle, Article 4(15) of the Paris Agreement explicitly requires states that enact climate mitigation ('response') measures to 'take into consideration ... the concerns of Parties with economies most affected by the impacts of response measures, particularly developing country Parties'.[45] This provision imposes an obligation of conduct, rather than of result, as it does not require parties to prevent or alleviate potential impacts of measures on developing countries, but only to 'take their concerns into consideration'. This wording reflects a compromise of the historical dispute within the UNFCCC negotiations on whether developing countries should be supported, or even compensated, for the economic impacts they may suffer as a result of developed countries' climate policies.[46] The latter was hesitant to acknowledge a general right of developing countries to financial support or compensation for such potential economic impacts, as it could lead to disproportional liabilities—for instance, requests to compensate the losses of oil-exporting developing countries due to decreased fossil fuel demand, caused by developed countries' stringent climate policies.

In 2018, the UNFCCC parties established a forum dedicated to fleshing out how negative economic consequences of response measures can be minimized for developing countries, while maximizing potential co-benefits.[47] So far, the work program, a minimum denominator of different political positions, has been framed mainly

---

[41] See, e.g., Peel 2017, pp. 1033–1036; Dobson and Ryngaert 2017, pp. 304–305.

[42] UNFCCC Article 3(3). See also Rio Declaration on Environment and Development in 'Report of the United Nations Conference on Environment and Development' UN Doc A/CONF.151/26 (vol I) (12 August 1992) (Rio Declaration) Principle 15; International Law Association, 'Resolution 2/2014, Declaration of Legal Principles Relating to Climate Change' (adopted 11 April 2014) Annex: Draft Articles, Article 7(A)2.

[43] See, e.g., Dobson 2018.

[44] IPCC 2021, para A3.

[45] Similarly, see UNFCCC Articles 4(8) and 4(10); Kyoto Protocol Articles 2(3) and 3(14).

[46] Chan 2016.

[47] UNFCCC 'Decision 7/CP.24, Modalities, Work Programme and Functions under the Convention of the Forum on the Impact of the Implementation of Response Measures' UN Doc FCCC/CP/

around capacity-building and the exchange of best practices for impact assessment exercises, as well as for the implementation of economic diversification and just transition policies in developing countries.[48] This type of support refers specifically to the impacts of climate response measures and is distinct from the hard obligation imposed on developed countries to financially assist developing ones with respect to climate mitigation and adaptation.[49]

Before proposing the expansion of the EU ETS to international shipping emissions, the European Commission assessed potential socioeconomic impacts at a global level and found that the scheme's effect on traded commodity prices would generally be negligible, as only goods with a very low weight-to-value ratio may be noticeably affected.[50] Nevertheless, this relatively small impact may have considerable consequences for certain developing countries, particularly the Least Developed Countries (LDCs) and Small Island Developing States (SIDS), that are economically dependent on EU imports and exports.[51] Relevant studies conducted by CE Delft and others had reached similar findings.[52] In the accompanying Impact Assessment, the Commission even acknowledged as a 'significant impact' that '[d]epending on its geographical scope, an extension of EU climate policies to maritime transport could impact third countries and trade and investment flows'.[53]

Despite being aware of these aspects, the Commission did not indicate how it intends to alleviate possible negative economic impacts of the scheme on affected countries. While no treaty provision imposes a hard obligation on the EU to do so, it could be argued that it did not sufficiently demonstrate that it took into consideration the concerns of vulnerable states, as established in Article 4(15) of the Paris Agreement, and particularly LDCs and SIDS that may be disproportionately affected by the extension of the EU ETS. This contention can be underpinned by the exceptional position that LDCs and SIDS hold within the UNFCCC regime. The fact that multiple provisions explicitly require consideration of their special circumstances could signify a special duty of care owed to LDCs and SIDS, particularly by developed countries, pursuant to the CBDR-RC principle.[54] Section 4 below examines how possible impacts of the scheme on vulnerable countries could be addressed in a non-discriminatory manner. Before doing so, it is first necessary to map the applicable jurisdictional boundaries in light of the law of the sea and general international law.

---

2018/10/Add.1 (15 December 2018). Co-benefits of the EU maritime climate policy could include, for example, an increase of ships' fuel-efficiency globally. See Bradford 2012, p. 31.

[48] UNFCCC Secretariat 2018, pp. 5–6; Chan 2016, p. 234.

[49] Paris Agreement Article 9(1).

[50] EU ETS Revision Impact Assessment Part 4, pp. 42–43.

[51] Ibid., pp. 42, 46.

[52] CE Delft et al. 2021, p. 16. See also Faber et al. 2009, pp. 25, 298–304.

[53] EU ETS Revision Impact Assessment Executive Summary, p. 3.

[54] See UNFCCC Articles 3(2), 4(8) and 4(9); Paris Agreement Articles 4(6), 9(4), 9(9), 11(1) and 13(3).

## 9.3 Jurisdiction for Expanding the EU ETS to International Shipping

### 9.3.1 *Law of the Sea*

The United Nations Convention on the Law of the Sea (UNCLOS) has been ratified both by the EU and by its Member States ('mixed agreement'), while most of its provisions reflect customary international law.[55] Often identified as a 'constitution for the oceans', the UNCLOS delimits states' maritime jurisdiction, and hence their capacity to prescribe and enforce their laws upon different actors in different areas of the sea. Since this capacity ultimately stems from the principle of sovereignty, it is severely limited with regard to activities that take place on the high seas, which is a maritime area that is subject to no state's sovereignty.[56] By contrast, the UNCLOS allows states to exercise broad jurisdictional powers over their ports and internal waters, which, similarly to land, are subject to the coastal state's full sovereignty.[57] As a result, it is widely accepted that customary international law does not recognize a general right of vessels to access ports.[58] States have a broad right to set conditions for entry into their harbors,[59] and by deciding to call at a port, ships voluntarily subject themselves to the applicable entry requirements.[60]

There are many examples of national or regional environmental regulations imposing port entry requirements that go beyond agreed IMO standards.[61] For instance, following the disastrous Exxon Valdez (1989) and Prestige (2002) oil spills, the United States and the European Community respectively restricted single-hull tankers from accessing their ports, regardless of flag.[62] The 'flag-neutral' application of port entry conditions is required by the UNCLOS, which prohibits all forms of discrimination against foreign ships.[63] This requirement serves the historical need

---

[55] United Nations Convention on the Law of the Sea (adopted 10 December 1982, entered into force 16 November 1994) 1833 UNTS 3 (UNCLOS). Pursuant to UNCLOS Article 1(2)(2) and Annex IX, Article 4(3), the rights and obligations of States under the treaty apply *mutatis mutandis* to the EU. See Paasivirta 2015, pp. 1046–1050. Regarding mixed agreements in EU external relations law, see Larik and Wessel 2020.

[56] UNCLOS Articles 87, 89 and 92.

[57] Ibid., Articles 25(2) and 211(3); *Case concerning Military and Paramilitary Activities in and against Nicaragua (Nicaragua v United States of America)* (Judgment) [1984] ICJ Rep 392, para 213. See Molenaar 2007, p. 227.

[58] Lowe 1977, p. 598; De La Fayette 1996, p. 1; Ryngaert and Ringbom 2016, p. 382.

[59] UNCLOS Articles 25(2) and 211(3).

[60] Crawford 2019, p. 449.

[61] See Molenaar 2007, pp. 231–232; Rodriguez 2012, pp. 40–41.

[62] Boyle 2006, p. 24.

[63] UNCLOS Article 227, but also Articles 24(1)(b), 25(3), 42(2), 52(2) and 234. UNCLOS imposes additional limitations on the exercise of port state jurisdiction, such as the requirements of good faith and non-abuse of rights, which are not analyzed here further. See Kotzampasakis 2023, pp. 34–35.

of global shipping for a regulatory environment that facilitates freedom of communications and unhindered trade.[64] Both the maritime EU ETS and FuelEU Maritime comply with the non-discrimination principle, as they apply equally to all ships, regardless of their country of registration (flag) or the nationality of their owners, managers, operators, or charterers.

In order to assess whether the EU acts within its jurisdictional boundaries, a distinction should be made between the prescriptive and the enforcement elements of port state jurisdiction. Prescriptive jurisdiction refers to 'the authority of a state to make the law applicable to persons, property, or conduct', while enforcement jurisdiction refers to 'the authority of a state to exercise its power to compel compliance with the law'.[65] Since the EU intends to enforce its maritime climate regulations within European ports, what merits further attention here is the permissible geographical radius of *prescriptive* port state jurisdiction.[66] For example, can a state's port entry requirements relate to ships' conduct on the high seas? This is a relevant question, as the extended EU ETS would partially cover emissions that occur beyond European territorial waters.

UNCLOS does not provide a definitive answer to this question. It clearly allows states to prescribe rules with regard to conduct in their internal waters, which are jurisdictionally assimilated to land, and in their territorial sea, without prejudice to the right of innocent passage.[67] However, from the exclusive economic zone (EEZ) and beyond,[68] UNCLOS deals with prescriptive port state jurisdiction only in connection to enforcement. While it does not preclude states from establishing port entry conditions in relation to ships' conduct beyond their territorial sea, it prevents them from undertaking in-port investigations and instituting proceedings related to extraterritorial vessel-source pollution, unless a breach of 'generally accepted international rules and standards' is suspected.[69]

Accordingly, States retain their broad right to altogether refuse port access to ships that do not comply with the prescribed entry conditions. This is perhaps counterintuitive, as a denial of entry can have more onerous economic implications for a shipping company than investigations and proceedings.[70] However, a denial of entry can be justified by the fact that it does not take away a benefit that ships are entitled to under international law, whereas investigations and proceedings require

[64] See Convention and Statute of the International Regime of Maritime Ports (adopted 9 December 1923, entered into force 26 July 1926) 58 LNTS 285, preamble.

[65] See Shaw 2017, p. 483; Crawford 2019, p. 440. The definitions above are excerpted from American Law Institute 2018, Sections 101(a) and 101(c).

[66] De Baere and Ryngaert 2013, p. 406.

[67] UNCLOS Articles 21(2), 24(1) and 211(4). Ringbom 2011, pp. 620, 623; McDorman 1997, p. 308.

[68] The EEZ is an area beyond and adjacent to the territorial sea in which the coastal state has sovereign rights and jurisdiction with regard to the purposes listed in UNCLOS Article 56.

[69] UNCLOS Articles 211(5)-(6), 218(1) and 220(1), in light of Articles 25(2), 211(3) and 230(1). See analysis in Kotzampasakis 2023, p. 33. See also Molenaar 2007, p. 237 and Ringbom 2011, pp. 626–627.

[70] Ryngaert and Ringbom 2016, p. 385.

stronger justification grounds, as they can lead to punitive actions, such as fines, cargo confiscation, or even vessel detention.[71]

In any case, the UNCLOS framework does not exhaustively determine the permissible reach of extraterritorial prescriptive port state jurisdiction.[72] The treaty's preamble explicitly confirms that 'matters not regulated by this Convention continue to be governed by the rules and principles of general international law'.[73] Therefore, in the absence of a *lex specialis*, the EU's jurisdiction to extend the EU ETS to international shipping needs to be assessed in light of general international law, which complements UNCLOS, and specifically in light of customary international law of state jurisdiction.

### *9.3.2 Customary International Law of State Jurisdiction*

Traditionally, States justify the exercise of prescriptive jurisdiction beyond their territory on the basis of one of the commonly accepted doctrines of extraterritorial jurisdiction: nationality, protection, effects, or universality.[74] The application of these customary jurisdictional bases is often challenging, due to a lack of consistent State practice and due to conflicting opinions about their status, scope, and interrelation.[75] This was also the case with the extension of the EU ETS to aviation, which triggered a debate on whether the full inclusion of intercontinental flights had a legitimate jurisdictional basis—and whether it constituted an extraterritorial exercise of prescriptive jurisdiction in the first place. From an EU law perspective, the dispute was settled in 2011, when the CJEU held that the EU ETS for aviation was not an extraterritorial measure but a purely territorial one, as it would apply only to planes physically located at European airports, subject to the full sovereignty of the respective State.[76]

Nevertheless, the Court's rationale has been criticized on the basis that, even if the measure is enforced territorially, the pricing of emissions beyond EEA territory still retains a *prescriptive* extraterritorial element that needs to be justified under international law.[77] On that premise, it has been considered whether the EU's jurisdiction to extend its ETS to intercontinental shipping could be based on the effects doctrine, which can justify the extraterritorial regulation of activities that produce (substantial) negative effects within the regulating state's territory.[78] This can arguably be the case for international maritime emissions, as they contribute significantly to

---

[71] Molenaar 2007, p. 229.

[72] Ringbom 2011, p. 628; Dobson and Ryngaert 2017, p. 299.

[73] See also UNCLOS Article 293(1).

[74] For definitions, see Crawford 2019, pp. 443–454.

[75] Ringbom 2011, p. 629; Kamminga 2008, p. 16.

[76] Case C-366/10, *Air Transport Association of America and Others* (*ATAA*), ECR I-13755, paras 124–130.

[77] Dobson and Ryngaert 2017, pp. 307–308; Ringbom 2011, p. 626.

[78] See Dominioni et al. 2018, p. 147.

climate change, whose catastrophic effects are increasingly felt throughout the globe, including Europe. Nevertheless, the effects doctrine has been mainly developed and applied in the area of anti-trust law. A strict application of the doctrine would require the EU to showcase a direct causal link between the maritime emissions to be covered and the damage inflicted upon its territory, which is scientifically difficult to establish due to the cumulative manner in which GHGs impact the Earth's climate system. It is apparent that the contended boundaries of traditional jurisdictional doctrines become even blurrier when confronted with the relatively novel, global, and complex threat of climate change.

To overcome this problem, the literature has resorted to a simplified, single jurisdictional requirement of a 'substantial and genuine connection' between the regulating state and the regulated subject matter.[79] This approach aims to strike a balance between competing jurisdictional claims, on the basis of good faith and reasonableness. While attempts have been made to analyze this 'balancing' or 'reasonableness' test into specific criteria, in the absence of a commonly agreed framework, jurisdictional conflicts tend to be resolved in state practice through political compromise.[80] The 'fifty-fifty' scope for intercontinental voyages could be perceived as a tacit political compromise of this type. As a 'half-step back' compared to the unsuccessful attempt to regulate emissions from all extra-EEA flights, it may signal the EU's effort to formulate a more balanced, politically acceptable solution for reducing shipping emissions.[81] In the meantime, the successful implementation of the EU MRV Shipping Regulation since 2018 has prepared the ground for the maritime ETS and FuelEU Maritime Regulation. Such instances of state practice might signify or catalyze normative developments with regard to the regulation of international maritime GHG emissions by port States.[82]

In light of the above, the breadth of port state jurisdiction, in combination with the legal duty of developed countries to lead action against climate change pursuant to the CBDR-RC, prevention, and precautionary principles, would support the existence of a substantial and genuine connection between the EU and its 50% share of intercontinental maritime emissions. The underlying basis beneath this connection could eventually be 'a novel ground of jurisdiction aimed at the protection of global public goods that are insufficiently protected by international solutions', as described by De Baere and Ryngaert.[83] The permissibility of safeguarding international community interests against common concerns is also the rationale behind the universality doctrine, mainly applied in the area of global criminal law.[84] In the

[79] Mann 1964, pp. 43–51; Crawford 2019, p. 461; Ringbom 2011, p. 632.

[80] International Law Association 1996, p. 523; De Baere and Ryngaert 2013, p. 402.

[81] The idea of a 'fifty-fifty' compromise approach in international climate policy can be traced back to 1999, when the EU, distrustful of the flexible emissions trading mechanisms of the Kyoto Protocol, had (unsuccessfully) proposed to quantitatively limit their use to 50%. See Woerdman 2004, p. 199.

[82] Whether 'unilateral' approaches crystallize as treaty or customary norms crucially depends on the reaction of other states, or lack thereof. Boyle 2006, p. 19.

[83] De Baere and Ryngaert 2013, p. 401.

[84] International Bar Association 2009, pp. 151–152.

absence of adequate international action on shipping emissions, the EU legislates as a custodian of a commonly agreed objective enshrined in treaties with virtually universal participation, namely the UNFCCC and the Paris Agreement.[85] The EU's exercise of jurisdiction is complemented by a strong territorial link, discerned from the application of the EU ETS and the FuelEU Maritime Regulation as port entry requirements and, consequently, the voluntary presence of the regulated entities (ships) within EU sovereign territory (ports). The fact that more than half of the ships performing intercontinental voyages are owned by EU-based entities[86] further strengthens the 'substantial and genuine connection' between the EU and 50% of its intercontinental maritime emissions.

Lastly, the establishment of a 'substantial and genuine connection' requires the consideration of competing jurisdictional claims by other states on the basis of comity and reasonableness.[87] While no such contradictory claims have been submitted to date, in the future other port states or flag states may be willing to enact regulations for reducing emissions from international shipping, in an attempt to reach Paris Agreement goals, like the EU does. An advantage of the 'fifty-fifty' scope is that it *a priori* administers the possibility of jurisdictional conflicts by providing non-European coastal states with equal jurisdictional space to regulate their corresponding 50% share of intercontinental emissions.[88]

The 'fifty-fifty' apportionment of jurisdictions is not an unknown concept in international law. In maritime delimitation, the border between the territorial seas of two opposite or adjacent coastal states is set at the median point between the respective baselines, unless agreed otherwise.[89] This even split reflects the principle of sovereign equality among states, which encompasses the notion that all states are equal before the law and have reciprocal obligations ('formal equality').[90] Nevertheless, particularly in the area of international environmental law, a gradual recognition of the structural inequalities between states has signaled departures from traditional 'formal equality' approaches to more Aristotelian perceptions of 'substantive equality', according to which the treatment of dissimilar situations should take existing differences into account.[91] The following section takes a closer look at

[85] UNFCCC Article 2; Paris Agreement Article 2(1). This hybrid jurisdictional basis can also contain elements of the protection doctrine, to the extent that the EU adopts measures to protect its own shipping industry and trade interests from ETS evasion strategies. See De Baere and Ryngaert 2013, pp. 401–402.

[86] See Commission (EU) '2020 Annual Report on CO2 Emissions from Maritime Transport' (Staff Working Document) SWD(2021) 228 final, 17 August 2021, p. 6.

[87] Crawford 2019, p. 441; International Bar Association 2009, p. 24.

[88] Similarly, Heine et al. 2015, p. 14, suggest that 'the EU only has a legitimate claim on half of the tax base', with each respective departure/destination state having the 'right to claim the other half'.

[89] UNCLOS Article 15.

[90] Simpson 2006, pp. 42–45.

[91] See Cullet 2006, pp. 307–309.

how this concept of 'formal equality', enshrined in the UNCLOS principle of non-discrimination, can be reconciled with the 'substantive equality' requirements, and thus the need for differential treatment, under the CBDR-RC principle of international climate law.

## 9.4 Reconciling Non-discrimination with the Need for Differential Treatment

The tension between the principles of non-discrimination and CBDR-RC has been documented as a major barrier to market-based emissions reduction measures for international shipping.[92] Many discussions at the IMO have focused on how the need for a uniformly applicable regulation can be reconciled with the need for differentiation that reflects developing countries' unequal historical responsibilities and economic capabilities regarding climate change. While the IMO GHG Strategy recognized 'the need to be cognizant of' the CBDR-RC principle,[93] so far there has been no agreement on how it should be operationalized.[94] Targeted financial support to developing countries, in particular SIDS and LDCs, possibly through redistribution of carbon pricing revenues, is an option supported by several delegations.[95] Yet, disagreement persists on the desirable scope and operation of such a financial mechanism, but also on whether exemptions for specific routes, trades, or regions should be introduced as an additional protective layer for vulnerable countries.

As IMO negotiations on market-based measures continue to progress slowly, the expansion of the EU ETS comes as a bottom-up effort to accelerate the alignment of international maritime emissions with the Paris Agreement goals. To that end, the EU places a share of international shipping under the absolute emissions cap of the EU ETS, while exerting political pressure for an agreement on an effective scheme with global coverage at the IMO.[96] The flag-neutral design of the maritime EU ETS, in the absence of nationality-based exemptions, ensures compliance with the UNCLOS principle of non-discrimination. At the same time, international climate law allows (or arguably even obliges) the EU to legislate for the reduction of maritime emissions, as long as it does so in a manner consistent with the CBDR-RC principle, which includes the consideration of impacts on vulnerable states.[97]

[92] See Psaraftis 2012, p. 231; Wu et al. 2022, p. 9.

[93] Initial IMO GHG Strategy, para 4.13. The same wording was preserved in the revised 2023 Strategy.

[94] See Kopela 2020, pp. 146–147.

[95] See IMO, 'Report of the Marine Environment Protection Committee on its Seventy-Eighth Session' MEPC 78/17 (24 June 2022) (IMO MEPC 78/17), paras 7.35–7.45.

[96] Heine et al. 2015, p. 44.

[97] Paris Agreement Article 4(15). The precise content and normative nature of CBDR-RC remain contested, both in relation to shipping and more generally. See IMO MEPC 78/17, paras 7.38–7.43; Scott and Rajamani 2012, pp. 477–479.

Looking back at the EU ETS aviation case, the Commission would consider exempting—from the scheme's broad coverage—flights from third countries, if they adopted regulations with 'an environmental effect at least equivalent to' the EU ETS.[98] Scott and Rajamani argued that by making the scope contingent on a uniform requirement of 'equivalent stringency', the EU disregarded the CBDR-RC principle.[99] The 'fifty-fifty' scope for intercontinental maritime transport may reflect the EU's effort to circumvent this problem by unconditionally ensuring that third countries are free to adopt climate measures in line with their self-perceived differentiated responsibilities and capabilities, as defined in their NDCs, with regard to their 50% share of shipping emissions from trade with Europe.[100] This is possibly what is implied with the remark that the 'fifty-fifty' scope is 'a practical way to solve the issue of' CBDR-RC.[101]

Yet, while the revised EU ETS Directive provides exemptions to alleviate possible socioeconomic impacts of maritime carbon pricing on certain regions of the EU,[102] the concerns of third states in similar or less advantageous positions have not been sufficiently considered. By applying a one-size-fits-all rule for intercontinental voyages, the EU still seems to disregard the diverse effects of its measure on different countries, especially vulnerable states that depend on maritime trade with Europe. The above calls for a more careful consideration of how elements of differentiation can be introduced to the maritime EU ETS, in line with the CBDR-RC principle, while preserving compatibility with the non-discrimination obligation under UNCLOS.

One way to operationalize the CBDR-RC in the maritime EU ETS would be to provide exemptions or uneven splits of regulatory scopes for routes between Europe and certain developing countries. For instance, the scope of the EU ETS could altogether exempt routes between Europe and Senegal, an LDC largely dependent on EU imports,[103] or alternatively be limited to e.g. 25%, so that Senegal can charge a carbon price on three-quarters of the emissions from its maritime trade with Europe. Additional income from emissions allowance auctions or carbon taxation could be used by the Senegalese state to alleviate the potential negative impacts of higher shipping costs on society or to finance climate mitigation and adaptation measures. Also, a relatively low carbon price applied by Senegal to emissions from maritime transport could contain an otherwise excessive increase in shipping costs (and thus the price of goods) for Senegalese households and industries.

---

[98] Directive 2008/101/EC of the European Parliament and of the Council of 19 November 2008 amending Directive 2003/87/EC so as to include aviation activities in the scheme for greenhouse gas emission allowance trading within the Community [2009] OJ L8/3, recital 17. See Scott and Rajamani 2012, p. 482. See also Martinez Romera and van Asselt 2015, pp. 277–278.

[99] Scott and Rajamani 2012, p. 483.

[100] This approach perhaps echoes the more flexible, bottom-up framework of the Paris Agreement in relation to differentiation. See Martinez Romera 2016, pp. 222–223.

[101] Directive (EU) 2023/959, recital 20.

[102] Ibid., notably, recitals 21 and 26.

[103] EU ETS Revision Impact Assessment Part 4, p. 45. Forty percent of Senegal's imports come from the EU.

Such route-based deviations would not infringe the UNCLOS principle of non-discrimination, as long as they apply equally to vessels of all nationalities. Nevertheless, varying the scope of GHGs covered under the EU ETS based on the place of departure or destination may raise problems under WTO law, as a potential violation of the most-favored-nation principle.[104] Moreover, a disadvantage of this approach is that it can incentivize the concentration of relatively old and inefficient ships on the (partially) exempted routes, which can exacerbate local pollution problems and create competitive distortions.[105] The incentive for investment in zero-carbon fuels and infrastructure in the respective port states would also be diluted.[106] Lastly, creating a patchwork of carve-outs or asymmetrical regulatory scopes would only further complicate an already convoluted regulatory landscape for international shipping.

A second, and probably preferable, way to operationalize CBDR-RC would be to use a share of maritime EU ETS revenues for providing assistance to the developing countries that are projected to be impacted by the scheme.[107] The EU could set up a support framework that further assesses the potential socioeconomic impacts of the scheme on developing countries, with a special focus on vulnerable LDCs and SIDS, and improves their capacity to maximize the positive and minimize the negative impacts, in line with the UNFCCC guidance documents.[108] While many (combinations of) different criteria and metrics could be used for allocating the revenues, this is bound to be a controversial exercise, prone to the risk of political recoil. Using an international fund or financial mechanism as an intermediary is seen as a way to mitigate this risk.[109]

Importantly, EU assistance should be provided *before* potential damages are felt and should not be conditional on requirements that overlook the CBDR-RC principle. For certain vulnerable countries, financial support and technical assistance from Europe is essential for implementing climate change adaptation and mitigation measures in the maritime sector, which can increase their resilience, diversify their economies, and reduce the costs incurred by maritime carbon pricing.[110] Such a targeted support framework would not only signal that the EU takes the concerns of vulnerable third countries into consideration, but could also contribute to the distinct, central obligation of developed countries to financially assist the mitigation and adaptation efforts of developing ones, particularly LDCs and SIDS.[111]

---

[104] GATT Article I(1). See Scott and Rajamani 2012, p. 490; Bartels 2012, pp. 447–449.

[105] Dominioni and Englert 2022, p. 16.

[106] Ibid., See also CE Delft et al. 2021, p. 77.

[107] Several studies have highlighted revenue sharing as a more suitable option for aligning the EU ETS with CBDR-RC. See Scott and Rajamani 2012, p. 492; Dominioni et al. 2018, p. 155; Heine et al. 2015, pp. 52–53; CE Delft et al. 2021, p. 17; Martinez Romera and van Asselt 2015, p. 282; Dominioni 2023.

[108] See UNFCCC Secretariat 2018.

[109] See Scott and Rajamani 2012, pp. 488–492; Martinez Romera and van Asselt 2015, p. 281.

[110] UNFCCC Secretariat 2018, p. 22. See also Pauw et al. 2019, p. 4; Dominioni and Englert 2022, p. 41.

[111] Paris Agreement Articles 9(1), 9(4) and 11(3). See also Pauw et al. 2019, p. 5.

## 9.5 Conclusion

The absence of references to international law from the EU climate policy expansion to international shipping leaves ample room for legal uncertainty. This chapter has analyzed certain international law aspects of that expansion. In light of the international climate law, the law of the sea, and customary international law of state jurisdiction, it can be concluded that the EU has sufficient legal grounds for regulating 50% of GHG emissions from incoming and outgoing extra-EEA voyages. However, the EU should be more vigilant about the applicable legal limitations when expanding its climate policy. Such limitations arise from the CBDR-RC principle of international climate law, which in this context requires the EU to more carefully consider the potential economic impacts of its extended carbon pricing regulation on vulnerable countries. In practice, this can be achieved by channeling a share of the allowance auctioning revenues from international shipping to a mechanism that assists targeted developing countries, particularly LDCs, and SIDS, to alleviate possible negative economic impacts, while maximizing associated climate benefits. In order to ensure efficacy, these arrangements should be inclusive and take place before the first financial obligations on shipping companies are imposed.

An advantage of the 'fifty-fifty' scope compared to a 100% emissions coverage is that it represents a more balanced claim of prescriptive extraterritorial jurisdiction. Does this necessarily mean that including 100% of emissions from extra-European voyages in the EU ETS would constitute a breach of international law? Against the background of the IMO's inaction on market-based measures, and with the climate clock ticking, it cannot be excluded that sufficient jurisdictional bases could be found to justify such full coverage. However, if other countries decide to regulate a share of intercontinental maritime emissions, the EU should provide them with the jurisdictional space to do so, without preconditions that neglect the CBDR-RC principle. In the end, the permissible limits of prescriptive EU climate change jurisdiction for intercontinental shipping may eventually be determined by the reaction of other states to the extension of the EU ETS. To reduce the risk of international backlash like in the case of aviation, the EU legislators opted for a politically safer albeit environmentally and economically sub-optimal solution. Yet, an explicit integration of international law in EU legislation with extraterritorial elements will not only contribute to its observance and development but can also reduce legal uncertainty and anchor the international political discourse to a more defined framework.

## References

American Law Institute (2018) Restatement (Fourth) of the Foreign Relations Law of the United States. American Law Institute Publishers, Philadelphia

Bartels L (2012) The WTO Legality of the Application of the EU's Emission Trading System to Aviation. EJIL 23:429–467

Bäuerle et al (2010) Integration of Marine Transport into the European Emissions Trading System - Environmental, economic and legal analysis of different options. https://www.umweltbundesamt.de/sites/default/files/medien/461/publikationen/3942.pdf Accessed 28 December 2022

Boyle A (2006) EU Unilateralism and the Law of the Sea. IJMCL 21:15–31

Bradford A (2012) The Brussels Effect. Northwestern University Law Review 107:1–68

Chan N (2016) The "New" Impacts of the Implementation of Climate Change Response Measures. RECIEL 25:228–237

Contini D and Merico E (2021) Recent Advances in Studying Air Quality and Health Effects of Shipping Emissions. Atmosphere https://doi.org/10.3390/atmos12010092

Crawford JR (2019) Brownlie's Principles of Public International Law. OUP, Oxford

Cullet P (2006) Differential Treatment in Environmental Law: Addressing Critiques and Conceptualizing the Next Steps. TEL 5:305–328

De Baere G and Ryngaert C (2013) The ECJ's Judgment in Air Transport Association of America and the International Legal Context of the EU's Climate Change Policy. European Foreign Affairs Review 18:389–409

De La Fayette L (1996) Access to Ports in International Law. IJMCL 11:1–22

CE Delft et al (2021) Study on Assessment of Possible Global Regulatory Measures to Reduce Greenhouse Gas Emissions from International Shipping. EU Publications Office, Luxembourg

Dobson NL (2018) Exploring the Crystallization of "Climate Change Jurisdiction": A Role for Precaution?. Climate Law 8:207–228

Dobson N and Ryngaert C (2017) Provocative Climate Protection: EU "Extraterritorial" Regulation of Maritime Emissions. ICLQ 66:295–333

Dominioni G (2023) Towards an equitable transition in the decarbonization of international maritime transport: Exemptions or carbon revenues? Marine Policy 154:105669

Dominioni G and Englert D (2022) Carbon Revenues from International Shipping: Enabling an Effective and Equitable Energy Transition. World Bank, Washington

Dominioni G, Heine D and Martinez Romera B (2018) Regional Carbon Pricing for International Maritime Transport: Challenges and Opportunities for Global Geographical Coverage. Carbon and Climate Law Review 12:140–158

European Environment Agency (2021) Rail and waterborne — best for low-carbon motorised transport. https://www.eea.europa.eu/publications/rail-and-waterborne-transport Accessed 28 December 2022

Faber J et al (2009) Technical Support for European Action to reducing Greenhouse Gas Emissions from International Maritime Transport. https://cedelft.eu/publications/technical-support-for-european-action-to-reducing-greenhouse-gas-emissions-from-international-maritime-transport/ Accessed 28 December 2022

Finska L and Ringbom H (2022) Regulation of GHGs from Ships - On the available discretion for regulatory solutions in a European and Finnish perspective. https://www.centrumbalticum.org/files/5185/Regulation_of_GHGs_from_ships_final.pdf Accessed 28 December 2022

Heine D et al (2015) A Regional Solution for a Transnational Problem? A Mechanism to Unilaterally Tax Maritime Emissions While Satisfying Extraterritoriality, Tax Competition and Political Constraints. Rotterdam Institute of Law and Economics, Rotterdam

Holz C, Kartha S and Athanasiou T (2018) Fairly Sharing 1.5: National Fair Shares of a 1.5 °C-Compliant Global Mitigation Effort. International Environmental Agreements: Politics, Law and Economics 18:117–134

Intergovernmental Panel on Climate Change (IPCC) (2021) Summary for Policymakers. In: Masson-Delmotte V et al (eds), Climate Change 2021: The Physical Science Basis, Working Group I Contribution to the Sixth Assessment Report. CUP, Cambridge

International Bar Association (2009) Report of the Task Force on Extraterritorial Jurisdiction. https://www.ibanet.org/MediaHandler?id=ECF39839-A217-4B3D-8106-DAB716B34F1E Accessed 28 December 2022

International Energy Agency (IEA) (2022), International Shipping. https://www.iea.org/reports/international-shipping Accessed 6 July 2023

International Law Association (1996) Final Report of the ILA's Committee on Extra-territorial Jurisdiction, Report of the Sixty-Seventh Conference. International Law Association, London
International Maritime Organization (IMO) (2021) Fourth IMO GHG Study 2020. IMO, London
International Transport Forum (2022) Carbon Pricing in Shipping. OECD Publishing, Paris
Kamminga MT (2008) Extraterritoriality. In: Peters A and Wolfrum R (eds), Max Planck Encyclopedia of Public International Law. OUP, online
Kopela S (2020) Climate Change and the International Maritime Organization. In: McDonald J, McGee J and Barnes R (eds), Research Handbook on Climate Change, Oceans and Coasts. Edward Elgar, Cheltenham pp 134–151
Kotzampasakis M (2023) Intercontinental shipping in the European Union emissions trading system: A 'fifty–fifty' alignment with the law of the sea and international climate law?. RECIEL 32:29–43
Larik J and Wessel RA (2020) Instruments of EU External Action. In: Wessel RA and Larik J (eds), EU External Relations Law: Text, Cases and Materials. Hart Publishing, Oxford pp 101–137
Lowe AV (1977) The Right of Entry into Maritime Ports in International Law. San Diego Law Review 14:597–622
Mann FA (1964) The Doctrine of Jurisdiction in International Law. AW Sijthoff, Leiden
Martinez Romera B (2016) The Paris Agreement and the Regulation of International Bunker Fuels. RECIEL 25:215-227
Martinez Romera B and van Asselt H (2015) The International Regulation of Aviation Emissions: Putting Differential Treatment into Practice. JEL 27:259–283
McDorman TL (1997) Port State Enforcement: A Comment on Article 218 of the 1982 Law of the Sea Convention. Journal of Maritime Law and Commerce 28:305–322
Molenaar EJ (2007) Port State Jurisdiction: Toward Comprehensive, Mandatory and Global Coverage. Ocean Development and International Law 38:225–257
Paasivirta E (2015) The European Union and the United Nations Convention on the Law of the Sea. Fordham International Law Journal 38:1045–1071
Pauw P, Mbeva K and van Asselt H (2019) Subtle Differentiation of Countries' Responsibilities under the Paris Agreement. Palgrave Communications https://doi.org/10.1057/s41599-019-0298-6
Peel J (2017) Climate Change. In: Nollkaemper A and Plakokefalos I (eds), The Practice of Shared Responsibility in International Law. CUP, Cambridge pp 1009–1050
Psaraftis HN (2012) Market-Based Measures for Greenhouse Gas Emissions from Ships: A Review. WMU Journal of Maritime Affairs 11:211–232
Ringbom H (2011) Global Problem—Regional Solution? International Law Reflections on an EU CO2 Emissions Trading Scheme for Ships. IJMCL 26:613–641
Rodriguez DP (2012) The Inclusion of Shipping in the EU Emission Trading Scheme: A Legal Analysis in the Light of Public International Law. Revista Catalana de Dret Ambiental 3:1–55
Ryngaert C and Ringbom H (2016) Introduction: Port State Jurisdiction: Challenges and Potential. IJMCL 31:379–394
Scott J and Rajamani L (2012) EU Climate Change Unilateralism. EJIL 23:469–494
Shaw MN (2017) International Law. CUP, Cambridge
Simpson G (2006) Great Powers and Outlaw States: Unequal Sovereigns in the International Legal Order. CUP, Cambridge
UNFCCC Secretariat (2018) Technical Papers on the Impact of the Implementation of Response Measures – Executive Summary. https://unfccc.int/documents/201299 Accessed 28 December 2022
United Nations Conference on Trade and Development (2022) Review of Maritime Transport 2022. United Nations Publications, New York

United Nations Environment Programme (2021) Emissions Gap Report 2021: The Heat Is On. https://www.unep.org/resources/emissions-gap-report-2021 Accessed 28 December 2022
Woerdman E (2004) The Institutional Economics of Market-Based Climate Policy. Elsevier, Amsterdam
Wu M et al (2022) Carbon Emission Trading Scheme in the Shipping Sector: Drivers, Challenges, and Impacts. Marine Policy https://doi.org/10.1016/j.marpol.2022.104989

**Manolis Kotzampasakis** is a PhD candidate at the University of Groningen, Netherlands. This chapter is a shorter and updated version of Kotzampasakis 2023.

# Chapter 10
# Pushing for Sustainability Abroad: EU Environmental Unilateralism and its Legal Limits

**Birgit Hollaus and Stefan Mayr**

## Contents



**Abstract** Its Treaties require the European Union (EU) to promote sustainable development in all its policies and activities, whether internal or external. In this regard, the EU increasingly relies on unilateral measures which build on a consumption-based approach to address sustainability concerns beyond its borders. Simply put, these instruments make market access—directly or indirectly—dependent on compliance with EU sustainability rules, aiming to steer the behavior of EU and non-EU market actors and thus allowing the EU to export its view on sustainability also to the rest of the world. However, as a series of recent conflicts highlight, such environmental unilateralism often finds its limits in international (trade) law. The present Chapter analyses these tensions in the context of the ongoing palm oil dispute between the EU and Indonesia in the World Trade Organization (WTO). It

B. Hollaus (✉) · S. Mayr
Institute for Law and Governance, WU Vienna University of Economics and Business, Vienna, Austria
e-mail: birgit.hollaus@wu.ac.at

S. Mayr
e-mail: stefan.mayr@wu.ac.at



R. A. Wessel et al. (eds.), *EU External Relations Law and Sustainability*, Global Europe: Legal and Policy Issues of the EU's External Action 4,
https://doi.org/10.1007/978-94-6265-655-0_10

shows that even "indirect sustainability rules" are prone to conflict with key WTO standards and that their fate depends on their justification, which ultimately proves difficult when global concerns are addressed unilaterally.

**Keywords** Integration principle · Sustainability · Environmental unilateralism · Environmental multilateralism · Environmental diplomacy · Palm oil · WTO · Green leader

## 10.1 Introduction: Promoting Sustainability in the EU and Beyond

Its Treaties expressly mandate the European Union (EU) to work towards the sustainable development of Europe[1] as well as of the wider world.[2] These treaty objectives translate into a legal requirement for the EU to take sustainability concerns into account, whether in its internal or external action. This observation is confirmed specifically for the environmental dimension of sustainability by Article 11 TFEU,[3] the so-called integration clause, and policy-specific treaty provisions that 'echo this horizontal commitment'.[4] The promotion of sustainability is thus a paramount EU concern, irrespective of policy area and its respective external or internal focus.

The strive for environmental sustainability is particularly pronounced in the EU's external policy.[5] Over the years, the EU has visibly strengthened its external environmental action, positioning itself as a "green leader" in environmental multilateralism who pushes for more ambitious action than its peers and seeks to lead by example.[6] Today, the narrative of the EU's green leadership extends to other policy areas, in which the EU, in pursuit of its wider constitutional objectives, seeks to integrate environmental, and more broadly, sustainability concerns in its multilateral efforts.[7] However, with multilateralism in crisis,[8] the EU considers itself increasingly forced to rely on unilateral solutions to address environmental concerns abroad instead ("environmental unilateralism").[9] Indeed, much of the implementation of the EU's

---

[1] Article 3(3) Treaty on European Union, opened for signature 13 December 2007 [2016] OJ C202/13, entered into force 1 December 2009 (TEU).

[2] Article 3(5) TEU (n 1); note the critique of the concept of sustainable development by Kotzé and Adelman 2023, p. 230 ff.

[3] Treaty on the Functioning of the European Union, signed on 13 December 2007 [2016] OJ C202/47, entered into force on 1 December 2009 (TFEU).

[4] Klamert 2019, mn 3.

[5] On the different dimensions of sustainability, e.g. Purvis, Mao and Robinson 2019.

[6] Nuanced, Kulovesi 2012.

[7] For trade, e.g. European Commission 2021a.

[8] See, Politico 2023. https://www.politico.eu/article/multilateralism-broken-diplomacy-united-nations-eu-global-governance/. Accessed 7 March 2024; United Nations 2023a. https://news.un.org/en/story/2023/09/1141262. Accessed 7 March 2024.

[9] See, e.g. van Zeben 2022, para 61; Hervé 2022.

flagship initiative, the European Green Deal, includes such unilateral solutions, which typically rely on the EU's market power to also steer the behavior of non-EU market actors towards compliance with its understanding of sustainability.[10]

Factually, the EU has often been quite successful with exporting its views on different sustainability issues by means of conditioning market access,[11] even though politically, this strategy has been criticized for essentially amounting to "regulatory globalization"[12] or even "regulatory imperialism".[13] However, as the EU focuses more and more on sustainability concerns arising outside the EU territory, today's environmental unilateralism faces distinct legal challenges as the respective instruments must be compatible with both EU law and relevant international law.[14] The EU's revision process of the Renewable Energy Directive (RED)[15] in relation to biofuels and the related ongoing disputes at the World Trade Organization (WTO)[16] serve as current examples illustrating that these legal challenges are not only real, but that they are also co-determining the EU's possibilities for working towards sustainability, irrespective of the instrument it uses.

Against this background, the current chapter explores the possibilities and constraints of the EU's environmental unilateralism for the purpose of promoting sustainable development outside its territory. In doing so, the chapter first zooms-in on the rise of environmental unilateralism in the EU. It outlines how the Treaties generally enable a unilateral approach to multilateral and global challenges such as sustainability before setting out how internal instruments address sustainability concerns beyond the borders of the EU. With this in mind, the contribution focuses on WTO law as a limit to the EU's environmental unilateralism and uses the current WTO dispute with Indonesia regarding palm oil-based biofuels as an example demonstrating such limitation.

---

[10] E.g. the carbon border adjustment mechanism (CBAM), established by Regulation (EU) 2023/956 of the European Parliament and of the Council of 10 May 2023 establishing a carbon border adjustment mechanism [2023] OJ L130/52.

[11] See, e.g. Dero-Bugny and Motte-Baumvol 2021, p. 47 f. Also, Öberg 2020; Gstöhl 2008.

[12] Scott 2014a, p. 88.

[13] De Ville et al. 2023, p. 16. Also, Financial Times 2023. https://www.ft.com/content/3cad4fab-cbdd-4ddb-a5a7-846c7de9c6e2. Accessed 7 March 2024; for a more positive view, Bradford 2020.

[14] See Vranes 2009, p. 176 f.

[15] Directive 2009/28/EC of the European Parliament and of the Council of 23 April 2009 on the promotion of the use of energy from renewable sources and amending and subsequently repealing Directives 2001/77/EC and 2003/30/EC [2009] OJ L140/16 (RED I); Directive (EU) 2018/2001 of the European Parliament and of the Council of 11 December 2018 on the promotion of the use of energy from renewable sources [2018] OJ L328/82 (RED II), last amended by Directive (EU) 2023/2413 of the European Parliament and of the Council of 18 October 2023 [2023] OJ L2023/2413 (RED III).

[16] *European Union—Certain measures concerning palm oil and oil palm crop-based biofuels*, WT/DS593 and *European Union and certain Member states—Certain measures concerning palm oil and oil palm crop-based biofuels*, WT/DS600.

## 10.2 EU Environmental Unilateralism: Internal Instruments and Rules on Sustainability Abroad

By its Treaties, the EU is called upon to integrate sustainability concerns in its internal and external action. However, the division between internal and external seems to fade as the EU relies more heavily on unilateral solutions to also address sustainability concerns abroad. While the reasons for such environmental unilateralism as well as its form of implementation may differ, the common denominator is clear: the effects of EU internal instruments beyond the borders of the EU. Unsurprisingly, however, these extraterritorial effects spark (legal) controversy.

### *10.2.1 Unilateral Solutions for Sustainability Concerns Beyond EU Borders*

The Treaties expressly make sustainability a concern to be taken account of in all policy areas. As the cross-cutting integration clause stipulates: "Environmental protection requirements must be integrated into the definition and implementation of the Union's policies and activities, in particular with a view to promoting sustainable development".[17] This provision can already be read to include both the internal and external dimensions of these substantive policy areas.[18] Such an understanding is confirmed by Article 21(3) TEU, which provides that the requirement to integrate sustainability concerns extends to the external aspects of *all* substantive policy areas,[19] be that the EU's external environmental policy or its common commercial policy. What Article 21(3) TEU does not clarify though, is how the EU should bring about sustainable development beyond its borders. As the example of the EU's environmental policy shows, solutions can be of a multi-, bi- or unilateral nature.

As part of what is deemed an ambitious environmental mandate,[20] Article 191 TFEU mentions the promotion of measures at the international level for dealing with worldwide or regional environmental pressures;[21] it is such environmental pressures that the Treaties most clearly understand as core sustainability concerns.[22] In order to pursue this mandate, the EU, necessarily, has the competence to act at the international level.[23] The EU can thus cooperate with third states as well as international organizations to develop solutions for common environmental concerns, or

---

[17] Article 11 TFEU (n 3).

[18] Marín Durán and Morgera 2012, p. 30.

[19] See Oeter 2013, mn 38, 43.

[20] See Morgera 2013.

[21] Article 191(1) 4th indent TFEU (n 3).

[22] E.g. Article 21(2)(f) TEU (n 1), Article 11 TFEU (n 3), both connecting environmental concerns to sustainable development. For the different dimensions of sustainability see above, n 5.

[23] The EU disposes of an explicit external environmental competence, based on Article 191(4) TFEU (n 3), which, however, is limited to mere cooperation. Its implicit external environmental

take part in efforts to that effect under the auspices of international organizations, for instance in the negotiations of multilateral agreements, the main instrument of international environmental cooperation.[24] However, in these international settings, bi- and specifically multilateral solutions are increasingly harder to achieve.[25] The price for reaching an agreement is often that the level of ambition falls victim to political concessions.[26]

Faced with a trade-off between ambition and pragmatism, unilateral solutions, i.e. solutions devised and implemented by the EU alone,[27] seem to become an increasingly attractive alternative for the EU to pursue its environmental goals.[28] At first glance, Article 191 TFEU may seem to stand in the way of such environmental unilateralism. However, while the provision may express a preference for multilateral solutions in face of global environmental pressures,[29] its wording ('promote') does not suggest that the EU is necessarily limited to bi- or multilateral solutions. Such a narrow understanding of Article 191 TFEU would ultimately be detrimental to the ambitious environmental mandate set out in the very same provision, particularly in light of the earlier mentioned factual and political difficulties arising in the context of cross-border environmental concerns. Consequently, the EU, as part of its external environmental policy, may principally pursue unilateral solutions to regional or worldwide environmental concerns. To this end, the EU can use internal instruments, i.e. instruments which are used for regulating internal policy matters, also in the context of EU external relations.[30] This view is well-grounded in the wording of Article 288 TFEU, which sets out different legal instruments at the EU's disposal in 'exerci[zing] [its] competence'. Accordingly, the provision is not limited to the exercise of internal competences but can extend to external competences,[31] whether explicit or implicit, as in the case of the external environmental competence.[32]

In practice, the use of internal instruments to devise unilateral solutions regarding sustainability concerns abroad is indeed on the rise. One factor that contributes to this development is the above-mentioned mounting difficulties in international cooperation. While the complexities of the subject, (historical) responsibility and (financial) capabilities have been core issues hampering multilateral negotiations in

---

competence flows from Article 192(1) TFEU (n 3) and is more encompassing, as argued by the Court, CJEU, *Cartagena Protocol*, Opinion, 6 December 2001, 2/00, para 43.

[24] Bruch and Broderick 2016, p. 36 ff.

[25] De Ville et al. 2023, p. 21 ff. Illustratively, e.g. Euractiv 2023. https://www.euractiv.com/section/economy-jobs/news/eu-australia-fail-to-agree-on-trade-deal. Accessed 7 March 2024.

[26] Scott 2020, p. 13.

[27] Note Boisson de Chazournes 2000, p. 316 f.

[28] See Bodansky 2000, p. 343 ff.

[29] Similarly, Article 21(2)(f) TEU (n 1) 'help develop' and Article 21(2)(h) TEU (n 1) "promote", for which Ramopoulos 2019, mn 7 highlights how Article 21(2) TEU reflects 'the Union's preference for a multilateral approach' to realising its objectives.

[30] Larik and Wessel 2020, p. 103.

[31] On the issue of addressing extraterritorial concerns by means of internal instruments, see below, Sect. 10.2.2.

[32] See above, n 23.

the past,[33] the recent geo-political developments—think multipolarity—and tensions are expected to impact more broadly on multilateralism and the quest for multilateral solutions to sustainability concerns.[34] In the past, stand-stills in multilateral negotiations have already led the EU to resort to unilateral solutions instead. In 2008, for example, following the continued delay to secure a global consensus on the issue of greenhouse gas emissions from international aviation, the EU decided to include such emissions by EU-administered aviation operators in its EU emissions trading scheme.[35] Even though this inclusion was later reversed following political and legal battles,[36] the reversal was made contingent on finding a multilateral solution.[37] Such a solution was ultimately found within the International Civil Aviation Organization (ICAO) by way of the compensation mechanism CORSIA (Carbon Offsetting and Reduction Scheme for International Aviation).[38]

Another factor that feeds into the reliance on internal instruments to address sustainability concerns abroad is the increasing awareness of the EU's own role in exacerbating the underlying issues.[39] Indeed, data demonstrates that EU consumption of raw materials and (consumer) goods is tied to negative effects on the environment outside its borders.[40] While this is not unique to the EU,[41] these effects are of a devastating scale,[42] which the EU cannot ignore—at least not morally or politically.[43] In the absence of a cooperative solution with the affected producer country or countries,

[33] Further, Zürn 2020, p. 119 ff.

[34] See e.g. United Nations 2023b. https://press.un.org/en/2023/gadis3727.doc.htm. Accessed 7 March 2024. EU Observer 2023. https://euobserver.com/green-economy/157630. Accessed 7 March 2024.

[35] Directive 2008/101/EC of the European Parliament and of the Council of 19 November 2008 amending Directive 2003/87/EC so as to include aviation activities in the scheme for greenhouse gas emission allowance trading within the Community [2008] OJ L8/3.

[36] See e.g. Denza 2012; Gattini 2012.

[37] Article 5 Decision No 377/2013/EU of the European Parliament and of the Council of 24 April 2013 derogating temporarily from Directive 2003/87/EC establishing a scheme for greenhouse gas emission allowance trading within the Community [2013] OJ L113/1; Article 28a(8) Regulation (EU) 421/2014 of the European Parliament and of the Council of 16 April 2014, of 16 April 2014 amending Directive 2003/87/EC establishing a scheme for greenhouse gas emission allowance trading within the Community, in view of the implementation by 2020 of an international agreement applying a single global market-based measure to international aviation emissions [2014] OJ L129/1.

[38] Article 28b Directive (EU) 2023/958 of the European Parliament and the Council of 10 May 2023, amending Directive 2003/87/EC as regards aviation's contribution to the Union's economy-wide emission reduction target and the appropriate implementation of a global market-based measure [2023] OJ L113/115.

[39] See Scott 2020.

[40] See e.g. Galli 2023.

[41] See WWF 2022, p. 66 f.

[42] See Scott 2019, p. 47 ff. Exact numbers may vary depending on the methodology used, see Tukker et al 2016, p. 172 f.

[43] Scott 2020, p. 15. Also legally speaking, consumption-related sustainability concerns abroad certainly fall within the ambit of the EU's mandate to work towards sustainability and should thus be understood as a clear responsibility of the EU.

unilateral measures in this area will usually have a strong market- and/or trade-component as they seek to reduce the ecological footprint of (consumer) products or raw materials being imported into the EU.[44] The new carbon border adjustment mechanism (CBAM),[45] for example, which supplements the EU's (internal) emission trading system has such a market component, as it requires EU importers of certain emission-intensive materials to buy emission certificates, equivalent to the amount of emissions generated during production, prior to their import.[46]

### *10.2.2 Internal Instruments and Rules on Sustainability Abroad*

Just as the reasons for the rise of unilateral solutions of the EU are different, so is the shape these solutions ultimately take. Whereas some unilateral solutions are modelled after international, not (yet) agreed solutions, others mirror existing, purely internal solutions while yet again others are uniquely devised. This diversity can also be seen in connection with rules incorporated in internal instruments that seek to specifically address sustainability concerns abroad ("sustainability rules"). What ties such rules together despite their diversity is that they usually build on the EU's market power to change and steer the behavior of non-EU market participants towards sustainability. To do so, such sustainability rules subject market access either directly or indirectly to compliance with EU sustainability criteria.[47] These differences consequently allow us to distinguish between direct and indirect sustainability rules. The former makes access to the EU market directly dependent on compliance with certain EU sustainability criteria ("direct sustainability rules"). The latter impose no direct restrictions on market access, but still have the potential to modify the conditions of competition in the relevant market to the detriment of imported products ("indirect sustainability rules"). Two examples shall illustrate this distinction.

#### 10.2.2.1 "Direct Sustainability Rules": The Case of the Anti-Deforestation Regulation

Preventing deforestation abroad has been on the EU's policy agenda since the 1990s. Its efforts in this area rely on different approaches, from engaging in partnerships with countries to support forest governance abroad,[48] to including sustainability chapters

---

[44] See Nessel 2023.

[45] Regulation (EU) 2023/956 of the European Parliament and of the Council of 10 May 2023 establishing a carbon border adjustment mechanism [2023] OJ L130/52.

[46] On the CBAM, see Patrick Abel in this volume.

[47] See Mignolli 2018, p. 245 f. Pointing out the decrease in the EU's market power—and concomitant normative power Gáspár-Szilágy 2022, p. 54.

[48] European Commission 2003.

in free trade agreements[49] and providing development assistance. However, starting with the EU's Timber Regulation,[50] also unilateral measures addressing international deforestation have become a firm part of the policy mix. The recently adopted Anti-Deforestation Regulation,[51] repealing the EU Timber Regulation,[52] once more relies on a unilateral approach, with its core rules making market access directly dependent on compliance with EU sustainability criteria. The Anti-Deforestation Regulation thus serves as a prime example for direct sustainability rules.

In 2008, the Commission set an EU objective to reduce gross tropical deforestation by 50% by 2020, and to halt global forest cover loss by 2030.[53] Within the framework of the Sustainable Development Goals (SDGs), also endorsed by the EU, that target was raised in 2015: deforestation globally should be halted by 2020.[54] Yet, with the deadline approaching, it became clear that the EU's objective could not be achieved with the existing efforts, let alone the SDG target.[55] In a 2019 communication, the Commission thus announced to take further steps to protect the world's forests, amongst others, by reducing the EU consumption footprint on land and encouraging the consumption of products from deforestation-free supply chains in the EU.[56] A later European Parliament resolution highlighted why this consumption-based approach was indeed a necessary step. In plain speech, the European Parliament found that the EU is 'indirectly involved in the deforestation and degradation of world forests [...] through the consumption of commodities associated with deforestation [in their countries of origin].'[57] And that consumption is not negligible, amounting in 2020 to about 10% of the global consumption in such commodities, not even counting their mere import into the EU.[58]

Building on the approach already taken in the EU Timber Regulation, the Commission published its proposal for a regulation to ensure deforestation-free supply chains

[49] On sustainability chapters, see Gracia Marín Durán and Saide Esra Akdogan in this volume.

[50] Regulation (EU) No 995/2010 of the European Parliament and of the Council of 20 October 2010 laying down the obligations of operators who place timber and timber products on the market [2010] OJ L295/23.

[51] Regulation (EU) 2023/1115 of the European Parliament and of the Council of 31 May 2023 on the making available on the Union market and the export from the Union of certain commodities and products associated with deforestation and forest degradation and repealing Regulation (EU) No 995/2010 [2023] OJ L150/206 (Anti-Deforestation Regulation).

[52] The Anti-Deforestation repeals the EU Timber Regulation by 30 December 2024.

[53] European Commission 2008, p. 3.

[54] United Nations General Assembly 2017, para 15.2.

[55] United Nations 2023. https://sdgs.un.org/goals/goal15#progress_and_info. Accessed 7 March 2024.

[56] European Commission 2019a, p. 7 ff.

[57] European Parliament 2020, para G.

[58] European Parliament 2024. https://www.europarl.europa.eu/legislative-train/theme-environment-public-health-and-food-safety-envi/file-eu-driven-global-deforestation. Accessed 7 March 2024.

in certain commodities and products in 2021.[59] In its adopted version, the Anti-Deforestation Regulation relates to certain products that have been fed with or have been made using certain commodities, which are associated with a risk of deforestation: cattle, cocoa, coffee, palm, rubber, soya and wood.[60] According to Article 3 Anti-Deforestation Regulation, neither such products nor commodities shall be supplied for distribution, consumption or use on the EU market, unless they are deforestation-free, have been produced in accordance with the relevant legislation of the country of origin and are covered by a due diligence declaration.[61] Natural or legal persons who still want to supply such products or commodities on the EU market—so-called operators—are required to fulfil the due diligence obligation set out in Article 8 of the Regulation in relation to all covered products supplied by each individual supplier.[62] In case an operator is established outside the EU, the first natural or legal person who supplies a relevant product or commodity to the EU market is required to fulfil the operator's obligations under the Regulation.[63]

In essence, the Anti-Deforestation Regulation introduces a *de facto* ban for deforestation-prone products or commodities. By defining when such a risk is negligible,[64] it defines criteria which certain products and commodities must fulfil in order to be exempted from that ban. Fulfilment of those criteria is thus rewarded with market access, the two are directly linked ("direct sustainability rules").

Formally, the Regulation may just oblige the usual subjects of EU secondary law: a natural or legal person with a link to the EU territory, either by place of residence, registered office, central headquarters or a permanent business establishment.[65] Factually, however, non-EU producers and exporters of deforestation-prone commodities are also affected by the EU's internal sustainability rules. If they do not comply with EU rules, they do lose access to the EU market. The Anti-Deforestation Regulation's sustainability rules thus clearly have extraterritorial effects[66]—and these effects are not unintended. The Commission proposal reveals as much when it underlines that, because deforestation has global impacts, it would be inappropriate for the Regulation to address only domestic products, but not products being imported into the EU.[67] It seems that the Commission felt required to

---

[59] European Commission 2021b.

[60] Article 1 Anti-Deforestation Regulation (n 51).

[61] Article 3(1) Anti-Deforestation Regulation (n 51).

[62] This due diligence obligation includes the collection of information, data and documentation (Article 9) that can be used to assess the risk of non-compliance of the relevant product (Article 10; risk assessment), such as the geo-localisation data of the production facility. If no negligible risk is identified, operators are obliged to take measures to minimise the risk, such as carrying out independent audits (Article 11). For the risk calculation, the Commission must carry out an assessment of countries by means of an implementing act (Article 29), in which the deforestation risk per country of production is identified (low, normal, high).

[63] Article 7 Anti-Deforestation Regulation (n 51).

[64] Article 2(26) Anti-Deforestation Regulation (n 51).

[65] Article 2(21) Anti-Deforestation Regulation (n 51).

[66] See Dero-Bugny and Motte-Baumvol 2021, p. 46.

[67] European Commission 2021b, p. 4.

justify the scope of application of the Regulation and perhaps even more broadly the external effects of the proposed measure.

#### 10.2.2.2 "Indirect Sustainability Rules": The Case of the RED

The promotion of renewable energy is a central part of the EU's efforts to reduce its domestic greenhouse gas emissions.[68] The Renewable Energy Directive (RED I) was designed to increase the share of renewable energy in EU energy consumption, amongst others, by pushing the use of biofuels in the fuel-dependent transport sector.[69] However, specifically crop-based biofuels—first generation biofuels—are associated with certain risks, such as *direct land-use change* (DLUC), a phenomenon that describes a situation in which the higher demand for crops—and perhaps a more lucrative market for energy crops—can provide an incentive for turning non-agricultural land into agricultural land.[70] DLUC is thus associated with the risk of losing valuable fallow and highly biodiverse land as well as important carbon emission sinks. Such a risk can materialize within the EU's territory or abroad, depending on the origin of the crop used.

Aware of the DLUC-risks associated with crop-based biofuels, already RED I included sustainability rules for biofuels and bioliquids. According to these rules, not all biofuels could be considered for the mandatory minimum share of energy from renewable sources established for the transport sector ("minimum share requirement"). Biofuels could only be counted towards that requirement if its raw material was not obtained from land of high biodiversity value, such as forests, land with high carbon stock, or peatland ("land criteria").[71] The recast of RED I, resulting in RED II, and the amendment of the latter by the so-called RED III, further strengthened these sustainability criteria.[72]

By establishing those sustainability criteria, the EU relies again on its market power to steer the behavior of market actors. Yet, it does so in a different way than in the context of the Anti-Deforestation Regulation. The RED II establishes sustainability rules to be fulfilled by energy crops only for the purpose of calculating and assessing the share of biofuels used in the transport sector. Thus, it does not make the overall access to the EU market dependent on fulfilment of those criteria. Energy crops harvested from land of high biodiversity value, for example, and biofuels made of them can still be imported into the EU. Indirectly, however, the sustainability rules may nevertheless hamper market access in the particular segment, the energy crop

---

[68] Note Kettner and Kletzan-Slamanig 2020, p. 145 f.

[69] Article 3(4) RED I (n 15).

[70] See Laborde (2011).

[71] Article 17(3), (4) and (5) RED I (n 15).

[72] For one, the RED II extended the existing land criteria to biomass fuels. For another, the RED II established producer country- and production-specific requirements that should minimise the risk of using unsustainable forest biomass production, see Article 29(6) and (7) RED II (n 15). With the amendment introduced by the RED III, the land criteria for agricultural biomass now also apply to forest biomass, Article 29(3), (4), (5) RED II (n 15), as amended by RED III (n 15).

market, which was deemed to be the driver of deforestation abroad. According to current Article 29(1) RED II, the sustainability rules apply irrespective of whether the raw material was cultivated inside or outside the territory of the EU, and thereby also to biofuels made from non-EU crops. In this respect, the rules included in the RED II were designed to have external effects, even though their formal addresses are the member states.

The matter became more difficult as evidence grew stronger that first generation biofuels can also be a driver for *indirect land-use change* (ILUC); a phenomenon, which occurs when the cultivation of crops for energy production displaces the existing cultivation of crops for food and feed purposes.[73] The European Parliament pointed out, when taking into account DLUC and ILUC effects, crop-based biofuels may paradoxically result in a net increase in greenhouse gas emissions.[74] The recast of RED I in 2018 thus sought to address the issue by supplementing the existing sustainability rules. First, RED II limited the contribution all crop-based biofuels can make to the—now increased—minimum share of renewable energy within the final consumption of energy in the transport sector.[75] Then, RED II introduced what has been termed as a "freeze and phase-out" in relation to biofuels produced from crops for which a significant expansion of the production area into land with high-carbon stock is observed ("high ILUC-risk biofuels").[76] For the purpose of the minimum share requirement, their share may principally not exceed the consumption levels of such biofuels in the member state in 2019 ("freeze"). Additionally, this limit will gradually decrease to zero by the end of 2030 ("phase-out"). The methodology for identifying biofuels covered by the freeze and phase-out, i.e. for which crops a significant expansion of the production area into land with high-carbon stock is observed, is set out in a Commission Delegated Regulation.[77] Biofuels produced from such crops would therefore be subject to the gradual phase out unless they meet the strict criteria for low ILUC-risk fuels.[78] The freeze and phase-out for high ILUC-risk biofuels continues to be part of the RED II, as recently amended by RED III.[79]

Obviously, RED II did not introduce an outright ban on high ILUC-risk biofuels similar to the Anti-Deforestation Regulation did for covered products. Its sustainability rules relate "only" to the calculation of the minimum share of renewable energy in the member states. Thus, member states were and are still free to allow the

[73] On the issue, e.g. Ahlgren and Di Lucia 2014.

[74] European Parliament 2016, paras 77–82.

[75] Article 25(1) RED II (n 15).

[76] Article 26(1) and (2) RED II (n 15).

[77] The review of the criteria on high ILUC-risk biofuels, including a decreasing trajectory for high ILUC-risk biofuels was expected for September 2023, Commission 2023, p. 2. However, at the time of writing, the review was not yet present nor adopted.

[78] Commission Delegated Regulation 2019/807 of 13 March 2019 supplementing Directive (EU) 2018/2001 of the European Parliament and of the Council as regards the determination of high indirect land-use change-risk feedstock for which a significant expansion of the production area into land with high carbon stock is observed and the certification of low indirect land-use change-risk biofuels, bioliquids and biomass fuels [2019] OJ L133/1 (Delegated Regulation).

[79] Article 26 RED II (n 15), as amended by RED III (n 15).

use of these biofuels. Yet, compared to RED I, the rules of RED II—now also part of RED III—are designed to have an even stronger indirect steering effect: they clearly signal that the energy crop market should not be expanding and that the market for deforestation-linked crops will vanish first. While those rules are addressed to the member states, the steering effect is meant to extend to non-EU crop-producers. After all, some of the crops included in the methodology for calculating high ILUC-risk are only produced outside the EU, most importantly palm.

### *10.2.3 Extraterritoriality, Extraterritorial Effects, and their Challenges*

While the reasons for environmental unilateralism, as well as its form of implementation may differ, the common denominator is clear: the aim to influence behavior outside the EU. That said, the above examples demonstrate that EU unilateral solutions to global sustainability concerns do not tend to technically apply beyond EU borders. The addressees of these obligations are the usual subjects of EU law—the member states, natural or legal persons with a link to the EU territory, either by place of residence, registered office, central headquarters or a permanent business establishment—and their immediate conduct. On a formal level, these instruments can thus not be understood as extraterritorial measures which would require justification under general international law.[80] Yet, it cannot be denied that sustainability rules are ultimately designed to have extraterritorial *effects*. This, essentially, is a result of the consumption-based approach often embodied in those rules.[81] When the EU assumes responsibility for the negative impacts of its domestic consumption abroad and addresses them, the respective rules inevitably have extraterritorial effects aimed to also influence the behavior of market actors outside the EU borders. In this way, such sustainability rules can be said to give rise to the territorial extension of the reach of EU law.[82]

Territorial extension and extraterritorial effects are legally not unproblematic. Much rather, they give rise to challenges under both EU and international law. EU law is perhaps the lesser hurdle. As Joanne Scott finds, there is 'no presumption *against* extraterritorial extension operating in EU law'.[83] Indeed, the Court of Justice of the European Union (CJEU) seems to accept that conduct, which takes place outside the EU, but has an impact on it, can still be regulated by EU law.[84] In other words, it accepts that such effects can be a legislative trigger under EU law.[85]

[80] Vranes 2009, p. 176 f.

[81] See Scott and Rajamani 2013, p. 213.

[82] On the concept of extraterritorial extension, Scott 2014a, p. 89 ff.

[83] Scott 2019, p. 36 (emphasis added).

[84] See CJEU, *Imperial Chemical Industries* v *Commission*, Judgment, 14 July 1972, 48/69, EU:C:1972:70, para 126.

[85] See Scott 2014b, p. 1356.

Yet, EU legislation still needs to establish a territorial connection. In the case of greenhouse gas emissions from international aviation, for example, the Court found such a connecting factor in the place of departure and arrival.[86]

Interestingly, an explanation why extraterritorial extension of EU law is possible is not found in EU law but in international law. In his *Ahlström* opinion, Advocate General (AG) *Darmon* refers to the *Lotus* judgement of the Permanent Court of International Justice (PCIJ) to highlight that considerations of the location of the effects as the basis of a state's jurisdiction was in conformity with international law and states that 'what is thus permissible for States must necessarily also be permissible for the [EU], as a subject of international law.'[87] However, while perhaps permissible, it is still possible that the exercise of this jurisdiction conflicts with obligations under international law, which the EU is bound to obey—as a matter of international *and* EU law.[88] Specifically, obligations under international trade law pose difficulties when devising unilateral sustainability rules.

## 10.3 Limits to EU Environmental Unilateralism: WTO Law, Extraterritorial Sustainability Concerns and the EU's Palm Oil Dispute

As discussed above, EU law does not *per se* limit the EU in devising unilateral rules on sustainability abroad. International law, however, is another matter. The EU's shift to a consumption-based approach to address sustainability concerns abroad can specifically get into conflict with the EU's obligations under international trade law. Thus, in designing such sustainability rules, the EU legislator has to keep an eye on these obligations, stemming, in particular, from WTO law.

As Eva Pander Maat shows in this volume, the EU institutions, indeed, tend to be very much aware of this fact.[89] However, despite their awareness, the occurrence of diplomatic controversies and, ultimately, even trade disputes cannot always be avoided.[90] The above-mentioned changes to the RED I rules regarding high ILUC-risk biofuels serve as a good example. Indonesia, the world's largest palm

---

[86] CJEU, *Air Transport Association of America and others*, Judgment, C-366/10, EU:C:2011:864, 21 December 2011, para 116 f.

[87] AG Darmon, *Ahlström*, AG Opinion, Joined cases 89, 104, 114, 116, 117 and 125-129/85, EU:C:1992:293, para 27.

[88] Note Article 3(5) TEU (n 1). On the notion of jurisdiction in international law, note Mignolli 2018, p 249 f.

[89] See Chap. 2 in this volume. The present authors have argued elsewhere that the freeze and phase-out in the RED II very much reflects the European Commission's attempt to avoid the pitfalls of WTO law, see Mayr et al. 2021.

[90] See, for example, Euractiv.com 2023. https://www.euractiv.com/section/global-europe/opinion/the-brief-clumsy-carbon-diplomacy/. Accessed 7 March 2024. Also, World Trade Organization Committee on Market Access 2022, with statements of EU trading partners as to their concerns in relation to EU initiatives, including the Anti-Deforestation Regulation and the CBAM.

oil producer, claims that the freeze and phase-out introduced in the RED II violates WTO law as it would unfairly single out and discriminate palm oil-based biofuels *vis-à-vis* other crop-based biofuels, e.g., biofuels on soybean basis. Consequently, Indonesia requested consultations, and when those consultations did not lead to the desired outcome, initiated dispute resolution proceedings in the WTO,[91] to which *inter alia* other palm-oil-producing countries, like Malaysia, Colombia or Costa Rica, joined as third parties.[92]

Due to the complexity of the legal and factual issues the dispute involves, the panel had to extend the standard six-month period for issuing its report twice so far. According to its latest communication of February 2023, the final report would not be issued before the third quarter of 2023.[93] However, by the end of February 2024, the panel had neither issued its report, nor announced when a decision could be expected. The complexity of the issues in dispute is further reflected in the extensive submissions of the parties, but also in the over 100 written questions the panel posed to the EU as part of the proceedings.[94]

The intricacies of the case raise a host of legal questions that are interesting in themselves. On a broader level, however, the palm oil dispute provides an illustrative example of the legal difficulties that can arise when aiming to address extraterritorial sustainability concerns through trade-related measures. While WTO law principally acknowledges that measures adopted in the public interest can be justified, it arguably sets clear limits to environmental unilateralism.

### *10.3.1 Indirect Sustainability Rules and their Trade Implications*

Indonesia's complaint essentially targets the freeze and phase-out of high ILUC-risk biofuels as devised in the EU's RED II.[95] As explained above, the RED II does not impose an import ban on high-ILUC risk biofuels. It also does not impose other direct limitations on their import, trade, or use in the EU. And yet, the measure has trade implications, which raise questions regarding its compatibility with key WTO rules.

---

[91] See WTO, *European Union—Certain measures concerning palm oil and palm crop-based biofuels*, Request for the establishment of a Panel by Indonesia, 24 March 2020, WT/DS593/9.

[92] Malaysia, in the meantime, has brought another palm-oil-related dispute before the WTO dispute settlement body, see WTO, *European Union and certain Member states—Certain measures concerning palm oil and crop-based biofuels*, WT/DS600.

[93] WTO, *European Union—Certain measures concerning palm oil and palm crop-based biofuels*, Communication from the Panel, 24 February 2023, WT/DS593/12.

[94] See European Union 2021c.

[95] The complaint was made before the adoption of RED III (n 15). Hence, the complaint relates to RED II in its initial version. While RED III did not alter the freeze and phase-out itself, it introduced some changes to rules in the broader context of the freeze and phase-out. Where it is necessary to refer to these changes as part of the argumentation, we will thus refer explicitly to the RED II, as amended by RED III.

Those trade implications, however, only come to light when reading all relevant EU measures together.

As may be recalled, the freeze and phase-out consists of a cap on the share of high ILUC-risk biofuels that count towards member states' renewable energy targets in the transport sector—principally "freezing" them at 2019 levels—and the gradual reduction of this limit to zero by 2030.[96] What qualifies as high-ILUC risk, however, is defined only in broad terms in the RED II, which speaks of a significant expansion of a crop's production area into land with high-carbon stock.[97] The concrete methodology for identifying such high ILUC-risk crops—and, by implication biofuels—is set out in a Delegated Regulation by the Commission, which specifies two criteria that determine what constitutes such a significant expansion. The first criterion concerns the average annual expansion of the global production area of the feedstock in general, and the second criterion concerns the share of such expansion into land with high-carbon stock.[98] That said, based on the figures provided in the Annex to the Delegated Regulation, and the formula set out therein, only biofuels produced from palm oil—an energy crop that is virtually *not* produced in the EU—qualify as a high ILUC-risk. Thus, even though the freeze and phase-out is origin-neutral on its face, the fact that it only affects biofuels produced from a crop exclusively produced outside EU territory implies trade repercussions and may thus raise issues under WTO law.

## 10.3.2 Indirect Sustainability Rules and their Discriminatory Effects

At the heart of Indonesia's case, there is a possible violation of the GATT's non-discrimination obligations set out in Articles I:1 and III:4 of the GATT.[99] According to the WTO's Appellate Body (AB), '[t]he essence of the non-discrimination obligations is that like products should be treated equally, irrespective of their origin'.[100] For that purpose, Article I:1 GATT sets out the most-favored nation (MFN) treatment obligation, prohibiting discrimination '*between* and *among* like products of different origins'.[101] It requires that an 'advantage, favor, privilege or immunity granted by any [Member] to any product originating in or destined for any other country shall be accorded immediately and unconditionally to the like product originating in or destined for the territories of all other [Members]'. Article III:4 GATT sets out the

[96] See Article 26(2) RED II (n 15).

[97] Ibid.

[98] Article 3 Delegated Regulation (n 78).

[99] For the TBT (Technical Barriers to Trade) agreement, see Mayr et al. 2021, p. 236 ff.

[100] WTO, *European Communities—Regime for the Importation, Sale and Distribution of Bananas*, Appellate Body Report, 9 September 1997, WT/DS27/AB/R, para 190 (*EC—Bananas III*).

[101] WTO, *European Communities—Measures Prohibiting the Importation and Marketing of Seal Products*, Appellate Body Report, 22 May 2014, WT/DS400/AB/R; WT/DS401/AB/R, para 5.82 (*EC—Seal Products*).

national treatment obligation, which prohibits discrimination of '*imported* products vis-à-vis like *domestic* products',[102] and requires that 'products of the territory of any [Member] imported into the territory of any other [Member] shall be accorded treatment no less favorable than that accorded to like products of national origin in respect of all laws, regulations and requirements affecting their internal sale, offering for sale, purchase, transportation, distribution or use'. While Articles I:1 and Articles III:4 GATT establish different non-discrimination standards, they both apply only to 'like products'.

#### 10.3.2.1 Likeness, Competitive Relationship, and the Question of a Market

Whether the GATT's non-discrimination standards apply depends, to begin with, on the 'likeness' of the products at issue. While this concept takes on different meanings in different contexts—the AB vividly speaks of an 'accordion of likeness'—the underlying question is largely similar, i.e. whether there exists a competitive relationship between the products at issue.[103] The freeze and phase-out distinguishes between biofuels produced from high ILUC-risk feedstock and biofuels produced from other crops, which are associated with no or only a lower ILUC-risk. The question is therefore whether biofuels produced from different raw materials are 'like products'. While the physical characteristics and qualities of products are a useful indicator for likeness,[104] the products' end-uses as well as consumers' habits and tastes are equally relevant.[105] The latter may also be influenced by process and production methods (PPMs) that do not affect the physical characteristics of a product (non-product-related PPMs),[106] such as upstream (in)direct deforestation. All these criteria need to be taken into account to establish whether (certain) crop-based biofuels are in a competitive relationship with each other and thus 'like' products.

In the dispute at hand, the likeness of palm oil-based and other crop-based biofuels is a highly contentious point. Indonesia, on the one hand, argues that palm oil-based biofuel is like other oil crop-based biofuel, in particular biofuel made from soybean and rapeseed, being the most commonly used oil-crops in biofuel production.[107] Consequently, these oil crops would compete with one another on the energy crop market established by the minimum share requirement in the EU's RED II.[108] The

---

102 *EC—Seal Products* (n 101) para 5.79.

103 See Van den Bossche, Zdouc 2022, p. 379 ff. In-depth on the concept of likeness, e.g., Vranes 2009, p. 191 ff.

104 WTO, *European Communities—Measures Affecting Asbestos and Asbestos-Containing Products,* Appellate Body Report, 5 April 2001, WT/DS135/AB/R, para 110 (*EC—Asbestos*).

105 Working Party on Border Tax Adjustments 1997, para 18.

106 Marín Durán 2015, p. 96. On PPMs, see also Charnovitz 2002; Vranes 2009, p. 320 ff; Sifonis 2018.

107 Indonesia 2021a, para 29.

108 See European Union 2021b, para 397.

EU, on the other hand, argues that Indonesia's like product analysis would be methodologically flawed since it focuses only on a specific subset of biofuels produced from food and feed crops and ignores other biofuels such as those made out of sunflower oil, waste materials (such as used cooking oil), or animal fats. Moreover, the EU contends that even if the products to be considered were only rapeseed oil, soybean oil and palm oil-based biofuels, these would in any case not meet the 'traditional' likeness criteria established in the WTO case law.[109] Finally, the EU also contests that the RED II actually establishes an energy crop market. While the Directive might create certain opportunities, whether and how these opportunities materialize would depend on the implementing measures of the EU member states.[110]

To a large extent, the likeness of the various oil crop-based biofuels—e.g., as regards their physical properties and end uses—appears to depend on how much of a zoomed-in or zoomed-out approach one takes. Be that as it may, the EU's argument regarding the absence of the creation of a relevant market seems not entirely convincing. Indeed, the EU member states can choose between different domestic measures to fulfil the minimum share requirement. This is because as an EU Directive, the RED II is binding as to the objectives prescribed but leaves the member states free to choose how to realize these objectives.[111] *Somehow*, however, they must incentivize the use of biofuels in the transport sector to meet their renewable energy targets, specifically as the share of biofuels has dropped in recent years.[112] For example, some EU member states, such as Austria, relied on a blending requirement for domestic fuel suppliers, simultaneously providing that high ILUC-risk biofuels cannot be counted towards the fulfilment of this requirement.[113] France, to give another example, pursued a different strategy, amending its Customs Code to incentivize the consumption of certain crop-based biofuels that could be counted towards the minimum share requirement. Palm oil was expressly excluded from that scheme.[114] While using different mechanisms, the member states' respective measures are a direct result of the minimum share requirement and the Directive's

[109] See European Union 2021a, paras 528 ff and 551 ff. At the EU level, the Court put forward the same argument when seeking to demonstrate that the Delegated Regulation does not directly concern the appellant, CJEU, *Lípidos Santiga v Commission*, Judgment, 21 October 2021, C-402/20 P, EU:C:2021:872, para 46. That, however, is a different (legal) standard than under WTO law.

[110] See European Union 2021c, para 398. Indeed, certain implementing measures of the member states have become subject of a later dispute, initiated by Malaysia, WTO, *European Union and Certain Member States—Certain Measures Concerning Palm Oil and Oil Palm Crop-Based Biofuels*, DS600.

[111] Prechal 2006, p. 71 ff.

[112] Eurostat 2023. https://ec.europa.eu/eurostat/statistics-explained/index.php?title=Renewable_energy_statistics#of_renewable_energy_used_in_transport_activities_in_2021. Accessed 7 March 2024.

[113] § 8(5) Austrian Fuel Regulation (*Kraftstoffverordnung 2012*) Federal Law Gazette II 398/2012, as amended by Federal Law Gazette II 452/2022.

[114] Article 266 quindecies of the French Customs Code. The French Conseil d'Etat has confirmed that all palm oil-based are excluded from this tax benefit, Conseil d'Etat, Decision, 24 February 2021, Nos 437277, 438782.

overall binding effects.[115] In this light, a minimum requirement, which drives the demand for biofuels, could thus be said to create an energy crop market, even when included in a Directive. In such a market, palm oil-based biofuels and other oil-crop-based biofuels—given their overall same end-uses[116]—can be considered to be in a competitive relationship.[117] This is in particular the case, since consumer preferences, such as deforestation-free production, seem not (yet) strong enough to affect the 'likeness' of those specific products.

#### 10.3.2.2 Different Faces of Discrimination

Assuming that palm oil-based and other oil crop-based biofuels are like products, the further requirements set out in Articles I:1 and III:4 GATT come into focus. Under Article I:1 GATT, the question is whether *any advantage* is granted to like products *immediately and unconditionally*. In WTO case law, the term advantage is generally understood broadly.[118] Thus, it seems that the eligibility for incentives seeking to ensure that the RED II's minimum share requirement can be such an advantage—an assumption for which Indonesia, in the eyes of the EU, did not provide enough proof.[119] The second question is whether such an advantage would be accorded 'immediately and unconditionally' to like products covered under Article I:1 GATT. In case of the RED II, this is not the case since palm oil-based biofuels would not be eligible for incentives aimed at ensuring the minimum share; this, in turn, would negatively impact the competitive opportunities for palm oil-based biofuels which are imported from Indonesia and a few other palm oil producing countries.[120] Thus, the EU's freeze and phase-out would conflict with the most-favored-nation requirement in Article I:1.

Article III:4 GATT provides that imported products shall be accorded treatment no less favorable than like domestic products in respect of all laws, regulations and requirements affecting their internal sale, offering for sale, purchase, transportation,

---

[115] Note, however, the CJEU's argument as regards the member state's discretion in implementing both the minimum share requirement and the directive. Specifically, it argues that the member states are still free to use even high-ILUC risk biofuels as part of their remaining energy mix, given that the share of sustainable biofuels would only have to rise to 32% by 2030 (now: 42.5%), CJEU, *Lípidos Santiga v Commission*, Judgment, 21 October 2021, C-402/20 P, EU:C:2021:872, para 42. This, however, would be counterproductive in view of not only the EU's own climate policies but also the member states'.

[116] On the differences in physical properties, see, for example, Canakci and Sanli 2008, p. 437.

[117] Whether cereal or sugar crop-based biofuels (bioethanol) and oil crop-based biofuels (biodiesel) are in a competitive relationship, and thus 'like', seems more doubtful.

[118] E.g., *EC—Bananas III* (n 100) para 206; WTO, *Canada—Certain Measures Affecting the Automotive Industry,* Appellate Body Report, 31 May 2000, WT/DS139/AB/R; WT/DS142/AB/R, para 79; WTO, *Columbia—Indicative Prices and Restrictions on Ports of Entry*, Panel Report, 27 April 2009, WT/DS366/R, paras 7.340 ff.

[119] European Union 2021a, paras 1102 ff.

[120] See *EC—Seal Products* (n 101) para 5.88 'detrimental impact on the competitive opportunities for like imported'. Similarly, Mitchell and Merriman 2020, p. 570.

distribution, or use. Again, WTO case law reveals a rather broad understanding of the key elements of the provision. Thus, for example, even though the EU's RED II rules have to be implemented by member states, they can arguably be considered as laws and regulations in the sense of Article III:4.[121] Moreover, these rules are likely to treat palm oil-based biofuels less favorably as they provide an incentive for fuel suppliers to not use them.[122] This is because, as mentioned above, the economic viability of biofuels depends largely on financial support or regulatory measures prescribing demand; but if palm oil-based biofuels do not count towards the minimum share, there is also no incentive to use them at a premium price. Importantly, when it comes to determining whether the EU rules result in less favorable treatment for palm oil-based biofuels, actual effects in the market do not need to be demonstrated by the complaining WTO member.[123] However, there has to be 'a genuine relationship between the measure at issue and the adverse impact on imported versus domestic like products'.[124] As mentioned earlier, while origin-neutral on its face, the freeze and phase-out *de facto* affects only imported products, since only palm oil-based biofuels are identified as high ILUC-risk by the EU rules, and oil palm is not grown in the EU. Against this background, it does not seem too far-fetched that the EU rules could be said to have an adverse impact on the competitive opportunities of a group of imported products *vis-à-vis* domestic like products.[125]

### *10.3.3 Justification is Key—But Does it Open the Door?*

The preceding section highlighted some key fault lines that may play a role for the assessment of the RED II's freeze and phase out under GATT non-discrimination standards. The WTO panel will have to address these and many more intricate issues in its report.[126] That said, what becomes clear even from this cursory analysis is that indirect sustainability rules, which do not *directly* link sustainability criteria to restrictions on market access, are nonetheless at risk of infringing the GATT's non-discrimination standards. This is the consequence of the rather broad understanding of the notions employed in the respective provisions, such as 'affect', which is driven—as is often underlined in the WTO case law—by the ultimate *telos* of the norms:

---

[121] WTO, *European Union and its Member States—Certain Measures Relating to the Energy Sector*, Panel Report, 10 August 2018, WT/DS476/R (appeal pending) paras 7.393–7.395.

[122] See *EC—Bananas III* (n 100) para 220 'effect on a product'.

[123] WTO, *United States—Tax Treatment for "Foreign Sales Corporations"—Recourse to Article 21.5 of the DSU by the European Communities*, Appellate Body Report, 14 January 2002, WT/DS108/AB/RW, para 215.

[124] WTO, *Thailand—Customs and Fiscal Measures on Cigarettes from the Philippines*, Appellate Body Report, 17 June 2011, WT/DS371/AB/R, para 134.

[125] See in more detail Mayr et al. 2021, p. 243 ff; Mitchell and Merriman 2020, p. 575 ff.

[126] E.g., in its complaint, Indonesia relies heavily on alleged violations of the Agreement on Technical Barriers to Trade (TBT Agreement). For an in-depth analysis of some of the complex legal questions arising in that context, see Mayr et al. 2021, pp. 233, 236 ff.

to ensure equality of competitive opportunities for like products.[127] Importantly, however, the *potential* incompatibility with the GATT's non-discrimination standards does not yet determine the fate of a measure such as the freeze and phase-out. This is because under Article XX GATT, WTO members can justify measures adopted in the pursuit of certain public interest objectives—though often only in theory.[128]

#### 10.3.3.1 Indirect Sustainability Rules Addressing Policy Objectives Abroad

Measures inconsistent with GATT rules can principally be justified if they genuinely address certain policy objectives, exhaustively listed in Article XX GATT, and meet the additional requirements set out in the *chapeau* of that provision. Some of these policy objectives seem principally suited to cover measures addressing concerns regarding the environmental dimension of sustainability. However, the fact that in the context of the freeze and phase-out the sustainability concerns addressed lie outside the EU's own territory arguably complicates matters.[129]

For one, the freeze and phase-out could be justified as a measure protecting human, animal or plant life or health based on Article XX(b) GATT. On a general level, it may not pose too much difficulty to argue that the freeze and phase-out is designed with that policy objective in mind.[130] The RED II is one of the EU's main instruments of climate policy, seeking to reduce greenhouse gas emissions and thereby limiting the detrimental effects of climate change on humans and the environment.[131] The freeze and phase-out itself is part of a recast that sought to increase these efforts, in particular by decarbonizing the transport sector—including from ILUC emissions.[132] However, Article XX(b) GATT sets quite a demanding standard, since the measure at issue must be 'necessary' to achieve the desired policy goal.[133]

The assessment of necessity involves the weighing and balancing of a number of factors, including the relative importance of the interests or values furthered by the measure at issue, the measure's contribution to achieving its objective, and its trade-restrictiveness.[134] In the present case, the perhaps trickiest part of this assessment

---

[127] *EC—Seal Products* (n 101) para 5.82.

[128] For an analysis of another unilateral environmental measure, the CBAM, under Article XX GATT see Patrick Abel in this volume.

[129] Devising an "extraterritoriality decision tree" within the framework of Article XX GATT, see Cooreman 2016, who emphasizes that even such a structured legal framework offers no one-size-fits-all blueprint (at p. 248).

[130] See WTO, *United States—Standards for Reformulated and Conventional Gasoline*, Appellate Body Report, 29 April 1996, WT/DS2/AB/R, para 6.21 (*US—Gasoline, AB Report*).

[131] E.g., recitals 2, 81 RED II (n 15); recitals 7-9 Delegated Regulation (n 78).

[132] See above, Sect. 10.2.2.2.

[133] On this criterion, see e.g., Regan 2007.

[134] WTO, *Brazil—Measures Affecting Imports of Retreaded Tyres*, Appellate Body Report, 3 December 2007, WT/DS332/AB/R, para 178 (*Brazil—Retreaded Tyres*).

is the contribution of the freeze and phase-out to achieving its objective, i.e. the reduction of global ILUC emissions. As ILUC cannot be measured or observed directly, demonstrating the contribution of the freeze and phase-out to the reduction of ILUC emissions—or, on an even broader scale, to mitigating climate change—is anything but trivial. That said, in WTO practice, the burden of proof is somewhat alleviated.

WTO case law has acknowledged in similar situations that there are other ways of establishing necessity, for example, by showing that the measure 'is apt to produce a material contribution to the achievement of its objective',[135] which can be demonstrated through 'quantitative projections in the future, or qualitative reasoning based on a set of hypotheses that are tested and supported by sufficient evidence'.[136] Still, quantifying the freeze and phase-out's contribution to reducing emissions from ILUC remains difficult. After all, ILUC is an indirect phenomenon, which is not immediately observable and measurable.[137]

In the dispute between Indonesia and the EU, Indonesia goes even a step further when it dismisses ILUC as 'a theoretical economic concept',[138] underlining that ILUC emissions cannot be quantified or measured at all.[139] By contrast, the EU argues that the criteria it uses—in particular the observable rates of expansion into areas of high carbon stock in combination with crop yield and productivity—are appropriate proxies for identifying crops that give rise to a high risk of ILUC emissions.[140] These criteria were deduced from available scientific evidence, amongst others a literature review and a separate global assessment of geodata,[141] and thus—in the eyes of the EU—are 'apt to make a material contribution' to the reduction of ILUC emissions by discouraging the use of high ILUC-risk biofuels. Indeed, while the evidence used by the EU may be disputed—Indonesia refers to it as a minority view[142]—it must be recalled that WTO members may very well rely on opinions other than the prevailing view when formulating their policies.[143] What the EU needs to demonstrate is that the evidence used is in itself conclusive and conclusively reflected in its choice of criteria for the freeze and phase-out.[144]

Even if the EU could demonstrate that the freeze and phase-out is a measure that is 'apt to make a material contribution' to the reduction of ILUC emissions, it would still be necessary to assess the availability of other, less trade-restrictive measures

---

[135] *Brazil—Retreaded Tyres* (n 134) para 151.

[136] Ibid.

[137] E.g. Ahlgren and Di Lucia 2014.

[138] Indonesia 2021a, para 233.

[139] Indonesia 2021b, para 136.

[140] European Union 2021c, para 112.

[141] European Commission 2019b, p. 6.

[142] See European Union 2021b, p. 26.

[143] E.g., WTO, *European Communities—EC Measures Concerning Meat and Meat Products (Hormones)*, Appellate Body Report, 16 January 1998, WT/DS26/AB/R; WT/DS48/AB/R, para 194; *EC—Asbestos* (n 104) para 178. Cf. also Ehlermann, Lockhart 2004, p. 519.

[144] See the EU's line of argumentation in this regard, European Union 2021b, para 79 ff.

that allow for the same level of protection.[145] In the dispute at hand, Indonesia points to a number of potential alternatives, which would essentially boil down to taking a domestic and cooperative approach instead of extraterritorial and unilateral one to the issue. Paradoxically, the complexity of determining causality and direct attribution of ILUC emissions also makes the claim that these alternatives offer the same level of protection difficult.[146] Moreover, the EU underlines that it is not 'constrained to addressing issues of global concern 'at home''.[147] Thus, if anything, Indonesia's proposal that the EU should limit the harvesting of forests for wood-based bioenergy to the EU territory, could be seen as a supplement to the measures addressing global ILUC emissions. Similarly, Indonesia argues that instead of the freeze and phase-out, the EU could provide Indonesia with financial and technical support to ensure the preservation of its forests, which is considered 'at best [as] a complementary measure' by the EU,[148] lacking a core preventative element. In case the EU succeeds in demonstrating that the proposed alternatives do not genuinely allow for the same level of protection or are not reasonably available, a *provisional* justification under Article XX(b) GATT seems possible.

For another, Article XX(g) GATT allows for the justification of measures that 'relate to the conservation of exhaustible natural resources if such measures are made effective in conjunction with restrictions on domestic production or consumption'. As the reference to 'natural resources' is understood evolutionary,[149] it arguably includes the global climate system and the Earth's atmosphere.[150] Thus, it is not too difficult to argue that the RED II's freeze and phase-out, which aims to promote the use of biofuels with a lower risk of ILUC emissions, and thereby secure greenhouse gas reductions, *relates to* this policy objective. Once again, an examination of the measure's actual effects is not required.[151] What Article XX(g) GATT does require, however, is that the burden of the restrictions relating to the conservation of natural resources is not externalized completely. Rather, such measures must be 'made effective in conjunction with restrictions on domestic production or consumption'.[152]

In the present case, this required 'even-handedness' of the freeze and phase-out raises questions. On a formal level, the RED II's requirement to ensure a minimum share of energy from renewable sources in the transport sector, is addressed to the

[145] On the trade-restrictiveness of the freeze- and phase-out, see above Sect. 10.3.2.1.

[146] See for alternative two, the extension of greenhouse gas emissions accounting to other agricultural products, European Union 2021c, para 297 ff and for alternative three, a tax on meat and dairy products, European Union 2021c, para 308 ff.

[147] European Union 2021c, para 293.

[148] European Union 2021c, para 305.

[149] WTO, *United States—Import Prohibition of Certain Shrimp and Shrimp Products*, Appellate Body Report, 12 October 1998, WT/DS58/AB/R, para 130 (*US—Shrimp*).

[150] See Mitchell, Tran 2010, p. 41.

[151] WTO, *China—Measures Related to the Exportation of Rare Earths, Tungsten and Molybdenum*, Appellate Body Report, 7 August 2014, WT/DS431/AB/R; WT/DS432/AB/R; WT/DS433/AB/R, para 5.113 (*China—Rare Earths*).

[152] Article XX(g) GATT. Cf. e.g., *China—Rare Earths* (n 151) para 5.93.

member states and domestic actors only.[153] However, based on its design and structure, considering, in particular, the formula and values laid down in the Delegated Regulation, the freeze and phase-out obviously has repercussions and affects (mainly) foreign producers and products. While the AB has clarified that the burden of conservation does not need to be distributed evenly between foreign and domestic producers or products, it has also emphasized that a measure imposing a 'significantly more onerous burden' on foreign producers or products is unlikely to meet the requirements set out in Article XX(g) GATT.[154] A central question in the context of subparagraph (g) is therefore whether the restrictions the freeze and phase-out imposed on domestic production or consumption are '"real" rather than existing merely "on the books"'.[155] If so, any disparate impact of the measure would not be enough to rule out its even-handedness and thereby its provisional justification under Article XX(g) GATT. However, in any case, any discriminatory effects in the application of the freeze and phase-out must then be taken into account in the context of the *chapeau* of Article XX.

#### 10.3.3.2 The Crux of the Matter—The Application of the Provisionally Justified Measure

Whereas the different subparagraphs of Article XX GATT are primarily concerned with the design of a measure, the provision's *chapeau* focuses on its application. By ensuring that a provisionally justified measure is not 'applied in a manner which would constitute a means of arbitrary or unjustifiable discrimination between countries where the same conditions prevail, or a disguised restriction on trade', Article XX GATT seeks to prevent the abuse or misuse of the exceptions granted.[156] This involves a highly complex legal analysis, which often really determines the legal fate of a measure under the GATT. For the freeze and phase-out in the RED II, the appraisal under the *chapeau* thus appears crucial.

Overall, an assessment of the requirements of the chapeau involves three questions: (1) whether the application of the measure results in discrimination; (2) whether the discrimination occurs between countries where the same conditions prevail; and (3) whether the discrimination is arbitrary or unjustifiable.[157] The first question was already briefly touched upon in the context of Article XX(g) GATT.[158] Indeed, the discriminatory effects resulting from the application of the freeze and phase-out can hardly be overlooked. The Delegated Regulation, determining the relevant numbers as well as the formula to calculate a high ILUC-risk, effectively singles out palm

[153] See above, Sect. 10.2.2.2.

[154] *China—Rare Earths* (n 151) para 5.143.

[155] *China—Rare Earths* (n 151) para 5.136.

[156] *Brazil—Retreaded Tyres* (n 134) para 227.

[157] E.g., *US—Shrimp* (n 149) para 150.

[158] See above, Sect. 10.3.3.1.

oil-based biofuels as the only oil crop-based biofuels which fall under the freeze and phase-out. Given that palm oil-based biofuels have been predominantly imported from certain WTO members, above all Indonesia and Malaysia, the application of the freeze and phase-out can be expected to have disparate economic impacts on *some* producer countries of oil crop-based biofuels, thereby resulting in discrimination. Indeed, data shows that palm oil consumption in the EU has dropped quite significantly from 2019 onwards,[159] with imports dropping by 19% between July 2022 and June 2023 alone.[160] This drop is partly to be attributed to the member states' rules implementing the freeze- and phase-out, specifically where member states opted for an early phase-out.[161]

If we accept that the application of the freeze and phase-out may entail certain discriminatory effects, the next question is whether this discrimination occurs between countries where the same conditions prevail. According to the AB, the relevant 'conditions' should be understood 'by reference to the applicable subparagraph of Article XX under which the measure was provisionally justified and the substantive obligations under the GATT [...] with which a violation has been found'.[162] In the analysis of subparagraphs (b) and (g) of Article XX GATT, ILUC emissions were the common denominator and are thus the relevant 'condition' to consider. In this regard, it is important to recall that the phenomenon of ILUC is not limited to certain countries, such as producer countries of (specific) (oil) crop-based biofuels. Rather, it is a global phenomenon.[163] Hence, it can be argued that the same conditions—i.e. a certain ILUC risk—prevail in any country and consequently that discrimination occurs between countries where the same conditions prevail.

Ultimately, the justification of the freeze and phase-out would then depend on the third analytical element under the *chapeau*, i.e. whether the application of the measure constitutes a means of arbitrary or unjustifiable discrimination, which 'bear[s] no rational connection to the [policy] objective' or 'would go against that objective'.[164] More specifically, the question would be whether there exists a rational connection between the freeze and phase-out of palm oil-based biofuels, but not other oil crop-based biofuels, and the reduction of ILUC emissions. The panel's answer to this question in the ongoing dispute would arguably depend on what it makes of the available scientific evidence on which the EU has based the criteria, which determine

[159] Statista 2024. https://www.statista.com/statistics/489370/palm-oil-consumption-european-union/. Accessed 29 February 2024.

[160] Oils and Fats International 2023. https://www.ofimagazine.com/news/eu-member-states-imported-significantly-less-palm-oil-in-2022-23. Accessed 29 February 2024.

[161] However, part of the drop seems also the result of exceptionally high palm oil prices in the analysed period, United States Department of Agriculture, Global Agricultural Information Network 2023, p. 30.

[162] *EC—Seal Products* (n 101) para 5.301.

[163] See also to this effect, European Commission 2019b.

[164] *Brazil—Retreaded Tyres* (n 134) para 227.

the existence of a high ILUC-risk and which effectively single out palm oil-based biofuels.[165]

Importantly, an additional hurdle could arise from the AB's ruling in *US—Shrimp*. In this case, the AB underlined that in face of transboundary or global environmental problems which 'call for concerted and cooperative efforts' of a large number of countries, WTO members had to make serious good faith efforts to negotiate and conclude bilateral or multilateral agreements.[166] Only where such efforts were unsuccessful, it seemed acceptable to the AB that WTO members would resort to unilateral measures. Whether the EU can be said to have made 'serious and good faith efforts' *prior* to adopting the freeze and phase-out is questionable. As far as can be seen it was really only *after* the adoption of the measure that the EU engaged more visibly in different dialogues with palm oil producing countries, as part of, for example, the Joint Working Group on Palm Oil, established under the auspices of the ASEAN-EU Ministerial Meeting[167] or—possibly—in the context of the ongoing negotiations of a trade agreement with Indonesia.[168]

Even more problematic, perhaps, seems that, in *US—Gasoline*, the AB judged a measure as unjustifiable under the *chapeau*, because the discrimination 'must have been foreseen, and was not merely inadvertent or unavoidable'.[169] In case of the freeze and phase-out in the RED II, such foreseeability can hardly be denied. A closer look at the EU legislative process leading to the RED II shows that the phase-out of high ILUC-risk biofuels was not part of the Commission's original proposal. It was only introduced following an amendment proposal by the European Parliament, which called for a swift phase-out of palm oil based biofuels by 2021.[170] However, within the Commission, this proposal raised 'substantive concerns, in terms both of compatibility with the EU's obligations in the [WTO] and of trade relations with palm oil producing countries'.[171] And even in the last phase of the legislative process, the trialogue, the Council was still concerned about the WTO compatibility of the suggested rule.[172] In light of this internal negotiation history, and the fact that the freeze and phase out *de facto* affects only palm oil-based biofuels, the understanding that the freeze and phase-out was tailored to target these biofuels does not seem too

---

165 See also the arguments submitted by the EU to the WTO panel in the dispute with Indonesia, European Union 2021a, para 1349.

166 *US—Shrimp* (n 149) paras 166–168.

167 For the third meeting, held in 2023, see Association of Southeast Asian Nations 2023. https://asean.org/joint-press-release-third-joint-working-group-meeting-on-palm-oil-between-the-european-union-and-relevant-asean-member-states/. Accessed 29 February 2024.

168 For an overview of the negotiations see European Commission 2024. https://policy.trade.ec.europa.eu/eu-trade-relationships-country-and-region/countries-and-regions/indonesia/eu-indonesia-agreement_en. Accessed 29 February 2024.

169 *US—Gasoline, AB Report* (n 130) paras 28–29.

170 European Parliament 2018, amendment 307.

171 Letter from European Commission Director-General for Trade, Jean-Luc Demarty to European Commission Director-General for Energy, Dominique Ristori (March 2018) (a redacted version of this letter is on file with the authors).

172 See Council of the European Union 2018, p. 5.

far-fetched. If so, however, the resulting discrimination can hardly be qualified as 'merely inadvertent or unavoidable'—and therefore may ultimately be unjustifiable under Article XX GATT.c

## 10.4 Conclusion and Outlook

Sustainability concerns—from climate change to biodiversity loss—are central concerns for international actors in this day and age. This includes the EU, which often presents itself as a "green leader". Indeed, the EU Treaties mandate the Union to address such concerns in all areas of its internal and external action. To fulfil this mandate, in practice, the EU also resorts to unilateral measures, whose reach extends beyond its borders. However, in its endeavor to achieve sustainability goals outside its borders, the EU must also observe the limits that international law—and in particular, WTO law—sets for such unilateral measures.

As the present contribution shows, the way such unilateral, often consumption-based measures work is likely to cause trade conflicts in which such measures are anything but easy to defend legally. Importantly, this is not only true of measures, which make the access to the EU market directly dependent on compliance with certain EU sustainability criteria ("direct sustainability rules"). Rather, it is also true for measures such as the freeze and phase-out in the RED II, which impose no direct restrictions on market access ("indirect sustainability rules"), but still have the potential to modify the conditions of competition in the relevant market to the detriment of imported products.

How the respective panels in the WTO disputes concerning the RED II will resolve the complex legal issues that these conflicts entail, remains to be seen. If the panels were to rule in favor of Indonesia and Malaysia, the EU principally has three options. First, the EU could comply with the rulings and adapt the RED II rules. Any such adaptation would have to comply with WTO rules and therefore be subject to the same considerations as outlined in this contribution for the original measure. Second, the EU could appeal an unfavorable panel report. However, in light of the wider WTO governance crisis, which has culminated in the absence of a functioning AB, such an appeal would go 'into the void',[173] thus postponing the legal resolution of the dispute well until after the AB crisis would finally be resolved. Third, the EU could opt not to appeal the report but to also not comply with it and accept any retaliatory measures imposed by Indonesia or Malaysia. In light of these three options, it appears that the EU may be caught between a rock and a hard place. It can either insist on working towards sustainability at home and abroad, if necessary, unilaterally. Or, it

[173] Pauwelyn 2019, p. 303 ff. In this regard, it should be noted that neither Indonesia nor Malaysia are parties to the Multi-Party Interim Appeal Arbitration Arrangement, which serves as an alternative for resolving appeals in WTO disputes in the absence of a functioning AB, cf. Geneva Trade Platform. 2024. https://wtoplurilaterals.info/plural_initiative/the-mpia/. Accessed 7 March 2024.

can remain committed to a rules-based trading system, even if this system may not allow it to pursue the policies it considers necessary for a sustainable future.

## 10.5 Postscript

Shortly after this contribution was finalised, the WTO panel in a parallel palm oil-related dispute, initiated by Malaysia against the EU and certain member states, issued its report.[174] The panel's lengthy and nuanced assessment as well as the dissent of one panellist demonstrate how complex and controversial the involved issues still are.

Overall, the panel report supports the understanding that WTO law does not pose an insurmountable hurdle to the freeze and phase-out of high ILUC-risk biofuels in the EU's RED II. While the panel considers the EU measure *inter alia* inconsistent with key GATT non-discrimination rules, it accepts the measure's possible justification under Article XX GATT. Indeed, the panel recognises that climate change is 'inherently global in nature' and therefore, that there is a 'sufficient nexus' between the EU territory and the objective of limiting the risk of ILUC emissions, even if the relevant agricultural activities are mostly expected to occur outside the territory of the EU.[175] For a majority of the panel the justification under Article XX GATT only fails because the freeze and phase-out has been administered in a manner that constitutes arbitrary or unjustifiable discrimination between countries where the same conditions prevail. More specifically, the majority criticises that the EU 'failed to conduct a timely review of the data used to determine which biofuels are high ILUC risk' as well as 'deficiencies in the design and implementation of the low ILUC-risk criteria and certification procedure' set out in the Delegated Regulation.[176] By contrast, one panellist, in a separate opinion, argues that the inconsistencies with WTO rules are not limited to certain deficiencies in the administration of the freeze and phase-out. Rather, in their view, Malaysia has provided 'sufficient evidence' to show that the objective of the freeze and phase-out 'includes an element of protectionism'.[177]

Against this backdrop, both sides involved in the dispute seem keen to present the complex decision as a win.[178] For now, whether Malaysia or the EU will appeal

[174] *European Union and certain Member States—Certain measures concerning palm oil and oil palm crop-based biofuels*, WT/DS600/R, Report of the Panel, circulated on 5 March 2024.

[175] *EU and Certain Member States—Palm Oil (Malaysia)* (n 184) paras. 7.309-7.314.

[176] *EU and Certain Member States—Palm Oil (Malaysia)* (n 184) para. 7.1097.

[177] *EU and Certain Member States—Palm Oil (Malaysia)* (n 184) para. 7.1443.

[178] Cf., on the one hand, the statements of Malaysia's Plantation and Commodities Minister Datuk Seri Johari Abdul Ghani, New Straits Times 2024. https://www.nst.com.my/business/corporate/2024/03/1022504/malaysia-wins-key-parts-eu-dispute-over-palm-oil-based-biofuels%C2%A0. Accessed 7 March 2024. Cf., on the other hand, the statement of EC's DG Trade, European Commission 2024. https://policy.trade.ec.europa.eu/news/wto-rules-renewable-energy-dispute-2024-03-05_en. Accessed 7 March 2024.

the panel report 'into the void' remains to be seen.[179] Otherwise, the report should be adopted by the WTO Dispute Settlement Body within the next two months and thereby become binding between Malaysia and the EU. Some uncertainty also remains with regard to the parallel dispute initiated by Indonesia, which serves as the reference point in this contribution. This dispute remains pending before the same panel. While the panel is not legally bound by its decision in the Malaysian case, both cases do raise similar legal issues. However, since Indonesia requested a two-month suspension of the proceedings as of 5 March 2024, these issues remain unresolved for now.

Beyond the individual dispute, the panel report in *EU and Certain Member States—Palm Oil (Malaysia)* arguably has broader relevance for the EU's (future) endeavours of pushing for sustainability abroad. The panel has indeed accepted a sufficient nexus between the EU territory and global environmental problems which allows principally justifying measures with extraterritorial effects. However, such measures, their design and application, are still subject to rigorous case-by-case scrutiny. Thus, even in light of the recent WTO panel decision, the devil remains in the detail.

## References


Ahlgren S, Di Lucia L (2014) Indirect Land Use Changes of Biofuel Production – A Review of Modelling Efforts and Policy Developments in the European Union. Biotechnology for Biofuels 7: 35, http://www.biotechnologyforbiofuels.com/content/7/1/35.

Bodansky D (2000) What's so Bad about Unilateral Action to Protect the Environment? European Journal of International Law 11:339–347.

Boisson de Chazournes L (2000) Unilateralism and Environmental Protection: Issues of Perception and Reality of Issues. European Journal of International Law 11:315–388.

Bradford A (2020) The Brussels effect: How the European Union Rules the World. Oxford University Press, Oxford.

Bruch C and Broderick J (2016) International Law and Processes. In: Paddock LRC, Glicksman RL and Bryner NS (eds) Decision Making in Environmental Law, Edward Elgar Publishing, Cheltenham, pp 35–51.

Canakci M, Sanli, H (2008) Biodiesel Production from Various Feedstocks and Their Effects on the Fuel Properties. Journal of Industrial Microbiology and Biotechnology 35:431–441.

Charnovitz S (2002) The Law of Environmental "PPMs" in the WTO: Debunking the Myth of Illegality. Yale Journal of International Law 27:59–110.

Cooreman B (2016) Addressing Environmental Concerns through Trade: A Case for Extraterritoriality? The International and Comparative Law Quarterly 65:229–248.

Council of the European Union (2018) General Secretariat Note of 4 June 2018 to Council, Doc 9287/18.

De Ville F, Happersberger S, Kalimo H (2023) The Unilateral Turn in EU Trade Policy? The Origins and Characteristics of the EU's New Trade Instruments. European Foreign Affairs Review 28(SI) pp 15–34.

Denza E (2012) International Aviation and the EU Carbon Trading Scheme: Comment on the Air Transport Association of America Case. European Law Review 37:314–326.


[179] Pauwelyn 2019. See above, 3.4.

Dero-Bugny D, Motte-Baumvol J (2021) Shaping EU External Relations Beyond Treaty-Making: The Scope of Extraterritorial EU Legislation and Its Enforcement Challenges. In: Douma WTh et al (eds) The Evolving Nature of EU External Relations Law, T.M.C Asser Press, the Hague, pp 41–60.

Ehlermann CD, N Lockhart (2004) Standard of Review in WTO Law. Journal of International Economic Law 7:491–521.

European Commission (2003) Forest Law Enforcement, Governance and Trade (FLEGT) - Proposal for an EU Action Plan, COM(2003) 251 final.

European Commission (2008) Addressing the Challenges of Deforestation and Forest Degradation to Tackle Climate Change and Biodiversity Loss, COM(2008) 645 final.

European Commission (2019a) Stepping up EU Action to Protect and Restore the World's Forests, COM(2019) 352 final.

European Commission (2019b) Report on the Status of Production Expansion of Relevant Food and Feed Crops Worldwide, COM(2019) 142 final.

European Commission (2021a) An Open, Sustainable and Assertive Trade Policy, COM(2021) 66 final.

European Commission (2021b) Proposal for a regulation on the making available on the Union market as well as export from the Union of certain commodities and products associated with deforestation and forest degradation and repealing Regulation (EU) No 995/2010, COM(2021) 706 final.

European Commission (2023) Report on the Exercise of the Power to Adopt Delegated Acts Conferred on the Commission Pursuant to Directive (EU) 2018/2001/EU on the promotion of the use of energy from renewable sources, COM(2023) 204 final.

European Parliament (2016) Resolution of 4 April 2017 on Palm Oil and Deforestation of Rainforests, 2016/2222(INI).

European Parliament (2018) Amendments adopted by the European Parliament on 17 January 2018 on the proposal for a directive on the promotion of the use of energy from renewable sources (recast), P8_TA(2018)0009.

European Parliament (2020) The EU's Role in Protecting and Restoring the World's Forests, P9_TA(2020)0212.

European Union (2021a) First written submission by the European Union in the World Trade Organisation Panel Proceedings EU – Palm Oil (Indonesia), DS 593. https://circabc.europa.eu/ui/group/cd37f0ff-d492-4181-91a2-89f1da140e2f/library/46ce6da8-f5e5-4c60-9f7f-34c8f3bf9595/details.

European Union (2021b) Responses by the European Union to Questions following the Panel's first substantive meeting with the parties in the World Trade Organisation Panel Proceedings EU – Palm Oil (Indonesia), DS 593, https://circabc.europa.eu/ui/group/cd37f0ff-d492-4181-91a2-89f1da140e2f/library/1d9f37dc-8aa9-4e98-948f-3d0ebb9cd853/details.

European Union (2021c) Second written submission by the European Union in the World Trade Organisation Panel Proceedings EU – Palm Oil (Indonesia), DS 593. https://circabc.europa.eu/ui/group/cd37f0ff-d492-4181-91a2-89f1da140e2f/library/1191532d-0d0f-4c99-8df1-6a20754cc43e/details.

Galli A (2023) EU-27 Ecological Footprint was Primarily Driven by Food Consumption and Exceeded Regional Biocapacity from 2004 to 2014. Nature Food 4:810–822.

Gáspár-Szilágy S (2022) The 'Palm Oil Wars' or How the EU's 'Inflated' Common Commercial Policy Might Need to Prioritize Its Non-trade Values. European Foreign Affairs Review 27:29–56.

Gattini A (2012) Between Splendid Isolation and Tentative Imperialism: The EU's Extension of its Emission Trading Scheme to International Aviation and The ECJ's Judgment in the ATA Case. International and Comparative Law Quarterly 61:977-991.

Gstöhl S (2008) The Internal Market's External Dimension. In: Pelkmans J, Hanf D, Chang M (eds) The EU Internal Market in Comparative Perspective: Economic, Political and Legal Analyses, Peter Lang, Brussels, pp 221–248.

Hervé A (2022) European Unilateralism as a Tool for Regulating International Trade: A Necessary Evil in a Collapsing Multilateral System. Fondation Robert Schuman European issues n°626.

Indonesia (2021) First written submission by Indonesia in the World Trade Organisation Panel Proceedings EU – Palm Oil (Indonesia), DS 593.

Indonesia (2021b) Responses by Indonesia to Questions following the Panel's first substantive meeting with the parties in the World Trade Organisation Panel Proceedings EU – Palm Oil (Indonesia), DS 593.

Kettner C, Kletzan-Slamanig D (2020) Is there Climate Policy Integration in European Union Energy Efficiency and Renewable Energy Policies? Yes, No, Maybe. Environmental Policy and Governance 30:141–150.

Klamert M (2019) Art 11 TFEU. In: Kellermann M, Klamert M, Tomkin J (eds) The EU Treaties and the Charter of Fundamental: Rights. A Commentary, Oxford University Press, Oxford, pp 386–387.

Kotzé LJ, Adelman S (2023) Environmental Law and the Unsustainability of Sustainable Development: A Tale of Disenchantment and of Hope. Law & Critique 34:227–248.

Kulovesi K (2012) Climate Change in EU External Relations: Please Follow My Example (or I Might Force You to). In: Morgera E (ed) The External Environmental Policy of the European Union: EU and International Law Perspectives, Cambridge University Press, Cambridge, pp 115–148.

Laborde D (2011) Assessing the Land Use Change Consequences of European Biofuel Policies. International Food Policy Research Institute, Washington.

Larik J, Wessel RA (2020) Instruments of EU External Action. In: Wessel RA, Larik J (eds) EU External Relations Law, 2nd edn. Hart Publishing, Oxford, pp 101–137.

Marín Durán G (2015) NTBs and the WTO Agreement on Technical Barriers to Trade: The Case of PPM-Based Measures Following US – Tuna II and EC – Seal Products. European Yearbook of International Economic Law, 6:87–136.

Marín Durán, Morgera E (2012) Environmental Integration in the EU's External Relations: Beyond Multilateral Dimensions, Hart Publishing, Oxford.

Mayr S et al. (2021) Palm oil, the RED II and WTO law: EU Sustainable Biofuel Policy Tangled up in Green?. RECIEL 30:155–293.

Mignolli A (2018) The European Union and Sustainable Development: A study on Unilateral Trade Measures, Edizione Nuova Cultura, Rome.

Mitchell AD, Merriman D (2020) Indonesia's WTO Challenge to the European Union's Renewable Energy Directive: Palm Oil and Indirect Land-Use Change. Trade, Law and Development 12:548–622.

Mitchell AD, Tran C (2010) The Consistency of the European Union Renewable Energy Directive with World Trade Organization Agreements: The Case of Biofuels. Renewable Energy Law and Policy Review 1:33–44.

Morgera E (2013) Ambition, Complexity, and Legitimacy of Pursuing Mutual Supportiveness Through the EU's External Environmental Action. In: van Vooren, B, Blockmans S, Wouters J (eds) The EU's Role in Global Governance: The Legal Dimension, Oxford University Press, Oxford, pp 195–208.

Nessel S (2023) Umwelt- und Konsumpolitik in der EU: Zum aktuellen Stand und zur Notwendigkeit einer realitätsnahen Fundierung. SWS-Rundschau [Social Science Review] 63:160–179.

Öberg ML (2020) Internal Market Acquis as a Tool in EU External Relations: From Integration to Disintegration. Legal Issues of Economic Integration 47:151–178.

Oeter S (2013) Article 21. In: Blanke HJ, Mangiameli S (eds) The Treaty on European Union (TEU): A Commentary, Springer, Berlin, pp 831–874.

Pauwelyn J (2019) WTO Dispute Settlement Post 2019: What to Expect? Journal of International Economic Law 22:297–321.

Prechal S (2006) Directives in EC Law, 2nd edn, Oxford University Press, Oxford.

Purvis B, Mao Y, Robinson D (2019) Three Pillars of Sustainability: In Search of Conceptual Origins. Sustainability Science 14:681–695 (2019).

Ramopoulos T (2019) Art 21 TEU. In: Kellermann M, Klamert M, Tomkin J (eds) The EU Treaties and the Charter of Fundamental: Rights. A Commentary, Oxford University Press, Oxford, pp 200–204.
Regan DH (2007) The meaning of 'necessary' in GATT Article XX and GATS Article XIV: the myth of cost-benefit balancing. World Trade Review 6:347–369.
Scott J (2014a) Extraterritoriality and Territorial Extension in EU Law. The American Journal of Comparative Law 62:87–125.
Scott J (2014b) The New "EU Extraterritoriality". Common Market Law Review 51:1343–1380.
Scott J (2019) The Global Reach of EU Law. In: Cremona M, Scott J (eds) EU Law Beyond EU Borders, Oxford University Press, Oxford.
Scott J (2020) Reducing the EU's Global Environmental Footprint. German Law Journal 21:10–16.
Scott J, Rajamani L (2013) Contingent Unilateralism: International Aviation in the European Emissions Trading Scheme. In: Van Vooren B, Blockmans S, Wouters J (eds) The EU's Role in Global Governance: The Legal Dimension, Oxford University Press, Oxford, pp 209–223.
Sifonis D (2018) Environmental Process and Production Methods (PPMs) in WTO Law, Springer International Publishing, Cham.
Tukker A et al. (2016) Environmental and Resource Footprints in a Global Context: Europe's structural Deficit in Resource Endowments. Global Environmental Change 40:171–181.
United Nations General Assembly (2017) Resolution 71/313. Work of the Statistical Commission pertaining to the 2030 Agenda for Sustainable Development, A/RES/71/313.
United States Department of Agriculture, Global Agricultural Information Network (2023) Biofuels Annual, E42023-0033. https://apps.fas.usda.gov/newgainapi/api/Report/DownloadReportByFileName?fileName=Biofuels%20Annual_The%20Hague_European%20Union_E42023-0033.pdf.
Van den Bossche P, Zdouc W (2022) The Law and Policy of the World Trade Organization, 5th edn. Cambridge University Press, Cambridge.
van Zeben J (2022) Environmental Law. In: Garben S and Gormley Laurence (eds), The Oxford Encyclopedia of EU Law, Oxford University Press, Oxford, https://doi.org/10.1093/law-oeeul/law-oeeul-e112.
Vranes E (2009) Trade and the Environment, Oxford University Press, Oxford.
Working Party on Border Tax Adjustments (1997) Report BISD 18S/97.
World Trade Organization Committee on Market Access (2022) Minutes of 18 and 19 October 2022, G/MA/M/77. https://docs.wto.org/dol2fe/Pages/SS/directdoc.aspx?filename=q:/G/MA/M77.pdf&Open=True.
WWF (2022) Living Planet Report 2022 – Building a nature-positive society, WWF, Gland.
Zürn M (2020) Multilateralism in Crisis: a European Perspective. In: Meyer T, de Sales Marques JL, Telò J (eds) Towards a New Multilateralism: Cultural Divergence or Political Convergence? 2nd edn. Routledge, London, pp. 119–131.

**Birgit Hollaus** is a postdoctoral Research and Teaching Associate at WU Vienna University of Economics and Business.

**Stefan Mayr** is a Senior Scientist and Reader in EU and International Law at WU Vienna University of Economics and Business.

# Chapter 11
# The EU's Evolving Approach to Environmental Provisions in Free Trade Agreements

**Gracia Marín Durán**

**Contents**



**Abstract** With the entry into force of the Lisbon Treaty, the promotion of sustainable development became a constitutional imperative for the EU in its relations with third countries. A practical manifestation of this has been the inclusion of environmental (and labour) provisions in the free trade agreements (FTAs) concluded by the EU since 2010, which have sought to advance a distinct 'promotional' model for regulating trade-environment (and trade-labour) linkages in FTAs. However, the more recent agreements with the United Kingdom and New Zealand signal a paradigm shift in the EU's approach to promoting environmental sustainability in FTAs. This chapter examines such an evolution in EU FTA practice with regards to both substantive environmental commitments and monitoring and enforcement mechanisms. In doing so, it discusses some of the key challenges that have emerged in terms of policy ambition and compliance.

**Keywords** Environmental sustainability · Sustainable development · Free trade agreements · Sanction-based enforcement · Level playing field · Effectiveness · Fairness

G. M. Durán (✉)
Faculty of Laws, University College London, London, United Kingdom
e-mail: gracia.marin-duran@ucl.ac.uk



R. A. Wessel et al. (eds.), *EU External Relations Law and Sustainability*, Global Europe: Legal and Policy Issues of the EU's External Action 4,
https://doi.org/10.1007/978-94-6265-655-0_11

## 11.1 Introduction

While there is a broad consensus that international trade should support sustainable development,[1] how exactly environmental (and social) concerns should be integrated into trade agreements has long been a matter of contention. For the European Union (EU), the promotion of sustainable development through its external trade policy became a constitutional imperative with the entry into force of the Lisbon Treaty.[2] A practical manifestation of this are the so-called 'Trade and Sustainable Development' (TSD) chapters, which have been a standard component of the comprehensive free trade agreements (FTAs) concluded by the EU with developed and developing countries since 2010. These TSD chapters are heralded as a central part of the EU's 'value-based' trade policy, where the Union is determined to use trade as a vehicle for promoting 'sustainable development worldwide'[3]—that is, not only in third countries but presumably also in the EU itself. In doing so, the EU has sought to depart from the pre-existent policy practice by the United States (US) and Canada, and to develop a distinct model for regulating trade-environment linkages in FTAs. In essence, the EU's approach has been more ambitious in terms of substantive commitments on environmental sustainability but less coercive with regards to their enforcement, notably by excluding the use of economic sanctions as a remedy in cases of infringement. However, after defending this 'promotional' model for a decade, post-2020 FTAs concluded by the EU are gradually marking a paradigm shift in relation to the enforcement of sustainability commitments.

This chapter traces the evolution of environmental provisions in EU FTAs, from the earlier new-generation agreements (starting with the 2010 EU-Korea FTA),[4] to the 2020 EU-UK Trade and Cooperation Agreement (TCA)[5] and the more recent 2023 EU-New Zealand FTA.[6] This comparative analysis seeks to pinpoint the main innovations with respect to both substantive commitments on environmental sustainability (Sect. 11.2.1) and institutional mechanisms for their implementation and enforcement (Sect. 11.2.2). The chapter also discusses the thorny issue of enforcing environmental provisions in EU FTAs through economic sanctions. It challenges the seemingly conventional wisdom that such sanction-based enforcement is warranted

[1] United Nations (UN) General Assembly, 'Transforming Our World: The 2030 Agenda for Sustainable Development' UN Doc/A/RES/70/1, 25 September 2015, paras 17.10–17.12.

[2] Treaty on the European Union (TEU), Articles 3(5) and 21; Treaty on the Functioning of the European Union (TFEU), Article 207(1).

[3] European Commission 'Communication on Trade for all – Towards a More Responsible Trade and Investment Policy', COM(2015)497 final, 14 October 2015, p. 15.

[4] Free Trade Agreement between the European Union and its Member States, of the one part, and the Republic of Korea, of the other part, signed on 6 October 2010, O.J. 2011 L 127/6 [EU-Korea FTA].

[5] Trade and Cooperation Agreement between the European Union and the European Atomic Energy Community, of the one part, and the United Kingdom of Great Britain and Northern Ireland, of the other part, O.J. 2021 L149/10, signed on 30 December 2020 [EU-UK TCA].

[6] Free Trade Agreement between the European Union and New Zealand, signed on 9 July 2023, O.J. 2024 L 1/229 [EU-New Zealand FTA].

from an effectiveness perspective, and argues that its implementation raises a number of legal challenges as well as serious equity concerns (Sect. 11.3).

## 11.2 Evolution of Environmental Provisions in EU FTAs

### *11.2.1 Substantive Commitments*

#### 11.2.1.1 TSD Chapters in New-Generation FTAs (2010–2020)

Beginning with the 2010 EU-Korea FTA,[7] EU free trade agreements[8] have consistently included dedicated TSD chapters, which contain a number of substantive and institutional provisions in respect of environmental (and labor)[9] protection. While there is some variation across agreements, all TSD chapters share three main types of substantive commitments.[10]

The first category is based on international environmental standards and may be defined as 'minimum-level' clauses. These demand each FTA party to effectively

[7] Technically, the 2008 EU-CARIFORUM Economic Partnership Agreement was the first to include legally-binding environmental provisions, but these differ in some respects from the dedicated TSD chapters in subsequent EU FTAs and will not be examined here.

[8] In this chapter, the term 'free trade agreement' is used for simplicity, even though EU agreements have different terminology. Trade Agreement between the European Union and its Member States, of the one part, and Colombia and Peru, of the other part, signed on 26 June 2012, O.J. 2012 L 354/21 [EU-COPE FTA]; Comprehensive Economic and Trade Agreement between Canada, of the one part, and the EU and its Member States, of the other part, signed on 30 October 2016, O.J. 2017 L 11/23 [EU-Canada FTA]; Agreement between the European Union and Japan for an Economic Partnership, signed on 6 July 2017, O.J. 2018 L330/3 [EU-Japan FTA]; Free Trade Agreement between the European Union and the Republic of Singapore, signed on 19 October 2018, O.J. 2019 L 294/3 [EU-Singapore FTA]; Free Trade Agreement between the European Union and the Socialist Republic of Vietnam, signed on 30 June 2019, O.J. 2020 L 63/3 [EU-Vietnam FTA]; Agreement amending the Association Agreement between the EU and its Member States, of the one part, and Mexico, of the other part, agreed in principle on 21 April 2018 [EU-Mexico FTA], https://policy.trade.ec.europa.eu/eu-trade-relationships-country-and-region/countries-and-regions/mexico/eu-mexico-agreement_en (accessed 27 March 2024); EU-Mercosur Trade Agreement, agreed in principle on 28 June 2019 [EU-Mercosur FTA], https://policy.trade.ec.europa.eu/eu-trade-relationships-country-and-region/countries-and-regions/mercosur/eu-mercosur-agreement_en (accessed 27 March 2024).

[9] This chapter focuses on environmental protection as an integral pillar of sustainable development. It does not deal with social protection due to space constraints.

[10] In addition to these core commitments, TSD chapters include provisions on specific environmental issues (e.g., trade and climate change, biodiversity conservation and illegal trade in endangered species, sustainable forest management and trade in illegally harvested timber, sustainable fisheries management, fossil fuel subsidy reform). There is disparity across EU FTAs on the content of these provisions and a detailed examination is outside the scope of this chapter. For an overview, see Velut et al. 2022, pp. 48–50.

implement in their domestic laws and practices the multilateral environmental agreements (MEAs) that they have each ratified.[11] These provisions add nothing substantively new, just incorporate pre-existing environmental commitments that are already binding on each FTA party under the relevant MEAs. Nonetheless, they are significant for promoting global environmental sustainability in a wide sense, even where bilateral trade and investment are not affected. In other words, they establish a minimum baseline of environmental protection that ought to be respected across each party's territory, irrespective of any trade (or investment) effects. However, the content of this minimum regulatory floor will vary from party to party, depending on the MEAs it has actually ratified. Unlike for social protection, there is no internally-agreed list of 'core' MEAs similar to the core labor standards enshrined in the International Labor Organization (ILO)'s Declaration on Fundamental Principles and Rights at Work and associated ILO Conventions,[12] which all FTA parties undertake to ratify and effectively implement.[13]

Nonetheless, there are some global MEAs with quasi-universal acceptance that have been ratified -and hence, ought to be effectively implemented- by all FTA parties, including in the areas of climate change (1985 Vienna Convention for the Protection of the Ozone Layer[14] and its 1987 Montreal Protocol;[15] 1992 United Nations Framework Convention on Climate Change[16] and its 2015 Paris Agreement[17]), biodiversity conservation (1973 Convention on International Trade in Endangered Species of Wild Fauna and Flora (CITES);[18] 1992 Convention on Biological Diversity[19]), and hazardous substances and waste (1989 Basel Convention on the Control of Transboundary Movements of Hazardous Wastes and their

[11] E.g., EU-Korea FTA, Article 13.5(2); EU-Singapore FTA, Article 12.6(2); EU-Canada FTA, Article 24.4; EU-Japan FTA, Article 16.4(2); EU-Mercosur FTA, Article 14.5(3).

[12] The relevant eight ILO Conventions are: Convention C87 on Freedom of Association and Protection of the Right to Organize, adopted 9 July 1948; Convention C98 on the Right to Organize and Collective Bargaining, adopted 1 July 1949; Convention C29 on Forced Labor, adopted 28 June 1930; Convention C105 on the Abolition of Forced Labor, adopted 25 June 1957; Convention C138 on Minimum Age for Admission to Employment, adopted 26 June 1973; Convention C182 on the Prohibition and Immediate Action for the Elimination of the Worst Forms of Child Labor, adopted 17 June 1999; Convention C100 on Equal Remuneration, adopted 29 June 1951; Convention C111 on Discrimination (Employment and Occupation), adopted 25 June 1958.

[13] E.g., EU-Singapore FTA, Article 12.3(3); EU-Japan FTA, Article 16.3(2); EU-COPE FTA, Article 269(3).

[14] Vienna Convention for the Protection of the Ozone Layer, adopted 22 March 1985, 1513 UNTS 293.

[15] Montreal Protocol on Substances that Deplete the Ozone Layer, adopted 16 September 1987, 1522 UNTS 3.

[16] United Nations Framework Convention on Climate Change, adopted 4 June 1992, 1771 UNTS 107.

[17] Paris Agreement (PA), adopted 12 December 2015, 3156 UNTS.

[18] Convention on International Trade in Endangered Species of Wild Fauna and Flora, adopted 3 March 1973, 993 UNTS 243.

[19] Convention on Biological Diversity, adopted 5 June 1992, 1760 UNTS 79.

Disposal;[20] 1998 Rotterdam Convention on the Prior Informed Consent Procedure for Certain Hazardous Chemicals and Pesticides in International Trade;[21] 2001 Stockholm Convention on Persistent Organic Pollutants[22]).

The second kind of commitments concern existing domestic environmental legislation more broadly (i.e., above and beyond that implementing international standards) and take the form of 'non-derogation and effective enforcement' clauses. These provide that FTA parties shall not waive or derogate, nor fail to effectively enforce, their environmental laws 'in a manner affecting [bilateral] trade or investment',[23] or in some agreements, 'to encourage [bilateral] trade or investment'.[24] Hence, these provisions do not prohibit any derogation or enforcement failure of domestic environmental standards, but only insofar as actual or intended effects on bilateral trade (or investment) can be shown, making a violation thereof harder to establish. In either case, the trade condition reflects the economically driven rationale behind these clauses. Their most immediate, direct goal is to avoid perceived 'unfair' competitive advantages that an FTA party could enjoy through undercutting its levels of environmental protection and thereby ensure a 'level playing field' in bilateral trade (and investment) relations—even if, *indirectly*, this may also preserve the global environment.

The third type of clauses relate to current and future levels of environmental protection in domestic laws but, unlike the other two sets of provisions, do not enshrine mandatory obligations. Each FTA party 'shall seek',[25] or 'shall strive',[26] to ensure that its laws and policies provide for 'high levels' of environmental protection and to continue to improve them over time. This aspirational and imprecise language allows for different levels of compliance and, arguably, renders these provisions difficult to implement and enforce in practice. However, they are not entirely meaningless: 'an overt weakening' of existing environmental laws 'could hardly be said to be consistent with striving to improve these standards'.[27]

---

[20] Basel Convention on the Control of Transboundary Movements of Hazardous Wastes and their Disposal, adopted 22 March 1989, 1673 UNTS 57.

[21] Rotterdam Convention on Prior Informed Consent for Certain Hazardous Chemicals and Pesticides, adopted 10 September 1998, 2244 UNTS 337.

[22] Stockholm Convention on Persistent Organic Pollutants, adopted 22 May 2001, 2256 UNTS 119.

[23] E.g., EU-Singapore FTA, Article 12.12.

[24] EU-Canada FTA, Articles 23.4–23.5; EU-Mexico FTA, Articles 27.2(3)–(5); EU-Mercosur FTA, Articles 14.2(3)–(5). Some FTAs combine both formulations, either making clear that 'encouragement' is dependent upon actual trade effects being shown [EU-Korea FTA, Article 13.7(2); EU-Japan FTA, Article 16.2(2)], or independently from each other [EU-COPE FTA, Articles 277(1)–(2)].

[25] E.g., EU-Korea FTA, Article 13.13; EU-Canada FTA, Article 24.3.

[26] E.g., EU-Singapore FTA, Article 12.2(2); EU Japan FTA, Article 12.2(1); EU-COPE FTA, Article 268.

[27] Bartels 2013, p. 308.

#### 11.2.1.2 Environment and Climate Chapter in EU-UK TAC (2020)

The TCA was bound to be different from all other EU FTAs since it governs the new relationship between the EU and a former Member State and, hence, was concluded in a very peculiar context of pre-existing deep levels of integration and regulatory convergence, including in the environmental field. Environmental provisions are mainly found in a novel title on 'Level Playing Field for Open and Fair Competition and Sustainable Development', which also includes rules on competition policy, subsidy control, taxation and social protection, and which proved particularly contentious to negotiate.[28] This seeks to address the EU's concerns over the UK's new 'freedom' to adopt divergent regulatory standards in the relevant areas and was set as a pre-condition for maintaining unfettered 'zero-tariff and zero-quota' UK access to the EU market.[29] As such, while the environmental commitments of the TCA are largely similar in some respects when compared to regular EU FTAs (e.g. the minimum-level clause requiring compliance with ratified MEAs),[30] they are more ambitious in others.

First, the EU and UK are prohibited from 'weaken[ing] or reduc[ing], in a manner affecting [bilateral] trade or investment' their respective levels of climate/environmental protection 'below the levels that [were] in place at the end of the transition period', including by failing to effectively enforce their climate/environmental laws.[31] This key non-regression obligation is broader in scope than the 'non-derogation and effective enforcement' clauses previously examined, in that it covers *any* weakening or reduction of domestic levels of environmental protection, and not just waivers or derogations from domestic environmental laws and sustained or recurring non-enforcement.[32] At the same time, a violation of the non-regression obligation in the TCA is similarly conditioned on demonstrating trade/investment effects. This means that a lowering from the common levels of climate/environmental protection that prevailed at the end of the transition period is permitted as long as it does not affect the economic level playing field between the EU and the UK.[33]

Second, the TCA provides for a 'rebalancing mechanism', which may be triggered in cases of divergence in future environmental (and other regulatory) standards between the EU and the UK, and which is unprecedented in EU FTAs. Either party may take rebalancing measures (including trade sanctions) if '*material impacts* on [bilateral] trade or investment … are arising as a result of *significant* divergences between the Parties'[34] in the areas of environmental and social protection, as well as subsidy control. Such measures 'shall be restricted with respect to their scope and

[28] EU-UK TCA, Title XI, Chapters 2–6; and for an overview, Peers 2022, pp. 62–66.

[29] Leonelli 2021, p. 614.

[30] EU-UK TCA, Article 400(2).

[31] EU-UK TCA, Article 391(2).

[32] Bronckers and Gruni 2021, p. 32.

[33] Leonelli 2021, p. 625.

[34] EU-UK TCA, Article 411(2) (emphasis added).

duration to what is strictly necessary and proportionate in order to remedy the situation',[35] and their adoption is subject to specific procedural requirements.[36] While the TCA thus provides for economic remedies should 'significant' divergences in EU/UK levels of environmental protection arise in the future, it sets a higher trade-related threshold ('material impacts') than that enshrined in the non-regression clause of the agreement. Hence, the burden of proof is likely to be even harder to discharge in practice for the party seeking to adopt such rebalancing measures.[37]

Third, the fight against climate change is elevated to constituting one of the 'essential elements' of the EU-UK partnership established by the TAC and any supplementing agreements, albeit the actual significance of this remains uncertain.[38] In principle, each party is obliged to 'refrain from acts or omissions that would materially defeat the object and purpose of the Paris Agreement'.[39] A breach of this essential obligation is subject to a special procedure potentially leading to fast-track termination or suspension of the TCA (or any supplementing agreement), partly or wholly.[40] However, in practice, the EU has only triggered this type of 'essential elements' clauses (concerning respect for democratic principles and human rights) on limited occasions under regular FTAs.[41]

#### 11.2.1.3 TSD Chapter in EU-New Zealand FTA (Post-2023)

At the time of writing, the EU-New Zealand FTA is the most recently concluded agreement incorporating a TSD chapter, and it followed the review process on the implementation and enforcement of these chapters launched by the Commission in June 2021.[42] As such, it presents some similarities and differences with previous regular FTAs. With regard to the former, the parties 'shall effectively implement' those MEAs that they have each ratified (minimum-level clause).[43] Likewise, each party 'shall strive to ensure that its relevant [laws] and policies provide for, and encourage, high levels of environmental [and labor] protection, and shall strive to improve such levels, [laws] and policies'.[44] The aspirational wording of this clause makes it legally weaker and harder to enforce in practice. But compared to other regular FTAs, one change is noticeable in terms of ensuring no back-pedaling in

---

35 Ibid.

36 Ibid., Article 411(3).

37 Leonelli 2021, pp. 632–633.

38 EU-UK TCA, Article 771.

39 Ibid., Article 764(1).

40 Ibid., Articles 772(1) and (4).

41 Peers 2022, pp. 53–54; Bartels 2013, pp. 299–305.

42 European Commission, 'Communication on the Power of Trade Partnerships: Together for Green and Just Economic Growth' COM(2022) 409 final, 29 June 2022 [Commission Communication 2022].

43 EU-New Zealand FTA, Article 19.5(2).

44 Ibid., Article 19.2(3).

existing levels of protection under domestic environmental laws. That is, the 'non-derogation and effective enforcement' clause[45] is reinforced by a broader and legally-binding non-regression obligation: '[a] Party shall not weaken or reduce the levels of protection afforded in its environmental law *in order to encourage* trade or investment'.[46] Here, a breach of the non-regression obligation is conditioned upon demonstrating an *intention* of encouraging trade (or investment), rather than an *actual* effect on trade (as under the EU-UK TCA), but it is unclear which one would be easier to establish in practice.

To recap, the evolution of substantive environmental provisions under EU FTAs since 2010 has mainly centered on strengthening the guarantees against a weakening of domestic environmental laws for competitive purposes, which is most evident in the TCA given the close economic interdependence and geographical proximity between the EU and the UK. As illustrated in Table 11.1, the broader reach of non-regression obligations under both the EU-UK TCA and EU-New Zealand FTA has been closely tied to proven impacts on bilateral trade or investment, reflecting the narrow focus of these clauses on safeguarding an economic level playing field.

## *11.2.2 Implementation and Enforcement Provisions*

### 11.2.2.1 New-Generation FTAs (2010–2020)

All EU FTAs concluded during the period 2010–2020 set up specific institutional mechanisms for the implementation and enforcement of TSD chapters. A specialized body, named Committee or Board on Trade and Sustainable Development and made up of senior representatives from each party, is assigned with the task of overseeing the implementation of the TSD chapter and guiding further bilateral cooperation in this area.[47] In addition to this joint inter-governmental body, each party is also required to have in place domestic consultative mechanisms (e.g. Domestic Advisory Groups), comprising a balanced representation of business, environmental and labor stakeholders, with a view to seeking their input on matters under the TSD chapter.[48] Bilateral consultative mechanisms are also foreseen in various forms (e.g. Civil Society Forums) for the parties to conduct regular dialogue with these stakeholders on the implementation of the TSD chapter.[49] Despite this emphasis on stakeholder participation in the monitoring of TSD chapters, there is no formal

[45] Ibid., Articles. 19.2(5)–(6).

[46] Ibid., Article 19.2(4) (emphasis added).

[47] E.g., EU-Korea FTA, Articles 13.12(2)–(3); EU-Singapore FTA, Articles 12.15(2)–(3); EU-COPE FTA, Articles 280; EU-Canada FTA, Articles 22.4.

[48] E.g., EU-Korea FTA, Articles 13.12(4)–(5); EU-Singapore FTA, Articles 12.15(5); EU-COPE FTA; Article 281; EU-Canada FTA, Article 24.13(5); EU-Japan FTA, Article 16.15.

[49] E.g., EU-Korea FTA, Article 13.13; EU-Singapore FTA, Article 12.15(4); EU-COPE FTA, Article 282; EU-Canada FTA, Article 22.5; EU-Japan FTA, Article 16.16.

**Table 11.1** Substantive environmental provisions in EU FTAs (2010–2023)

| Agreement | Minimum-level clause | High-level clause | Non-derogation/ non-enforcement clause | Non-regression clause | Rebalancing clause |
|---|---|---|---|---|---|
| 2010–2020 FTAs | Effective implementation of ratified MEAs | 'High levels' of protection in domestic environmental laws | No derogation or lack of enforcement of domestic environmental laws | Not included, or legally weak[50] | Not included |
| | Mandatory obligation | Aspirational commitment, violation difficult to establish | Mandatory obligation | | |
| | Not conditioned on trade/ investment related effects | | Violation conditioned on actual/intended effects on trade or investment | | |
| EU-UK TCA (2020) | Same as above | Same as above | Same as above | No weakening of domestic environmental laws | Rebalancing measures may be adopted if 'significant' divergences in domestic environmental laws |
| | | | | Mandatory obligation | Discretionary right |
| | | | | Violation conditioned on actual/intended effects on trade or investment | Adoption of rebalancing measures conditioned on 'material impacts' on trade/ investment |
| EU-New Zealand FTA (2023) | Same as above | Same as above | Same as above | Same as EU-UK TCA | Not included |

requirement upon the parties to follow up on the submissions received, and the operation of these mechanisms has met considerable criticism for lack of accountability and transparency, as well as different degrees of inclusiveness.[51]

[50] Where included in previous FTAs, non-regression clauses were significantly weaker, using aspirational rather than mandatory language: Leonelli 2021, pp. 622–624.

[51] For a careful analysis, see Drieghe et al. 2022, pp. 586–92; Prévost and Alexovicova 2019, pp. 244–48.

Disputes concerning any matter arising under the TSD chapter may only be resolved through its specific dispute settlement procedures, and recourse to the general dispute settlement mechanism is explicitly excluded in most EU FTAs[52] (with the exception of the EU-CARIFORUM EPA).[53] Pursuant to this self-contained system of dispute settlement, the parties are first required to seek a mutually satisfactory resolution of the matter through governmental consultations.[54] Where these initial consultations do not lead to a satisfactory resolution of the dispute, any party may refer the matter to a Panel of Experts, composed of members with expertise in trade, environmental and labor issues. The Panel is to issue a final report within the established timeframe, with findings as to whether there has been a failure to comply with the relevant obligations and (non-binding) recommendations for the resolution of the matter, which is to be made public unless the parties agree otherwise. In its deliberations, the Panel may seek information and advice from (*inter alia*) multilateral environmental organizations and private stakeholders. The follow-up to the Panel's report is to be monitored by the joint Board or Committee on Trade and Sustainable Development, and private stakeholders may submit observations in this regard.[55]

While stakeholder involvement is thus favored in the dispute settlement process, there is no formal requirement to act upon any such submission. At the domestic level, the EU has taken some steps to improve stakeholder involvement in the enforcement of TSD chapters, including the creation of a 'Single Entry Point' (SEP) in November 2020 under the remit of the Chief Trade Enforcement Officer (CTEO) in the Commission's Directorate General for Trade. The SEP provides a centralized contact point for EU-based stakeholders (e.g. industry associations, trade unions or non-governmental organizations) to lodge individual or collective complaints (including through Domestic Advisory Groups) on violations of TSD commitments, which are handled by the CTEO. However, stakeholder use of this new system to raise TSD-related concerns has been limited to date.[56]

[52] E.g., EU-Korea FTA, Article 13.16; EU-Singapore FTA, Article 12.16(1); EU-COPE FTA, Article 285(5); EU-Canada FTA, Article 24.16; EU-Japan FTA, Article 16.17(1); EU-Mexico FTA, Article 27.15; EU-Mercosur FTA, Article 14.15(5).

[53] The EU-CARIFORUM EPA differs from other EU FTAs in that the regular dispute settlement procedure applies to matters arising out of the TSD chapter, although trade sanctions are ruled out for breaches of trade/environment and trade/labor provisions [Article 213(2)]. However, this remedy carve-out does not apply to violations of environmental and labor standards set out in the investment chapter of the agreement, which may thus be subject to trade sanctions (Articles 72 and 73).

[54] E.g., EU-Korea FTA, Article 13.14; EU-Singapore FTA, Articles 12.16(2)–(6); EU-COPE FTA, Article 283; EU-Canada FTA, Article 24.14; EU-Japan FTA, Articles 16.17(2)–(5); EU-Mexico FTA, Article 27.16; EU-Mercosur FTA, Article 14.16.

[55] E.g., EU-Korea FTA, Article 13.15; EU-Singapore FTA, Article 12.17; EU-COPE FTA, Articles 284–285; EU-Canada FTA, Article 24.15; EU-Japan FTA, Article 16.18; EU-Mexico FTA, Article 27.17; EU-Mercosur FTA, Article 14.17.

[56] Commission Communication 2022, p. 9. See also the amendments to the EU Trade Enforcement Regulation: Regulation (EU) 2021/167 of the European Parliament and of the Council of 10

These institutional provisions reflect the EU's promotional approach towards compliance with environmental commitments in 2010–2020 FTAs, relying on inter-party dialogue and third-party adjudication, but ruling out the possibility of imposing economic sanctions (in the form of fines or withdrawal of trade concessions) in cases of non-compliance with the report of the Panel of Experts. Against this backdrop, the EU-UK TCA and the EU-New Zealand FTA signal a shift in the EU's approach to the enforcement of environmental provisions.

#### 11.2.2.2 EU-UK TCA (2020)

The EU-UK TCA largely retains the model of providing dedicated institutional mechanisms in relation to both the implementation of environmental provisions (overseen by the joint Trade Specialized Committee on Level Playing Field for Open and Fair Competition and Sustainable Development)[57] and dispute settlement (involving inter-governmental consultations and independent review by a Panel of Experts),[58] but introduces two main innovations. First, it provides more detailed rules on the compliance stage following the issuance of the Panel report, including possible review by the Panel of any measures taken by the defaulting party to address non-conformity with the agreement.[59] Second, temporary remedies for breach under the general dispute settlement rules are available in disputes concerning the non-regression clause[60]—but *not* other environmental provisions of the TCA.[61] In cases where the Panel finds a violation of the non-regression obligation, the disputing parties may agree on compensation.[62] Failing this, and where no measure is taken to comply with the Panel report, the winning party may retaliate against the losing party through the suspension of trade concessions.[63] There are, however, limits to such trade retaliation which are similar to those found under the dispute settlement rules of the World Trade Organization (WTO), including that it shall be proportionate.[64]

It is important to note that this trade retaliation as a means of inducing compliance with the non-regression obligation is distinct from the trade measures that may be adopted under the rebalancing mechanism of the TCA discussed earlier. The latter

---

February 2021 concerning the exercise of the Union's rights for the application and enforcement of international trade rules, O.J. 2021 L49/2, para 10.

[57] EU-UK TCA, Article 8(j).

[58] EU-UK TCA, Articles 408–409.

[59] Ibid., Article 409(18). See, however, Article 409(9) suggesting that, if the Panel makes specific recommendations for the resolution of the matter, these are non-binding on the defaulting party: '[f]or greater certainty, the Parties share the understanding that if the Panel makes recommendations in its report, the respondent Party does not need to follow these recommendations in ensuring conformity with this Agreement.'

[60] Ibid., Article. 410(2).

[61] Ibid., Article 409(19).

[62] Ibid., Article 749(1).

[63] Ibid., Articles 749(2)–(3).

[64] Ibid., Article 749(5); and for discussion, Peers 2022, pp. 73–78.

are not temporary remedies (or countermeasures) for a prior breach of TCA environmental obligations, but just rebalancing measures to offset the (material) impacts on bilateral trade/investment that may result from future (significant) divergences in EU/UK levels of environmental protection.[65] Notably, such regulatory divergence may take the form of one party raising its level of environmental protection vis-à-vis that existing at the end of the transition period, while the other does not. Yet, none of these actions would itself be inconsistent with the TCA.

#### 11.2.2.3 EU-New Zealand FTA (Post-2023)

The EU-New Zealand FTA is broadly similar to other regular FTAs when it comes to monitoring the implementation of TSD commitments (by a joint specialized Committee on Trade and Sustainable Development)[66] and stakeholder involvement mechanisms at the domestic and transnational levels.[67] However, it is the first agreement to incorporate the 'more assertive' approach to enforcement proposed by the Commission following the 2021 review process of TSD chapters. Breaking with previous practice, the EU-New Zealand FTA brings disputes arising under the TSD chapter into the general dispute settlement mechanism established by the agreement.[68] This sets out the two-stage process of inter-governmental consultations and Panel review procedures in substantially more detail when compared to the dedicated method for settling TSD disputes under previous FTAs. But the key difference between the two mechanisms lies in what happens at the compliance stage, after infringement findings have been made in the Panel's final report. Under the EU-New Zealand FTA, there is no ambiguity that the Panel's findings and recommendations in relation to TSD provisions are legally binding, and the defaulting party shall promptly take the necessary measures to comply with them.[69] The specialized Committee on Trade and Sustainable Development is responsible for monitoring the implementation of such compliance measures.[70] If the parties disagree over the existence or consistency of these measures with the relevant provisions of the agreement, independent review by the Panel is available.[71] In cases where compliance is not reinstated within the arranged period of time,[72] the general temporary remedies (i.e., compensation and trade retaliation) are only applicable to one specific breach of environmental provisions -namely, 'the Party complained against failed to refrain

[65] EU-UK TCA, Article 411.

[66] EU-New Zealand FTA, Articles 24.4.1(e) and 24.4.6.

[67] Ibid., Article 24.6 (on Domestic Advisory Groups) and 24.7 (on Civil Society Forum).

[68] Ibid., Article 26.2.

[69] Ibid., Article 26.13.1.

[70] Ibid., Article 26.13.3(b).

[71] Ibid., Article 26.15.2.

[72] Ibid., Article 26.14.

from any action or omission that *materially defeats the object and purpose* of the Paris Agreement' (PA clause).[73]

This provision is puzzling for a number of reasons. First, the Commission fails to explain why it considers that the use of trade sanctions is appropriate as a remedy of last resort to foster compliance with the Paris Agreement, but presumably not for other MEAs.[74] This is surprising because the imposition of trade sanctions is not available as a penalty for non-compliance under the Paris Agreement itself, nor more generally under MEAs with few exceptions (e.g. CITES and Montreal Protocol).[75]

Second, it is unclear what the terms 'materially defeats the object and purpose' of the Paris Agreement actually mean. The central purpose of the Paris Agreement is stipulated in its Article 2 as 'strengthen[ing] the global response to the threat of climate change', and further defined in more specific aims for mitigation ('holding the increase in the global temperature to well below 2 °C above pre-industrial levels and pursuing efforts to limit the temperature increase to 1.5 °C above pre-industrial levels'),[76] adaptation ('increasing the ability to adapt to the adverse impacts of climate change and foster climate resilience and low greenhouse gas emissions development')[77] and finance ('making finance flows consistent with a pathway towards low greenhouse emissions and climate-resilient development').[78] However, which individual actions (or omissions) would materially defeat such an object and purpose is far less clear. For instance, in relation to mitigation, each party is under an obligation (of conduct) to 'prepare, communicate and maintain' nationally determined contributions (NDCs) 'with a view to achieving' the long-term temperature goal of the agreement.[79] Yet, there are no mandatory obligations on the substance or stringency of such NDCs—just a normative expectation that each successive NDC represents a 'progression' and reflects the 'highest possible ambition', which each party is left to determine 'in light of different national circumstances'.[80] Arguably, a party's complete failure to submit an NDC, or full inaction towards complying with its self-determined mitigation commitments therein, would likely meet the threshold of materially defeating the object and purpose of the Paris Agreement. But the key question is which other actions (or omissions) may also meet that standard.[81] The

---

[73] Ibid., Article 26.16.2(a)(ii) (emphasis added).

[74] Commission Communication 2022, p. 11.

[75] UN Environment Programme, Compliance Mechanisms under Selected Multilateral Environmental Agreements, 2007, pp. 118–119 (Table 3.5).

[76] Article 2.1(a) PA.

[77] Article 2.1(b) PA.

[78] Article 2.1(c) PA.

[79] Articles 3 and 4(2) PA.

[80] Article 4(3) PA; and for discussion, Rajamani 2016, pp. 500–501; Voigt and Ferreira 2016, pp. 295–297.

[81] A combined reading of Articles 19.6.2 and 19.6.3 of the EU-New Zealand FTA suggests it goes beyond effective implementation of NDCs.

term '*materially* defeating' sets a high bar,[82] and is likely to be confined to serious violations of provisions that are essential to the accomplishment of the object and purpose of the Paris Agreement.[83]

Third, from a global governance standpoint, a question arises as to whether it would be appropriate for an FTA Panel to determine when a party's action (or inaction) materially frustrates the object and purpose of the Paris Agreement, without any guidance from the multilateral climate regime on the matter. In principle, it is true that an FTA Panel may request information from the relevant MEA bodies when deciding bilateral disputes concerning compliance with MEAs.[84] But in practice, such advice is likely to be quite limited in the context of the Paris Agreement. This is because the key oversight mechanism established under this agreement (the so-called 'Global Stocktake') is only authorized to assess collective progress towards meeting the global warming targets, thereby insulating individual parties from any assessment as to the adequacy of their mitigation action under NDCs.[85] In this regard, climate-related commitments are in a distinct position when compared to core labor standards included in TSD chapters, where an FTA Panel may rely on the fact that the ILO supervisory system does monitor the application of ratified conventions in individual members.[86]

## 11.3 Towards a Harder Sanction-based Enforcement?

As exposed in the previous section, recent EU FTAs have marked a significant shift in the approach towards the enforcement of environmental commitments at two levels. First, in institutional terms, the EU-New Zealand FTA abandons the separate procedures for settling disputes concerning the TSD chapter and fully integrates them into the general dispute mechanism applicable to the whole agreement. This alignment of dispute settlement arrangements is a welcome step, as it remedies the fact that the earlier TSD-dedicated dispute settlement procedures lacked detailed rules on the compliance stage. However, it is important that the specificities raised by trade-and-environment disputes continue to be duly accounted for in the general dispute settlement procedures—e.g. in terms of ensuring subject-matter expertise in

---

[82] Arguably higher than the obligation in Article 18 of the Vienna Convention on the Law of Treaties (VCLT), adopted on 23 May 1969, 1155 UNTS 331, not to defeat the object and purpose of a treaty prior to its entry into force.

[83] A parallel may be drawn with 'material break' under the VCLT as a ground for terminating (in whole or in part) a treaty and its definition in Article 60(3)(b). On this high threshold, see *Gabcikovo-Nagymaros Project (Hungary/Slovakia)*, Judgment, [1997] ICJ Reports, p. 7.

[84] EU-New Zealand FTA, Article 26.21(3).

[85] Article 14 PA. This is complemented by a transparency framework (Article 13) and a compliance mechanism that is facilitative, non-adversarial and non-punitive in character (Article 15). For an overview of this oversight system, see Rajamani 2016, pp. 502–505.

[86] ILO Constitution, Article 22 (reporting procedure) and Articles 24–34 (representation and complaints procedures).

the Panel's composition and consultation with competent monitoring bodies under relevant MEAs.[87]

More problematic is the second innovation, which concerns the use of economic sanctions as a temporary remedy in cases of non-compliance with *selected* (not all) environmental provisions—i.e. the non-regression obligation in the EU-UK TCA and the PA clause in the EU-New Zealand FTA. The Commission justifies this move as a response to the open public consultation (OPC) it conducted,[88] as well as the independent comparative study on TSD provisions in FTAs it requested,[89] as part of the 2021 review of TSD chapters.[90] Yet, a closer look at these two documents does not reveal a clear-cut case in favor of introducing economic sanctions as an enforcement tool for environmental commitments. Notably, among the 71 total stakeholders (largely EU-based) participating in the OPC, the majority of trade unions (11 respondents in total) and non-governmental organizations (20 respondents in total) were in favor of sanctions, whereas most business associations (26 respondents in total) were against them and public authorities (4 respondents in total) were split on the question.[91] These responses reflect a continued absence of consensus on a sanction-based enforcement model for TSD chapters, just as in the previous OPC run by the Commission in 2017, which had then led it to conclude that a move towards such an approach was 'impossible'.[92] Similarly, the independent comparative study does not take a clear position on the suitability of compliance approaches, and notes that 'even for sanction-based enforcement models like in Canada and the US, cooperation remains the watchword for the implementation of TSD provisions'.[93] Therefore, the real motivations behind the Commission's change of perspective on the issue of sanctions remain obscure.

Nonetheless, it is true that the EU's traditional cooperative approach to compliance with TSD provisions in earlier FTAs has attracted criticism in academic and institutional circles for being too 'soft' and ineffective in enhancing global environmental governance. In particular, the European Parliament[94] and some scholars have argued in favor of a 'harder' enforcement of TSD commitments, including by adding economic sanctions as a remedy of last resort in far-reaching cases of

---

[87] This has been the case in the EU-New Zealand FTA, Articles 26.7(3) and 26.21(3).

[88] LSE Consulting 2021.

[89] Velut et al. 2022.

[90] Commission Communication 2022, p. 1.

[91] LSE Consulting, pp. 4 and 10.

[92] European Commission, 'Non-Paper on Feedback and Way Forward on Improving Implementation and Enforcement of Trade and Sustainable Development Chapters in EU Free Trade Agreements', 18 February 2018 [Commission Non-Paper 2018], p. 3.

[93] Velut et al. 2022, p. 18.

[94] See e.g., European Parliament, Resolution on Human Rights and Social and Environmental Standards in International Trade Agreements of 25 November 2010, 2009/2219(INI), para 22(a); and Resolution on Implementation of the 2010 Recommendations of Parliament on Social and Environmental Standards, Human Rights and Corporate Responsibility of 5 July 2016, 2015/2038(INI), paras 21–22.

non-compliance.[95] The underlying assumption behind these calls seems to be that such retaliatory economic measures can be more effective at inducing compliance with TSD commitments by recalcitrant trade partners where dialogue and cooperation have failed to do so. Such an assumption, however, is questionable on several grounds.

First, the availability of sanctions as a remedy does not necessarily mean that environmental (or labor) complaints will be pursued more frequently to the phase of dispute settlement under FTAs. In fact, FTA practice thus far suggests otherwise. While the EU has only triggered once dispute settlement procedures (with no economic sanctions) under TSD chapters in 13 years of practice (i.e. the 2018 labor dispute under the EU-Korea FTA),[96] this is not less than what the US and Canada have done to enforce environmental and labor commitments through the dispute settlement mechanisms (with economic sanctions) under their respective FTAs in 30 years of practice since the North American Free Trade Agreement came into force (i.e. equally one labor arbitration so far under the US-CADR FTA).[97] Second, empirical evidence on the presumed compliance-inducing effect of economic sanctions is scant and at best mixed. Even if one looks at the WTO system where practice has been more extensive when compared to FTAs, trade retaliation has been requested and authorized only in a handful of WTO cases (12 disputes as of December 2020), actually implemented in an even smaller number of cases, and led to a certain degree of compliance with the condemnatory WTO ruling in some (e.g. *US-FSC*) but not all (e.g. *EC-Hormones*) instances.[98] Conversely, the EU's traditional cooperative approach turned out to be successful in the 2018 EU-Korea labor dispute. As showcased in the Commission's 2021 FTA Implementation Report, within one month of the issuance of the Panel of Experts' final report in January 2021, Korea took concrete steps to address infringement findings through amendments to its domestic labor legislation and the ratification of 3 of the 4 pending fundamental ILO Conventions.[99]

But leaving aside the issue of effectiveness, enforcing environmental commitments through economic sanctions raises a number of legal and policy challenges. First, as the Commission has itself recognized, most EU negotiating partners 'would

[95] See notably, Bronckers and Gruni 2021; Mazzotti 2021. This view is not unanimously shared in the literature. For a contrary view, see e.g. Prévost D and Alexovicova 2019; Hradilovà and Svoboda 2018.

[96] *Panel of Experts Proceedings Constituted under Article 13.15 of the EU-Korea Free Trade Agreement—Report of the Panel*, 20 January 2021.

[97] *In the matter of Guatemala—Issues relating to the Obligations under Article 16.2.1(a) of the CAFTA-DR—Final Report of the Panel*, 14 June 2017.

[98] In other cases, the mere threat of trade retaliation may have been enough to induce the withdrawal of the WTO-inconsistent measure: e.g. *United States—Certain Country of Origin Labelling Requirements, Recourse to Article 22.6 Arbitration Report, WT/DS384/DS386/ARB*, 7 December 2015.

[99] European Commission, 'Report on the Implementation and Enforcement of EU Free Trade Agreements'.

COM(2021) 654 final, 27 October 2021, p. 18. See also follow-up in TSD Committee: https://trade.ec.europa.eu/doclib/docs/2021/may/tradoc_159567.pdf.

not accept a broad scope [of TSD chapters] combined with trade sanctions.'[100] Hence, there seems to be an unavoidable trade-off between the breadth of environmental obligations and their enforceability through economic sanctions. Looking at EU and US FTA practice, trade retaliation is only available as a remedy for breaches of a narrow set of environmental obligations, and subject to high thresholds.[101] As seen above, a violation of the non-regression clause in the EU-UK TCA is conditioned on demonstrating trade/investment effects, whereas the PA clause in the EU-New Zealand FTA is confined to actions or inactions that materially frustrate the object and purpose of the Paris Agreement. In both cases, the burden of proof will be hard for a complaining party to discharge and may render these clauses very difficult to enforce in practice.[102]

Second, and in line with customary international law,[103] FTAs will often require that economic sanctions (or countermeasures) be proportionate—that is, they 'shall not exceed the level equivalent to the nullification or impairment caused by the violation'.[104] This proportionality requirement thus limits the degree of intensity of the retaliatory response, and demands that the level of injury caused by the violation be determined in order to calculate the appropriate amount of trade retaliation that may be applied by the offended treaty partner. This determination, however, can be particularly complex in the case of the PA clause: how should we measure injury suffered by an *individual* State resulting from interference by another State with the *global* commons (i.e. the Earth's climate)? Bronckers and Gruni suggest that it is not necessary to assess proportionality in this context in purely quantitative terms, and that qualitative factors may also be considered (e.g. importance of the interest protected by the rule infringed or the seriousness of the breach).[105] While this is certainly true under customary rules of State responsibility,[106] this may not be allowed under EU FTAs which explicitly require the level of trade retaliation to be equivalent to the 'nullification and impairment caused by the violation'.[107] This follows closely the language used in the WTO Dispute Settlement Understanding and has been more narrowly interpreted as involving economic harm.[108] Hence, assuming an action (or inaction) by party A is found to materially defeat the object

---

[100] Commission Non-Paper 2018, p. 3.

[101] An exception here is Article 26.16.2(a)(i) EU-New Zealand FTA, which makes trade retaliation available as a temporary remedy for breaches of the minimum-level clause in respect of core labour standards (not subject to a trade/investment condition).

[102] On the trade-effects condition in the non-regression clause in the US-Guatemala labour dispute, see Marín Durán 2020, pp. 1064–1065.

[103] International Law Commission (ILC), Articles on State Responsibility with Commentaries 2001, Article 51.

[104] See e.g. EU-UK TCA, Article 749(5); EU-New Zealand FTA, Article 26.16.5.

[105] Bronckers and Gruni 2021, p. 42.

[106] ILC 2001, p. 135, noting that Article 51 explicitly refers to 'taking into account the gravity of the internationally wrongful act and the rights in question'.

[107] See e.g. EU-UK TAC, Article 749(5); EU-New Zealand FTA, Article 26.16.5.

[108] WTO Understanding on Rules and Procedures Governing the Settlement of Disputes, 14 April 1994, Article 22.4.

and purpose of the Paris Agreement, how may we determine the level of economic injury suffered by party B?

Third, and perhaps most importantly, a shift towards a sanction-based enforcement of environmental obligations raises fundamental questions of equity between FTA partners. As has been well documented in the WTO context, trade sanctions are inherently inequitable as an enforcement tool where significant disparities in market size and economic power exist between the disputing parties. This is undoubtedly the case with most FTAs concluded by the EU(27), being the world's third-largest economy. The reason for this lies, essentially, in the asymmetric capacity to actually use trade retaliation as means to induce compliance with treaty commitments. Typically, for a small-market country seeking to retaliate against an economically powerful country, trade sanctions often result in 'shooting oneself in the foot' (due to increased import prices) while inflicting little economic harm—and hence, retaliatory pressure—on the offending party.[109] Overall, there is yet to be one example of 'David vs Goliath' compliance-inducing retaliation in the WTO. For our purposes, this means that introducing trade sanctions to enforce TSD commitments would, in practice, translate into an imbalanced one-way enforcement mechanism in favor of the EU in most FTAs. Regrettably, these equity concerns have been largely ignored by the Commission in its 2021 TSD review process, as well as in most scholarly debates on the issue of enforceability. Proponents of sanction-based enforcement are yet to explain how it can be reconciled, in practice, with their formal acceptance that compliance with environmental provisions in EU FTAs is a reciprocal matter and should go in both directions (i.e., not only by third countries, but equally by the EU and its Member States).

## 11.4 Conclusion

This chapter has examined the evolution in the EU's approach to environmental provisions in FTAs, at both substantive and institutional levels. Two main trends have emerged from this analysis. First, with regards to the substance of these provisions, we have witnessed a firming up of non-regression clauses, which prohibit the lowering of environmental protection levels under domestic laws for competitive purposes. This has been most noticeable in the EU-UK TCA, which is hardly surprising in light of the pre-existing high degree of economic interdependence and geographical proximity between the EU and its former Member State. However, the formulation of this non-regression obligation has not been consistent across EU FTAs, raising important questions of interpretation and application. Notably, what is the difference, if any, between *actual* and *intended* effects on bilateral trade/investment as a threshold condition for establishing a breach of these clauses? Similarly, the PA clause which has been included in more recent FTAs (either as an 'essential element' of the EU-UK TCA, or as an obligation in the climate provisions of the EU-New Zealand FTA, both

---

[109] See further Marín Durán 2020, pp. 1059–1062 and references to WTO case law therein.

enforceable through trade sanctions) leaves ambiguous which actions (or inactions) may *materially defeat* the object and purpose of the Paris Agreement. FTA parties should use the joint institutional mechanisms to provide further guidance on these standards.

Second, the EU seems now ready to embrace the idea that trade retaliation should be available as a remedy to counter far-reaching violations of selected environmental obligations. While this move towards more coercive means of enforcement may be welcomed in some quarters, this chapter has argued that the case for trade sanctions is at best dubious from an effectiveness perspective, while such retaliatory measures raise a number of legal and policy concerns. Most significant among them are issues of equity, which have been largely neglected in the scholarly and institutional debates in EU circles. Trade sanctions are clearly inequitable as an enforcement tool where economic imbalances exist between trading partners, with the 'stick' being a real option only for the economically powerful side—in most instances, the EU—regardless of its own compliance record. As such, a sanction-based enforcement mechanism is at odds with the proclaimed 'partnership' spirit of TSD chapters, whereby compliance with sustainability commitments should not be one-directional but genuinely a two-way street (i.e. by EU/Members States as much as by third countries). Against this background, it is not surprising that the more assertive approach to enforcement proposed by the Commission in the 2021 TSD review and first implemented in the EU-New Zealand FTA has not been followed in the EU-Chile FTA, which was concluded in December 2022.[110] This raises questions as to whether a 'one-size-fits all' approach to enforcing environmental obligations is suitable—and will be acceptable—to all EU FTA partners.

## References

Bartels L (2013) Human Rights and Sustainable Development Obligations in EU Free Trade Agreements, *Legal Issues of Economic Integration* 40:297–314

Bronckers M and Gruni G (2021) Retooling the Sustainability Standards in EU Free Trade Agreements, *Journal of International Economic Law* 24:25–51.

Drieghe L et al (2022) Participation of Civil Society in EU Trade Policy Marking: How Inclusive is Inclusion? *New Political Economy* 27:581–596.

Hradilovà K and Svoboda O (2018) Sustainable Development Chapters in EU Free Trade Agreements: Searching for Effectiveness, *Journal of World Trade* 52:1019–1042

Leonelli GC (2021) From Extra-Territorial Leverage and Transnational Environmental Protection to Distortions of Competition: The Level Playing Field in the EU–UK Trade and Cooperation Agreement, *Journal of Environmental Law* 33:611–638

Marín Durán G (2020) Sustainable Development Chapters in EU Free Trade Agreements, *CMLRev* 57:1031–1068

[110] EU-Chile Advanced Framework Agreement, signed 9 December 2022, https://policy.trade.ec.europa.eu/eu-trade-relationships-country-and-region/countries-and-regions/chile/eu-chile-agreement/text-agreement_en (accessed 27 March 2024). This follows the 'promotional' approach similar to that in earlier 2010–2020 FTAs, discussed in Sect. 11.2.2.1.

Mazotti P (2021) Stepping Up the Enforcement of Trade and Sustainable Development Chapters in the European Union's Free Trade Agreements: Reconsidering the Debate on Sanctions. https://www.europeanlawinstitute.eu/fileadmin/user_upload/p_eli/YLA_Award/Submission_ELI_Young_Lawyers_Award_Paolo_Mazzotti_2021.pdf Accessed 27 March 2024

LSE Consulting (2021) Open Consultation on the Trade and Sustainable Chapters Review. https://policy.trade.ec.europa.eu/development-and-sustainability/sustainable-development/sustainable-development-eu-trade-agreements_en#tsd-review-2021 Accessed 27 March 2024

Peers S (2022) So Close, Yet So Far: The EU/UK Trade and Cooperation Agreement, *CMLRev* 59:49–80

Rajamani L (2016) Ambition and Differentiation in the 2015 Paris Agreement: Interpretative Possibilities and Underlying Politics, *International and Comparative Law Quarterly* 65: 493–514

Prévost D and Alexovicova I (2019) Mind the Compliance Gap: Managing Trustworthy Partnerships for Sustainable Development in the European Union's Free Trade Agreements, *International Journal of Public Law and Policy* 6:236–269

Velut JB et al (2022) Comparative Analysis of Trade and Sustainable Development Provisions in Free Trade Agreements. https://policy.trade.ec.europa.eu/development-and-sustainability/sustainable-development/sustainable-development-eu-trade-agreements_en#tsd-review-2021 Accessed 27 March 2024

Voigt C and Ferreira F (2016) Dynamic Differentiation: The Principles of CBDR-RC, Progression and Highest Possible Ambition in the Paris Agreement, *Transnational Environmental Law* 5: 285–303

**Gracia Marín Durán** is a Professor of International Economic Law in Faculty of Laws of University College London, London, United Kingdom.

# Part III
# Environmental Diplomacy and Legal Frameworks

# Chapter 12
# EU Sustainable Water Diplomacy: Security Challenges in EU External Relations Law

**Luisa Jarchow and Luigi Lonardo**

**Contents**



**Abstract** The imperative of water security (defined by the UN as 'the capacity of a population to safeguard sustainable access to adequate quantities of acceptable quality water') looms large in the quest for the European Union's (EU) sustainable water management, both within its borders and beyond. As climate change continues, the EU faces growing threats to water security, stemming from increased freshwater variability, scarcity, and declining freshwater quality. This chapter delves into the largely uncharted territory of security considerations underpinning sustainable water management as they play out in EU external relations law. Of particular concern is in fact the international dimension of the water security challenge, necessitating cross-border cooperation with third-party nations. Tensions and conflicts in shared

L. Jarchow (✉)
University of Oxford, Oxford, United Kingdom
e-mail: jarchow.luisa@gmail.com

King's College London, London, United Kingdom

L. Lonardo
Centre for European Integration, University College Cork, Cork, Ireland
e-mail: llonardo@ucc.ie

Adjunct Faculty, Paris School of International Affairs, Sciences Po, Paris, France

 
R. A. Wessel et al. (eds.), *EU External Relations Law and Sustainability*, Global Europe: Legal and Policy Issues of the EU's External Action 4,
https://doi.org/10.1007/978-94-6265-655-0_12

transboundary basins, lakes, and aquifers serve as stark reminders of the importance of cooperative approaches. The Chapter explores the security facets of freshwater and the legal instruments available to the EU to face them. To that aim, this Chapter undertakes an analysis of the interplay between water security and sustainability, especially focusing on the concepts of "conflicts" and "migration," exemplified by situations in the Nile and Tigris-Euphrates basins. It delves into the complex web of EU external relations instruments, encompassing treaty law, secondary law, international law, and policy instruments, highlighting that the legal framework differs significantly depending on whether the situation falls partially within or entirely outside the EU's territorial boundaries. The Chapter concludes by emphasizing the critical role the EU must play in fostering resilience and cooperation in water governance, as climate change-induced unpredictability continues to reshape global water dynamics.

**Keywords** Sustainability · Water diplomacy · Water security · Sustainable Development Goals · EU environmental law · EU water law

## 12.1 Introduction

The EU needs water security (defined in the next section) to achieve the sustainable—or integrated—management of water resources within and outside of the EU.[1] Water security, and thus sustainable (or, as the EU says, integrated) water management, is increasingly being threatened by climate change, which is causing increased freshwater variability and scarcity, as well as deteriorating freshwater quality. The security underpinnings of sustainable water management are underexplored as far as EU law is concerned—and this is what we set out to do in this chapter. The main challenge that the EU currently faces in terms of water security is the international nature of this issue and the resulting need for transboundary cooperation with third countries to solve it. Conflict in transboundary basins, lakes and aquifers have been brewing as states who share these resources fail to cooperate between themselves. But global interconnectedness means that crises in one place affect other places around the world, and issues with water management and governance outside the EU can also have grave consequences for the Union's security (every actor, in sum, has a "selfish" reason to cooperate to address water insecurity).

This Chapter explores the security dimension of freshwater[2] (and the legal instruments available to the EU) because it is a necessary condition for sustainable freshwater management within and outside of the EU. The link between water security and sustainability was identified by the EU in the Council Conclusions on Water Diplomacy:

[1] These key terms are defined in Sect. 12.2 of this chapter.

[2] Water that is not salty, found in lakes, rivers, aquifers and so on.

> EU water diplomacy must aim at facilitating the prevention, containment and resolution of conflicts, contributing to the equitable, sustainable and integrated management of water resources from source to sea, and promoting resilience to climate change impacts on water. Cooperation on water must be harnessed to promote regional integration, and address political instability.[3]

The EU has adopted legal acts for the management of freshwater within its territory (e.g. the 2000 Water Framework Directive), and while there are international measures in place (e.g. 1992 UNECE Water Convention; 1997 UN Watercourses Convention), the application of freshwater governance and management mechanisms is not as established outside of the EU. It is the argument of this Chapter that EU external relations law will need to engage with international freshwater governance, and that it is already well-equipped to do so. Having instituted sound water management and governance on the EU's territory, the EU must now look toward regions which are experiencing water security issues (e.g. transboundary conflict, ineffective transboundary cooperation, pollution from natural or anthropogenic sources, prolonged drought, catastrophic flooding), such as the Middle East, Northeast Africa, and the Horn of Africa, to improve water security there (which is also in the EU's own self-interest) and therefore sustainability everywhere (which is in everyone's interest). With an interdisciplinary approach merging law and water security studies, this chapter will first analyze the two main channels through which water insecurity can pose a threat to the EU's sustainable water management, namely "conflicts" and "uncontrolled migration" (Sect. 12.2).[4] These are illustrated with reference to the situations in the Nile basin (Sect. 12.2.1) and the Tigris-Euphrates basin (Sect. 12.2.2). The security-sustainability nexus of freshwater management has been underexplored in EU external relations law, with the 2018 Council Conclusions on Water Diplomacy being one of the few (soft law) instruments in existence. Hence, Sect. 12.3 will reconstruct the intricate web of EU external relations instruments that are in place to tackle these issues. This web is the result of treaty law, secondary law, international law, and policy instruments. When it comes to external relations competences of freshwater management, the legal framework appears to differ depending on whether the situation falls partially inside the EU's territory (Sect. 12.3.2), or entirely outside (Sect. 12.3.3).

---

[3] Council of the European Union 2018, para 5.

[4] These two channels are those identified by EU institutions, see next section.

## 12.2 The Relationship Between Water Security and Sustainability in EU Foreign Policy

Water security is a complex phenomenon, including dimensions of national, economic and human security, that for a long time escaped a precise definition at the international level[5] (inter alia, because "it is not just the *absence* of water but also its *presence* that can be a threat,"[6] and so water security cannot simply be defined as water scarcity). In 2013, UN-Water[7] created a working definition, removing any uncertainties about and reservations preventing the use of the term.[8] Water security is:

> The capacity of a population to safeguard sustainable access to adequate quantities of acceptable quality water for sustaining livelihoods, human well-being, and socio-economic development, for ensuring protection against water-borne pollution and water-related disasters, and for preserving ecosystems in a climate of peace and political stability.[9]

Water security therefore, is recognized as a precondition for sustainable water management in international documents. While there is no commonly accepted working definition for sustainable water management, the EU has referred to Integrated Water Resource Management (IWRM) in the 2018 Council Conclusions on Water Diplomacy,[10] combining the terms "sustainable and integrated" when referring to water management, and directly emphasizing its "commit[ment] to promoting Integrated Water Resources Management".[11] Short of a definition, the Council laid out all dimensions of IWRM, citing "political and security, developmental, humanitarian and environmental"[12] challenges. In light of the EU's interchangeable use of *sustainable* and *integrated* water management, in this chapter we will adopt the Global Water Partnership's widely accepted IWRM definition:

> Integrated Water Resources Management (IWRM) is a process which promotes the coordinated development and management of water, land and related resources in order to maximize economic and social welfare in an equitable manner without compromising the sustainability of vital ecosystems and the environment.[13]

---

[5] The most comprehensive definition was produced by Grey and Sadoff in 2007 and defined water security as "the availability of an acceptable quantity and quality of water for health, livelihoods, ecosystems and production, coupled with an acceptable level of water-related risks to people, environments and economies," p. 548.

[6] Grey and Sadoff 2007, p. 547.

[7] UN-Water is an interagency coordination mechanism for water and sanitation work, which is made up of UN members and international organizations that the UN recognizes as partners.

[8] UN-Water 2013, p. 2.

[9] UN-Water 2013, p. vi.

[10] Council of the European Union 2018.

[11] Ibid. para 14.

[12] Ibid. para 22.

[13] Global Water Partnership 2011.

With relation to security, the question is how likely conflict[14] over water resources is. While scholarly opinions are divided, with Neo-Malthusians citing the advent of water wars[15] and liberal institutionalists believing in the triumph of cooperation over conflict,[16] EU policymakers lean toward the former. The High Representative Josep Borrell stated that "there is unfortunately every reason to fear that this type of conflict will multiply in the future. (...) Europe and its neighborhood are very directly concerned by the water scarcity issue and the geopolitical tensions it can generate."[17] In the event of such water-related conflict, peace and stability in the affected regions will be compromised, leading to increased migration within the region, migration to surrounding regions, and "intensif[ied]" migration "to reach Europe."[18] Uncontrolled migration to Europe has been called a "direct implication" of water-related conflict and is expected to "further strain the relationships between Member States" because of allocation decisions and increased pressure on water resources, which can challenge sustainable water management.[19] This does not appear to be an unfounded fear, as water insecurity has already forced an estimated 100 million individuals globally to migrate internally and externally, and this number is likely to surpass 216 million by 2050.[20]

The EU is thus committed to "promot[ing] regional integration" in river basins, lakes and aquifers beyond its borders to fight the specter of water insecurity-induced instability in its surrounding regions that may threaten sustainable water management within its borders.[21] Specifically, para 11 of the Council Conclusions on Water Diplomacy states that "the EU stands ready to work in partnership with others to promote collaborative and sustainable water management, encouraging and supporting regional and international cooperation."[22] While it may not be possible to fully get rid of this security issue, as climate change causes unpredictable variability and scarcity, the EU's efforts to promote cooperation on water abroad should aim to build resilient governance and management institutions, as well as strong relationships between basin states[23] to withstand the destabilizing effects of water insecurity. The details of this will be explored in the following sections.

---

[14] Conflict will be interpreted as "situations of tension and hostility that may threaten international peace and security," see Vinogradov et al. 2003, p. 25.

[15] Furlong et al. 2006; Gleditsch and Brochmann 2012; Gleick 1993, p. 79; Homer-Dixon 1994, p. 8; Starr 1991, p. 17.

[16] De Stefano et al. 2012; Dinar et al. 2010, p. 2568; Kalbhenn 2011, p. 715; Mitchell and Hensel 2007, p. 721; Tir and Stinnett 2012, p. 211; Zawahri and Mitchell 2011, p. 841.

[17] EEAS 2020.

[18] Council of the European Union and European Commission 2008, p. 6.

[19] Council of the European Union 2018, para 2; European Parliament 2021, p. 20.

[20] World Bank 2021, p. 80.

[21] Council of the European Union 2018, para 5.

[22] Ibid. para 11.

[23] A basin state is a state which has an area of a certain river basin on its territory.

### *12.2.1 The Nile Basin*

The Nile is the world's longest river, stretching a vast 6,550km from central Africa to the Mediterranean, crossing 11 basin states on its northbound journey. Its water however, is monopolized by only two of them, the downstream states Egypt and Sudan, on the basis of colonial era treaties that have been dismissed as invalid by the remaining basin states, as will be shown below. Since 2011 the situation has become even more volatile, as Ethiopia started unilaterally building the Grand Ethiopian Renaissance Dam (GERD), which Egypt and Sudan perceive as a threat to their water security. In the absence of a basin-wide treaty regarding the management and governance of the Nile waters, and as Ethiopia continues to act unilaterally in the filling of the GERD, conflict has become a real possibility.[24] On July 19th 2020, the day that the first filling of the GERD was completed without an agreement between the three states, High Representative Josep Borrell stated that "the situation along the Nile is particularly worrying (…) and the EU became engaged [a few weeks earlier] because the concerned countries – Egypt, Sudan and Ethiopia – are all of strategic importance for us."[25] Ethiopia's plan to fill the GERD's reservoir by 2023, instead of slowly over a decade as proposed by Egypt and Sudan so as not to threaten their water security,[26] is creating insecurity that extends beyond the Nile Basin to the EU, as conflict in the Nile Basin can lead to uncontrolled migration, including "through Northern Africa to reach Europe."[27] The World Bank has estimated that by 2050 water insecurity will lead 105 million people in Sub-Saharan Africa and North Africa to migrate.[28] The EU has consequently recognized the Nile basin dispute as a security issue,[29] because it is threatening political stability in an already volatile region, which could lead to conflict and uncontrolled migration flows.

Political stability in Ethiopia, Egypt and Sudan are key to the EU's interest of "address[ing] political instability" in the wider region.[30] EEAS Spokesperson Peter Stano, has expressed that "the EU is ready to play a more active role, if useful and desirable to all parties, putting forward its rich experience in the management of shared water courses."[31] He went on to assert that cooperation between the basin

---

[24] European Parliament 2021, p. 16; Kantemnidis D (2021) Europe can help prevent a 'water war' over Ethiopia's Nile dam https://www.friendsofeurope.org/insights/europe-can-help-prevent-a-water-war-over-ethiopias-nile-dam/ Accessed 7 September 2022.

[25] EEAS 2020.

[26] Bearak M, Raghavan S (2020) Africa's largest dam powers dreams of prosperity in Ethiopia —and fears of hunger in Egypt https://www.washingtonpost.com/world/interactive/2020/grand-ethiopian-renaissance-dam-egypt-nile/ Accessed 8 September 2022.

[27] Council of the European Union and European Commission 2008, p. 6.

[28] World Bank 2021, p. xxii.

[29] Council of the European Union 2018, para 8.

[30] Council of the European Union 2018, para 5.

[31] Stano 2021.

states "could be turned into an opportunity for many people," highlighting "predictability" and "water security" as benefits.[32] These *many people* would certainly also include the EU, as cooperation and stability in the basin would lessen the threat to water resources that conflict and uncontrolled migration pose. The Nile basin dispute taken aside, stability in the region is already threatened by internal instability in the three basin states, and a study by the European Parliament has found that "instability in one [of the basin states] tends to have regional repercussions."[33]

In Ethiopia, widespread violence, mainly associated with the Tigray conflict,[34] is complicating humanitarian relief efforts for the 20 million in need.[35] The humanitarian crisis was caused by drought, seasonal flooding and violence that led to food insecurity, a staggering increase of internally displaced persons, increased outward migration and even more violence.[36] Further downstream, Sudan is experiencing similar issues, with over a quarter of its population—14.3 million—in need of humanitarian assistance as a result of flooding, hunger, disease, conflict and economic crisis.[37] A combination of rising internal displacement and increased migration inflows from Sudan's neighbors, have put further pressure on water resources and led to an overall deterioration of violence and conflict in Sudan.[38] Egypt's sources of internal instability largely differ to her humanitarian crisis-stricken upstream neighbors. At the Nile's delta, the people are struggling with repression and unresolved economic and social problems. While the election of president al-Sisi in 2014 brought steady economic growth,[39] large parts of the population feel that their economic and social grievances have been neglected in the economic development of the past few years, causing "anger, frustration, and discontent" to spread.[40] If the political instability of the basin states continues to deteriorate, this will become highly problematic for the EU, who sees all three states as crucial to peace and stability in the region. Indeed, the EU has described Ethiopia as "one of the EU's important partners on the

---

[32] Ibid.

[33] European Parliament 2021, p. 16.

[34] The Tigray conflict is a war between the federal government of Ethiopia and the Tigray People's Liberation Front (TPLF) which took power in 1991 through a war which toppled the former military regime. Human rights and democracy suffered under the TPLF regime and in 2018 Abiy Ahmed Ali, a member of the Oromo ethic group, came to power through widespread protests against the TPLF-led government. His reforms were ill-received by the Tigrayans and the situation escalated into a civil war in November 2020.

[35] De Waal A (2022) Ethiopia civil war: Why fighting has resumed in Tigray and Amhara https://www.bbc.com/news/world-africa-62717070 Accessed 8 September 2022; OCHA 2022a.

[36] Mutahi B (2020) Ethiopia's Tigray crisis: How the conflict could destabilise its neighbours https://www.bbc.com/news/world-africa-54904496 Accessed 11 September 2022; OCHA 2022a; Walsh D, Dahir A L (2022) Why Is Ethiopia at War With Itself? https://www.nytimes.com/article/ethiopia-tigray-conflict-explained.html Accessed 11 September 2022.

[37] OCHA 2022b.

[38] Ibid.

[39] IMF 2022.

[40] Dentice G, Melcangi A (2019) Challenges for Egypt's fragile stability https://www.atlanticcouncil.org/blogs/menasource/challenges-for-egypt-s-fragile-stability/ Accessed 7 September 2022.

African continent, active in regional peace and security."[41] In Sudan, the EU donated "€52 million in humanitarian assistance in 2021."[42] An effort, which is in line with their interests to curb conflict and migration in the region, as their humanitarian assistance was designed to "support the most vulnerable households — internally displaced and refugee families and host communities."[43] As for its Southern Neighborhood Partner Egypt, the EU has high expectations in the shape of a stabilizing role, extending to the Middle East. Stability in Egypt is especially important because it controls one of the entry points to the Mediterranean—the Suez Canal—where any trade coming from Asia to the EU must pass. The EU's virtual water trade with Egypt through agricultural products is also of importance[44] and most recently Egypt has started to export natural gas in cooperation with Israel to the EU to replace part of what Russia used to supply (El Safty and Rabinovitch 2022), keeping in step with their joint "Partnership Priorities" of "guarantee[ing] stability on both sides of the Mediterranean" as well as reinforcing cooperation in matters of security and migration.[45]

While the existing instability in Ethiopia, Egypt and Sudan and the resulting outmigration already point toward an increased likelihood of conflict and uncontrolled migration,[46] we must also consider the role of climate change and the strained relations between the basin states. In its sixth assessment report, the IPCC confirmed an "increased risk of conflict" in the Nile basin because of "the impact of climate change on water resources [increasing] tensions, particularly in the absence of strong institutional capacity."[47] With higher temperatures leading to increasing evaporation,[48] more variability in precipitation patterns,[49] and extreme flooding and droughts,[50] the IPCC has predicted, with "high confidence," that in regions already under water stress, "the number of people being water stressed will increase by 50% to 100%" by mid-century.[51] Furthermore, the Nile's annual flow pattern is already "highly seasonal," discharging 80% of its annual flow between the months of August and October[52] and with increased variability as a result of climate change and the higher demands of bigger populations and growing economies,[53] this will lead to more uncertainty and tensions between the basin states.

---

[41] EEAS 2021a.

[42] EEAS 2021c.

[43] Ibid.

[44] EEAS 2021b.

[45] Ibid.

[46] European Parliament 2021, p. 13.

[47] IPCC 2022, p. 618.

[48] Barnes 2017, p. 2; Beyene et al. 2010, p. 433; Enyew et al. 2014, p. 1.

[49] Barnes 2017, p. 2; Goulden et al. 2009, p. 809

[50] Beschorner 1992, p. 46; Conway 2004, p. 111; Kimenyi and Mbaku 2015, p. 34.

[51] IPCC 2022, p. 2293; Munia et al. 2020.

[52] Henshaw and Paisley 2013, p. 61.

[53] EEAS 2020.

Indeed, Egypt has repeatedly threatened to go to war over the Nile. In 1980, President Anwar el-Sadat declared that "If Ethiopia takes any action to block our right to the Nile waters, there will be no alternative for us but to use force."[54] Current president al-Sisi proclaimed the unimpeded flow of the Nile to his nation as "a matter of life and death"[55] and in 2013 Egyptian politicians were shown live on TV "proposing military action to stop the construction of the dam."[56] Most recently, following Ethiopia's decision to conduct the second filling of the GERD without an agreement with Egypt and Sudan, the two downstream states conducted military exercises.[57] The combination of internal instability, climate change, and existing tensions between the basin states, make conflict a real possibility in the absence of a basin-wide agreement on the management and governance of the Nile waters and cooperation on the filling and operating of the GERD. It is thus important to explore the obstacles to cooperation that can be found in the past and present.

For almost a century, Egypt and Sudan have been monopolizing the Nile waters on the basis of the colonial-era 1929 Anglo-Egyptian[58] and 1959 Egypto-Sudanese[59] treaties, which allocated the entire Nile flow between the two states and forced the upstream states to renounce their claims to the water of the Nile. The 1929 Anglo-Egyptian Treaty was signed by Britain on behalf of its Nile basin colonies and newly independent Egypt. Upon independence, the former British colonies rejected the validity of this treaty, including Sudan, which signed a successor treaty with Egypt in 1959 to replace the 1929 Treaty.[60] To the present day, the treaties have been posing an obstacle to cooperation, as they are not representative of contemporary customary international water law[61] and rejected by the other basin states, while Egypt and Sudan cling to their validity.

The upstream states argue that they are not bound by the 1929 Treaty by virtue of the Nyerere doctrine which gives newly independent states the choice of whether to

[54] Kimenyi and Mbaku 2015, p. 99.

[55] Abdulrahman S A (2018) Agreements that favour Egypt's rights to Nile waters are an anachronism https://theconversation.com/agreements-that-favour-egypts-rights-to-nile-waters-are-an-anachronism-103353 Accessed 7 September 2022.

[56] Laccino L (2018) Grand Ethiopian Renaissance Dam And The Nile Water Dispute https://www.ibtimes.co.uk/grand-ethiopian-renaissance-dam-nile-water-dispute-1597589 Accessed 6 September 2022.

[57] Kantemnidis D (2021) Europe can help prevent a 'water war' over Ethiopia's Nile dam https://www.friendsofeurope.org/insights/europe-can-help-prevent-a-water-war-over-ethiopias-nile-dam/ Accessed 7 September 2022.

[58] 1929 Treaty. Officially called 'Exchange of Notes between His Majesty's Government in the United Kingdom and the Egyptian Government in regard to the Use of the Waters of the Nile for Irrigation Purposes,' signed in Cairo on 7 May 1929.

[59] 1959 Treaty. Officially called 'Agreement between the Republic of the Sudan and the United Arab Republic for the Full Utilization of the Nile Waters,' signed in Cairo on 8 November 1959. Taken from United Nations Treaty Series 1963, No. 6519, pp. 63–76, English translation. Registered by the United Arab Republic on 7 February 1963.

[60] Sanchez and Gupta 2011, p. 366.

[61] Article 4(1) United Nations Convention on the Law of the Non-Navigational Uses of International Watercourses (1997) United Nations, Treaty Series, vol. 2999.

accede to colonial-era treaties.[62] In the case of the 1959 Treaty, the upstream states argue that they are not bound by the treaty because they never signed or validated it.[63] Article 34 of the 1969 *Vienna Convention on the Law of Treaties* confirms this argument, stating that "a treaty does not create either obligations or rights for a third State without its consent." Egypt and Sudan however, point to the doctrine of universal succession to validate the 1929 Treaty[64] and emphasize their "natural and historical rights" to the Nile waters as established in the 1929 Treaty[65] to validate their existing uses of the Nile water since the 1959 Treaty. It must be noted that Egypt and Sudan have a strong argument in favor of their existing uses, as the Nile is Egypt's only source of freshwater[66] and Sudan's alternative freshwater sources are "limited" with "large variations in annual flow."[67] As a result, the downstream states feel threatened by any attempts to form basin-wide agreements which could change their allocated share of the Nile water. The attempt by Ethiopia and the upstream states to form a new basin-wide agreement—the Common Framework Agreement[68]—which would recognize the rights of all basin states, was consequently perceived as a threat to Egypt and Sudan's water security and failed because the basin states could not agree on the definition of water security in Article 14(b) of the agreement. Egypt and Sudan wanted it to confirm their existing uses and historical rights as expounded in the colonial-era treaties, and the upstream states wanted it to lead to a more equitable distribution of the Nile waters.[69] The two downstream states felt their concerns regarding the Common Framework Agreement confirmed when Ethiopia started unilaterally building the GERD in 2011. In 2015 however, they took a seemingly important step toward cooperation, when the three basin states signed the *Agreement on Declaration of Principles On The Grand Ethiopian Renaissance Dam Project*, which stated that Ethiopia, Egypt and Sudan would "agree on guidelines and rules on the first filling [...] and annual operation of the GERD".[70] Unfortunately, Ethiopia disregarded her previous commitment and instead proceeded to unilaterally fill the reservoir of the GERD without an agreement between the three states,[71] causing tensions to skyrocket and creating a security issue for the region and ultimately the EU.

---

[62] Collins 2000, p. 275; Crawford 2012, p. 438.

[63] Abseno 2013, p. 194.

[64] Salman 2013, p. 19.

[65] 1929 Treaty, note 2, para 4.

[66] CIA 2021.

[67] FAO 2015, p. 6.

[68] CFA. Officially called 'Agreement on the Nile River Basin Cooperative Framework'

[69] Barnes 2017, p. 13; Mekonnen 2010, p. 422.

[70] Article 5(2) Agreement on Declaration of Principles between The Arab Republic of Egypt, The Federal Democratic Republic of Ethiopia And The Republic of the Sudan On The Grand Ethiopian Renaissance Dam Project (2015).

[71] Abdelhadi M (2020) Nile dam row: Egypt fumes as Ethiopia celebrates https://www.bbc.co.uk/news/world-africa-53573154 Accessed 7 September 2022.

To achieve security from this threat, the EU and Nile basin states have the same objectives, as for the basin states they lead to cooperation and water security and for the EU they lead to stability in the region, which will make conflict, migration, and thus a threat to EU water security less likely, and sustainable water management more easily achievable. If the EU were to achieve its desired effect of Water Diplomacy, it would take the shape of cooperation between Egypt, Ethiopia and Sudan on the remaining filling and future operation of the GERD; as well as a basin-wide water management and governance agreement.

### *12.2.2 The Tigris Euphrates Basin*

About 4500 years ago, the first and only full-scale "war over water" broke out in the Tigris-Euphrates basin between the city-states of Umma and Lagash.[72] In recent times, the water of the two rivers has once again become a source of conflict between the present-day basin states of Turkey, Syria and Iraq. The Tigris and Euphrates rivers originate in Turkey, the upstream state which has claimed the Tigris and Euphrates as "Turkish river[s]" and consequently monopolized their waters.[73] From southeast Turkey, the Tigris flows through Iraq, while the Euphrates flows through Syria before it reaches Iraq, where the two rivers join together to flow into the Persian Gulf. Both Syria and Iraq have been largely dependent on the goodwill of Turkey because the upstream state has most control over the flow of the rivers as a result of the large-scale dam project Güneydoğu Anadolu Projesi (GAP). This "integrated and multi-sector socioeconomic development project,"[74] according to some commentators, has "negatively impacted" the relations between the basin states and brought them to the brink of conflict twice.[75] Climate change has also contributed to tensions and will continue to increase variability and scarcity in the Tigris-Euphrates basin, with scientists predicting "unprecedented super- and ultra-extreme heatwaves,"[76] extreme droughts[77] and in the worst-case, that the two rivers "will disappear this century."[78]

Facing these extremes, the European Parliament expects increasingly limited access to water "to have highly destabilizing results,"[79] and already in 2008 the European Commission highlighted conflict in the region as a security issue, by stating that "tensions over access to water are almost certain to intensify in this region leading

[72] Dinar 2014, p. 217.

[73] Chalaby and Majzoub 1995, quoted in: Daoudy 2009, p. 373.

[74] Ministry of Industry and Technology 2015.

[75] Dohrmann and Hatem 2014, p. 583.

[76] Zittis et al. 2021, p. 1.

[77] Al-Ansari et al. 2013, p. 667.

[78] Pearce F (2009) Fertile Crescent Will Disappear This Century https://www.newscientist.com/article/dn17517-fertile-crescent-will-disappear-this-century/ Accessed 17 September 2022.

[79] European Parliament 2021, p. 12.

to further political instability with detrimental implications for Europe."[80] The EU's security has already once been impacted by instability in the basin stemming from the Syrian civil war, which led to mass migration to Europe.[81] In fact, it has often been argued that an underlying cause of the Syrian civil war was a shortage of freshwater,[82] which will be discussed below.

Climate change's impact on variability and scarcity in the basin, ongoing and past water disputes between the basin states, and long-standing political tensions, all point toward conflict resurfacing in the basin, especially in the absence of a basin-wide treaty on the management and governance of the Tigris-Euphrates waters. First, it is important to explore the role of water in the Syrian civil war to highlight the threat of water scarcity and variability to stability in the region. Next, cooperation and conflict in the Tigris-Euphrates basin, as well as the basin states' relationships will be discussed to highlight the volatility of the basin in the absence of a basin-wide treaty. Lastly, climate change's impact on the basin will be laid out, allowing us to draw conclusions on the likely future of conflict in the basin.

In 2021, the European Parliament recognized the role of water in the Syrian civil war and the consequent mass migration to Europe, by citing "the long drought that preceded the civil war" and "the Islamic State['s] violent control over water."[83] More specifically, the three-year drought that led to "empty irrigation canals and dried-up wells" is argued to have been the "underlying reason behind the social discontent that unleashed the civil war in Syria in 2011."[84] From winter 2007 onwards, Syria was hit by the most extreme 3-year drought since records started in 1940.[85] The country was already suffering from water scarcity, caused by an overexploitation of groundwater reserves,[86] as well as a decline in Euphrates flow from reduced rainfall and increasing water diversions and extractions by upstream Turkey,[87] but the drought heightened this scarcity and "led to the death of 85 percent of livestock (…) and widespread crop failure."[88]

This hydrological disaster forced an estimated 1.5 million Syrians to migrate from rural to urban areas such as Aleppo and Damascus, which were already struggling with migration influxes from Iraq of the same magnitude since 2002.[89] In fact, Syria's urban population had increased by almost 50% between 2002 and 2010, from

[80] Council of the European Union and European Commission 2008, p. 7.

[81] Abel 2019.

[82] Kelley et al. 2015, p. 3241; Pearce 2019, p. 130; Gleick 2014, p. 331.

[83] European Parliament 2021, pp. 12, 19.

[84] Pearce 2019, p. 130.

[85] Trigo et al. 2010, p. 1255.

[86] Kelley et al. 2015, pp. 3241-42.

[87] Pearce 2019, p. 130; Al-Ansari et al. 2013, p. 667.

[88] El-Geressi Y (2020) Climate Change, water woes, and conflict concerns in the Middle East: A toxic mix https://www.earthday.org/climate-change-water-woes-and-conflict-concerns-in-the-middle-east-a-toxic-mix/ Accessed 16 September 2022

[89] Kelley et al. 2015, p. 3242.

8.9 to 13.8 million,[90] "priming the populace for concentrated, large-scale political unrest."[91] Throughout the civil war, ISIS gained control of vital drinking water, irrigation and hydroelectricity infrastructure: the Syrian Tabqa Dam and the Iraqi Fallujah Dam on the Euphrates; and the Iraqi Mosul Dam on the Tigris. In 2013, when ISIS took the Syrian Tabqa Dam, they wanted to maximize energy production which led to dangerously low levels of water in the reservoir. When ISIS realized this, they drastically reduced the flow out of the turbines, leading to extreme electricity shortages in Aleppo. Upon seizing the Iraqi Fallujah Dam in 2014, ISIS stopped the Euphrates water from flowing downstream, which caused the reservoir to overflow eastward as far as Baghdad.[92] ISIS' brief control of the Mosul Dam in 2014 was possibly the most dangerous episode, as the dam has been designated "the most dangerous dam in the world" because of its unstable foundations.[93] If it were to break, a flood wave would "swallo[w] everything in its path for more than a hundred miles," flooding Mosul within three hours and reaching Baghdad within four days, still almost five meters high, killing an estimated 1.5 million people on its destructive path.[94]

In 2021, the European Parliament highlighted "the Islamic State['s] violent control over water" and the previously-described "weaponiz[ing] [of] natural resources by seizing or destroying water supplies and infrastructure" as contributing to the instability that caused the migration flows to Europe.[95] The possibility of non-state actors taking over strategic Tigris-Euphrates basin infrastructure in the future and creating instability which causes conflict and uncontrolled migration is real, creating a security threat not only for the basin states, but once again also for the EU.

In absence of a basin-wide treaty on water management and governance, the relations between Turkey, Syria and Iraq have been predominantly conflictual, punctuated by short periods of cooperation that produced some bilateral agreements. The most intense crises that brought the basin states to the verge of conflict happened in 1974 and 1990. So far, the EU's role in the basin's cooperation and conflict "has been strikingly hands-off" according to the European Parliament.[96] Before the crisis of 1974, Turkey and Syria had started individual building works on the Euphrates—the GAP and the Tabqa dam respectively. The push to utilize Turkey's water resources in the Southeastern Anatolia region for irrigation and electricity developed over the

---

[90] Ibid.

[91] El-Geressi Y (2020) Climate Change, water woes, and conflict concerns in the Middle East: A toxic mix https://www.earthday.org/climate-change-water-woes-and-conflict-concerns-in-the-middle-east-a-toxic-mix/ Accessed 16 September 2022.

[92] Pearce 2019, p. 156.

[93] Ibid. p. 157.

[94] Filkins D (2016) A Bigger Problem than ISIS? https://www.newyorker.com/magazine/2017/01/02/a-bigger-problem-than-isis Accessed 18 September 2022.

[95] European Parliament 2021, p. 12.

[96] Ibid., p. 23.

1950s and 1960s into the GAP,[97] in an effort to modernize the country[98] and reduce dependency on oil imports by increasing hydropower generation.[99]

As a response to the upstream constructions, Syria started its own development works on the Euphrates and built the Tabqa dam.[100] In 1974, Syria and Iraq came close to conflict over the simultaneous filling of the Turkish Keban dam and the Syrian Tabqa dam during a period of drought. The filling of both reservoirs significantly reduced the flow of the Euphrates waters to Iraq, which led to hostilities being exchanged,[101] Syria and Iraq sending troops to their shared border, and Iraq threatening Syria to bomb the Tabqa dam if they did not reinstate the usual flow pattern.[102] Mediation from Saudi Arabia and the Soviet Union, as well as a break in the prevailing drought prevented an armed conflict between Syria and Iraq.[103] Despite a period of heightened tensions after the crisis, the basin states were able to reach some minor bilateral agreements. In 1987, Turkey and Syria agreed that Turkey must "release a yearly average of more than 500 cubic meters per second at the Turkish-Syrian border" and in exchange the two states would cooperate on "border security" regarding Kurdish separatists.[104] Iraq was unhappy about being left out of these talks and consequently became "less cooperative on border security," a grievance that had previously enabled closer relations between Turkey and Iraq.[105] Nevertheless, in 1989 a second bilateral agreement was reached, when Iraq and Syria agreed to divide whichever amount of water flowed downstream from the Turkish border by 58% and 42% respectively.[106] In 1990, another crisis emerged when Turkey diverted the flow of the Euphrates during the filling of the Atatürk dam. The diversion of the Euphrates brought Turkey's power over the management of the Euphrates to Syria's attention, which suffered from reduced energy production at the Tabqa dam and less water for irrigation and drinking purposes as a result.[107] Despite coming dangerously close to the verge of conflict, the basin states could still not come together in a basin-wide treaty. In fact, Turkey was one of the only three states who refused to sign the *United Nations Convention on the Law of the Non-navigational Uses of International Watercourses* in 1997,[108] which did not bode well for cooperation in the basin. While some bilateral cooperation persisted throughout the early 2000s, basin-wide cooperation has not progressed.

[97] Ministry of Industry and Technology 2015.

[98] Conde 2016.

[99] Kibaroglu 2014, p. 154.

[100] Varsamidis 2010, p. 91.

[101] Wolf 1994, p. 29.

[102] Varsamidis 2010, p. 92.

[103] Daoudy 2009, p. 372.

[104] Beschorner 1992, p. 40.

[105] Ibid. p. 38.

[106] Daoudy 2009, p. 373.

[107] Beschorner 1992, p. 41.

[108] Carius et al. 2005, p. 19.

The main obstacles to cooperation in the form of a basin-wide treaty remain the basin states' opposing views toward international water law. More specifically, it was previously mentioned that Turkey considers the Tigris and Euphrates as "Turkish river[s]," invoking the Harmon Doctrine[109] of absolute territorial sovereignty according to which "a state is free to dispose, within its territory, of the waters of an international river in any manner it deems fit, without concern for the harm or adverse impact that such use may cause to other riparian states."[110] Downstream states Syria and Iraq however, have been advocating for the international customary water law principles of equitable and reasonable use, and the obligation not to cause significant harm,[111] as well as their "acquired rights."[112] Accordingly, Turkey has not been eager to sign a basin-wide treaty on the management and governance of the Tigris-Euphrates basin, as it would likely have to make concessions to Syria and Iraq in regards to quantity and quality of the flows, and only *might* gain more stability in the region.

Without a major breakthrough in basin-wide cooperation, the effects of climate change on the Tigris-Euphrates basin are likely to have grave consequences for stability in the basin. Once the cradle of civilization, the fertile crescent may disappear within the century.[113] Water supplies are already running much lower than a century ago, when uses were not as established and ground water resources were not yet being overexploited.[114] Furthermore, scientists predict that droughts are likely to get worse and last longer, perhaps even become a permanent feature in the basin.[115] There is no doubt however, that surface and groundwater resources will decline in quantity and quality. Regarding quality, Iraq has already lost large areas of fertile land due to "excessive salinity."[116] At the same time, the GAP has not even come to conclusion: in 2005 Turkey decided to greatly expand the GAP to include "22 dams, 19 hydroelectric plants, and extensive irrigation systems on the Tigris and Euphrates Rivers."[117] Upon completion, the combined magnitude of GAP building works on the Tigris and Euphrates by Turkey would gravely impact the water quantity and quality that reaches Syria and Iraq, leading to further tensions and inhibiting cooperation. Electricity generation from the dams in Syria and Iraq will also likely decline with decreased quantities of water flowing in through the Tigris and Euphrates from Turkey to the downstream states, which is likely to create further instability in the basin. Facing the threat of renewed conflict and uncontrolled migration from

[109] Dohrmann and Hatem 2014, p. 575.

[110] Salman 2007, p. 627.

[111] Daoudy 2009, p. 375.

[112] Jouejati 1996, pp. 134–136.

[113] Pearce F (2009) Fertile Crescent Will Disappear This Century https://www.newscientist.com/article/dn17517-fertile-crescent-will-disappear-this-century/ Accessed 17 September 2022.

[114] Al-Ansari et al. 2013, pp. 667, 676; Kelley et al. 2015, p. 3241–42.

[115] Pearce F (2009) Fertile Crescent Will Disappear This Century https://www.newscientist.com/article/dn17517-fertile-crescent-will-disappear-this-century/ Accessed 17 September 2022.

[116] Varsamidis 2010, p. 97.

[117] Pool and Grover 2008, p. 373.

the Tigris-Euphrates basin, it is clear that the EU needs to take action to persuade Turkey, Syria and Iraq to reach a basin-wide agreement which is resistant to conflict and adaptable to the uncertainties of climate change. What remains to be determined is how the EU can use its experience and influence to achieve this.

## 12.3 How EU External Relations Law Can Tackle the Water Security-sustainability Nexus

The web of EU instruments to ensure water security consists of the EU Treaties, secondary law, international law, and policy instruments. As far as Treaty law is concerned, Article 3(5) mentions the protection of fundamental rights (and access to water is one), derived from the right to an adequate standard of living under Article 11(1) of the International Coventant on Economic, Social, and Cultural Rights as objective of the EU external action; Article 21(2)(d) TEU also lists sustainability (environmental, social, and economic) as such an objective; Article 191(1) TFEU lists the protection of the environment and the prudent and rational utilization of natural resources as an objective of the Union. As far as secondary law is concerned, the Water Framework Directive imposes an obligation for Member States to coordinate with non-EU countries for the management of water basins extending beyond EU boundaries (provided that there is an element within EU territory).[118]

As far as international agreements are concerned, the EU and Member States are parties to the 1992 UNECE Convention on the Protection and Use of Transboundary Watercourses and International Lakes; the 1997 Convention on the Law of the Non-navigational Uses of International Watercourses; and conflict prevention and water sustainability are also among the objectives of the Cotonou-agreement between the EU and African, Caribbean and Pacific group of States.[119] As far as policy documents are concerned, the 2018 Council Conclusions on Water Diplomacy set as objective 'facilitating the prevention, containment and resolution of conflicts, contributing to the equitable, sustainable and integrated management of water resources from source to sea, and promoting resilience to climate change impacts on water', including through cooperation to promote regional integration and address political instability (outside the EU). The EU is also committed to the 2030 Sustainable Development Goals agreed at UN level, among which there is 'to ensure availability and sustainable management of water and sanitation for all' (Goal 6). The 2022 EU Strategic Compass, which sets out a strategy for the security and defense of the block, recognizes the impact of competition for natural resources for the EU's security landscape, and the previous "version", the 2016 EU Global Strategy also stressed the need to

[118] Article 13(2) Water Framework Directive 2000/60/EC of 23 October 2000 establishing a framework for Community action in the field of water policy.

[119] Article 11 and 23(e) Partnership agreement between the members of the African, Caribbean and Pacific Group of States of the one part, and the European Community and its Member States of the other part, signed in Cotonou on 23 June 2000 [2000] OJ L317/3.

improve resilience in areas surrounding Europe in relation to issues such as water scarcity.

### *12.3.1 The External Dimension of EU Water Legislation*

Many authors have stressed that EU constitutional objectives and values are (binding) legal standards for the EU's conduct, both in its internal[120] action as well as its external one.[121] Reference is typically made to Article 2 TEU, Article 3(5) TEU (which includes, as we have seen, the protection of human rights and sustainable development of the planet) as well as to Article 21(1) TEU. To these, one probably ought to add the objective of "the sustainable development of Europe" (Article 3(3) TEU), and economic, social, and environmental development (Article 21(2)(d) TEU). This is because they are operationalized in EU policies which, albeit "soft law" instruments, are part of the rules-based order of the EU: in other words, an EU institution is expected not to deviate from the commitments it makes, even when these are purely political. They are also binding because international agreements to which the EU is a party (such as the 1992 UNECE Convention on the Protection and Use of Transboundary Watercourses and International Lakes or the 1997 Convention on the Law of the Non-navigational Uses of International Watercourses) or other secondary law (such as the Water Framework Directive) give expression to them.

For the purposes of EU law, the regulation of water is not an external relations competence, but falls instead under environmental policy. Environmental policy is defined in Title XX of the TFEU (Articles 191–193 TFEU). Of special relevance here is Article 191(4) TFEU, which reads: 'Within their respective spheres of competence, the Union and the Member States shall cooperate with third countries and with the competent international organizations. The arrangements for Union cooperation may be the subject of agreements between the Union and the third parties concerned. The previous subparagraph shall be without prejudice to Member States' competence to negotiate in international bodies and to conclude international agreements.' Article 192(1) and (2) TFEU empower the EU to achieve the objectives which are listed in Article 191(1) TFEU, namely: preserving, protecting and improving the quality of the environment, protecting human health, the prudent and rational utilization of natural resources, and so on. Water is directly or indirectly relevant to all of the objectives of the environmental policy as defined in Article 191(1) TFEU.

When it comes to external relations competences of freshwater management, the legal framework appears to differ depending on whether the situation falls partially inside the EU's territory (Sect. 12.3.2), or entirely outside (Sect. 12.3.3).

---

[120] Scheppele et al. 2021.

[121] Kassoti and Wessel 2022; Sommermann 2013, p. 160; Cannizzaro 2021.

### 12.3.2 *Freshwater Partially Inside the EU and Partially Outside*

Pursuant to Article 3(2) TFEU, the EU has capacity to conclude an international agreement when EU legislation says so, or 'in so far as its conclusion may affect common rules or alter their scope'. Common rules already in place at EU level are found in the Water Framework Directive, which is the most comprehensive piece of legislation enacted by the EU on the theme. In a nutshell, it establishes the EU's model for freshwater management: by river basin (the natural geographical and hydrological unit) instead of according to administrative or political boundaries. It also establishes rules for the protection of water from pollution, with different regimes for surface water and groundwater; as well as for the protection of the ecology of the basin and quantity (the "security" of supply element). Further, the Directive introduces the regulation of water pricing as an incentive for the sustainable use of water resources. EU legislation relevant to freshwater also includes the environmental quality standards directive, the groundwater directive, and the floods directive. Article 13(3) of that Directive imposes an obligation for Member States to coordinate with third countries to produce a single river basin management plan 'in the case of an international river basin district extending beyond the boundaries of the Community, (...) and, where this is not possible, the plan shall at least cover the portion of the international river basin district lying within the territory of the Member State concerned'. Article 13(3) Water Framework Directive therefore creates (implicit) competence for the EU to enter international agreements for the management of rivers with parts outside the EU's boundaries. The geopolitical importance of these commitments cannot be overstated, as some river basins will interest both Russia and an EU Member State, for example, and thus cooperation between Member State and Russian authorities is needed.

The international dimension of water legislation beyond EU territory is also important in so far as so-called "mixed" international agreements (those to which both the Union and the Member States are parties), form an integral part of the EU legal order.[122] It follows that the Court of Justice of the European Union has jurisdiction to monitor whether a Member State complies with certain provisions of international agreements.[123]

### 12.3.3 *Freshwater Entirely Outside the EU's Territory*

What was presented so far applies to freshwater which is partially in the EU's territory, and partially outside. What about freshwater entirely outside the EU territory?

[122] Case 181/73 Haegeman, para 5.

[123] Case C-213/03 Syndicat professionnel coordination des pêcheurs de l'étang de Berre et de la région v Electricité de France.

First, with regard to environmental protection more broadly, the Court held in the *Slovak Bear* case that the EU has shared competence to conclude international agreements pursuant to the above-mentioned Article 191(4), read in conjunction with Article 192 TFEU.[124] The fact that the competence is shared is relevant because it means that the EU and the Member States may conclude mixed agreements, with the consequences outlined in the previous section.

Second, and focusing on freshwater security, since access to water is a fundamental human right, the EU is bound to promote it. This flows from Article 3(5) TEU, which states that "in its relations with the wider world and it may be read into Article 11(1) of the International Covenant on Civil and Political Rights (...) the EU shall contribute to the protection of fundamental rights". This means that the EU can conclude international agreements with third countries for the protection of access to water—an objective for the pursuit of which the attainment of water security is instrumental. The main multilateral forum of relevance here is the United Nations, where climate change and water security have featured for well over a decade in the agenda of the UN Security Council. For example, in 2017 the UNSC recognized the adverse effects of climate change on freshwater (the Lake Chad Basin).[125] In relation to this, one ought to mention project Frexus on Improving security and climate resilience in a fragile context through the Water-Energy-Food Security Nexus in Niger River basin and Lake Chad Basin, which is led by the Niger Basin Authority and co-funded by the EU.

Third, the EU can act under its development cooperation policy (Article 208 TFEU), one objective of which is the promotion of peacekeeping and conflict prevention.[126] Article 208(2) TFEU refers to the attainment of objectives approved in the context of the UN and other international organizations. These include the 2030 Agenda for Sustainable Development. The Agenda, originally adopted in 2015 by the UN General Assembly, is an important political objective of the international community. And while it is not legally binding on the EU or its Member States, it identifies the (shared) political priorities of the block. As the name itself implies, the Agenda is a plan for action, setting objectives (the Sustainable Development Goals, SDGs) to be achieved (it identifies 17 of them). Of particular relevance here is Goal 6, ensure availability and sustainable management of water and sanitation for all. The EU committed to implement the SDGs in all its policies,[127] which include, of course, external relations. In the context of the predecessor of the SDGs, the Millennium Development Goal targets, the EU and the African Union started cooperating

[124] Case C-240/09 Slovak Bears, para 35.

[125] UNSC Resolution 2349.

[126] Article 3(13) Regulation (EU) No 231/2014 of the EP and of the Council of 11 March 2014 establishing an Instrument for Pre-accession Assistance [2014] OJ L77/11; Article 11 Partnership agreement between the members of the African, Caribbean and Pacific Group of States of the one part, and the European Community and its MS, of the other part, signed in Cotonou on 23 June 2000 [2000] OJ L317/3.

[127] As affirmed in the conclusions of the Council of the EU 'A sustainable European future: The EU response to the 2030 Agenda for Sustainable Development' and in the European Commission's reflection paper 'Towards a Sustainable Europe by 2030'.

through the EU-Africa Strategic Partnership on Water Affairs and Sanitation (now Africa-EU Water Partnership Project) to contribute to achieving water supply, sanitation and water resources management. This includes the financing by the EU of programmes in the regions referred to in the previous sections of this chapter. For example, the Nile Basin Initiative, long supported also by German funds, sees the contribution of the EU as well.[128]

Fourth, insofar as there are aspects of security in freshwater management, these would most likely fall under the Common Foreign and Security Policy (CFSP). For example, a study prepared for the European Parliament[129] shows acute awareness of the need for preparedness of EU Common Security and Defense Policy (CSDP) missions or military operations in the regions of Sahel and Horn of Africa. These two regions are particularly sensitive to the scarcity of water and associated risks. Even though the authors of the study correctly point out that there is a CSDP 'blind spot' because 'missions have deployed to climate-stressed areas, but such stresses have not been the reason for their deployment', one ought to point out that the EU's experience with CSDP operations and missions in the Sahel (and, albeit in a different context, in the Horn of Africa) are evidence of the existence of EU foreign policy tools in this domain.

Understanding whether EU action should fall under EU environmental policy, development, or Common Foreign and Security Policy is a classic question of legal basis. The issue would be solved by the center of gravity test. In Case C-300/89, *Titanium Dioxide*, the Court held that a dual legal basis is not possible when they involve two different procedures for the adoption of an act. In that case, the Court held that the correct legal basis was (now) Article 114 TFEU and not (now) Article 192(1) TFEU for a measure approximating national rules concerning production conditions which were liable to lead to distortions of competition, even if the measure also related to environmental protection. Since CFSP has a requirement of unanimity in the majority of cases, and since the Court held in Case C-36/98, *Spain v Council*, para 52 that the unanimity requirement of Article 192(2) only applies to the quantitative and not qualitative aspects of water management, the issues with the correct identification of the legal basis persist.

## 12.4 Conclusion

As conflicts like the ones in the Nile basin and the Tigris-Euphrates basin start to emerge, water is set to have a central place in the future landscape of international security. This fact, which derives from the issues caused by climate change, global inequalities, and current water management, is generally acknowledged both in academic writings and in policy recommendations, but it is not central to policy formulation.

128 EEAS 2017.

129 Meyer et al. 2021.

But while the centrality of water security is clear on paper, it is not yet key in EU policy documents. The 2018 Council Conclusions on Water Diplomacy are an important milestone in this direction, but they are only the beginning of the EU's engagement with water security and sustainable management. In 2021, the EU reaffirmed this commitment to enhanced EU engagement in water diplomacy with the Council Conclusions on Water in the EU's External Action. While the content of these Council Conclusions did not add any significantly different insights compared to the 2018 Council Conclusions on Water Diplomacy, they "reaffirm[ed] the EU's diplomatic engagement on water, especially transboundary water cooperation, as a tool for peace, security and stability, calling for their further implementation."[130] The 2018 and 2021 Council Conclusions have already led to some tangible action by the EU in regard to the two river basins discussed in this chapter. In 2021 for example, representatives from EU member states, Turkish and Iraqi experts, and international organizations assembled for a joint working group to discuss the issues that have inhibited successful cooperation in the basin, as well as the "EU's possible role in facilitating capacity building, knowledge sharing, and greater cooperation over transboundary river governance in the region."[131] At COP28 in December 2023 the EU also made a commitment for water diplomacy efforts in the Nile basin by signing a Joint Declaration to enter into a Partnership for Water Resilience with Egypt. This partnership aims to increase water security, "foster regional dialogues and policy exchanges," enhance "technological, scientific and capabilities" in the region, and to set a "precedent for similar partnerships across Africa."[132] Most recently it has also been announced that the European commission will launch an "Initiative on Water Resilience" to "build a shared understanding of global water stress challenges and strategize a water-resilient future."[133] While the EU is clearly still in the early stages of its engagement in water diplomacy, these initiatives pave the way for more engagement and the possibility of a clear legal framework in the future.

Indeed, as far as the legal framework is concerned, there are also reasons for optimism. This Chapter showed that the EU legal framework equipped it well to tackle the challenges deriving from water security and therefore relating to what the EU calls sustainable and integrated water management. In particular, the Chapter showed that the EU has power to conclude international agreements for the management of freshwater entirely outside the EU's territory, even though the classic "EU external relations law" difficulty in identifying the correct legal basis remains. In the absence of decisive case law, these issues will have to be further developed and only further practice can provide a better foundation on how the EU will operate in these matters.

---

130 Council of the European Union 2021, para 4.

131 Aamer 2021.

132 Press and information team of the EEAS Delegation to Egypt 2023.

133 Ibid.

## References

Aamer F (2021) Joint Working Group on International and EU Water Diplomacy—In Focus: The Euphrates-Tigris River Basin. https://www.stimson.org/2021/joint-working-group-on-international-and-eu-water-diplomacy-in-focus-the-euphrates-tigris-river-basin/ Accessed 2 March 2024

Abel G J et al (2019) Climate, conflict and forced migration, Global Environmental Change 54:239-249

Abseno M M (2013) The influence of the UN Watercourses Convention on the development of a treaty regime in the Nile River basin, Water International 38:192-203

Al-Ansari N et al (2013) Management of Water Resources in Iraq: Perspectives and Prognoses, Engineering 5:667-684

Barnes J (2017) The future of the Nile: climate change, land use, infrastructure management, and treaty negotiations in a transboundary river basin, WIREs Climate 8:1-18

Beschorner N (1992) The Nile basin, Adelphi Papers 32:45-61

Beyene T, Lettenmaier D, Kabat P (2010) Hydrologic impacts of climate change on the Nile River Basin: implications of the 2007 IPCC scenarios, Climate Change 100:433-461

Cannizzaro E (2021) The Value of the EU International Values. In: Douma, Eckes, van Elsuwege, Kassoti, Ott, Wessel, Cannizzaro (eds), The Evolving Nature of EU External Relations Law. Asser Press

Carius A et al (2005) Cooperation on Turkey's transboundary waters. Status Report commissioned by the German Federal Ministry for Environment.

Chalaby H, Majzoub T (1995) Turkey, the waters of the Euphrates, and public international law. In: Allan A J et al, Water in the Middle East: Legal, Political and Commercial Implications. Tauris Academic Studies, London

CIA (2021) Egypt. The World Facebook. https://www.cia.gov/the-world-factbook/countries/egypt/ Accessed 15 September 2022

Collins R O (2000) In Search of the Nile Waters, 1900-2000. In: Erlich H, Gershoni I (eds), The Nile: Histories, Cultures, Myths. Lynne Rienner Publishers, London

Conde G (2016) Water and Counter-Hegemony: Kurdish Struggle in the Tigris and Euphrates in Turkey, Revista de Paz y Conflictos 9:43–58

Conway D (2004) From headwater tributaries to international river: Observing and adapting to climate variability and change in the Nile basin, Global Environmental Change 15:99–114

Council of the European Union (2018) Council Conclusions on Water Diplomacy https://data.consilium.europa.eu/doc/document/ST-13991-2018-INIT/en/pdf Accessed on 5 September 2022

Council of the European Union (2021) Council Conclusions on Water in the EU's External Action. Accessed on 2 March 2024

Council of the European Union, European Commission (2008) *Climate change and international security: paper from the High Representative and the European Commission to the European Council*. https://doi.org/10.2860/50106 Accessed 15 September 2022

Crawford J (2012) Brownlie's Principles of Public International Law. Oxford University Press, Oxford

Daoudy M (2009) Asymmetric Power: Negotiating Water in the Euphrates and Tigris, International Negotiation 14:361–391

De Stefano L et al (2012) Climate change and the institutional resilience of international river basins, Journal of Peace Research 49:193–209

Dinar A, Blankespoor B, Dinar S, Kurukulasuriya P (2010) Does precipitation and runoff variability affect treaty cooperation between states sharing international bi-lateral rivers?, Ecological Economy 69:2568-2581

Dinar S (2014) Transboundary Water Regime. In: Morin J F, Orsini A (eds), Essential Concepts of Global Environmental Governance. Routledge

Dohrmann M, Hatem R (2014) The Impact of Hydro-Politics on the Relations of Turkey, Iraq, and Syria, Middle East Journal 68:567-583

EEAS (2017) The EU supports inclusive and cooperative management of the Nile Basin for all Nile countries. https://www.eeas.europa.eu/node/30919_en Accessed 22 November 2022
EEAS (2020) The Nile and beyond: geopolitics of water. https://www.eeas.europa.eu/eeas/nile-and-beyond-geopolitics-water_en Accessed 10 September 2022
EEAS (2021a) The European Union and Ethiopia https://www.eeas.europa.eu/sudan/european-union-and-sudan_en?s=122 Accessed 8 September 2022
EEAS (2021b) The European Union and Egypt https://www.eeas.europa.eu/sudan/european-union-and-sudan_en?s=122 Accessed 8 September 2022
EEAS (2021c) The European Union and Sudan https://www.eeas.europa.eu/sudan/european-union-and-sudan_en?s=122 Accessed 8 September 2022
Enyew B, Van Lanen H, Van Loon A (2014) Assessment of the impact of climate change on hydrological drought in Lake Tana Catchment, Blue Nile Basin, Ethiopia, Journal of Geology & Geosciences 3:1-17
European Parliament (2021) *In-Depth Analysis: Preparing the CSDP for the new security environment created by climate change* https://www.europarl.europa.eu/RegData/etudes/IDAN/2021/653639/EXPO_IDA(2021)653639_EN.pdf Accessed 18 September 2022
FAO (2015) Country Profile – Sudan: FAO AQUASTAT Reports. https://www.fao.org/publications/card/fr/c/I9808EN/ Accessed 6 September 2022
Furlong K, Gleditsch N P, Hegre H, Lacina B, Owen T (2006) Conflicts over shared rivers: Resource scarcity or fuzzy boundaries?, Political Geography 25:361-382
Gleditsch N P, Brochmann M (2012) Shared rivers and conflict e A reconsideration, Political Geography 31:519-527
Gleick P H (1993) Water and Conflict: Fresh Water Resources and International Security, International Security 18:79-112
Gleick P H (2014) Water, drought, climate change, and conflict in Syria, Weather Climate Society 6:331–340
Global Water Partnership (2011) What is IWRM? https://www.gwp.org/en/GWP-CEE/about/why/what-is-iwrm/ Accessed 27 November 2022
Goulden M, Conway D, Persechino A (2009) Adaptation to climate change in international river basins in Africa: a review, Hydrological Sciences Journal 54:805-828
Grey D, Sadoff C W (2007) Sink or Swim? Water security for growth and development, Water Policy 9:545-571
Henshaw T W, Paisley R K (2013) Transboundary governance of the Nile River Basin: Past, present and future, Environmental Development 7:59-71
Homer-Dixon T F (1994) Environmental Scarcities and Violent Conflict: Evidence from Cases, International Security 19:5-40
IMF (2022) Arab Republic of Egypt: Country Data. https://www.imf.org/en/Countries/EGY Accessed 9 September 2022
IPCC (2022) Impacts, Adaptation and Vulnerability—Working Group II Contribution to the Sixth Assessment Report of the Intergovernmental Panel on Climate Change. https://report.ipcc.ch/ar6/wg2/IPCC_AR6_WGII_FullReport.pdf Accessed 20 September 2022
Jouejati M (1996) Water Politics as High Politics: The Case of Turkey and Syria. In: Barkey H J (ed), Reluctant Neighbor: Turkey's Role in the Middle East. United States Institute for Peace, Washington DC
Kalbhenn A (2011) Liberal Peace and Shared Resources—A Fair-Weather Phenomenon?, Journal of Peace Research 48:715–735
Kassoti E, Wessel R (2022) The Normative Effect of Article 3(5) TEU: Observance and Development of International Law by the European Union. In García Andrade P (ed), Interacciones entre el derecho de la Union Europea y el Derecho Internacional Publico. Tirant lo Blanc
Kelley C et al (2015) Climate Change in the Fertile Crescent and Implications of the Recent Syrian Drought, PNAS 112:3241-3246
Kibaroglu A (2014) An analysis of Turkey's water diplomacy and its evolving position vis-à-vis international water law, Water International 40:1–15

Kimenyi M, Mbaku J (2015) Governing the Nile River Basin: The Search for a New Legal Regime. Brookings Institution Press

Mekonnen D Z (2010) The Nile Basin Cooperative Framework Agreement negotiations and the adoption of a 'water security' paradigm: flight into obscurity or a logical cul-de-sac?, European Journal of International Law 21:421–440

Meyer C et al (2021) Preparing the CSDP for the new security environment created by climate change

Ministry of Industry and Technology (2015) History of GAP—Southeastern Anatolia Project Regional Development Administration. http://www.gap.gov.tr/en/history-page-3.html Accessed 18 September 2022

Mitchell S M, Hensel P R (2007) International institutions and compliance with agreements, American Journal of Political Science 51:721–737

Munia H A et al (2020) Future transboundary water stress and its drivers under climate change: a global study, *Earths Future* 8:1-21

OCHA (2022a) Ethiopia: Situation Report https://reports.unocha.org/en/country/ethiopia Accessed 12 September 2022

OCHA (2022b) Sudan: Situation Report https://reports.unocha.org/en/country/sudan/ Accessed 12 September 2022

Pearce F (2019) When the Rivers Run Dry: Water – the Defining Crisis of the Twenty-First Century. Granta, London

Pool, A M, Grover V I (2008) GAPs in the Dialogue of Governance: Conflicting Ideologies of Development in Turkey. In: (ed) Grover V I, Water: Global Common and Global Problems. Science Publishers

Press and information team of the EEAS Delegation to Egypt (2023) New Milestone in EU-Egypt Relations: A Partnership for Water Resilience. https://www.eeas.europa.eu/delegations/egypt/new-milestone-eu-egypt-relations-partnership-water-resilience_en Accessed 2 March 2024

Salman S M A (2007) The Helsinki Rules, the UN Watercourses Convention and the Berlin Rules: Perspectives on International Water Law, International Journal of Water Resources Development 23: 625-640

Salman S M A (2013) The Nile Basin Cooperative Framework Agreement: a peacefully unfolding African spring?, Water International 38:17-29

Sanchez N, Gupta J (2011) Recent changes in the Nile Region may create an opportunity for more equitable sharing of the Nile River waters, Netherlands International Law Review 58:363-385

Scheppele K L, Kochenov D V, Grabrowska-Moroz (2021) EU Values Are Law, after All: Enforcing EU Values through Systemic Infringement Actions by the European Commission and the Member States of the European Union, Yearbook of European Law 39:3–121

Sommermann K P (2013) Article 3 TEU. In: Blanke H J and Mangiameli S (eds), The treaty on European Union (TEU): a commentary. Springer

Stano P (2021) Grand Ethiopian Renaissance Dam: Statement by the Spokesperson on the Announcement of the Second Filling. https://www.eeas.europa.eu/eeas/grand-ethiopian-renaissance-dam-statement-spokesperson-announcement-second-filling_en Accessed 12 September 2022

Starr J (1991) Water wars, Foreign Policy 82:17–36

Tir J, Stinnett D M (2012) Weathering climate change: Can institutions mitigate international water conflict?, Journal of Peace Research 49:211-225

Trigo R M, Gouveia C M, Barriopedro D (2010) The intense 2007–2009 drought in the Fertile Crescent: Impacts and associated atmospheric circulation, Agricultural Meteorology 150:1245–1257

UN-Water (2013) Water Security and the Global Water Agenda—A UN-Water Analytical Brief. https://www.unwater.org/sites/default/files/app/uploads/2017/05/analytical_brief_oct2013_web.pdf Accessed 27 November 2022

Varsamidis A (2010) An Assessment of the Water Development Project (GAP) of Turkey: Meeting its Objectives and EU Criteria for Turkey's Accession, Naval Postgraduate School

Vinogradov S, Wouters P, Jones P (2003) Transforming Potential Conflict into Cooperation Potential: The Role of International Water Law. UNESCO, PCCP Publications

Wolf A T (1994) A Hydropolitical History of the Nile, Jordan and Euphrates River Basins. In: Biswas A (ed), International Waters of the Middle East. Oxford University Press, Oxford

World Bank (2021) Groundswell Part 2: Acting on Internal Climate Migration. https://hdl.handle.net/10986/36248 Accessed 24 February 2024

Zawahri N, Mitchell S M (2011) Fragmented Governance of International Rivers: Negotiating Bilateral versus Multilateral Treaties, International Studies Quarterly 55:835–858

Zittis et al (2021) Business-as-usual will lead to super and ultra-extreme heatwaves in the Middle East and North Africa, Climate and Atmospheric Science 20:1-9

**Luisa Jarchow** is a University of Oxford alumna in Water Science, Policy and Management, and a King's College London alumna in International Relations. She specializes in transboundary river basin conflict and cooperation, including international water law.

**Luigi Lonardo** is a Lecturer in EU Law and Acting Director, Centre for European Integration, University College Cork. Adjunct Faculty, Paris School of International Affairs, Sciences Po.

# Chapter 13
# The Role of Climate and Energy Diplomacy in Promoting Sustainability in EU External Relations Law

**Tuula Honkonen**

## Contents




**Abstract** Environmental and climate diplomacies have become key external policy agendas for the European Union (EU). Climate change policy and governance are major fields where the EU seeks to broaden its external influence. EU climate diplomacy policy can be divided into three overlapping spheres: participation in global climate negotiations; promotion of ambitious domestic climate legislation and policies beyond the EU borders; and initiatives for climate cooperation and partnerships with third countries. As a framework for these actions, the EU has developed a solid climate diplomacy policy: a series of Council Conclusions on the theme have been adopted since 2008. Interestingly, energy diplomacy was coupled with climate diplomacy in the Council Conclusions in 2021, and the linkage was further strengthened in the new set of Council Conclusions in 2023. The, by now, rather well-established EU climate and energy diplomacy policy is part of the external dimension of the European Green Deal (EGD). It also underlines the inherent nexus thinking that is



T. Honkonen (✉)
Center for Climate, Energy and Environmental Law (CCEEL), University of Eastern Finland, Joensuu, Finland
e-mail: tuula.honkonen@uef.fi





R. A. Wessel et al. (eds.), *EU External Relations Law and Sustainability*, Global Europe: Legal and Policy Issues of the EU's External Action 4,
https://doi.org/10.1007/978-94-6265-655-0_13

part and parcel of the EGD. The climate and energy diplomacy policy has called for a stronger mainstreaming of climate and energy diplomacy considerations in all areas of EU external action, including its foreign and security policy and development cooperation.

**Keywords** EU external relations law · Sustainability · Climate governance · Climate diplomacy · Energy diplomacy · European green deal

## 13.1 Introduction

EU environmental policy is not confined to the Member States and the EU as an institution, but it also has a solid external dimension.[1] This does not consist of haphazard spill-over effects, although such also sometimes materialize, but is generally a result of determined policymaking with the intention of shaping environmental policies, practices and conditions in non-EU countries.[2] The EU is keen to profile itself as a world leader in the field of environmental policy, which translates into efforts to show a good example, spread good practices and raise the level of ambition of environmental policymaking in other countries and regions. The intended external effects may be targeted at all countries or specified EU partners in specific fields and projects.

Climate change is never only a domestic matter. There may be domestic climate policies in place, even very ambitious ones, but their effectiveness is limited if other countries or regions do not follow suit. What is needed is the art of skillful negotiation, persuasion, trust-building, showing a good example, joint projects and cooperation with mutual and global gains. In two words: climate diplomacy. Climate diplomacy comprises the means to advance climate action beyond a country's own borders.

Climate change policy and governance is a major field where the EU seeks to broaden its external influence. In concrete terms, two interrelated key areas are often distinguished in this respect: the EU's climate diplomacy within the United Nations Framework Convention on Climate Change (UNFCCC)[3] and beyond; and EU support for and capacity-building in developing countries,[4] and partnerships with other countries as well. According to another definition, EU climate diplomacy can, in its broadest sense, be divided into two separate, yet interrelated spheres: activities in multilateral settings; and bilateral cooperation with third countries.[5] The ideas behind these are basically the same. Of course, 'bilateral' can also mean a somewhat

[1] Key literature on EU external environmental policy includes Morgera 2012; Adelle et al. 2018; Biedenkopf and Groen 2021; Wouters et al 2021.

[2] Biedenkopf and Groen 2021, p. 33.

[3] United Nations Framework Convention on Climate Change, New York 1992, 31 ILM, p. 849.

[4] Dupont et al. 2018, p. 108.

[5] Earsom and Delreux 2021, p. 716. See also Torney and Cross 2018; Biedenkopf and Petri 2021. Literature on EU climate diplomacy has largely focused on the EU role and action within the UNFCCC negotiations (see, e.g., van Schaik 2010; Bäckstrand and Elgström 2013; Delreux 2021;

larger set of countries, but it is still a distinct track from the global climate change regime.

EU external climate policy is usually largely about promoting the EU's internal climate policies beyond its borders. Overall, the tools of the EU external climate policy can be divided into those that create incentives and disincentives for external actors, and those that are founded on dialogue and persuasion.[6] The former could be established through conditional funding schemes, for instance, whereas the latter is more about the EU leading by example and showing the way towards ambitious climate policy goals. In any case, the internal and external dimensions of the EU climate and energy policies are closely intertwined.[7]

One could always argue about the extent to which the "externalization" of a climate policy is, indeed, an intentional and strategic action by the EU or more akin to a by-product or "natural" domestic norm or policy diffusion. Either way, the observed climate diplomacy action by the EU has relied on two main components: the credibility of domestic action, and the EU regulatory and market power.[8] This is a strong combination that gives the EU a significant advantage as an actor seeking to effectively implement climate diplomacy. The setting also works vice versa: the EU's global and regional leadership aspirations have been an important driver for ambitious domestic policy development. Consequently, it has been argued that the distinction between EU exemplary and diplomatic leadership has become blurred and the separation of EU domestic and international/external climate policy elusive.[9] This has become evident within the global climate change negotiations where the EU has arguably shifted from a one-sided leadership strategy to a "leadiator" role, whereby it does not try to convert other countries to follow its strong climate policy actions, but it acknowledges the geopolitical changes in the negotiation setting and adopts a more nuanced and delicate strategy of coalition-building and bridge-building activities.[10] It could be said that this approach is now shared in both dimensions of the EU external climate policy: that which is focused on the multilateral setting under the UNFCCC and the one focusing on bilateral activities and partnerships.

This Chapter examines the role that climate and energy diplomacy plays in EU external relations law. The Chapter builds on a broad definition of climate and energy diplomacy, comprising EU action within both global climate governance and bilateral and regional partnerships. However, emphasis will be placed on the legal and policy basis of the EU climate diplomacy, followed by an examination of how climate diplomacy has been integrated as part of different areas of EU external relations law and policies. The latter study is limited to the most traditional fields of EU

---

Earsom and Delreux 2023); the partnership type of climate and energy diplomacy action by the EU has been subject to academic studies mainly more recently.

[6] Biedenkopf and Groen 2021, pp. 38–41.

[7] Dupont et al. 2018, pp. 119–120.

[8] Oberthür and Dupont 2021, p. 1109.

[9] Ibid., p. 1109.

[10] Bäckstrand and Elgström 2013.

external relations, i.e. foreign and security policy; human rights policy; and development policy. Thereafter, the chapter discusses some cross-cutting themes for the EU climate and energy diplomacy policy, in particular, the ever more frequently emerging nexus thinking in the EU external relations. The Chapter ends with some concluding remarks.

## 13.2 EU Climate and Energy Diplomacy Policy

This section examines the general basis of the EU climate and energy diplomacy policy, its development over the years and the current status, including the relation with the European Green Deal (EGD).[11]

### *13.2.1 Policy Development on EU Climate Diplomacy*

Environmental and climate diplomacies have become key diplomatic agendas for the EU.[12] This stems from various sources, including the heightened ambition of the EU's domestic environmental and climate policies and legislation, the search for a leadership role in multilateral arenas dealing with key global environmental problems, and the increased recognition of linkages between environmental crises and relations between states and regions.

The starting point for the EU climate diplomacy policy, specifically labeled as such, was the first Council Conclusions[13] on climate diplomacy that was adopted in 2011.[14] The document divided the concept of climate diplomacy into three "strands": supporting UNFCCC negotiations; promoting low-emission and resilient transformations in third countries; and analyzing and reducing climate fragility risks in order

[11] The European Green Deal, Communication from the Commission to the European Parliament, the European Council, the Council, the European Economic and Social Committee and the Committee of the Regions, COM(2019) 640 final.

[12] Minas and Ntousas 2018; Earsom and Delreux 2023. Also, ecological diplomacy has been suggested as a policy for the EU to pursue (Lazard and Youngs 2021), but it would be, in many ways, a much broader term.

[13] Council Conclusions is a document that expresses the EU's political commitment in selected matters. General Secretariat of the Council 2020. It also sheds light on the EU's motives and aims with regard to specific policy fields. However, Conclusions does not amount to a legally binding document. The main shortcoming of Conclusions has been that it often says relatively little about the tangible methods of implementation to be used. When Conclusions are not supported by any kind of action plan, the practical significance of the document may remain rather low. However, in case of climate and energy diplomacy, the most recent Council Conclusions shows a turn in the approach.

[14] Council conclusions on EU Climate Diplomacy, Foreign Affairs Council meeting, Brussels, 18 July 2011 (hereinafter 2011 Council Conclusions). See also European External Action Service (EEAS) and European Commission 2011.

to foster peace and international security.[15] The approach very well illustrates the EU stance towards climate diplomacy, then and still today.

The Council adopted the next Conclusions on climate diplomacy in 2013,[16] 2015[17] and 2016.[18] They largely followed the approach of the 2011 document, but were gradually developing towards more coherent and ambitious policy guidelines, calling for "a more pro-active and targeted EU climate diplomacy agenda",[19] recognizing "the need to further step up EU climate diplomacy efforts"[20] and "the importance of mainstreaming climate diplomacy into the priority agendas of the EU and Member States".[21] The EU was keen to deploy a more strategic approach to climate diplomacy with a specific toolbox to be developed.[22]

Adoption of the Paris Agreement in December 2015 clearly sparked intensified efforts within the EU to grab momentum and invest in reinforced climate diplomacy policymaking. The EEAS and Commission prepared a Climate Diplomacy Action Plan[23] that was adopted in 2015. The Plan outlined concrete actions to implement the objectives laid down in the Council Conclusions. In addition, the 2016 Council Conclusions included in an annex separate "Elements for climate diplomacy action in 2016". These were presented as action points for the EU under three strands that could be taken as developed versions from those presented in the 2011 Conclusions. The new strands were: "Continue to advocate climate change as a strategic priority in diplomatic dialogues, public diplomacy and external policy instruments"; "Support implementation of the Paris Agreement, in the context of low-emission and climate-resilient development"; and "Increase efforts to address the nexus between climate, natural resources, prosperity and stability."

Apart from the aftermath of the UNFCCC Paris Conference, no separate EU-wide action plan documents on climate diplomacy have been adopted. This is not to be interpreted as due to lack of interest by the EU, but it is more likely to be a consequence of the Council Conclusions documents having become more detailed and practice-oriented since then.

A second more active phase in the policy development of EU climate diplomacy could be said to have started around 2018. At that time, the European Parliament passed a Resolution on climate diplomacy.[24] In the Resolution, the Parliament

---

[15] 2011 Council Conclusions; Tänzler et al 2021, p. 3.

[16] Council conclusions on EU Climate Diplomacy, Foreign Affairs Council meeting, Luxembourg, 24 June 2013 (hereinafter 2013 Council Conclusions). See also EEAS and European Commission 2013.

[17] Council conclusions on EU Climate Diplomacy, Foreign Affairs Council meeting, Brussels, 20 July 2015, 11095/15.

[18] Council Conclusions on European Climate Diplomacy after COP21: Foreign Affairs Council meeting, Brussels, 15 February 2016.

[19] 2013 Council Conclusions, para 6.

[20] Ibid., para 7.

[21] Ibid., para 8.

[22] EEAS and European Commission 2013.

[23] European Commission and EEAS 2015, Action Plan for Climate Diplomacy, January 2015.

[24] European Parliament Resolution of 3 July 2018 on Climate Diplomacy, 2017/2272 (INI).

highlighted, among other things, the importance of mainstreaming climate across different foreign policy issues and the need for strengthening the EU capacity for climate diplomacy.[25]

The next Council Conclusions on climate diplomacy were adopted in 2018[26] and 2019.[27] They were built around several dimensions: strengthening of international climate ambition and multilateralism; the EU's long-term strategy on climate change; the peace and security aspect of climate change; and further enhancement of the EU's international climate cooperation and support. The Conclusions clearly signaled the EU's commitment to global action to deal with climate change and EU leadership. The domestic climate policy ambitions were present as well. Finally, the EU external relations were accounted for in terms of the peace and security aspect and the EU's climate cooperation and support beyond its borders. However, the Conclusions did not envision any significant new openings to the EU climate diplomacy action.

### *13.2.2 Integrating Energy into EU Climate Diplomacy*

After 2019, there were plans to renew the Council Conclusions on Climate Diplomacy,[28] but, instead, the Council adopted Conclusions on Climate and Energy Diplomacy[29] in 2021. They were developed to explicitly deliver on the external dimension of the EGD, adopted by the Council on the same date. In this context, it was recognized that an ambitious internal climate policy of the EU will not be enough to achieve the long-term goals of the Paris Agreement. The goal of the EU climate and energy diplomacy is to pursue global transition towards climate neutrality, while ensuring the EU's own resilience and competitiveness in a shifting security and geopolitical environment.[30]

2021 was the first time that energy diplomacy was coupled with climate diplomacy in the Council Conclusions. This can be considered to be quite a natural development, especially given the intertwined relationship between the two policy areas and the (geo)political challenges involved.[31] The addition of the energy diplomacy element to the EU climate diplomacy policy principally brings the goal of accelerating the global energy transition to the fore in the EU external action on climate and energy. However, the 2021 Conclusions state that the implementation of the energy transition

---

[25] Ibid., paras 16–33 and 38.

[26] Council Conclusions on Climate Diplomacy, Foreign Affairs Council meeting, Brussels, 26 February 2018, 6125/18, Annex.

[27] Council Conclusions on Climate Diplomacy, Foreign Affairs Council meeting, Brussels, 18 February 2019, 6153/19, Annex.

[28] See Draft Council Conclusions on Climate Diplomacy, 20 January 2020, 5033/20, Annex.

[29] Council Conclusions on Climate and Energy Diplomacy—Delivering on the external dimension of the European Green Deal, 25 January 2021, 5263/21, Annex.

[30] Ibid., para 1.

[31] On the development of EU energy diplomacy and its external dimensions as well as its linkage with EU climate diplomacy, see, e.g., Barra and Svec 2018.

goal should also ensure affordability, safeguarding the environment and achieving the Sustainable Development Goals (SDGs).[32,33] Thus, energy and climate appear to be on an equal footing in this context.

Overall, the Conclusions call for further reinforcement of the external dimension of the EGD.[34] From a practical diplomacy action perspective, the Conclusions state that the EU will continue to place climate action and the energy transition at the top of the agenda in its bilateral engagement with third countries.[35]

It is also interesting that the 2021 Conclusions place some emphasis on climate and energy security aspects as part of the related diplomacy efforts. The Council states the already previously recognized need to "strengthen and mainstream work on the climate and security nexus".[36] In addition, the EU and Member States' foreign and security policy "will systematically consider climate and environmental factors and risks, and will work with partners, including the UN, to develop conflict prevention measures" in this respect.[37] Hence, it could be argued that the 2021 Conclusions did not only integrate the energy dimension to the EU climate diplomacy policy but also strengthened the security and conflict prevention elements therein.

The Council adopted yet another Conclusions on EU climate diplomacy in February 2022.[38] They were published in the aftermath of the 26th UNFCCC Conference of the Parties (COP) and specifically tailored for "accelerating the implementation of Glasgow outcomes". From the EU external relations perspective, the Conclusions do not offer much new: the Council reaffirmed the high level of ambition of the EU's climate diplomacy and called for concrete actions.[39]

A new set of Council Conclusions on climate and energy diplomacy were adopted in March 2023.[40] With the title "Bolstering EU climate and energy diplomacy in a critical decade", the Conclusions was prepared in response to the Russian invasion of Ukraine and the resultant multidimensional crisis that had hit Europe and the world as large. The tone and approach differ somewhat from the previously adopted Conclusions. The document begins by stressing the EU's determination to engage and work with partners worldwide through its climate and energy diplomacy.[41] It then continues by highlighting the EU's commitment to implement the Paris Agreement,

---

32 Transforming our world: The 2030 Agenda for Sustainable Development, UNGA Res. 70/1 of 25 September 2015.

33 2021 Council Conclusions, para 9.

34 Ibid., para 20.

35 Ibid., para 19.

36 Ibid., para 4.

37 Ibid.

38 Council conclusions on EU Climate Diplomacy: accelerating the implementation of Glasgow outcomes, 21 February 2022, 6120/22 (hereinafter 2022 Council Conclusions).

39 Ibid., para 1.

40 Council conclusions on Climate and Energy Diplomacy "Bolstering EU climate and energy diplomacy in a critical decade", 9 March 2023, 7248/23, Annex (hereinafter 2023 Council Conclusions).

41 Ibid., para 1.

provide support to the most vulnerable and increase collective climate finance.[42] In the field of energy, the Conclusions stresses increased cooperation with countries that rely heavily on fossil fuels[43] and the need to forge new energy partnerships, while safeguarding the EU's own resilience and competitiveness and domestic resources.[44] Finally, the conflict-aggravating nature of environmental problems is recognized by the Council, which highlights increasing risks to human, state and regional security and the risk of intensified dimensions of fragility.[45]

What is new and remarkable with the 2023 Conclusions as a whole, is the level of determination of the envisioned EU action and the steps to ensure a regular follow-up. The Council invites the High Representative of the Union for Foreign Affairs and Security Policy, the Commission and all Member States to strengthen EU climate and energy diplomacy "as a political priority".[46] This is to take place through intensified coordination, information exchange and strengthening of the EU Delegations and Member States' embassies, and relevant EU and international networks and working groups. In addition, the Council encourages EU and Member States' climate outreach missions and regional initiatives, including joint ones.[47] These are much-needed concrete action points for implementing the goals enshrined in the Conclusions. In addition, the Council takes up the task of regular follow-up on the joint work to coordinate and enhance the EU's climate and energy diplomatic impact. Finally, the Council invites the High Representative and the Commission to strengthen their capacity dedicated to EU climate and energy diplomacy.[48] With these actions, it becomes clear that climate and energy diplomacy, indeed, becomes elevated to a priority position in the EU agenda. It also appears that the Council has been determined to ensure that the Conclusions document will not remain a collection of dead letters but that its objectives will be effectively implemented by the different EU actors and institutions.

In addition to specific Council Conclusions, climate and energy diplomacy is being advanced in the EU also through other means. The Green Diplomacy Network, created in the early 2000s, has served to coordinate climate diplomacy initiatives across the EU and Member States. The Network collects together the EEAS, EU delegations and Member State embassies around the world. These actors coordinate the implementation of EU priorities in the area of green diplomacy. The Network is a vehicle for bilateral outreach, partnerships with third countries and information-gathering.[49]

---

[42] Ibid.

[43] Para 20.

[44] Para 26.

[45] Para 21.

[46] Para 43.

[47] Ibid.

[48] Ibid. Initiation of a process for regular reporting on the EU's climate diplomacy activities and its achievements, as well as its shortcomings was proposed already by the EP in its Resolution in 2018 (para 26).

[49] See also, e.g., Biedenkopf and Petri 2021; Earsom and Delreux 2023, pp. 201–202.

### 13.2.3 Implementing the EGD

The EU has had climate-related legislation for a long time.[50] Until relatively recently, however, the legislation did not have any explicit link to the EU's climate diplomacy policy. A turn came when the EGD was introduced in 2019. Political agreement on it was reached in 2021. The EGD is a comprehensive climate change legislation package for the EU. In practice, it is a strategy and roadmap on how the EU will reach its climate targets, most importantly climate neutrality in 2050.

While the EGD is majorly a domestic strategy and legislative package, it has important explicit and implicit effects on EU policies beyond its borders. "The EU as a global leader", part three of the EGD document, testifies the key external element of the EGD. It envisions the EU as a global leader in climate policy and names EU priorities related to, inter alia, the Paris Agreement and continued and strengthened engagement with international partners bilaterally and in regional contexts. This is also necessary for reaching the goals of the EGD, as it is recognized that "[t]he environmental ambition of the Green Deal will not be achieved by Europe acting alone".[51]

Climate and energy diplomacy is part of the external dimension of the EGD. The EGD calls for intensified efforts "to mobilize all diplomatic channels both bilateral and multilateral";[52] for forging "green alliances" through "diplomatic and financial tools" with Africa, Latin America, the Caribbean, Asia and the Pacific;[53] and for more integrated thinking about climate and interrelated policies such as environmental, security and defense issues.[54] Nevertheless, neither climate nor energy diplomacy is specifically mentioned in the EGD document.

Implementation of the external dimension of the EGD can be said to largely amount to another diplomacy policy for the EU: the "green deal diplomacy" as called for in the EGD document.[55] Accordingly, green deal diplomacy is implemented through the EU and the Member States setting a credible example and convincing and supporting actors and countries outside the EU to take up their share of promoting more sustainable development.[56] The EU is to integrate climate diplomacy with other external policy areas and bring EGD into all EU and Member States diplomatic channels as well as bilateral partnerships with third countries.[57] Furthermore, it has been recognized that EU foreign and security policy plays a key role in the efforts of green deal diplomacy, given the geopolitical dimension of climate change.[58] So far,

---

[50] See, e.g., Woerdman et al. 2021.

[51] COM(2019) 640 final, p. 1.

[52] Ibid., p. 20.

[53] Ibid., p. 21.

[54] Ibid.

[55] P. 20.

[56] Ibid.

[57] Ibid., and Teevan et al. 2021, p. 8.

[58] See Borrell 2019.

the EU has not developed any additional strategy for the realization of green deal diplomacy, although a need for such has been recognized in the literature.[59]

## 13.3 Climate and Energy Diplomacy in EU External Relations Law

### 13.3.1 The Legal and Policy Basis

The EU climate and energy diplomacy policy reflects the aims defined in the Treaty of the European Union.[60] These general aims include peace, the sustainable development of Europe and a high level of protection and improvement of the quality of the environment, among other things.[61] Furthermore, in its external action, the EU must "preserve peace, prevent conflicts and strengthen international security", "help develop international measures to preserve and improve the quality of the environment and the sustainable management of global natural resources, in order to ensure sustainable development" and "promote an international system based on stronger multilateral cooperation and good global governance".[62] Regarding the EU's environmental policy, one of its goals is to contribute to "promoting measures at international level to deal with regional or worldwide environmental problems, and in particular combating climate change".[63] Taken together, environmental protection and sustainable development strongly feature as key objectives of the EU. In the same vein, EU external action is based on conflict prevention and developing international measures to ensure environmental protection and sustainable development. The EU's environmental policy also has objectives that cross the Union borders, and climate change has been named as a priority area in this respect. The idea and practices of climate and energy diplomacy, therefore, fit very well into the general frameworks for the EU and the goals of its external and environmental policies.

The word "diplomacy" binds climate diplomacy closely to foreign and security policy. Illustratively, the main purpose of diplomacy has been defined in literature as "to enable states to secure the objectives of their foreign policies without resort to force".[64] This also applies to climate and energy diplomacy, which can be classified as an indirect objective of the EU's external policy and which the EU seeks to advance through amicable measures. In this study, climate and energy diplomacy is

59 Petri 2020.

60 Consolidated version of the Treaty on European Union (TEU), O.J. C 326/13, 26.10.2012.

61 Ibid. Article 3.

62 Ibid. Article 21.

63 Consolidated version of the Treaty on the Functioning of the European Union (TFEU), O.J. C326/47, 26.10.2012, Article 191(1), fourth indent.

64 Berridge 2005, p. 1.

considered to have a broader field of application than being merely part of the EU's foreign policy.

The EU has not formally addressed the role of climate in its external action, other than what has been specified in the series of Council Conclusions concerning climate, and later also energy diplomacy. Specific Council Conclusions on climate in the EU's external action have not been adopted, contrary to what has been done in the area of water diplomacy, for instance.[65] However, Council conclusions on strengthening the external dimension of the EU energy policy[66] exist, but they were adopted in 2011. An initial impetus for combining climate and energy diplomacy was provided in 2015 when the Commission stated that the external dimension of the new Energy Union includes coordinating energy and climate diplomacy.[67]

In the following, brief overviews are provided on the role of climate and energy diplomacy in some of the key policy fields of EU external action.

### *13.3.2 Foreign and Security Policy*

The EU founding treaties specify that in its external actions, the Union "shall define and pursue common policies and actions, and shall work for a high degree of cooperation in all fields of international relations".[68] Furthermore, it shall contribute to peace, security and sustainable development of the Earth, and it shall uphold and promote its values and interests in its external relations.[69] These principles and objectives of its external action also provide the framework for the EU's Common Foreign and Security Policy (CFSP).[70]

While the EU does not possess a specific strategy for its environmental foreign policy, a trend towards strategic thinking in that field can be discerned.[71] Climate does not explicitly feature in the founding treaties' provisions on the EU's external action, nor concerning its foreign and security policy.

The Global Strategy for the European Union's Foreign and Security Policy (EUGS),[72] adopted in 2016, provides more information on the EU's shared interests and principles and jointly agreed upon priorities in its external action. The document was designed to be thematically global and so it encompasses not only the traditional

---

[65] Council Conclusions on Water in the EU's External Action, 19 November 2021, 14108/21.

[66] Council conclusions on strengthening the external dimension of the EU energy policy, Transport, telecommunications and Energy Council meeting, 24 November 2011.

[67] A Framework Strategy for a Resilient Energy Union with a Forward-Looking Climate Change Policy, COM(2015) 80 final.

[68] Article 21.2 TEU.

[69] Articles 3 and 21 TEU.

[70] Article 24.2 TEU.

[71] Schunz 2019.

[72] Shared Vision, Common Action—A Stronger Europe: A Global Strategy for the European Union's Foreign and Security Policy, 2 June 2016.

domains of foreign policy but also the external dimensions of literally all EU policies.[73] As regards to the theme of the present chapter, two main observations can be made from the EUGS: (1) climate considerations have been effectively mainstreamed into the Strategy; and (2) energy and climate are entangled in the Strategy.

Concerning the first point, the word 'climate' features 24 times in the EUGS text. As part of strategic action for global governance, it is stated that the EU will lead by example, by implementing its commitments on sustainable development and climate change. Furthermore, the EU will increase climate financing, promote climate mainstreaming in multilateral fora, and work for clean energy cost reductions.[74] As regards looking at energy and climate issues together, the Strategy acknowledges that energy and environmental sector reform policies can assist partner countries along a path of energy transition and climate action. Through such efforts, the EU encourages positive developments in both the energy sector and climate change mitigation and adaptation.[75] At the same time, however, the EU also envisions risks that climate change and energy insecurity may endanger the security of the EU.[76]

A later assessment of the EUGS has brought the climate security dimension more firmly to the fore, seeing the EU place more emphasis on the climate-security nexus in the future. Full utilization of the relevant EU toolbox could arguably contribute to conflict prevention and peace-building in the context where climate change, environmental degradation and food and water insecurity have severe conflict multiplying effects.[77] The approach fits well with the traditional view of the EU's common foreign and security policy.

The EEAS produced a working document "The EU Concept for an Integrated Approach on Climate Change and Security"[78] in 2021, with the aim of establishing a framework to address the climate, peace and security nexus in the EU external action. Consequently, the impact of the relevant EU action should be increased, and the climate and security nexus be addressed in all relevant EU instruments and policies underpinning its peace and security approach.[79] The Concept is set to complement the external dimension of the EGD, particularly in the field of crisis and conflict management.[80]

The 2022 Council Conclusions on EU climate diplomacy affirmed that climate diplomacy remains a core component of EU foreign and security policy. The document also highlighted that climate diplomacy is an increasingly important, joint priority in the dialogues and cooperation of the EU and its Member States with third countries and regions.[81]

---

[73] Tocci 2020, p. 178.

[74] EUGS, p. 40.

[75] Ibid., p. 27.

[76] Ibid., p. 9.

[77] EEAS 2019, p. 26.

[78] Concept for an Integrated Approach on Climate Change and Security, EEAS(2021)770.

[79] Ibid., p. 3.

[80] Ibid.

[81] 2022 Council Conclusions.

Moreover, it is notable that climate change is recognized as a security risk and source of instability by the EU, in the EGD document, for instance.[82] Consequently, the EU has stated that climate policy implications should become an integral part of the EU's thinking and action on external issues, also in the context of the common security and defense policy.[83] The Green Deal is an excellent vehicle for mainstreaming climate and energy diplomacy into the core policy areas of the EU. According to recent literature, the broader implications of climate change and climate policy for the EU's external relations and foreign policy at large have been slow to move into political focus.[84] The Green Deal package, and the subsequent geopolitical developments with their considerable ramifications, have certainly increased the leverage of climate and energy diplomacy as a security matter for the EU.

An interesting recent development within the EU Member States is the establishment of an inclusive "Group of Friends for an Ambitious EU Climate Diplomacy" in October 2022, ahead of UNFCCC COP-27. It is a group of, originally, 11 EU Member States[85] which aims to promote the placement of climate at the heart of the EU's foreign and security policy. More precisely, the group is committed to advance all of the previously identified elements of climate diplomacy as part of the EU foreign and security policy: enhance the nexus between climate, peace and security within the EU foreign policy; to accelerate a Paris-aligned global energy transition by actively supporting the implementation of the external dimension of the Green Deal and by showing leadership; and to ensure the effective mainstreaming of climate action across all areas of EU foreign and security policy.[86]

In the light of the above, it is apparent that climate figures strongly in the EU's foreign and security policy.[87] It has been reported that foreign services have been increasingly involved and coordinated across the EU for enhanced climate diplomacy.[88] This is very well in line with the idea that climate diplomacy includes a peace and security aspect. Through the EUGS, the EU has launched a shift from a traditional approach to foreign and security policy towards one that is characterized by greater diversity in which climate and climate diplomacy have a recognized and more prominent role.

---

82 Ibid., p. 5.

83 COM(2019) 640 final, p. 21.

84 Oberthür et al. 2022, p. 5.

85 Denmark, Finland, France, Germany, Ireland, Latvia, Lithuania, Luxembourg, the Netherlands, Spain, Sweden. The Group is open to all EU Member States who want to foster a more ambitious EU Climate Diplomacy.

86 Statement by the Group of Friends for Ambitious EU Climate Diplomacy in connection with the publishing of the synthesis report of the United Nations Intergovernmental Panel on Climate Change (IPCC). www.auswaertiges-amt.de/en/newsroom/news/-/2588800, accessed 13 July 2023.

87 See also EEAS 2019, p. 40.

88 Oberthür and Dupont 2021, p. 1108.

### 13.3.3 Human Rights Policy

The EU shall promote human rights as one of its general aims[89] and as a central objective of its external action.[90] The EU is, thus, strongly committed to protecting and advancing human rights both internally and in its policies towards third countries.

The current EU priorities in human rights policy making, in relation to its external relations, are recorded in the EU Action Plan on Human Rights and Democracy 2020–2024.[91] Accordingly, the EU supports measures to address the high risk and serious impacts of climate change, environmental degradation and biodiversity loss on the exercise of human rights, such as rights to life, health, safe drinking water and sanitation, food, adequate housing and standard of living, including for climate-induced displaced people.[92] The EU is also committed to strengthening the linkage between human rights and the environment, including climate change in its external action. In practical terms, the EU provides support to activities to raise awareness of the human rights impact and implications of climate change and environmental degradation.[93] Moreover, the EU will continue to strengthen the links between human rights, security, environmental degradation and climate in policy dialogues, conflict prevention, development and humanitarian actions, as well as in disaster risk reduction strategies.[94]

Despite the numerous stated commitments and recognized policy linkages to promote the climate—human rights nexus in its external relations, the EU does not have specific human rights guidelines related to climate. Such guidelines have been created for other types of human rights, such as safe drinking water and sanitation and rights of the child.[95] However, it could be argued that those are established human rights, whereas climate change does not have an associated human right of its own.

The 2021 Council Conclusions on Climate and Energy Diplomacy recognize the need for human rights to be systematically integrated into EU climate action and energy diplomacy.[96] The 2022 Conclusions repeat the same, while also emphasizing gender equality and participatory rights of women and the youth.[97] The 2023 Conclusions also express the EU's commitment to respect and promote numerous human rights when taking action to address climate change.[98] The statement in the 2021

---

[89] Article 2 TEU.

[90] Articles 3(5) and 21 TEU.

[91] EU Action Plan on Human Rights and Democracy 2020–2024, Commission and High Representative of the Union for Foreign Affairs and Security Policy, Joint Communication, JOIN(2020) 5 final.

[92] Ibid., p. 13.

[93] Ibid., p. 15.

[94] Ibid., p. 21.

[95] EEAS, EU Human Rights guidelines, www.eeas.europa.eu/eeas/eu-human-rights-guidelines_en, accessed 13 July 2023.

[96] 2021 Council Conclusions, para 19.

[97] 2022 Council Conclusions, para 10.

[98] 2023 Council Conclusions, paras 41–42.

Conclusions is particularly significant in that it does not speak in abstract about the recognition of or respect for human rights, but it stresses the need to systematically integrate human rights into EU climate and energy diplomacy action.

### 13.3.4 Development Policy

As defined in the founding treaties, the primary objective of EU development cooperation is the reduction and, in the long term, the eradication of poverty.[99] This goal has to be pursued through all EU development policy action.[100] The goal could, in theory, be considered to be a mismatch with climate diplomacy objectives because a reduction of poverty is often likely to lead to increased greenhouse gas emissions due to an increased consumption of goods and services. However, an increased wealth of people and societies can also help them move towards using new and more sustainable energy sources, for instance, which results in fewer emissions. The conflict potential of climate change is also likely diminished by reduced poverty because people can then afford to be better prepared for the negative effects of climate change, for instance. Consequently, it can be said that poverty eradication is generally in line with the broad goals of climate, and, especially, energy, diplomacy. The issue of energy poverty was specifically addressed in the 2021 Council Conclusions on climate and energy diplomacy.[101]

Besides a reduction and eradication of poverty, the goals of the EU development policy include promotion of security and stability and fostering sustainable development in accordance with the European consensus on development,[102] adopted in 2017. The document contains the EU's vision and action framework for development cooperation and explicitly addresses climate change, committing the EU to integrate environment and climate change throughout its development cooperation strategies.[103]

The Council Conclusions on climate and energy diplomacy do not address the EU's development policy to a significant extent. This could be considered to be a missed opportunity, given the potential of directly employing climate and energy diplomacy in the EU development policy activities. Nevertheless, a gradual mainstreaming of climate and energy cooperation into the EU's development policy has taken place.

---

[99] Article 208 TFEU; Article 3(5) TEU.

[100] In a strict sense, development policy is part of the EU's Common Foreign and Security Policy. However, here those two are treated separately.

[101] 2021 Council Conclusions, p. 12.

[102] The new European consensus on development—our World, our Dignity, our Future, Joint statement by the Council and the representatives of the governments of the Member States meeting within the Council, the European Parliament and the Commission, OJ C210/1, 27.6.2017.

[103] Ibid., section 2.2.

EU policy in the field of development cooperation must also be conducted within the framework of the principles and objectives of the EU's external action.[104] As described earlier, the EU's objectives in this field include the prevention of conflicts and contribution to the preservation of peace and sustainable development on a global scale, which reflect elements of climate diplomacy. Yet, action and instruments for development cooperation are generally driven by the priorities of partner countries, which, it has been argued, provide limited opportunities for the EU to put forward its own policy priorities.[105] However, climate change and energy are issues that can be expected to appear on the agendas of basically all developing countries nowadays.

Overall, development policy offers one of the most tangible ways to advance the climate and energy diplomacy agenda in relations between the EU and third countries. At the same time, it also offers a means for the EU to promote sustainable development beyond its borders.

### *13.3.5 EU and Climate Diplomacy Under the UNFCCC*

Global climate negotiations and policymaking under the UNFCCC are an integral element of the EU's climate diplomacy. Work under the global climate agreements has been in a key position in all of the Council Conclusions on climate and energy diplomacy adopted so far.

The EU is, together with the Member States, a party to the international climate agreements. The EU speaks with one voice throughout the negotiations. This requires a great deal of internal coordination, time and effort. Nevertheless, it is a great opportunity to bring forward the EU climate and energy diplomacy agenda in global climate policymaking.

Over the years, the EU has acted as a leader in the global climate negotiations, and it aspires to continue in that role.[106] Besides pursuing an ambitious and forward-looking stance in the negotiations, the EU seeks to also lead by example. This is what is stressed in the EU's climate diplomacy policy as well.[107]

Showing a good example, the exercise of persuasion and provision of support are the cornerstones of the EU's climate diplomatic action under the global climate negotiations. This has led to a situation where the EU is seen as exercising both exemplary and diplomatic leadership under the UNFCCC.[108] Ambitious domestic climate legislation and policy is a key element to the EU's climate diplomacy efforts at the level of the global negotiations. In a way, the EU attempts there to externalize

---

[104] Article 208 TFEU.

[105] Kettunen et al. 2019, p. 15.

[106] All Council Conclusions on climate (and energy) diplomacy adopted so far speak for that. See also, e.g., Wurzel et al. 2017.

[107] See, e.g., para 1 of the 2021 Council Conclusions.

[108] Oberthür and Dupont 2021, pp. 1099–1101.

its existing regulatory framework to the rest of the world.[109] In the EGD, the issue has been formulated in somewhat softer terms so that the EU mobilizes, through its influence, expertise and financial resources, its neighbors and partners to "join it on a sustainable path".[110]

## 13.4 EU Action on Climate and Energy Diplomacy in Practice

The EU's institutional and administrative structure principally provides sectoral support to climate and energy diplomacy action.[111] Climate issues have a Directorate-General (DG) of their own under the European Commission, DG Climate Action, and the EU's energy policy lies under the mandate of DG for Energy. The Conclusions on climate (and energy) diplomacy have been prepared under and adopted by the Foreign Affairs Council. Climate diplomacy, as an issue, belongs to the mandate of the EEAS which is also responsible for the implementation of the EU's development policy (together with the DG for International Partnerships) and the EU's external human rights policy. The EU's Common Foreign and Security Policy is led by the High Representative of the Union for Foreign Affairs and Security Policy, assisted by the EEAS and EU Special Representatives (EUSRs), nine of them in total at the time of writing. The EGD, for its part, is such a multi-faceted package that its implementation is distributed to a large number of EU institutions, their tasks depending on the themes at hand.

Whether in the name of climate (and energy) diplomacy or without the explicitly expressed linkage, the EU has initiated numerous partnerships, policy initiatives and development projects that have had a focus on climate and energy cooperation in third countries and/or on strengthening regional stability through climate action. Partnership-based cooperation on climate and energy diplomacy has a benefit over pure investment projects or development assistance that it is flexible and engages the partners on a more equal basis. It, therefore, fits very well with the idea of diplomatic action.

The number and level of intensity of partnership-based climate initiatives is likely to only increase, given the felt urgency of the need to tackle the climate and energy crises. Recent EU policy and legislative developments also point in the same direction; for instance, the EGD called for continued and strengthened engagement with international partners bilaterally and in regional contexts, with an emphasis on, among others, Africa and the EU neighborhood.[112] Furthermore, the High Representative of the European Union for Foreign Affairs and Security Policy,

---

[109] Schunz 2019; Delreux 2021, p. 270; Çelik 2022, p. 1021.

[110] COM(2019) 640 final, p. 2.

[111] As has been stated by Bremberg and Bunse: "Siloed approaches remain the norm." Bremberg and Bunse 2023, p. 4.

[112] COM(2019) 640 final, pp. 20–21.

Josep Borrell, has stated that "[s]trengthening climate diplomacy means working in partnership."[113] Partnerships are also at the center of the recently introduced "Team Europe" approach[114] in the EU's external development policy action. Its current portfolio largely focuses on green deal diplomacy related projects.

Below, an example is provided on the current framework of EU external action in Africa which bears very close relevance to climate and energy diplomacy.

The EU—Africa partnership forms a very good example of how the EU seeks to carry out coordinated climate and energy diplomacy action with third countries at the regional level. The EU has implemented climate-related projects in Africa for a rather long time; in recent years, the energy element has come more strongly to the forefront.[115] In addition to development projects focusing on energy access and infrastructure, for instance, the climate and energy diplomacy related EU—African cooperation includes trade arrangements, technical and technological cooperation, research and innovation projects etc.

Already, the 2016 EUGS recognized the need to mobilize investments in Africa to support sustainable development by linking policies and blending development efforts with energy and climate, among others.[116] The 2021 Council Conclusions on climate and energy diplomacy set a priority for the EU to "develop a strategic and coordinated approach on the energy partnership with Africa".[117] Furthermore, the document foresaw the launching of an African Union—EU Green Energy Initiative,[118] which should support universal sustainable energy access ambition for Africa.[119] While the energy partnership will benefit African countries in terms of strengthening energy institutions and frameworks and contribute to climate change mitigation, it will also assist the EU in transitioning away from carbon dependency.[120] Ideally, the cooperation initiatives would also align with the objectives of the EU's development and human rights policy as well as contribute to peace, security and sustainable development as per the Union's foreign and security policy objectives.

---

[113] Borrell 2019.

[114] Team Europe consists of the EU, its Member States as well as the European Investment Bank and the European Bank for Reconstruction and Development. The idea is to pool resources and expertise to make development projects more impactful.

[115] For instance, the Africa-EU Energy Partnership was created in 2000 to improve access to secure, affordable and sustainable energy for both continents. The partnership was renewed in 2022. "A partnership for green transition and energy access" is one of the key areas of the EU strategy with Africa. See Towards a Comprehensive Strategy with Africa, High Representative of the Union for Foreign Affairs and Security Policy, Joint Communication to the European Parliament and the Council, JOIN(2020) 4 final.

[116] EUGS, p. 36.

[117] 2021 Council Conclusions, p. 12.

[118] For more information, see EU-Africa: Global Gateway Investment Package—Green Energy Initiative. ec.europa.eu/commission/presscorner/api/files/attachment/874173/GG_Africa_Green_and_Sustainable_Energy.pdf.pdf, accessed 13 July 2023. The Initiative was formally proposed by the EU in 2022.

[119] Ibid.

[120] Dennison and Engström 2023, p. 3.

The EU's climate and energy related projects in and partnerships with African countries and regions are not, as a rule, explicitly labeled as tools to promote climate and energy diplomacy. This is sensible, given the many (geo)political sensitivities involved and the practice-oriented flavor of the cooperative projects.

The Just Energy Transition Partnership with South Africa, launched at the UNFCCC COP-26 in 2021, is a prime example of a new kind of climate diplomacy initiative in which the EU is involved. The initiative is a partnership between South Africa and the EU, France, Germany, the United Kingdom and the United States. The aim is to accelerate the decarbonization of Africa's highest emitter of greenhouse gasses. In practice, the Partnership mobilizes a significant amount of financing to support South Africa in its transition to a low emission economy and to support South African workers and affected communities during the transition.

The Partnership can be regarded as one of the first larger deliverables of the call made in the EGD to develop a stronger green deal diplomacy and to build green alliances.[121] The Partnership has been described as "a good illustration of what we would label the 'coming of age' of Europe's climate diplomacy."[122] The statement appears to be justified, given that the Partnership has adopted a just transition perspective that is more comprehensive than traditional development assistance or climate finance. In addition, the Partnership involves a large donor group, promoting a global approach to climate diplomacy partnerships.

## 13.5 Perspectives on Sustainability, Energy Diplomacy and Nexus Thinking

The EU does not have an official policy document on sustainability or sustainable development of its external policies. The EU Strategy for Sustainable Development[123] was adopted in 2001. Much later, the EU action and policymaking on sustainability was connected with the implementation of the globally agreed SDGs. Their implementation is to be integrated into all EU policymaking, both internal and external.[124] The EGD is also highly relevant for EU policy-making on sustainable development: it is an integral part of the Commission's strategy to implement the United Nations 2030 Agenda and the SDGs.[125]

Mention of the SDGs has been included in all Council Conclusions on climate (and energy) diplomacy adopted since 2015. However, it was not until 2023 that

---

[121] Petriccione and Van Kemseke 2022. https://www.friendsofeurope.org/insights/the-coming-of-age-of-europes-climate-diplomacy/, accessed 13 July 2023.

[122] Ibid.

[123] A Sustainable Europe for a Better World: A European Union Strategy for Sustainable Development, Communication from the Commission, COM/2001/0264 final.

[124] Delivering on the UN's Sustainable Development Goals—A comprehensive approach, Commission staff working document SWD(2020) 400 final.

[125] COM(2019) 640 final, p. 2.

the SDG connection was given a bit more elaborate treatment in the Conclusions document: reference was made to the urgent need to deliver on energy poverty and universal energy access in line with the SDG 7.[126] Given that the SDGs were agreed on already in 2015 and that they should, as a rule, be integrated into all EU policy-making, explicit mainstreaming has been rather slow in coming in the area of the EU's climate and energy diplomacy policy.

As regards the EU energy policy and its external dimension, it has been explicitly at the fore since 2011 when the first Council conclusions on strengthening the external dimension of the EU energy policy were adopted.[127] Relevant Conclusions were adopted also in 2013,[128] which sought to respond to the challenge of diversifying the EU's energy supply and reducing external energy dependency.[129]

Earlier, the EU objectives for energy diplomacy were not very clear and the actions focused on internal coordination and building on the format of energy partnerships and dialogues.[130] However, more recently, the EU has adopted a strategic approach to multilateral energy governance.[131]

The energy diplomacy element was formally added to the EU approach to climate diplomacy in 2021 when the joint Council Conclusions were adopted. This natural policy turn can be said to have followed the objective on synergies and joint priorities presented in the 2016 EUGS document. Moreover, the Strategic Approach to Resilience in the EU's external action, adopted in 2017, underlines the link between global energy security and the EU's climate change policy, and the need to accelerate the global energy transition towards carbon neutrality. It was concluded that EU energy and climate diplomacy aims at promoting this approach.[132] In a way, it is interesting that the common goals of the EU energy and climate diplomacies were recognized in the context of defining the EU approach to resilience in its external action, well before formally combining the two policy areas in the EU external policy.

Climate and energy represent, perhaps, even the most obvious theme for a nexus approach for the EU. Overall, nexus thinking has become prominent in EU policy-making in recent years, especially concerning external relations.[133] It can be considered to result from attempts to abandon a silo approach in the EU policy-making and to address linkages more effectively between issues.

The 2023 Conclusions on climate and energy diplomacy specifically adopts a nexus approach. The document underlines "the importance of integrating the climate,

---

[126] 2023 Council Conclusions, para 38.

[127] Council of the EU, Council conclusions on strengthening the external dimension of the EU energy policy, 3127thCouncil meeting (Energy items), 2011.

[128] Conclusions of the European Council, 22 May 2013, EUCO 75/1/13.

[129] Ibid., para 6.

[130] Petri 2020, p. 3.

[131] See, e.g., Siddi and Kustova 2021.

[132] A Strategic Approach to Resilience in the EU's external action, High Representative of the Union for Foreign Affairs and Security Policy, Joint Communication to the European Parliament and the Council, JOIN(2017) 21 final, p. 13.

[133] See, e.g., Leal-Arcas and Alvarez Armas 2018.

peace and security nexus in EU's external policy and actions" including, inter alia, in "conflict prevention, development cooperation, climate finance and climate diplomacy including dedicated water diplomacy".[134] The Council further invited the High Representative to "mainstream the climate, peace and security nexus, and issue a timely warning and analysis on climate related risks."[135] So far, the Council Conclusions and the EU action on climate and energy diplomacy have focused on the peace and security nexus, but there has also been action, through trade agreements, in relation to the climate-energy-trade nexus,[136] on the climate-energy-development nexus[137] and on the climate-energy-water nexus,[138] among others, in EU external relations in recent years. In this light, it is fair to say that, at the same time, that the climate and energy diplomacy is becoming an ever more prominent guideline in the EU external relations, it is also gaining more ground as a result of a broadening of its scope and "entering new waters",[139] mainly through the nexus approach.

## 13.6 Conclusion

The EU likes to profile itself as a world leader in ambitious climate policymaking. This has translated into action to externalize its climate and energy policy and legislation, by both showing a good example and action to spread good practices and raise the level of ambition of climate and energy policymaking in other countries and regions.

Climate and energy diplomacy is a way to carry out the external dimension of EU climate and energy policy and legislation. At the same time, it provides the framework and means for mainstreaming climate considerations into EU external action in other fields. In addition, climate and energy diplomacy is a defining factor in the EU action under international climate negotiations. Consequently, it can be said that the EU climate and energy diplomacy works in different spheres: in a global setting through climate negotiations; through an application and promotion of EU domestic legislation and policies beyond the Union borders; and through specific cooperation and partnerships with third countries.

Climate and energy diplomacy has become a prominent element, and even a key priority during the last few years, of the EU external policy agenda. The Council

[134] 2023 Council Conclusions, para 21.

[135] Ibid., para 21.

[136] Leal-Arcas and Alvarez Armas 2018.

[137] See Council Conclusions on Energy and Development 2018, 8954/18.

[138] See, e.g., the Team Europe Initiative on Water-Energy-Climate Change in Central Asia, which was launched in 2022. The EU has a rather long history in water-related cooperation in Central Asia, and the water-energy-climate nexus has now been made explicit there. This is certainly beneficial, from diplomatic and conflict prevention perspectives, for EU action in the region. The need to coordinate with energy and water diplomacy policies was recognized by the European Parliament in its 2018 Resolution on climate diplomacy. Para 23.

[139] Petriccione and Van Kemseke 2022.

has adopted a series of Conclusions on climate, and more recently on climate and energy diplomacy. A clear development trend can be discerned in the documents over the years: the Conclusions have gradually developed towards more coherent and ambitious policy guidelines, calling for stronger mainstreaming of climate and energy diplomacy considerations in all areas of EU external action and increasingly addressing the peace and security aspect of climate and energy diplomacy.

The formal integration of energy into the EU climate diplomacy policy since 2021 has brought the goal of accelerating the global energy transition to the fore in the EU external action on climate and energy. It has also strengthened the security and conflict prevention elements therein. These elements are clearly visible in the latest set of Council Conclusions on climate and energy diplomacy, adopted in 2023, where the EU showed determination to continue strong climate action and cooperation and forge new energy partnerships.

The energy aspect also strongly features in the "green deal diplomacy", as called for in the context of the EGD. Climate and energy diplomacy are, indeed, tools to implement the external element of the EGD. In addition, aspects of climate and energy diplomacy are nowadays present in the key policy fields of EU external action. For instance, climate security and strong linkages between climate and energy issues are important ingredients of the current EU foreign and security policy, not least due to the recent (geo)political developments and energy crisis. A gradual mainstreaming of climate and energy cooperation into the EU's development policy has also taken place.

Nexus-thinking has become increasingly prominent in EU policy making in recent years, especially concerning external relations. This trend has become highly visible in the climate and energy diplomacy area, and it seems clear that the EU will put more emphasis on the climate-security nexus in the future. The 2023 Council Conclusions took the nexus approach quite far, not only underlining the peace and security nexus vis-à-vis climate and energy diplomacy, but also other combinations such as those related to development, trade and water.

The Russian aggression against Ukraine, which started in early 2022, has reinforced the close link between climate and energy diplomatic action in the EU agenda. Consequently, energy security has become a topical issue, which has ramifications, on one hand, on the EGD and all the related climate and energy legislation and policies within the EU, and on the key external policies with third countries, on the other hand. EU diplomatic action in the multilateral setting within the UNFCCC negotiations has not seen big changes in recent years. In contrast, the externalization of domestic climate and energy policies has intensified, and smaller setting or bilateral partnerships and initiatives in third countries have increasingly adopted climate and energy diplomatic elements. The approach is in line with the key goals of EU external action, namely conflict prevention and development of international measures to ensure environmental protection and sustainable development.

## References

Adelle C et al (eds) (2018) European Union External Environmental Policy. Rules, Regulation and Governance Beyond Borders. Palgrave Macmillan, Cham

Barra M and Svec M (2018) Reinforcing Energy Governance under the EU Energy Diplomacy: A Proposal for Strengthening Energy Frameworks in Africa, European Journal of Risk Regulation 9: 245–267

Berridge G (2005) Diplomacy: Theory and Practice. Palgrave Macmillan, Basingstoke

Biedenkopf K and Groen L (2021) External EU environmental policy. In: Jordan A and Gravey V (eds), Environmental Policy in the EU. Actors, Institutions and Processes. Routledge, London, pp. 33–50

Biedenkopf K and Petri F (2021) The European External Action Service and EU Climate Diplomacy: Coordinator and Supporter in Brussels and Beyond, European Foreign Affairs Review 26: 71–86

Borrell J (2019) The EU Green Deal—A Global Perspective, a blog post. eeas.europa.eu/eeas/eu-green-deal-%E2%80%93-global-perspective_en, accessed 11 July 2023

Bremberg N and Bunse S (2023) Climate, Peace and Security in a Changing Geopolitical Context: Next Steps for the European Union, SIPRI policy brief. www.sipri.org/publications/2023/sipri-policy-briefs/climate-peace-and-security-changing-geopolitical-context-next-steps-european-union, accessed 13 July 2023

Bäckstrand K and Elgström O (2013) The EU's role in climate change negotiations: from leader to 'leadiator', Journal of European Public Policy 20: 1369–1386

Çelik F B (2022) The EU's different faces in climate diplomacy: leadership, interests, and responsibilities, Journal of European Integration 44: 1019–1039

Delreux T (2021) The EU in international environmental negotiations. In: Jordan A and Gravey V (eds), Environmental Policy in the EU. Actors, Institutions and Processes. Routledge, London, pp. 259–275

Dennison S and Engström M (2023) Decarbonization Nations: How EU Climate Diplomacy Can Save the World, European Council on Foreign Relations policy brief. ecfr.eu/wp-content/uploads/2023/05/Decarbonisation-nations-How-EU-climate-diplomacy-can-save-the-world.pdf, accessed 13 July 2023

Dupont C et al (2018) Climate Change: Adapting to Evolving Internal and External Dynamics. In: Adelle C et al (eds) (2018) European Union External Environmental Policy. Rules, Regulation and Governance Beyond Borders. Palgrave Macmillan, Cham, pp. 105–124

Earsom J and Delreux T (2021) Evaluating EU responsiveness to the evolution of the international regime complex on climate change, International Environmental Agreements: Politics, Law and Economics 21: 711–728

Earsom J and Delreux T (2023) One Big Conversation: The EU's Climate Diplomacy across the International Regime Complex on the Paris Agreement Negotiations, Journal of Common Market Studies 61: 198–214

EEAS (2019) EUGS three years on, looking forward. www.eeas.europa.eu/sites/default/files/eu_global_strategy_2019.pdf, accessed 11 July 2023

EEAS and European Commission (2011) Joint Reflection Paper. Towards a Renewed and Strengthened EU Climate Diplomacy, Brussels, Belgium, 9 July 2011

EEAS and European Commission (2013) EU Climate Diplomacy for 2015 and Beyond. Reflection Paper, Brussels, 26 June 2013

General Secretariat of the Council (2020) Council conclusions and resolutions. www.consilium.europa.eu/en/council-eu/conclusions-resolutions/#. Accessed 9 November 2022

Kettunen M et al (2019) Reinforcing Environmental Dimensions of European Foreign and Security Policy. ieep.eu/uploads/articles/attachments/9219f70d-8d00-4635-ad7e-a5430a40070b/think2030_-_environmental_security_-_dec_2018_-_final.pdf?v=63717390338. Accessed 9 November 2022

Lazard O and Youngs R (eds) (2021), The EU and Climate Security: Toward Ecological Diplomacy, Carnegie Endowment for International Peace, Washington

Leal-Arcas R and Alvarez Armas E (2018) The climate–energy–trade nexus in EU external relations. In: Minas S and Ntousas V (eds) (2018) EU Climate Diplomacy. Politics, Law and Negotiations, Routledge, Abingdon, pp. 153–168
Minas S and Ntousas V (eds) (2018) EU Climate Diplomacy. Politics, Law and Negotiations, Routledge, Abingdon
Oberthür S and Dupont C (2021) The European Union's international climate leadership: towards a grand climate strategy?, Journal of European Public Policy 28: 1095–1114
Oberthür S et al (2022) European Foreign Policy in a Decarbonising World. Challenges and Opportunities, Routledge, Abingdon
Petri F (2020) Revisiting EU Climate and Energy Diplomacy: A Starting Point for Green Deal Diplomacy?, Egmont European policy brief No. 65. aei.pitt.edu/103670/1/EPB65%2Dfinal.pdf, accessed 11 July 2023
Petriccione M and Van Kemseke P (2022) The coming of age of Europe's climate diplomacy, a Friends of Europe insight. www.friendsofeurope.org/insights/the-coming-of-age-of-europes-climate-diplomacy/, accessed 13 July 2023
Schunz S (2019) The European Union's environmental foreign policy: from planning to a strategy?, International Politics 56: 339–358
Siddi M and Kustova I (2021) From a liberal to a strategic actor: the evolution of the EU's approach to international energy governance, Journal of European Public Policy, 28: 1076–1094
Teevan C et al (2021) The Green Deal in EU foreign and development policy, European Centre for Development Policy Management briefing note No. 131. ecdpm.org/work/the-green-deal-in-eu-foreign-and-development-policy, accessed 11 July 2023
Tocci N (2020) Resilience and the role of the European Union in the world, Contemporary Security Policy, 41: 76–194S
Torney D and Davis Cross MK (2018) Environmental and Climate Diplomacy: Building Coalitions Through Persuasion. In: Adelle C et al (eds), European Union External Environmental Policy. Rules, Regulation and Governance Beyond Borders, Palgrave Macmillan, Cham, pp. 39–58
Tänzler D et al (2021) EU climate change diplomacy in a post-Covid-19 world, European Union, Brussels. www.europarl.europa.eu/RegData/etudes/STUD/2021/653643/EXPO_STU(2021)653643_EN.pdf, accessed 11 July 2023
van Schaik L (2010) The Sustainability of the EU's Model for Climate Diplomacy. In: Oberthür S and Pallemaerts M (eds), New Climate Policies of the European Union: Internal Legislation and Climate Diplomacy, ASP VUBPress, Brussels, pp. 251–280
Woerdman E et al. (eds) (2021), Essential EU Climate Law, Edward Elgar, Cheltenham
Wouters J et al (2021) External Environmental Policy. In: Wouters J et al (eds), The Law of EU External Relations. Cases, Materials, and Commentary on the EU as an International Legal Actor, Oxford University Press, Oxford, pp. 316–358
Wurzel R K W et al (eds) (2017) The European Union in international climate change politics: Still taking a lead?, Routledge Abingdon

**Tuula Honkonen** is a Senior Lecturer of International Environmental Law at the University of Eastern Finland, Center for Climate, Energy and Environmental Law.

# Chapter 14
# EU External Relations Law and Sustainability: Exploring the Potential for Engaging China

**Amon Elpídio da Silva**

## Contents



**Abstract** This Chapter examines the relationship between European Union (EU) external relations law, sustainability, and the role of China in advancing global sustainable development goals. The Chapter analyzes the EU's approach to engaging China in its external relations and explores the potential for collaboration on sustainability issues. Additionally, it identifies key challenges and opportunities in the EU-China relationship concerning sustainability, offering insights into the legal and policy dimensions. By focusing on the specific dynamics between the EU and China, this chapter contributes to the academic discourse on EU external relations law. The main findings highlight the potential of EU-China collaboration on sustainability while emphasizing the need to strengthen the legal framework, address geopolitical tensions, enhance climate change cooperation, promote regulatory convergence, and balance economic and environmental interests for a more sustainable relationship.

**Keywords** EU-China collaboration · Geopolitical tensions · Competing interests · Legal framework · Sustainable relationship · Future direction

---

A. Elpídio da Silva (✉)
Federal University of Minas Gerais, Belo Horizonte, Brazil
e-mail: amonelpidio@ufmg.br



R. A. Wessel et al. (eds.), *EU External Relations Law and Sustainability*, Global Europe: Legal and Policy Issues of the EU's External Action 4,
https://doi.org/10.1007/978-94-6265-655-0_14

## 14.1 Introduction

EU external relations law is the legal framework guiding the European Union's external action and interactions with non-EU Member States and international organizations.[1] It encompasses a wide range of areas, including trade, diplomacy, development cooperation, human rights, and environmental policies. The external relations of the European Union (EU) are rooted in principles such as promoting peace, democracy, and the rule of law while advancing its economic and political interests globally.

Sustainability has emerged as a critical global priority, emphasizing the need to address environmental, social, and economic challenges to ensure the well-being of present and future generations. It entails promoting practices that maintain ecological balance, preserve natural resources, foster social equity, and promote economic prosperity. The urgency of addressing sustainability challenges, such as climate change, biodiversity loss, and resource depletion, has elevated the importance of integrating sustainable development principles into various policy domains, including international relations.

The relationship between the EU and China is multifaceted and complex, spanning trade, investment, diplomacy, and cooperation on global issues. China is the EU's second-largest trading partner, and the EU is China's largest trading partner. The EU and China engage in political dialogues and collaborate on various issues, including climate change, sustainable development, and global governance.

Cooperation between the EU and China on sustainability holds significant importance for several reasons. Firstly, it contributes to advancing global sustainable development goals, such as those outlined in the United Nations' Sustainable Development Goals (SDGs). Collaboration between these major global actors enhances the overall effectiveness of global efforts to address environmental challenges.

Secondly, EU-China cooperation is instrumental in tackling climate change. Both entities play crucial roles in combating climate change, and collaboration between them is necessary for implementing the Paris Agreement, sharing best practices in renewable energy, promoting energy efficiency, and facilitating the transition to a low-carbon economy.

Thirdly, promoting sustainable trade and investment is essential. By strengthening sustainability considerations in EU-China trade and investment relations, responsible business practices can be encouraged, environmental protection can be supported, and social and labor rights can be upheld. Enhanced cooperation on sustainability standards, responsible supply chains, and corporate social responsibility can contribute to sustainable economic growth.

Lastly, biodiversity conservation and environmental protection are shared concerns for the EU and China. Collaborative efforts can address challenges related to illegal wildlife trade, promote conservation measures, and strengthen environmental governance.[2]

---

[1] Wouters et al. 2021, p. 26.

[2] Massé et al. 2020, p. 25.

However, the EU-China relationship on sustainability also faces challenges. Differences in regulatory frameworks between the EU and China can create obstacles in aligning policies and practices. Competing economic and geopolitical interests may hinder cooperation on sustainability issues, necessitating a delicate balance. Building trust and ensuring transparency between the EU and China is crucial for effective cooperation on sustainability, addressing concerns related to information sharing, compliance, and enforcement. Achieving policy coherence within the EU and between the EU and China is another challenge, as it requires aligning diverse policy areas, such as trade, energy, and environmental policies, with sustainability objectives.[3]

With that, understanding the background and significance of EU External Relations Law, sustainability, and the EU-China relationship provides valuable insights into the complexities of promoting sustainability in the context of EU-China relations. Overcoming challenges and fostering effective cooperation can pave the way for a more sustainable future.

The primary objectives of this Chapter are to examine and analyze the relationship between EU External Relations Law, sustainability, and the EU-China relationship; assess the significance of EU External Relations Law in promoting sustainability in the context of EU-China relations; and, identify the challenges and opportunities in integrating sustainability into the EU-China relationship.

## 14.2 EU External Relations Law and Sustainability: An Overview

The legal framework of EU External Relations Law provides the basis for the European Union's engagement with other countries and international organizations on various fronts, including trade, diplomacy, development cooperation, human rights, and environmental policies.[4] However, while the EU has demonstrated a commitment to sustainability in its external actions, significant gaps still need to be addressed to achieve greater equity and balance.

One of the key gaps in the legal framework of EU External Relations Law lies in the lack of enforceability of sustainability provisions. While the EU has incorporated sustainability objectives into its trade agreements and other external instruments, the mechanisms for ensuring compliance and accountability remain limited.[5] This undermines the effectiveness of sustainability provisions, as they often lack teeth and fail to produce tangible outcomes.

Furthermore, there is a need for greater coherence and integration of sustainability considerations across different policy areas within the EU's external relations. While progress has been made in certain domains, such as climate change and human

[3] Christiansen and Maher 2017, p. 128.

[4] Craig and Búrca 2020, p. 367; Larik and Wessel 2020.

[5] Morgera et al. 2020, p. 16.

rights, there is a lack of comprehensive integration of sustainability principles into all aspects of EU External Relations Law.[6] This fragmentation limits the ability of the EU to effectively address sustainability challenges and promote a holistic approach to sustainable development.

Another critical gap is the limited focus on social and environmental justice in the EU's external actions. While the EU has emphasized economic development and environmental protection, there is a need to ensure that sustainability efforts also prioritize social equity and environmental justice.[7] This requires considering the interests and needs of marginalized communities, both within the EU and in partner countries, and addressing the unequal distribution of resources and benefits.

Additionally, the legal framework should take into account the differentiated responsibilities and capacities of different countries. Ensuring equity and balance[8] in EU External Relations Law means recognizing the varying levels of development, resources, and capabilities among partner countries. This requires tailoring sustainable development approaches to the specific circumstances and needs of different countries, rather than applying a one-size-fits-all approach.

Moreover, the legal framework should promote meaningful participation and inclusion of civil society organizations, local communities, and affected stakeholders in decision-making processes.[9] Greater transparency and public engagement are essential to ensure that sustainability objectives align with the interests and aspirations of those directly affected by EU external actions.

To address these gaps and achieve greater equity and balance, several measures are required. Strengthening the enforcement mechanisms of sustainability provisions in trade agreements and other external instruments must enhance accountability. This needs to include mechanisms for monitoring compliance, dispute settlement, and targeted sanctions for non-compliance.

Promoting policy coherence across different areas of EU External Relations Law can be achieved by integrating sustainability considerations into all aspects of the legal framework.[10] This requires close coordination and cooperation among different EU institutions and policy areas, ensuring that sustainability objectives are mainstreamed and pursued consistently.

Furthermore, the EU should prioritize social and environmental justice by adopting a rights-based approach to sustainability. This entails recognizing and addressing the social and environmental impacts of EU external actions, promoting inclusive and participatory decision-making processes, and ensuring that the benefits of sustainable development are shared equitably among all stakeholders.

Recognizing the differentiated responsibilities and capacities of different countries is crucial. The EU should adopt a more nuanced and context-specific approach to sustainable development, taking into account the varying needs and capacities of

[6] Eeckhout and Lopez-Escudero 2019, p. 94.

[7] European Commission 2023c, para 9.

[8] European Union 2010, p. 82.

[9] Wessel 2011, p. 622.

[10] Estrada Canãmares 2016, p. 34.

partner countries.[11] This can be achieved through targeted capacity-building support, technology transfer, and financial assistance to facilitate sustainable development pathways.

While the legal framework of EU External Relations Law demonstrates a commitment to sustainability, these critical gaps need to be addressed to achieve more equity and balance. Strengthening enforceability, enhancing policy coherence, prioritizing social and environmental justice, and recognizing differentiated responsibilities are essential steps toward a more equitable and balanced approach to sustainability.[12] It is believed that by addressing these gaps, the EU can play a more transformative role in promoting sustainability globally while upholding principles of equity and justice.

The EU's commitment to sustainable development influences its external relations and has implications for its engagement with other countries, including China. The EU's dedication to sustainable development is reflected in its policies, strategies, and frameworks, as exemplified by the European Green Deal launched in 2019.[13] This comprehensive initiative aims to transform the EU into a climate-neutral and environmentally sustainable economy by 2050. The EU's commitment to sustainability permeates its external relations, shaping its cooperation with China and other partners.

In the context of EU-China cooperation on sustainability, several past initiatives and agreements demonstrate the EU's commitment and its implications for the bilateral relationship. These initiatives include the EU-China 2020 Strategic Agenda for Cooperation,[14] which prioritized sustainability as an area for collaboration. The EU-China Partnership on Urbanization,[15] established in 2012, focused on sustainable urban development. Additionally, the EU-China High-Level Environment and Climate Dialogue[16] serves as a platform for discussing common challenges and enhancing cooperation on sustainability.

Joint statements and declarations between the EU and China have reaffirmed their commitment to sustainable development. For example, the EU-China Summit Joint Statement in 2019[17] emphasized the importance of addressing climate change, protecting biodiversity, and implementing the Paris Agreement. Furthermore, the EU incorporates sustainability provisions in its trade and investment agreements, including those with China, covering areas such as environmental standards and corporate social responsibility.

These past initiatives and agreements underscore the EU's commitment to sustainability and its implications for EU-China cooperation. The EU seeks to engage China in promoting sustainable development, addressing climate change, and advancing

---

[11] Dzhuraev and Muratalieva 2020, p. 8.

[12] European Commission (2023d), p. 5.

[13] European Commission 2019a, para 2.

[14] European Union 2013, p. 2.

[15] European Union 2012, para 2.

[16] European Commission 2023a, para 1.

[17] European Commission 2019b, para 3.

environmental protection through various channels, including dialogues, partnerships, and the integration of sustainability provisions in trade and investment agreements.

The EU's commitment to sustainable development in its external relations implies a shared vision with China on addressing global challenges and achieving sustainable growth.[18] It fosters opportunities for knowledge sharing, capacity building, and technology transfer. Moreover, it promotes the adoption of sustainable practices, standards, and regulations, which can lead to economic and environmental benefits for both the EU and China.

However, challenges remain in fully realizing the implications of the EU's commitment to sustainable development in its engagement with China.[19] These challenges include ensuring the effective implementation of sustainability provisions, addressing differences in regulatory frameworks, enhancing policy coherence, and addressing issues of transparency, accountability, and enforcement.

The EU's focus on sustainability is evident in various initiatives, dialogues, and agreements highlighting the shared vision for addressing global challenges and promoting sustainable growth. Nevertheless, further efforts are required to address challenges and fully capitalize on the potential of EU-China cooperation on sustainability.

## 14.3 China's Role in Global Sustainability Efforts

China recognizes sustainable development as a national priority, aiming for a balance between economic growth and environmental protection. The government promotes green growth, circular economy, and resource efficiency. It has made significant investments in renewable energy, becoming a global leader in solar and wind power. China also emphasizes poverty alleviation and social development to ensure inclusive growth and benefits for all segments of society.[20]

Despite China's progress in sustainable development and climate change mitigation, several gaps and challenges remain. Air and water pollution continue to pose significant challenges, requiring stronger environmental regulations and pollution control measures.[21] Environmental governance needs improvement in terms of enforcement and monitoring. Balancing economic growth with sustainability objectives is a challenge, and efforts are needed to align economic development with cleaner energy sources and sustainable consumption patterns. Enhancing transparency and data sharing is crucial for accurate monitoring and reporting. Further international cooperation is needed to address global climate challenges and support developing countries in their climate mitigation and adaptation efforts.

[18] European Commission 2023b, para 4.

[19] European Commission 2022a, para 1.

[20] Liu et al. 2021, p. 4.

[21] Gibson 2018, para 4.

Addressing these gaps requires ongoing efforts, including strengthening environmental regulations, promoting sustainable practices, enhancing transparency, and deepening international collaboration. China must continue its commitment to sustainable development and climate change mitigation while addressing these challenges to achieve a more sustainable future domestically and globally.

The Belt and Road Initiative (BRI), launched by China in 2013, is an ambitious infrastructure and connectivity project that aims to enhance regional integration and economic cooperation across Asia, Latin America, Africa, and beyond.[22] The BRI encompasses a vast network of transportation, energy, and telecommunication projects, including railways, ports, pipelines, and power plants.[23] While the BRI presents significant economic opportunities, its impact on sustainability has been a subject of debate. The BRI includes investments in renewable energy projects such as solar and wind farms. These projects can help reduce carbon emissions, promote clean energy generation, and contribute to global efforts in combating climate change.[24] Through promoting trade, investment, and connectivity, the BRI has the potential to stimulate economic growth and reduce poverty in participating countries.[25] Enhanced economic development can lead to improved livelihoods and social well-being.

The BRI's large-scale infrastructure projects can have significant environmental impacts, including habitat destruction, deforestation, and increased carbon emissions.[26] Careful environmental assessments, mitigation measures, and adherence to international environmental standards are necessary to minimize negative ecological consequences. BRI's projects may result in population displacements, social disruptions, and cultural challenges in local communities. Ensuring inclusive and participatory decision-making processes, respecting the rights and interests of local communities, and conducting thorough social impact assessments are essential for mitigating potential negative social and cultural impacts.

Also, the BRI presents both opportunities and challenges for sustainability. While it can contribute to infrastructure development, renewable energy, poverty reduction, and knowledge sharing, careful planning, environmental safeguards, social inclusiveness, and transparency are vital to mitigate potential negative impacts. By addressing these challenges and adopting a sustainable approach, the BRI can play a crucial role in fostering regional connectivity, economic development, and environmental sustainability.

---

22 McBride et al. 2023, p. 1.

23 Wang 2023, paras 1, 3.

24 Ibid.

25 McBride et al. 2023, para 27.

26 Teo et al. 2019, pp. 6-7.

## 14.4 EU-China Cooperation on Sustainability: Challenges and Opportunities

The differences in regulatory frameworks and standards between the EU and China present challenges to their cooperation and trade relations. These disparities stem from variations in legal systems, cultural norms, historical contexts, and economic priorities. Overcoming these differences requires a multifaceted approach that involves dialogue, harmonization efforts, mutual recognition, and the development of common standards.

The EU and China have distinct regulatory frameworks that govern various sectors, including trade, investment, product safety, intellectual property, and environmental protection. These frameworks reflect the unique legal traditions and policy priorities of each jurisdiction. The EU operates within a framework characterized by a high level of regulatory harmonization, emphasizing consumer protection, environmental sustainability, and social welfare.[27] In contrast, China's regulatory framework is evolving, influenced by a mix of socialist principles,[28] market-oriented reforms,[29] and a focus on economic development.[30]

The disparities in regulatory frameworks and standards create challenges in terms of market access, trade barriers, and the protection of public health and safety. Regulatory divergence can lead to discrepancies in product specifications, testing requirements, certification procedures, and labeling standards. These discrepancies not only impede market entry for businesses but also raise concerns regarding consumer protection, environmental sustainability, and fair competition.

To address these challenges, the EU and China can engage in dialogue and regulatory cooperation. Establishing a structured dialogue mechanism can facilitate mutual understanding, identify areas of convergence, and address divergences constructively. Regulatory cooperation initiatives, such as the EU-China Regulatory Cooperation Project,[31] can promote information exchange, capacity building, and the alignment of regulatory approaches.

Furthermore, the harmonization of regulatory frameworks and standards is crucial for reducing trade barriers and enhancing compatibility between the EU and China. This process involves identifying common goals, principles, and best practices to align regulations and facilitate trade while safeguarding public interests. It requires the identification of areas where regulatory convergence is feasible and beneficial for both parties.

Mutual recognition agreements (MRAs) can play a vital role in overcoming differences in regulatory frameworks. MRAs establish mechanisms for recognizing and accepting each other's conformity assessment procedures, certification, and

[27] Inês 2020, p. 2.

[28] Zhang 2010, p. 7.

[29] Yang 2020, p. 8.

[30] Ibidem, p. 27.

[31] European Commission 2023e, para 2.

testing results.[32] By recognizing the equivalence of standards and procedures, MRAs enhance market access and reduce the need for redundant testing and certification, thus facilitating trade and ensuring the safety and quality of goods and services.

Developing common standards is another effective approach to bridging regulatory gaps between the EU and China. Common standards provide a foundation for shared understanding and facilitate market integration.[33] Collaborative efforts to develop international standards can draw upon the expertise and experiences of both parties, leading to higher-quality standards that are more globally accepted.

Overcoming differences in regulatory frameworks and standards between the EU and China requires a long-term commitment, political will, and mutual trust. It necessitates recognizing the importance of regulatory cooperation for trade facilitation, consumer protection, and sustainable development. By engaging in dialogue, harmonizing regulations, fostering mutual recognition, and developing common standards, the EU and China can enhance their cooperation, strengthen economic ties, and address the challenges posed by regulatory disparities.

The differences in regulatory frameworks and standards between the EU and China present challenges to their cooperation and trade relations. Overcoming these differences requires a multifaceted approach, including dialogue, harmonization efforts, mutual recognition, and the development of common standards.[34] By embracing regulatory cooperation, both parties can facilitate market access, enhance trade flows, and ensure the protection of public health, safety, and the environment.

Balancing economic and environmental interests in EU-China relations presents a complex and multifaceted challenge that requires navigating competing priorities, addressing environmental concerns, and fostering sustainable development. The relationship between economic growth and environmental sustainability is crucial for both parties and achieving a harmonious balance necessitates a comprehensive and integrated approach.

The economic dimension of EU-China relations is marked by extensive trade and investment flows, with both entities being major economic players on the global stage. China is the EU's principal partner for the import of goods,[35] and the EU is one of China's largest trading partners.[36] This economic interdependence presents opportunities for collaboration and mutual benefits. However, it also raises concerns about the environmental impact of economic activities, including carbon emissions, resource depletion, and pollution.

Environmental interests are of increasing significance in the EU-China relationship, as both entities recognize the urgency of addressing climate change, biodiversity loss, and other environmental challenges.[37] The EU has been positioning itself as a global leader in environmental protection, promoting sustainability, and pursuing

---

[32] OECD 2023, para 1.

[33] Egan 2001, p. 265.

[34] Brown 2016, p. 4.

[35] European Commission 2022c, para 5.

[36] People's Republic of China 2023, para 1.

[37] Casarini 2006, 25.

ambitious climate goals.[38] China, as well, has demonstrated a growing commitment to environmental stewardship, exemplified by its efforts to transition to a low-carbon economy, reduce pollution, and invest in renewable energy.[39]

Balancing economic and environmental interests requires a multi-pronged approach. First and foremost, it necessitates promoting sustainable trade and investment practices that prioritize environmental considerations.[40] This can be achieved through the development and implementation of green standards, eco-labeling schemes, and sustainable procurement policies. Aligning regulatory frameworks to ensure compatibility and prevent regulatory arbitrage is crucial in fostering a level playing field.

Furthermore, integrating environmental considerations into bilateral investment agreements and trade agreements can help ensure that economic activities meet high environmental standards. Provisions for environmental impact assessments, sustainable resource management, and biodiversity conservation can be included to safeguard the environment while promoting economic cooperation.

At the international level, the EU and China can collaborate to address global environmental challenges through multilateral frameworks such as the United Nations Sustainable Development Goals,[41] the Paris Agreement,[42] and the Convention on Biological Diversity.[43] Joint efforts in climate change mitigation, ecosystem conservation, and biodiversity protection can amplify their impact and showcase leadership in sustainable development.

However, challenges persist in balancing economic and environmental interests. These include different levels of economic development, variations in environmental regulations, and concerns over competitiveness and market access. Overcoming these challenges requires trust, transparency, and mutual understanding. Both China and the European Union must engage in constructive dialogue, respect each other's development priorities, and work towards win-win solutions that promote sustainable development and environmental protection.

Enhancing cooperation on climate change mitigation and adaptation is a pressing priority for both, the EU, and the international community.[44] The EU has positioned itself as a global leader in climate action, emphasizing the importance of addressing climate change through collective efforts.[45] However, critical aspects remain in the EU's approach to climate cooperation, including challenges related to ambition, policy coherence, financial support, and equitable burden-sharing.

With regard to climate change, the EU has demonstrated its commitment to mitigate it by setting ambitious targets and adopting comprehensive policies. The EU's

[38] European Commission 2019a, para 2.

[39] The World Bank 2022, para 1.

[40] Miles 2013, p. 119.

[41] United Nations 2023, para 1.

[42] United Nations Framework Convention on Climate Change 2016, para 1.

[43] United Nations 1992, p. 3.

[44] Harris, 2007, p. 151.

[45] Wurzel and Connelly 2010, p. 285.

long-term goal is to achieve climate neutrality by 2050, and it has implemented various measures to reduce greenhouse gas emissions, transition to renewable energy sources, and improve energy efficiency.[46] The EU has also played a prominent role in international climate negotiations, advocating for the Paris Agreement and urging other nations to enhance their climate ambitions.[47]

While the EU's efforts are commendable, critical aspects arise when considering the level of ambition. Some argue that the EU's current targets may not be sufficient to limit global warming to well below 2 °C, as outlined in the Paris Agreement.[48] There is a need for the EU to continuously reassess and strengthen its targets to align with the latest scientific evidence and the urgency of the climate crisis.

The EU's climate policies span across different sectors and levels of governance, involving complex interactions between member states and EU institutions. This leads us to make reflections on the necessity to consider policy coherence as a critical aspect in enhancing cooperation on climate change. Achieving policy coherence within the EU and in international cooperation is essential for the effective implementation and coordination of climate actions.[49] Strengthening the alignment between EU policies and climate objectives, including those related to trade, agriculture, and energy, is necessary to ensure a consistent and integrated approach.

Nevertheless, financial support is a key point when States are urged to make efforts in climate cooperation. The EU has committed to providing financial assistance to developing countries for climate change mitigation and adaptation efforts through mechanisms such as the Green Climate Fund.[50] However, the adequacy and predictability of financial support may be a target to debate. Ensuring sufficient and predictable funding is essential for enabling developing countries, particularly the most vulnerable, to effectively address climate change impacts and implement adaptation measures.

At the same time the EU has made substantial efforts to reduce its own emissions,[51] global climate goals cannot be achieved without the active participation of all countries. It is fundamental to ensure that the burden of climate action is distributed fairly and takes into account different countries' historical emissions, development needs, and capacity limitations. Addressing issues of equity and common but differentiated responsibilities is vital for building trust, enhancing cooperation, and fostering a sense of global solidarity in tackling climate change.

EU's approach to its external relations has some aspects to improve including reassessing and strengthening climate targets to align with the latest scientific evidence, ensuring policy coherence across different sectors and levels of governance, providing sufficient and predictable financial support to developing countries,

---

[46] European Commission 2020, para 1.

[47] Rayner and Jordan 2016, p. 13.

[48] Delreux and Ohler 2019, p. 13.

[49] Ibid.

[50] European Commission 2022b, para 25.

[51] European Council 2023, para 3.

and promoting equitable burden-sharing among nations.[52] By addressing these critical aspects, the EU can play a more effective and influential role in global climate cooperation, contributing to a sustainable and resilient future for all.

## 14.5 Legal and Policy Perspectives on EU-China Collaboration

Legal and policy perspectives on EU-China collaboration encompass a range of issues, including the analysis of existing legal instruments and agreements. Such analysis is essential for understanding the framework within which the EU-China collaboration operates and identifying critical aspects that may impact the effectiveness and sustainability of this collaboration. This discussion will focus on key legal instruments and agreements, essential aspects of their implementation, and challenges that may hinder EU-China collaboration.

One of the primary legal instruments guiding EU-China collaboration is the Commission's paper on *A maturing partnership—shared interests and challenges in EU-China relations*, which was published in 2003 and has evolved to cover various areas of cooperation, including trade, investment, climate change, and environmental protection.[53] This partnership provides a framework for dialogue and cooperation, aiming to enhance mutual understanding, deepen economic ties, and address common challenges.

Another important instrument with regard to the EU-China collaboration framework is the already mentioned *EU-China 2020 Strategic Agenda for Cooperation*, which sets out specific objectives and actions in areas such as trade and investment, connectivity, sustainable development, and people-to-people exchanges.[54] While this document outlines important areas of cooperation, critical aspects arise concerning the implementation and effectiveness of these objectives and actions.

Even so, one critical aspect of EU-China collaboration is the need for a level playing field and fair competition. Challenges related to market access, investment barriers, intellectual property rights protection, and state subsidies can hinder fair competition and create imbalances in economic relations.[55] Ensuring a transparent and predictable legal and regulatory environment is crucial for promoting a level playing field and fostering trust between the EU and China.

Despite that, the alignment of legal frameworks and standards is another point that remains to be solved. Differences in regulatory approaches, technical requirements, and product standards can create barriers to trade and impede effective collaboration.[56] Harmonization and mutual recognition of standards and regulations are vital

[52] Gallas 2023, paras 1, 3.

[53] European Commission 2003, pp. 6-7.

[54] European Union 2013, p. 3.

[55] Song, 2012, p. 26.

[56] Pan 2012, p. 49.

for enhancing cooperation, facilitating market access, and promoting a sustainable and integrated trading relationship.

In what touches on the protection of goods, intellectual property rights (IPR) is a matter which has gained strength in debates through the years, also in relation to EU-China relations. The EU has expressed concerns about the enforcement of IPR in China, including issues related to counterfeiting, piracy, and technology transfer.[57] Addressing these concerns and ensuring effective IPR protection matters for fostering innovation, encouraging technology transfer, and promoting a conducive environment for collaboration in research and development.

Legal and policy perspectives on EU-China collaboration involve the analysis of existing legal instruments and agreements. Addressing challenges related to implementation and enforcement is vital for promoting effective collaboration and achieving the objectives outlined in the legal instruments and agreements. By recognizing and addressing these critical aspects, the EU and China can foster a more robust and sustainable collaborative relationship.

A proposal of a comprehensive bilateral investment agreement (BIA) would be an exciting start for enhancing EU-China collaboration. Negotiations for such an agreement have been ongoing, aiming to establish a more predictable and transparent legal framework for investment between the two parties.[58] A robust BIA can address critical aspects, such as market access, investment protection, and dispute resolution mechanisms, providing greater certainty and stability for investors.

When discussing EU-China relations, the challenges to an agreement usually are human rights, democracy, and environmental protection. While China is investing billions to expand its BRI, the EU has made several declarations showing concerns on the topic.[59] The same happens to human rights and the lack of 'alignment' from China.[60]

Geopolitical competition and differing strategic priorities can impede cooperation and undermine efforts to develop a mutually beneficial relationship. For instance, the EU's alignment with other global powers, such as the United States, on various geopolitical issues may create friction with China, leading to complexities in finding common ground and shared objectives.[61]

Moreover, the economic interests of both the EU and China often diverge, presenting another significant hurdle to collaboration. Their status as major economic powers engenders interdependence and competition simultaneously. The prevailing trade imbalances, market access restrictions, concerns about intellectual property rights, and the use of state subsidies introduce frictions that challenge the establishment of a level playing field and fair competition. The challenge lies in reconciling these economic interests with the need to protect domestic industries, and foster an equitable trading relationship.

---

[57] Vogt 2012, p. 68.

[58] Houde and Yannaca-Small 2004, p. 4.

[59] Grieger 2021, p. 11.

[60] Human Rights in China 2016, p. 9.

[61] Marcin 2023, p. 1.

The principle of reciprocity constitutes yet another obstacle to EU-China collaboration. The EU champions the notion of equal market access and fair treatment, but concerns persist regarding the imbalance in market openness between the two parties.[62] Addressing this issue necessitates fostering a more balanced and reciprocal economic relationship, which, in turn, demands a commitment to fair competition and the creation of an environment that supports businesses from both sides.

To forge a more sustainable and balanced EU-China collaboration, it is vital to chart future directions that tackle these challenges head-on. One way to achieve this is by fostering constructive dialogue and engagement, enabling open discussions on sensitive issues while identifying areas of mutual interest. Enhanced transparency and trust-building measures, coupled with regular high-level dialogues, can facilitate a better understanding and management of geopolitical tensions and competing interests. To promote shared values and principles, including respect for human rights, EU-China collaboration should emphasize dialogue, engagement, and the integration of human rights considerations into all aspects of cooperation. Adherence to international norms and standards should be a shared commitment, serving as a unifying factor for collaboration.

Additionally, both parties should actively seek opportunities for cooperation in areas of mutual interest, such as climate change, sustainable development, and global governance. This can contribute to building trust and achieving shared goals while emphasizing the importance of multilateral cooperation within the United Nations framework to address global challenges and identify common solutions.

## 14.6 Conclusion

This Chapter has explored various aspects of EU-China collaboration in the context of sustainability, focusing on sustainable development, climate change, the Belt and Road Initiative (BRI), regulatory frameworks, and balancing economic and environmental interests in the context of EU external relations law.

It has become evident that EU-China collaboration holds great potential for addressing common sustainability challenges. The EU has demonstrated a commitment to sustainable development and has actively engaged with China on various initiatives and agreements. However, critical aspects and challenges persist, including the need to strengthen the legal framework, address geopolitical tensions and competing interests, enhance cooperation on climate change mitigation and adaptation, and overcome differences in regulatory frameworks and standards.

Several recommendations can be made for future EU-China sustainability collaboration. Efforts should be made to enhance bilateral agreements and investment treaties, ensuring enforceability, transparency, and compliance mechanisms. Dialogue, engagement, and the pursuit of shared objectives can help manage and overcome geopolitical tensions while promoting reciprocity, fair competition, and

[62] Hanemann and Huotari 2018, p. 22.

balanced economic relations. Furthermore, the EU and China should deepen their collaboration on climate change mitigation and adaptation, aligning their targets and policies, enhancing financial support, and promoting technology transfer. Harmonization and mutual recognition of standards should be pursued to facilitate trade and investment and promote a level playing field.

These findings have several policy implications. Policymakers would have to prioritize strengthening the legal framework for EU-China collaboration, ensuring coherence across different sectors and levels of governance. Efforts should be made to balance economic and environmental interests, emphasizing the importance of sustainability in trade and investment relations. Policy actions should also consider the geopolitical context and competing interests, aiming for constructive engagement and finding common ground on shared challenges.

Further research is needed to delve deeper into specific areas related to EU-China sustainability collaboration. This includes examining the effectiveness of existing legal instruments and agreements, analyzing the implementation and impact of sustainability initiatives, and exploring the role of non-state actors in promoting sustainable development. Additionally, research can focus on the implications of geopolitical tensions on sustainability collaboration, the potential for technology transfer and innovation, and the social and environmental impacts of the Belt and Road Initiative.

EU-China sustainability collaboration thus presents both opportunities and challenges. Strengthening the legal framework, addressing geopolitical tensions, enhancing climate change cooperation, promoting regulatory convergence, and balancing economic and environmental interests are crucial for fostering a more sustainable and balanced relationship. Policymakers should consider these findings and recommendations to inform their actions, while researchers should continue exploring the complexities of EU-China sustainability collaboration to contribute to the advancement of knowledge and policy in this area.

## References

Borell J (2020) China carbon neutrality in 2060: a possible game changer for climate. https://www.eeas.europa.eu/eeas/china-carbon-neutrality-2060-possible-game-changer-climate_en Accessed on 26 July 2023

Brown K (2016) China EU Relations: Where to Now?. v. 11:2016–2017

Casarini N (2006) The evolution of the EU-China relationship: from constructive engagement to strategic partnership. European Union Institute for Security Studies. Paris

Craig P and Búrca G (2020) EU Law: Text, Cases, and Materials. Oxford University Press. Oxford

Christiansen T, Maher R (2017) The rise of China—challenges and opportunities for the European Union. Asia Europe Journal, 15, 121–131.

Delreux T and Ohler F (2019) Climate policy in European Union politics. EBOR https://doi.org/10.1093/acrefore/9780190228637.013.1097

Dzhuraev E and Muratalieva N (2020) The EU Strategy on Central Asia. https://osce-academy.net/upload/file/16168.pdf Accessed on 26 July 2023

Eeckhout P and Lopez-Escudero M (2019) The European Union's External Action in Times of Crisis. Bloomsbury Publishing. Oregon

Egan M (2001) Constructing a European market: standards, regulation, and governance. Oxford University Press. Oxford

Estrada Canãmares M (2016) Coherence in EU external relations and the law: the case of the CFSP-development cooperation nexus in the Union's action in Somalia. EBOR https://doi.org/10.2870/348432

European Commission (2003) A maturing partnership—shared interests and challenges in EU-China relations. https://eur-lex.europa.eu/LexUriServ/LexUriServ.do?uri=COM:2003:0533:FIN:EN:PDF Accessed on 26 July 2023

European Commission (2019a) Annex to the Communication from the Commission to the European Parliament, the European Council, the Council, the European Economic and Social Committee and the Committee of the Regions, The European Green Deal. https://eur-lex.europa.eu/legal-content/EN/TXT/?uri=CELEX%3A52019DC0640 Accessed on 26 July 2023

European Commission (2019b) EU-China Summit Joint Statement. https://www.consilium.europa.eu/media/39962/eu-china-summit-joint-statement-20190409-final.pdf Accessed on 26 July 2023

European Commission (2020) 2050 long-term strategy. https://climate.ec.europa.eu/eu-action/climate-strategies-targets/2050-long-term-strategy_en#:~:text=The%20EU%20aims%20to%20be,AgreementEN%E2%80%A2%E2%80%A2%E2%80%A2 Accessed on 26 July 2023

European Commission (2022a) EU-China Relations, https://www.eeas.europa.eu/sites/default/files/documents/EU-China_Factsheet_01Apr2022.pdf Accessed on 26 July 2023

European Commission (2022b) International climate finance. https://climate.ec.europa.eu/eu-action/international-action-climate-change/international-climate-finance_en Accessed on 26 July 2023

European Commission (2022c) International trade in goods by partner. Focus on EU trade in goods—an overview. https://ec.europa.eu/eurostat/statistics-explained/index.php?title=International_trade_in_goods_by_partner#Focus_on_EU_trade_in_goods_-_an_overview Accessed on 26 July 2023

European Commission (2023a) Readout of the Fourth EU-China High-Level Environment and Climate Dialogue. https://climate.ec.europa.eu/news-your-voice/news/readout-fourth-eu-china-high-level-environment-and-climate-dialogue-2023-07-04_en Accessed on 26 July 2023

European Commission (2023b) The European Union and China. https://www.eeas.europa.eu/delegations/china/european-union-and-china_en?s=166 Accessed on 26 July 2023

European Commission (2023c) Sustainable development. https://policy.trade.ec.europa.eu/development-and-sustainability/sustainable-development_en Accessed on 09 October 2023

European Commission (2023d) A Union standing firm and united. https://commission.europa.eu/system/files/2022-10/com_2022_548_3_en.pdf Accessed on 09 October 2023

European Commission (2023e) International Aspects. https://single-market-economy.ec.europa.eu/sectors/toys/international-aspects_en Accessed on 09 October 2023

European Council (2023) Climate change: what the EU is doing. https://www.consilium.europa.eu/en/policies/climate-change Accessed on 26 July 2023

European Union (2010) Institutional balance of the European Union. https://eur-lex.europa.eu/LexUriServ/LexUriServ.do?uri=OJ:C:2010:212E:0082:0093:EN:PDF Accessed on 09 October 2023

European Union (2012) EU-China Partnership on Urbanization. https://energy.ec.europa.eu/system/files/2014-10/12_cities_0.pdf Accessed on 26 July 2023

European Union (2013) EU-China 2020 Strategic Agenda for Cooperation. https://www.eeas.europa.eu/delegations/china/eu-china-2020-strategic-agenda-cooperation_en?s=166 Accessed on 26 July 2023

Gallas D (2023) Why Europe is now in a hurry to re-engage with Brazil and Latin America. https://www.bbc.com/portuguese/articles/c4n49ervdelo Accessed on 26 July 2023

Gibson C (2018) Water pollution in China is the country's worst environmental issue. https://borgenproject.org/water-pollution-in-china Accessed on 26 July 2023

Grieger G (2021) Towards a joint Western alternative to the Belt and Road Initiative?. https://www.europarl.europa.eu/RegData/etudes/BRIE/2021/698824/EPRS_BRI(2021)698824_EN.pdf Accessed on 26 July 2023

Hanemann T and Huotari M (2018) EU-China FDI: Working towards reciprocity in investment relations. Mercator Institute for China Studies. Berlin

Harris P G (2007) Europe and global climate change: Politics, foreign policy and regional cooperation. Edward Elgar Publishing. UK

Houde M F and Yannaca-Small K (2004) Relationships between international investment agreements. EBOR https://doi.org/10.1787/171461325566

Human Rights in China (2016) The China challenge to international human rights: what's at stake?. https://www.ohchr.org/sites/default/files/Documents/HRBodies/UPR/NGOsMidTermReports/HumanRightsChina_2.pdf Accessed on 26 July 2023

Inês C, Guilherme P L, Esther M G, Swantje G, Stephen H, & Lars H (2020) Regulatory challenges and opportunities for collective renewable energy prosumers in the EU. EBOR https://doi.org/10.1016/j.enpol.2019.111212

Larik J, and Wessel R A (2020), *EU External Relations Law: Text, Cases and Materials.* Hart Publishing. Oxford.

Liu B, Wang T, Zhang J, Wang X, Chang Y, Fang D, & Sun X (2021) Sustained sustainable development actions of China from 1986 to 2020. EBOR https://doi.org/10.1038/s41598-021-87376-8

Marcin C G (2023) China-US global rivalry and the EU [What Think Tanks are thinking]. https://www.europarl.europa.eu/RegData/etudes/BRIE/2023/749803/EPRS_BRI(2023)749803_EN.pdf Accessed on 26 July 2023

Massé F, Dickinson H, Margulies J, Joanny L, Lappe-Osthege T, Duffy R (2020) Conservation and crime convergence? Situating the 2018 London illegal wildlife trade conference. Journal of Political Ecology. EBOR https://doi.org/10.2458/v27i1.23543.

McBride J, Berman N, Chatzky A (2023) China's Massive Belt and Road Initiative. https://www.cfr.org/backgrounder/chinas-massive-belt-and-road-initiative Accessed on 26 July 2023

Miles, K. (2013) The origins of international investment law: empire, environment and the safeguarding of capital. Cambridge University Press. UK

Morgera E, Geelhoed M, Ntona M (2020) European Environmental Law. In: Techera E, Lindley J, Scott K, Telesetsky A (eds), Routledge Handbook of International Environmental Law. Routhledge. London, pp 233-248

OECD (2016) China's climate change combat. https://www.oecd.org/environment/china-climate-change-combat.htm Accessed on 26 July 2023

OECD (2023) Mutual recognition agreements (MRAs) https://www.oecd.org/gov/regulatory-policy/irc-mras.htm Accessed on 26 July 2023

Pan C (2012) Problematizing "Constructing Engagement" in EU-China Policy. In: Vogt R (ed.), Europe and China: Strategic Partners or Rivals?. Hong Kong University Press. Hong Kong. pp 37-58

People's Republic of China (2023), Table of total value of major countries (regions) of import and export commodities in December 2022 (dollar value) http://www.customs.gov.cn/customs/302249/zfxxgk/2799825/302274/302275/4794352/index.html Accessed on 26 July 2023

Rayner T, & Jordan A (2016) Climate change policy in the European Union. EBOR https://doi.org/10.1093/acrefore/9780190228620.013.47

Song X (2012) Challenges and opportunities in EU-China relations. In: Vogt R (ed.), Europe and China: Strategic Partners or Rivals?. Hong Kong University Press. Hong Kong. pp 19-36

Teo H C, Lechner A M, Walton G W, Chan F K S, Cheshmehzangi A, Tan-Mullins M, & Campos-Arceiz A (2019) Environmental impacts of infrastructure development under the belt and road initiative. EBOR https://doi.org/10.3390/environments6060072

The World Bank (2022) China's Transition to a Low-Carbon Economy and Climate Resilience Needs Shifts in Resources and Technologies. https://www.worldbank.org/en/news/press-release/2022/10/12/china-s-transition-to-a-low-carbon-economy-and-climate-resilience-needs-shifts-in-resources-and-technologies Accessed on 09 October 2023

United Nations (2023) 17 Goals to Transform Our World https://www.un.org/sustainabledevelopment Accessed on 26 July 2023

United Nations Framework Convention on Climate Change (2016) The Paris Agreement. https://unfccc.int/process-and-meetings/the-paris-agreement Accessed on 26 July 2023

United Nations (1992) Convention on biological diversity. https://www.cbd.int/doc/legal/cbd-en.pdf Accessed on 26 July 2023

Vogt R (2012) Limitations for Europe's Partnership with China. In: Vogt R (ed.), Europe and China: Strategic Partners or Rivals?. Hong Kong University Press. Hong Kong. pp 59-81

Wang C N (2023) China Belt and Road Initiative (BRI) Investment Report 2022. https://greenfdc.org/china-belt-and-road-initiative-bri-investment-report-2022 Accessed on 26 July 2023

Wessel R A (2011) The legal framework for the participation of the European Union in international institutions. In: Oberthür S, Jørgensen K E, Shahin J (eds), The Performance of the EU in International Institutions. Routhledge. London. pp 23-37

Wouters J, Hoffmeister F, De Baere G, Ramopoulos T. (2021) The law of EU external relations: cases, materials, and commentary on the EU as an international legal actor. Oxford University Press. Oxford

Wurzel R & Connelly J (2010) The European Union as a leader in international climate change politics. Routledge. NY

Yang C (2020) Policies, regulatory framework and enforcement for air quality management: The case of China. EBOR https://doi.org/10.1787/7d1d1a82-en

Zhang M (2010) The socialist legal system with Chinese characteristics: China's discourse for the rule of law and a bitter experience. Temp. Int'l & Comp. v. 24.

Zhang R & Hanaoka T (2021) Deployment of electric vehicles in China to meet the carbon neutral target by 2060: Provincial disparities in energy systems. EBOR https://doi.org/10.1016/j.resconrec.2021.105622

**Amon Elpídio da Silva** is a L.L.M. in European and International Law, Universität des Saarlandes, a Ph.D. Student in Law at the Federal University of Minas Gerais, and a Researcher within the Jean Monnet Centre of Excellence at the Federal University of Minas Gerais.

# Chapter 15
# Crafting Global Green Diplomacy: The Strategic Narrative of the European Union

**Jamile Bergamaschine Mata Diz and Júlia Péret Tasende Tarsia**

## Contents



**Abstract** This Chapter examines the pivotal role of strategic narrative in the European Union's external action. By deftly articulating values and interests, the European Union's strategic use of narrative shapes global diplomacy, projecting its vision for cooperation and advancement. Serving as an embodiment of soft diplomacy, this narrative fosters collaboration and positions the European Union as a guiding force. However, divergent narratives among international actors, exemplified by Brazil's perspective on Amazon protection, can impede its effectiveness. Balancing the narrative with a nuanced recognition of diverse viewpoints is crucial for successful diplomacy, as the European Union navigates complexities to uphold its values and interests in agreements such as the European Union-Mercosur Association Agreement.

---

This chapter is published as partial results of a master's Research Dissertation written by Júlia Péret Tasende Tarsia entitled "European Union And Global Environmental Governance: Strategic Narratives And The Defense Of The Amazon Rainforest" into the post-graduate program at the Law Faculty—Federal University of Minas Gerais.

---

J. B. M. Diz (✉)
Law School in UFMG, Federal University of Minas Gerais (UFMG), Belo Horizonte, Minas Gerais, Brazil
e-mail: jbmatadiz@yahoo.com.br

J. P. T. Tarsia
UFMG Jean Monnet Center of Excellence, Federal University of Minas Gerais (UFMG), Belo Horizonte, Minas Gerais, Brazil

Pompeu Fabra University, Barcelona, Spain

J. P. T. Tarsia
e-mail: julia.tasende01@estudiant.upf.edu

 
R. A. Wessel et al. (eds.), *EU External Relations Law and Sustainability*, Global Europe: Legal and Policy Issues of the EU's External Action 4,
https://doi.org/10.1007/978-94-6265-655-0_15

**Keywords** Amazon rainforest · Brazil · Environmental politics · European Union · External action · Green diplomacy · Strategic narratives

## 15.1 Introduction

The scholarly pursuit of studying the EU's external relations has evolved into an independent discipline, with the primary objective of comprehending how the EU operates on the global stage, including the timing and legal foundations underlying its actions.[1] The focus extends beyond investigating the EU's interactions with specific States or regions; it also delves into understanding the EU's contribution to global sustainable development, safeguarding human rights, addressing peace and security concerns, and fostering democracy on a global scale.

Given the *sui generis* nature of the EU as an innovative actor in international politics, various theories and theses have emerged from these studies, shedding light on how the EU shapes, collaborates, and integrates into the global political landscape. In the 1970s, Duchêne[2] characterized the EU's international identity as a "civilian power," emphasizing its ability to exert influence through non-military means, particularly economic avenues. Hanns Maull[3] revisited this concept, asserting that the EU's foreign policy role is constrained by values and principles, allowing it to wield influence in the pursuit of a civilization of international order.

Manners[4] famously introduced the term "normative power Europe" (NPE) to describe the EU's external action, emphasizing its non-military and non-purely economic nature, operating through the power of norms to shape actors' behavior. Diez[5] later revisited this idea, highlighting the significance of discourse on NPE in the EU's external construction. In a more recent study, Wessel[6] explored the normative transformation in EU external relations, observing a departure from its legal framework in favor of "softer" arrangements, such as instruments of soft law, rather than traditional international agreements.

Bergamaschine and Araújo[7] delved into the extraterritorial impact of internal EU legislations designed to combat deforestation, emphasizing the nuanced and indirect manifestation of the EU's extraterritorial effect, distinct from a direct claim. And, as well-known, Bradford,[8] in her work on regulations shaping the international business environment and elevating global standards, coined the term "Brussels Effect".

[1] See Duchêne 1972; Maull 2007; Manners 2002; Mata Diz and Paiva Araújo 2021; Wessel 2021.

[2] Duchêne 1972.

[3] Maull 2007.

[4] Manners 2002.

[5] Diez 2005.

[6] Wessel 2021.

[7] Mata Diz and Paiva Araújo 2021.

[8] Bradford 2020.

Indeed, the EU has been considered an "umbrella term for a set of external policies, instruments and actors across a vast range of substantive domains" as an international actor in the past, which is constructed as a "piecemeal construction of political and legal developments pushed forward by geopolitical and socioeconomic *stimuli*".[9] The EU external relations and action is a complex field that embraces different and numerous actors, institutions, norms, principles, discourses, laws, and actions—internally and externally—to exert influence on the world stage.

This Chapter will examine one of the tools that the EU embraces in its external action: the use of strategic narratives. The use of narratives and communication in political science is not new. The way in which individuals and groups exchange information and meaning, and the impact of these exchanges on society, has been subject to research for centuries. In the field of international law and geopolitics, the study of discourse to construct meaning and shape perceptions is overall important and is often used to analyze power relationships.

To better analyze this concept, this Chapter will focus on the EU's use of Strategic Narratives concerning the protection of the Brazilian Amazon rainforest. This theme has been under spotlight in recent years considering that, after 20 years of negotiation, the EU-Mercosur Association Agreement (EUMAA) failed to reach a ratification. Although there are several reasons for the difficulty to establish a trade agreement between the EU and Mercosur, one that is pointed out by European discourses with some emphasis in the environmental deregulation and deforestation in Brazil, specially concerning the Amazon rainforest.[10]

Given its importance to a global matter—climate change and sustainable development—the Amazon rainforest has constantly been the focus of global attention and, indeed, Brazil has been failing to maintain its environmental commitments under the Paris Agreement. In 2004, Brazil committed to the National Policy on Climate Change to reduce its yearly overall deforestation rate to 3.925 $km^2$ by 2020. Between 2004 and 2014 the deforestation rate in the Brazilian Amazon was reduced by 82%.[11] However, analysis of satellite data provided by the Brazilian National Institute for Space Research (INPE) reveals a 34% increase in deforestation within the Amazon region from August 2019 to July 2020, encompassing an expansive area exceeding 9200 $km^2$, compared to the previous reporting period's 6800 $km^2$. This alarming trend persisted in 2021, witnessing a further rise as a total of 13,038 $km^2$ succumbed to deforestation during the year. Notably, between 2022 and 2023, the deforested area amounted to approximately 9000 $km^2$—equivalent to the dimensions of the Republic of Cyprus—within a single year.[12]

It is important to emphasize that from the perspective of global politics, the Amazon is a symbol of cross-borders activism, and, because of that, it is currently portrayed in an environmentalist perspective internationally that overshadows the local conflicts surrounding it.· In the region, there are indigenous people fighting for

[9] Van Vooren and Wessel 2014, p. 65.

[10] European Parliament 2020.

[11] PRODES 2023.

[12] INPE 2023.

their right to land, criminal networks, national and international NGOs, civilians, and companies with commercial interest in the region.[13] The matter of the Amazon involves a historically complex process with local, national, and international actors, all of them with different and unequal narratives and interests revolving around the Amazon rainforest. This Chapter endeavors to elucidate the role of the EU external action as one of these actors.

The EU has consistently positioned itself as a leader in the pursuit of global sustainable development and has advocated for international collaboration to safeguard global and tropical forests. However, this narrative is juxtaposed by the perspective espoused by Brazil, which critiques the presence of "international greed" surrounding the Amazon.[14]

The following section will commence by explicating the concept of strategic narratives in the geopolitical domain, followed by a brief exposition of the EU's narrative regarding global forest protection. Subsequently, attention will be directed towards the Brazilian counter-narrative, accompanied by an exploration of the current intricacies entwined within the process of ratifying the EUMAA.

## 15.2 Strategic Narratives in the International Sphere

According to Castells,[15] power relies on the control of communication and, therefore, the most fundamental form of power lies in the ability to shape the human mind. This can be done through coercion, but not only, as it is not sufficient to stabilize domination. The ability to build consent is also essential to enforce the rules that govern the institutions and organizations of society, which is a result of processes of compromise between conflicting social and political actors who mobilize their interests and values through communication. The point of Castell's analysis is that communication changes how power works.

One of the ways that communication can change power relations in the field of International Relations and International Public Law, is using narratives. According to Miskimmon, O'Loughlin and Roselle,[16] the way narratives are formed and projected in a communication environment explains major dynamics in international affairs. That is because narratives are central to human relations and, consequently, can be used to shape the behavior of others, in a strategic manner.

Firstly, it is important to highlight some aspects of narrative study. The use of narrative is not similar to the construction of political propaganda or national branding. It

[13] Zhouri 2004.

[14] Brasil 2008a, b; Agência Senado 2005.

[15] Castells 2009.

[16] Miskimmon et al. 2013.

is the recognition of a fundamental shift, and especially so, in relatively open societies, of how power, communication, influence, and decision-making have spread, and how complex they have become.[17]

Shenhav[18] sets out two strategies for defining the concept of narrative. First, narratives need to have a time sequence. In the words of Prince,[19] narrative is "*the representation of at least two real or fictive events or situations in time sequence, neither of which presupposes or entails the other*". Second, Shenhav points out other characteristics added to this basic idea of chronological narrative, such as causality, resolution, and the idea of "non randomness", i.e., that events are connected in a non-random way.

According to Miskimmon et al.,[20] a narrative entails an initial situation or order, a problem that disrupts that order, and a resolution that tries to re-establish that order. Moreover, a narrative structure embraces the existence of actors; events, plot, and time; and a setting or space. The actors are central to the meaning of a narrative, as they are the ones to frame their own character and the character of others, by selecting and highlighting some facets of their history of actions to promote a particular interpretation of the issue. Nonetheless, the theme of the given narrative, the issue, can be framed by various actors in different ways. Narratives also carry an implicit sense of time and space. They have an interpretation of the beginning, middle and end of a given issue. This definition is more complete, as it allows a way to identify actors and actions, while still recognizing the importance of temporality. In international relations, actors usually carry long-term narratives about themselves and about the international system.[21] For that reason, this research adopts this understanding of narrative.

To better understand the use of strategic narratives in international politics, it is also important to differentiate narrative and discourse. Both are means of communication, and both can shape our understanding of the world and influence our participation in social and political processes. Discourse, unlike narrative, does not feature a transformation that takes actors from a *status quo* to another, as narratives do. Miskimmon et al. defines discourse as the "*raw material of communication*", that is, bodies of knowledge about science, law, history, and theology that actors plot into given narratives. Hence, discourses have an embedded sense of historicity in their construction.[22] A discourse analysis can explain why and how certain ideas come to have significant power in shaping political practices. Therefore, discourses have a structuring effect upon the use of narratives. As Cristian Niţoiu identified,[23] "*narratives are the result of the overlapping of various discourses*" and can provide a clearer picture of the processes behind the influence of discourses on political reality.

[17] Wallström 2008.

[18] Shenhav 2006.

[19] Prince 1982, p. 4.

[20] Miskimmon et al. 2014.

[21] Miskimmon et al. 2013.

[22] Krzyżanowski 2009.

[23] Niţoiu 2014, p. 242.

Accordingly, only the discourses that are institutionalized through social, historical, and political practice come to be part of a narrative.

In the international political field, narratives are dynamic, and are social products based on State interactions both within their own society and with external parties. In their seminal work "Forging the World: Strategic Narratives and International Relations", Miskimmon, O'Loughlin, and Roselle[24] present a detailed examination of the application of narrative research in the field of international relations. Their concept of "*strategic narratives*" centers on the ways in which these narratives are formed, projected, and received. The authors delve into the process of narrative formation, specifically examining the role of political actors in constructing such narratives. This leads to an analysis of the projection and contestation of narratives within the global public sphere and, consequently, how these narratives are received and understood by individual actors and states in international relations.

Therefore, strategic narratives are representations of a sequence of events and identities, a communicative tool through which political actors attempt to give meaning to a certain issue to achieve political objectives. The creation of strategic narratives by political actors can be used to construct a shared meaning of the past, present and future of international events and, with this, political actors can then "*extend their influence, manage expectations and change the discursive environment in which they operate*".[25]

In this context, the significance of narratives becomes pronounced for the EU, particularly within the context of a shifting international order characterized by mounting pressures related to relative decline and power transitions. This is evident through the articulation of forward-looking strategic narratives that elucidate the EU's self-perception in the context of this evolving global order. A noteworthy illustration of this phenomenon is the EU's commitment to adopting a distinctive stance in addressing the repercussions of climate change on international security. This commitment is underscored by the EU's acknowledgment of its pivotal role in development, global climate policy, and its extensive array of tools and instruments at its disposal. For example, the EU has committed itself to assume a "*unique position*" to respond to the impacts of climate change on international security, given its "*leading role in development, global climate policy and the wide array of tools and instruments as its disposal*".[26]

As mentioned, a narrative structure embraces the existence of actors; events, plot, and time, and a setting or space. In this study, the EU itself is the main *actor*, which means it is divided into different EU actors, such as the European Parliament, the Council, the European Commission, the European External Action Service, EU Delegations, and individual EU Member States with their own governments.

EU foreign policy towards global environmental protection and sustainable development is the *action* taken by these actors, which take the form of a set of norms and agreements with partner States—in this study specifically, to Brazil. These actions

[24] Miskimmon et al. 2017.

[25] Miskimmon et al. 2013, p. 3.

[26] European Union 2008, p. 2.

are taken in several international *settings*, such as the Paris Agreement, Agenda 21, Agenda 2030 and, more regionally, the European Green Deal, which are complemented and permeated by principles of Public International Law, such as the principles of sustainable development, ecological proportionality, precaution, and common but differentiated responsibilities. In addition to this broader setting, the foreign policies of the EU concerning its relations with Brazil (the strategic partnership between the two actors), and its attempts to diminish deforestation rates around the globe, are of paramount importance. All these settings shape a space where the world is understood as one that "*grows through interdependence and globalization, with prospects for cooperation in pursuit of common goals*".[27]

The *conflict* or *action* presents itself when the EU, which portrays itself as the leader in the search for global sustainable development and environmental protection, expresses its concern about the way that Brazil has been managing the Amazon rainforest, and the consequential increase of deforestation in the region. The conflict in this matter arises from the fact that Brazilian authorities and society are distrustful of European interests in the region, and therefore do not receive the narrative from the EU well, as it will be shown.

This is a future-oriented narrative, with a clear cause-effect element. The *solution* of the conflict, given by the EU, would guarantee a better world for future generations, decreasing global climate change and protecting the diverse fauna and flora in the region. Miskimmon et al.[28] argues that strategic narratives come in three main forms that are inextricably linked:

(1) Strategic narratives about the international system, articulating how the principal actor conceives their understanding of international order. An example would include narratives from different countries about the Cold War.
(2) Narratives employed by political actors seeking to influence the development of policies. Here, the narrative sets out why a policy is needed and desirable, and how it would be successfully implemented.
(3) Strategic narratives of identity, which is how a political actor seeks to project their identity in international affairs. These define the story of a political actor, which leads to better understanding what values and goals it has and, consequently, the possible terms of policy.

As defended by Miskimmon et al., an actor able to align the three main forms of narrative has a greater chance of influence in the Global Governance scenario. In the next section, we will analyze how this is done by the current EU narrative towards global environmental protection and forestry protection, represented by the European Green Deal and the European Commission Communication on Stepping up EU Action Towards the Protection of the World Forests.[29]

---

[27] Roselle et al. 2014, p. 75.

[28] Miskimmon et al. 2013.

[29] European Commission 2019a.

## 15.3 The Narrative of Global Leader for Environmental Protection

There is some debate about how the EU persists in the global political scenario. In that context, Wessel[30] defines the EU as an entity which interacts with third countries and international organizations in distinguished ways from its Member States, with a "*stand alone identity composed of values, interests, and policies which it seeks to define and promote internationally as its own*". This promotion is done through a series of external policy instruments; political, economic, commercial, and financial: the EU is an "umbrella term" for external policies, instruments, and actors, across a range of substantive domains.

Manners[31] advocates that the EU's external behavior is founded on a set of core values that are inscribed into its treaties, acting as a "normative power". The NPE thesis advocates that the EU's action on the global governance scenario is based on a setting of standards that are considered "normal" in world politics, and that this idea of normality is mainly defined by EU's "core" norms globally. Manners identifies five core norms that comprise the EU's communitarian and political actions globally: peace, liberty, democracy, rule of law and respect for human rights. Added to these five core norms, Manners suggest four "minor" norms within the constitution and practices of the EU: social solidarity, anti-discrimination, good governance, and sustainable development.

In fact, these values are explicitly shown in the Lisbon Treaty, which, in its second article, presents a general provision about EU foreign policy:

> In its relations with the wider world, the Union shall uphold and promote its values and interests and contribute to the protection of its citizens. It shall contribute to **peace, security, the sustainable development of the Earth, solidarity and mutual respect among peoples, free and fair trade, eradication of poverty and the protection of human rights, in particular the rights of the child, as well as to the strict observance and the development of international law**, including respect for the principles of the United Nations Charter.[32]

Through this EU narrative about its values, it is possible to link the international position of the EU to its use of strategic narrative. In fact, Manners do suggest that EU's normative power stems from six factors, shaping norm diffusion in IR: (i) *contagion*, through an unintentional diffusion of ideas from the EU to other political actors; (ii) *informational diffusion*, as the result of a range of strategic communications; (iii) *procedural diffusion*, which involve the institutionalization of the relationship between the EU and a third party; (iv) *transference*, when the diffusion takes place with the exchange of goods, trade, aid or technical assistance with third parties; (v) *over diffusion*, which occurs as a result of the physical presence of the EU in third states and international organizations; and, finally, (vi) *cultural filter*, based on the interplay between the construction of knowledge and the creation of social and political identity.

---

[30] Wessel 2020, p. 60.

[31] Manners 2002.

[32] European Union 2007.

Through Manners' thesis, it is possible to denote that the actions of the EU in the global scenario are guided by a set of norms and principles, which reflect the EU's normative identity. And, especially through *informational diffusion* and *cultural filters*, it is possible for an international political actor to establish the terrain on which policy discussions take place. Subsequently, as the arguments of Miskimmon et al. and Roselle et al. have shown, this can be done by using strategic narratives.

Niţoiu[33] wrote about how the EU constructed different narratives concerning its role in the global scenario and mapped some influences on policy outcomes derived by EU narratives. One of the narratives is about the EU being a promoter of well-being for people around the world. This narrative is based on the self-established commitment of the EU to tackle the increase of climate change through its external relations. Climate change has been framed as a threat to the well-being of people around the world due to its environmental, economic, and social impact,[34] and the EU's official narrative puts the blame on other world powers such as the United States (US) or China, for their unwillingness to make environmental concessions.[35] Niţoiu highlights the way the European narrative has been crafted in opposition to the actions of the US, emphasizing, for example, the European Union's willingness to participate in multilateralism and to engage with matters considered "low politics", secondary to "hard" issues in international politics, such as economics and security. In the context of this narrative, the lack of international progress in environmental policy, exemplified by the unsuccessful outcome of the Copenhagen Conference of 2009 and the World Trade Organization's (WTO) Doha Development of 2001, is validated through the attribution of blame to specific parties, notably the US and China, who are deemed resistant to complying with the rigorous environmental standards proposed by the EU.

It is evident the EU's prominent role in global environmental politics, particularly following the surge in global interest and concern for environmental protection since the 1970s. Starting from 1987, the pursuit of policies aimed at ensuring a robust level of environmental protection emerged as one of the rapidly expanding domains within (then) European Community legislation. A pivotal juncture in this trajectory is often traced back to the Stockholm Conference in 1972, which is recognized as a seminal moment. This conference not only marked the initiation of the European Community's environmental initiatives but also catalyzed the establishment of dedicated environment ministries in several member states.[36]

The EU External Policy towards environmental protection is set out on several principles that guides its law-making and international relations. These principles are mainly presented in Article 191(2) of TFEU: a high level of environmental protection, principles of prevention, precaution, rectification at the source and the polluter pays. Those are applied in the "regional and worldwide environmental problems" according to Article 191(1), which indicates that the EU also takes measures targeting the

[33] Niţoiu 2014.

[34] European Commission 2009.

[35] European Parliament 2011; Simon 2024.

[36] Vogler 2005.

environment beyond its borders, just as its Member States - limited to the international law imposition on extraterritorial application of domestic environmental measures.

Currently, the EU is signatory of several environmental agreements, ranging from regional conventions on seas and watercourses to global agreements, such as the conventions on Stratospheric Ozone, on Biodiversity (CBD) and Desertification, and, of course, the UN Framework Convention on Climate Change (UNFCCC). Furthermore, in alignment with the UN 2030 Agenda,[37] the Paris Agreement,[38] and the post-2020 global framework of the UN Convention on Biological Diversity,[39] both the EU and its Member States have committed to provide support to developing nations. This support is framed by the principles of equity and common but differentiated responsibilities and respective capabilities. These principles, delineated in the Rio Declaration on Environment and Development (1992), emphasize the imperative for states to collaborate in a global partnership aimed at conserving, protecting, and restoring the Earth's ecosystem. This collaborative approach underscores the recognition that developed countries bear a responsibility in the international pursuit of sustainable development, acknowledging the impact their societies exert on the global environment, along with the financial and technological resources at their disposal.

In recent years, notable advances have been made towards environmental policy integration, particularly in the energy sector, as evidenced by the simultaneous progress of the EU's climate and energy package and the roadmap for transitioning towards a competitive low-carb economy by 2050.[40] Finally, in December 2019, the Commission launched the European Green Deal, which should help to focus EU policies on making Europe the first climate-neutral continent in the world.

In the European Green Deal, it is possible to denote the three forms of strategic narrative perpetrated by the EU on global environmental protection, as coined by Miskimmon et al.[41] Firstly, it is possible to perceive the construction of a narrative about the international system, and how the EU (the main actor) conceives an understanding of international order, praising ways for Global Environmental Governance, global cooperation, and multilateral framework:

> The environmental ambition of the Green Deal will not be achieved by Europe acting alone. The drivers of climate change and biodiversity loss are global and are not limited by national borders. The EU can use its influence, expertise and financial resources to mobilize its neighbors and partners to join it on a sustainable path. The EU will continue to lead international efforts and wants to build alliances with the likeminded. It also recognizes the need to maintain its security of supply and competitiveness even when others are unwilling to act.[42]

In this document presented by the EU, a paradigm emerges wherein the origins of climate change and biodiversity loss transcend the confines of national boundaries.

---

[37] United Nations 2015.

[38] UNFCC 2015.

[39] UNEP 2022.

[40] European Commission 2019b.

[41] Miskimmon et al. 2013.

[42] European Commission 2019b, p. 1.

This conception is intertwined with the imperative of forging collaborative partnerships, thus delineating an international framework wherein the diffusion of innovative politics transpires through interconnected networks, comprising decision-makers operating on different levels.

More than forging partnerships, the EU has put in action unilateral measures to influence the global market and other States to reduce deforestation through Regulations—especially Regulation (EU) No. 995/2010, and Regulation (EU) No. 1169/2011, as they seek to encourage the consumption of products from deforestation-free supply chains, and facilitate the process for all the chain of production. The use of internal policy to curb failed international environmental attempts to curb climate change is exemplified by the Regulation 2023/1115 of the European Parliament and of the Council that entered into force in June of 2023. Under the Regulation, any operator or trader who places commodities (soy, beef, palm oil, wood, cocoa, coffee, rubber and other derived products) on the EU market, must be able to prove that the products do not originate from deforestation or forest degradation.

The narrative put forth by the EU, in this sense, fundamentally perceives a transformative transition, denoted as a paradigm shift from "government to governance".[43] In this redefined landscape, emphasis is laid upon the intricate involvement of multipartite policy networks and advocacy coalitions, characterized by international actors engaging in concerted efforts, to address the highly complex problem that is climate change and biodiversity loss. Within the new governance arrangement, the traditional form of sovereign authority is revisited, operating within a limited subject-matter sphere and at spatial scale, with its boundaries defined by the nature and scale of the problem to be addressed.[44] Through the propagation of this narrative, the EU aspires to establish a standardized arena within global politics, affording it an expanded sphere of influence and maneuverability with other international actors, acting through its "normative power".

The same narrative is seen in the Communication Stepping up EU Action to Protect and Restore the World's Forests:

> There is no one-size-fits-all solution. Fighting deforestation and attaining sustainable forest management are complex challenges. Solutions need to be specific to each country and region, with an overall two-fold objective of protecting existing forests, especially primary forests, and significantly increasing sustainable, biodiverse forest coverage worldwide. It is clear that the EU by itself cannot reverse the trend of deforestation. It needs to be part of a global alliance.[45]

Applying its values as standard for multilateral and bilateral agreements, the EU also intends to influence the development of policies in the international sphere and, at the same time, projects its identity as a "global leader" for environmental protection:

> The EU will continue to promote and implement ambitious environment, climate and energy policies across the world. It will develop a stronger 'green deal diplomacy' focused on convincing and supporting others to take on their share of promoting more sustainable

[43] Rosenau and Czempiel 1992; Rhodes 1997.

[44] Karkkainen 2004.

[45] European Commission 2019c, p. 4.

> development. By setting a credible example, and following-up with diplomacy, trade policy, development support and other external policies, the EU can be an effective advocate.[46]

Here, the narrative paints the EU as the leader and the example that will ensure that partners increase their effort and implement nationally determined contributions and ambitious long-term strategies. It is a strategic narrative of identity that leads to how the EU wants to be perceived as a political actor with its own values and goals that should be shared by its partners for the protection of common global good. For that reason, it is, at the same time, a narrative that seeks to influence the development of policies, as it sets out why this cooperation and leadership are needed and desirable, and how they would be successfully implemented (with diplomacy, trade policy, development support and other means).

Therefore, the EU narrates that it holds a leadership position in the global fight against climate change. This objective has been pursued through two primary means: firstly, by striving to enhance the multilateral framework for mitigating climate change, and secondly, by setting a positive example with its own internal climate policies and laws.[47]

Another way that the EU prioritizes climate change matters in its external relations, is using bilateral and inter-regional cooperation. Kulovesi[48] has observed how the insertion of sustainable development and environmental cooperation clauses has become a normative practice. Kulovesi argues that the EU is progressively leveraging from its political and economic influence to promote its climate change objectives via bilateral and inter-regional external relations. It is clear, then, that the EU uses strategic narrative to position itself as a rule maker and a central actor to preserve environmental principles in the face of the crisis of climate change.

A way that the EU prioritizes climate change matters in its external relations and, therefore, "leads by example", is through the use of bilateral and inter-regional cooperation. As Farrell notes, since 1995 a "human rights clause is standard for all EU treaties with third countries and regions". In the same way, inclusion of environmental clauses and policies is nowadays a part of EUs bilateral and inter-regional cooperation.[49] In fact, this is established by Articles 11 and 21 of the TFEU, which obliged the EU political institutions to integrate environmental protection requirements into all EU policies and activities. Therefore, EU Free Trade Agreements (FTAs) include Trade and Sustainable Development (TSD) chapters, with provisions on environmental protection, climate change, biodiversity and forests, including the obligation to follow multilateral environmental agreements.

Strategizing in international relations, represents a complex endeavor that considers the intricacies of the world, and situates an agent within the vast dynamic global landscape by crafting a comprehensive vision and contemplating long-term

---

46 European Commission 2019c, p. 20.

47 Durán 2020.

48 Kulosevi 2013.

49 Farrel 2005, p. 275; Afionis and Stinger 2014, p. 50.

outcomes.[50] The centrality of strategy to foreign policy is evidenced through the analysis of various contexts, such as those within the EU and environmental contexts. For example, EU foreign policy has been depicted as the area of European politics that "*is directed at the external environment with the objective of influencing that environment and the behavior of other actors within it, in order to pursue interests, values and goals*".[51]

For example, in the European Commission document "Stepping up EU Action to Protect and Restore the World's Forests" the EU calls for a global alliance to reverse the trend of deforestation, developing partnerships with other countries to reduce pressures on forests. In fact, this has been the focus to ensure the final ratification of the EUMAA. For the first time during the negotiations, environmental concerns became the central political issue, with claims that the expansion of cultivated area (by forest clearing) would give South American agriculture an advantage over European farmers and harm the environment.[52]

In a later publication explaining the content of the EUMAA, the European Commission points out that the agreement is also about setting international and disseminating EU norms.[53] In 2020, the EU launched a resolution on the implementation of the common commercial policy, in which it was emphasized that the EUMAA cannot be ratified as it stands, given the continued Brazilian policy at the time of deflecting responsibility for the deforestation of the Amazon.[54] In addition, parliaments of France, Austria, Belgium, Ireland, and the Netherlands voted against the ratification, given the weak enforceability of the TSD chapter, built on dialogue alone.[55]

This chapter does not intend to do a viability study of the impact of the EU environmental narrative and the EUMAA. Determining the effects and underlying reasons behind a narrative can present a challenge, since the true intention behind a political narrative is not always evident, and researchers must take into consideration various factors to gain a deeper understanding of the strategy behind a given narrative. Additionally, political actors themselves may not have a clear strategy or defined goals. The difficulty in analyzing narratives in the realm of international relations is also rooted in the complexity of establishing clear cause-effect relationships within a global and historical context, which may necessitate oversimplifying connections to demonstrate the influence of a narrative.[56]

Although narratives cannot capture the political reality, this does not mean that all narratives are equally true or false, they are constructed based on a certain political

---

[50] Schunz 2017.

[51] Keukeleire and Delreux 2022, p. 1.

[52] Nolte and Ribeiro 2021.

[53] European Commission 2019d.

[54] European Parliament 2019.

[55] Recently the European Commission introduced the possibility of applying sanctions for enforcement of TSD. The reader can learn about the problems of enforcement of TSD chapters in this book in Said Esra Akdogan chapter.

[56] Miskimmon et al. 2017.

perspective.[57] This is of paramount importance when analyzing political discourses on a democratic scope and, specifically, when analyzing political discourses and actions taken through Global Governance, since it embraces different positions from different actors. In this way, the theme of the narrative itself can be framed by various actors in certain ways and, consequently, contribute to the construction of a shared meaning about certain events.

For this reason, it is interesting to denote that, while the EU may have constructed a narrative aimed at portraying itself as a leader of environmental protection and sustainable practices, the reception of this narrative by other countries, such as Brazil in the context of the Amazon rainforest, has not always been positive.

## 15.4 The EU Narrative Towards the Amazon Rainforest and the EU-Mercosur FTA

The Amazon rainforest, more than a tropical forest, is a complex socio-political system, divided by nine nations and hundreds of ethnicities, with national and international interests surrounding it. Its story and role in the globalized world are told in a fragmented manner by different voices, such as indigenous communities, civil society, academics, Federal Governments, scientists, NGOs, international organizations, States, etc. For that reason, everything that is studied and told about the *Amazônia* is surrounded by relativism and often is presented without any knowledge of its history and current reality. In Marcio Souza words: "*of all the ways that the Amazon suffers, the most slippery is exactly this one of the hasty opinions. At the same time, it is the simplest to overcome*".[58]

Until a few years ago, the general belief was that there was no possible human development into the forest until the arrival of European colonization, besides few indigenous tribes that lived on a small subsistence economy. However, in the last 20 years, with the advancement of Amazon archaeology, it was proved that when the Europeans arrived in the 16th century, the Amazônia was already habited by complex hierarchical societies of high demographic density.[59] For millennials, the Amazonian people developed a culture nominated as "tropical jungle culture" (*cultura da selva tropical* in Portuguese), with its own power, language, religion, social and economic systems. Therefore, it is important to denote for this research that the Amazônia was not—and still isn't—a demographically empty space, and it is intrinsically intertwined with hundreds of different societies and cultures.[60]

It is important to highlight the culture that is part of the Amazon, since it is a space with a lot of global visions attributed by European colonizers. Since the colonization, the forest is seen by outsiders as a productive unit and an economical frontier, creating

[57] Shenhav 2006.

[58] Souza 2019, p. 23.

[59] Neves et al. 2021.

[60] Souza 2019.

an arena of dispute for control between national power and international interests. Into this matter, the Amazon is narratively presented today as the hope of humanity to "solve" the climate and environmental crises. Indeed, forest protection is important not only for global climate concerns, but also for the conservation of the Amazonian culture. For that reason, the Amazon became a symbol for biodiversity governance and for the way nature is represented in the political and economic spheres. Currently, main environmental discussions go through the management of the South American rainforest, giving the Amazonia a strategic importance in political narratives worldwide.[61] This narrative, however, as it will be shown, is permeated with controversy given the historical international greed and interest towards the forest.[62]

In recent years, the EU has expressed concerns about the protection of the Amazon rainforest due to its critical role in global biodiversity, climate regulation, and carbon sequestration, as the issue even became a matter of debate at the European Parliament.[63] The Amazon Forest, the biggest tropical forest in the world, has already lost up to 20% of its original area since the 1970s, and deforestation has been rising in recent years.[64]

To ensure that the EU environmental objectives would be respected, the environmental concern narrative has been linked to the trade negotiations of the EUMAA. Certainly, interregional negotiations are often overshadowed by internal problems and challenges in both regions, as it is complex to reach a unified position on either side, especially regarding trade liberalization. However, when considering the EUMAA within the context of environmental protection, it becomes evident that while the aspiration to safeguard the environment is commendable, it is accompanied by underlying and pronounced economic motivations—the articulation of this strategic narrative by the EU is intertwined with substantial economic interests and the influence exerted by the European agribusiness lobby, particularly prominent within the beef and ethanol sectors.[65]

This convergence of ecological concerns and economic imperatives has led to significant criticism from Brazil, manifesting its opposition aimed at the EU approach to the EUMAA negotiations and provisions. In the beginning of 2023, the EU presented an additional instrument to address the environmental concerns in the Agreement.[66] The document was not well received by Mercosur, as it included sanctions for alleged non-compliance, such as the limitation of Brazilian products or the suspension of quotas. President Lula (2022–2026) called the document as "inadmissible", "offensive" and a "threat", while discoursing in the Summit for a New Global Financial Compact.[67]

---

[61] Bezerra 2015.

[62] Toledo 2012; Souza 2019.

[63] European Parliament 2020.

[64] INPE 2023.

[65] Karatepe et al. 2020.

[66] European Union 2023.

[67] CNN Brasil 2023.

This opposition to external interference in the Amazon can also be pointed as a strategic narrative, which highlights the suspicions of the Northern interest in the forest and calls for threats to Brazilian sovereignty to the forest. This narrative is based on an "anti-internationalization" discourse, against the international pressure towards the Brazilian government about the sustainability of the Amazon.

The use of this narrative can be traced back to the military dictatorship period in Brazil (1964–1988). It was during this period that the Amazon Cooperation Treaty (Tratado de Cooperação Amazônica—TCA) (1978) was signed, with the objective of environmental and geographical integration for "brotherhood" between the "neighboring Nations": Bolivia, Brazil, Colombia, Ecuador, Guiana, Peru, Suriname and Venezuela. The regional integration defended the principles of territorial sovereignty and, at the same time, aimed for the construction of a regional identity against the extra-regional international threat. That is, the territorial auto determination of Latin American countries would strengthen the region against the "imperialist greed".[68]

At the end of the 1980s, as the "internationalization" of environmental matters arose more intensely—given the Stockholm Declaration in 1972 and international environmental discussions, international pressure on the Brazilian government to change its destructive management of the Amazon started to become more visible, leading to crescent suspicious about international interest in the region.[69] The sustainable maintenance of forests became the central subject in the environmental global articulation, drawing attention from environmentalist groups from the United States and Europe. There was pressure not only from governments, but from transnational environmental groups that actively denounced unsustainable projects financed by multilateral banks, such as the World Bank and the Inter-American Development Bank.[70]

However, as for most of its colonial and post-colonial history, foreigners have regarded the Brazilian Amazon as a natural resource to serve a range of economic, political, and geopolitical interests, this sudden interest in "saving" the Amazon was not well received. The recurrent interference of foreign States, as well as an unequal exchange between them and Brazil, have been a core element of the history of Amazon since the colony period.

Through the economic and political domination by the "centers of power" (i.e., Europe),[71] there was an imposition of a paradigm of territory and natural resource use in which progress was seen as an endless growth and prosperity based on the exploitation of natural resources. In the Amazon, this paradigm took form in several "cycles" of natural resource exploration and exploitation, which gave rhythm to the development of the region.[72] Moreover, the Amazon Forest has been, since the 19th century, the main target for biopiracy in Brazil,[73] through the illegal exploitation of natural

[68] Júnior 2012.

[69] Toledo 2012.

[70] Zhouri 2006.

[71] Becker 1988; Bunker 2003.

[72] De Sartre and Taravella 2009.

[73] Toledo 2012.

resources, and the use of indigenous tribes' knowledge, without permission from the State or without returning any profit back to them. One famous example was when Henry Wickham, an English explorer, smuggled "seringueira" tree seeds—popularly known as rubber tree—in 1876. The British empire initiated a highly productive rubber plantation in Malaysia, causing a collapse of the Amazonian economy at the time, that relied on the "cycle of rubber", initiated by the industrialization of the United States and Europe.[74]

A more recent example, in the beginning of the century, is related to the "cupuaçu" fruit, when it was found that the fruit had been registered as a trademark in Europe, United States and Japan, by the Japanese company Japan's Asahi Foods and the US company Cupuaçu international, without paying the Brazilian State their right for the patent.[75]

On one hand, this induces an understandable concern by the Brazilian government to the international interests on the Amazon rainforest. But, on the other hand, this could be resolved with a national political strategy on international relations linked to the exploitation of the forest resources, although Brazil chooses not to invest sustainably in the region. In reality, this led to the adoption of a historical "anti-internationalization" discourse in Brazil, well exemplified over the years. In a discourse made by President José Sarney (1985–1990) in a reunion of the presidents of countries in the Amazon region in Manaus May 6th 1989, it is possible to denote the narrative against international interest over the forest:

> For our countries, the Amazon was a fundamental part of our national identity. It cost blood, struggle, tenacity, and heroism of our ancestors. It is a unique characteristic of our geographies. The Amazon has been the object of universal fascination. Its mysteries, vastness, and mythical dimensions have captivated imaginations around the world. And, we cannot deny, also, greed. Since the early days of the political configuration of South America, the Amazon has been a target of ambition. The thesis of its internationalization is not new and will not disappear. It emerges and resurfaces, taking on stronger or weaker contours depending on our ability to react. **If we falter, if we lack the determination to preserve the Amazon and maintain its sovereignty, it will be destroyed like all the lands that fell under the action and force of empires.**[76]

This anti-internationalist political narrative is seen by the different National Defense Policy of Brazil (Política Nacional de Defesa—PND) throughout the years, pointing to the Amazon as a concern for public security. Under the government of Luiz Inácio Lula da Silva (2003–2011), in the Decree Nº 5.484 from 2005, the PND stated that "*the Brazilian Amazon, with its great potential for mineral wealth and biodiversity, is the focus of international attention*". In the part of "Strategic Orientations", it presents the Amazon as a priority area for National Defense and puts as a necessity to present strategic actions to strengthen the military presence in the region, to defend its natural resources and the environment.[77]

---

[74] Merson 2000.

[75] Schmidlechner 2003.

[76] Brasil 2008b, p. 82, translated.

[77] Brasil 2005.

In April of 2005, the Federal Senate convened a public audience titled "The Internationalization of the Amazon: Real Risk, or Unfounded Fear",[78] during which examples of the international greed towards the forest were presented. One of the examples given at the event involved declarations made by the former European Union Commissioner, Pascal Lamy, advocating that the Amazon should not be regarded as a "nation asset", but rather as a matter of "collective governance".

During the XX National Forum in 2008, Lula gave a similar message to the international community: "*the world needs to understand that the Brazilian Amazon has an owner. The owners of the Amazon are the Brazilian people: they are the indigenous, the rubber tappers, the fishermen. But so are we*". Furthermore, he expressed amusement about the international cooperation regarding the Forest: "*it is very amusing that the countries responsible for 70% of the planet's pollution now keep an eye on the Amazon in South America, as if the responsibility were solely ours to do what they did not do throughout the entire last century*".[79]

In July of 2020, Bolsonaro government presented its PND, with the same discourse surrounding the Amazon. It is presented that "*the Amazon, like the South Atlantic, is also an area of geostrategic interest for Brazil. The protection of biodiversity, mineral and water resources, in addition to energy potential, in Brazilian territory is a priority for the country*", and it continued saying that:

> (...) [the] exploration and socio-economic development of the Amazon, in a sustainable way, will continue to be vital for national integration, **demanding an increase in the capacities to provide security and sovereignty, intensifying the military presence and the effective action of the State, preventing exogenous entities from influencing the local communities**. For the expansion of this security, it is essential to strengthen cooperation and integration with other Amazonian countries.[80]

As De Sartre and Taravella[81] point out, this denunciation of the foreign interest is also a "strategic card" in the power game that surrounds the narrative arena of the Amazon Forest. This strategy is also seen currently during the President's Lula government, but in different terms, more open to regional cooperation for tackling deforestation and illegal logging. In January 2023 at the 7th summit of the CELAC, President Lula said that international cooperation from outside the region is welcome, but it should be the regional countries that should lead, in a sovereign manner, the initiatives to care for the Amazon, valuing the Amazon Cooperation Treaty Organization.[82]

This proposal was reinforced by the launch of the new Amazon Security and Sovereignty Plan (Decree Nº 11.614/2023),[83] that creates measures to combat land grabbing, illegal mining, and logging, as well as deforestation and enhanced environmental protection. It is possible to affirm, then, that Lula welcomes international

---

78 Agência Senado 2005.

79 Brasil 2008a, translated.

80 Brasil 2020, p. 33, translated.

81 De Sartre and Taravella 2009.

82 Verdélio 2023.

83 Brasil 2023b.

help in the battle to reduce deforestation but insists in Brazil's and other Amazon countries' sovereignty over the forest.

In the Declaration of Belém (2023), a document created by the Amazon Cooperation Treaty Organization and signed by Brazil, it reads that the Organization condemns the "*proliferation of unilateral trade measures based on environmental requirements and norms which constitute trade barriers (...)*" and "*threats, aggression and unilateral coercive measures that are opposed to the international law*".[84]

Hence, there emerges a discernible inclination within the Brazilian government to foster heightened international cooperation concerning the preservation of the Amazon Forest. This inclination, however, appears to be more inclined towards regional cooperation, while concurrently diverging from the environmental protection methodologies advocated by the EU narrative.

To facilitate a more seamless negotiation process surrounding the ratification of the EUMAA, it is imperative for the EU to conscientiously account for Brazil's historical narrative of resistance against internationalization and its associated implications. Declarations of imposing sanctions and assertions of the Amazon's status as a "shared global asset" directly challenge the sovereignty-centric narrative upheld by countries with jurisdiction over the Amazon.

Some authors defend that blaming the EUMAA for contributing to Amazon deforestation is an overestimation of its potential impact.[85] As defended by Malamud (2022), "not engaging with Mercosur will not solve deforestation and will give more space to other global actors to set their agenda with less attention to sustainability", such as China, which already occupies the position of biggest trade partner to Latin America. Not signing the EU-Mercosur Agreement will not reduce the already existent rate of deforestation in the forest and will deprive the EU of leverage against Brazil to protect the forest.[86]

Anticipating the ratification of the EUMAA proves challenging, contingent upon various factors and a convergence of opportune political, economic, and geopolitical conditions between the two blocs. A crucial consideration involves formulating a shared agenda that accommodates the divergent characteristics of the blocs and devising adaptable methodologies that incorporate the parties' diversity. Consequently, both parties should aspire to establish an agenda that comprehensively encompasses all activities emanating from the Agreement.[87]

The significance of Brazil in the Agreement between the EU-Mercosur is multifaceted and pivotal to the overall success and impact of this diplomatic and economic alliance. Brazil holds a prominent position in South America, contributing significantly to its economic prowess and geopolitical influence. The sheer size of Brazil's economy, coupled with its abundant natural resources and the sovereignty over 60%

---

[84] Brasil 2023a.

[85] Grieger 2020; Nolte and Ribeiro 2021; Malamud 2022.

[86] Nolte and Ribeiro Neto 2021.

[87] Souza Costa and Pereira 2020; Mata Diz 2021.

of the Amazon rainforest, positions the country as a critical player in the dynamics between the negotiations for the EU-Mercosur Agreement.

Given the increasing concern over climate change and biodiversity loss, Brazil's commitment to sustainable practices and its role as a custodian of the Amazon rainforest carry immense weight in the ratification of the Agreement. If both parties align preferences and recalibrate their objectives, the EUMAA might be a platform for more profound collaborative efforts between the EU and Brazil to address environmental issues, with responsible resource management and biodiversity conservation.

A central element of the ongoing discourse within the intricate relationship between the EU and Brazil pertains to the multifaceted dimensions of Amazonia management. Unquestionably, one of the contemporary and crucial foundations influencing the dynamics between these entities is the administration of the Amazon rainforest. This pivotal aspect has emerged as a focal point, revealing a profound dichotomy in perspectives concerning environmental obligations and responsibilities.

## 15.5 Conclusion

This Chapter focused on one of the instruments used by the EU in its external action. Strategic narrative is shown as a pivotal instrument in shaping the trajectory of the EU's external action, wielding a profound influence on its diplomatic pursuits and engagement with international partners. This narrative serves as a potent tool through which the EU articulates its values, interests, and aspirations on the global stage, effectively conveying its vision for global cooperation and collective progress. By deftly articulating its stance on critical issues like sustainable development and environmental protection, the EU's strategic narrative not only shapes its identity but also elevates its position as a global influencer.

The strategic narrative employed by the EU as a leader on global environmental protection serves as an embodiment of soft diplomacy. It becomes a tool not only to promote the EU's interests, but also to project its core values onto the international stage, creating a common playing field for like-minded nations and actors, fostering an environment of global cooperation. This approach not only offers a platform for dialogue, but also positions the EU as the leader to guide other countries by example, according to its own values.

This Chapter intended to demonstrate that this strategic narrative can be hindered by divergent narratives among different international actors. The case of the Brazilian perspective on international interest over the Amazon Forest protection underscores this complexity. As the EU seeks the ratification of the European Union and Mercosur Association Agreement, it must demonstrate a nuanced understanding of Brazil's historical perspective of "anti-internationalization". The EU's assertions of global environmental leadership and declarations of tropical forests as common global goods might inadvertently challenge the sovereignty-centered narrative championed by not only Brazil, but also other Amazon countries. Balancing its strategic narrative with a recognition of this viewpoint is essential for the EU to foster a conducive atmosphere for negotiation, overcome impasses, and ultimately achieve more successful diplomatic outcomes while preserving its own values and interests.

## References

Afionis, S., & Stringer, L. C. (2014). The environment as a strategic priority in the European Union–Brazil partnership: Is the EU behaving as a normative power or soft imperialist? International Environmental Agreements: Politics, Law and Economics, 14, 47–64.

Agência Senado. (2005). Comissão discute internacionalização da Amazônia. Senado Federal. Retrieved from https://www12.senado.leg.br/noticias/materias/2005/04/07/comissao-discute-internacionalizacao-da-amazonia.

Becker, B. K. (1988). Significância contemporânea da fronteira: Uma interpretação geopolítica a partir da Amazônia Brasileira. Fronteiras. Brasília: Editora UnB.

Bezerra, J. (2015). The Brazilian Amazon: Politics, Science and International Relations in the History of the Forest. Springer.

Bradford, A. (2020). The Brussels Effect: How the European Union Rules the World. Oxford University Press.

Brasil. (2023a). Belém Declaration. The Leaders of State Parties to the Amazon Cooperation Treaty (ACT), meeting in the city of Belém do Pará, on August 9, 2023.

Brasil. (2023b). Decreto Nº 11.614, de 21 de julho de 2023. Institui o Plano Amazônia: Segurança e Soberania – Plano Amas.

Brasil. (2005). Decreto Nº 5.484, de 30 de junho de 2005. Aprova a Política de Defesa Nacional, e dá outras providências.

Brasil. (2008a). Luíz Inácio Lula da Silva (2007–2011). Discurso do Presidente da República, Luiz Inácio Lula da Silva, durante cerimônia de abertura do Instituto Nacional de Altos Estudos—INAE. Retrieved from http://www.biblioteca.presidencia.gov.br/presidencia/ex-presidentes/luiz-inacio-lula-da-silva/discursos/2o-mandato/2008/26-05-2008-discurso-do-presidente-da-republica-luiz-inacio-lula-da-silva-durante-cerimonia-de-abertura-do-xx-forum-nacional-do-instituto-nacional-de-altos-estudos-2013-inae/view.

Brasil. (2008b). Presidente (1985–1990: José Sarney). Discursos Selecionados do Presidente José Sarney. Fundação Alexandre de Gusmão. Brasília.

Brasil. (2020). Política Nacional de Defesa, Estratégia Nacional de Defesa.

Bunker, S. G. (2003). Matter, space, energy, and political economy: The Amazon in the world-system. Journal of World-Systems Research, 9(2), 219–258.

Castells, M. (2009). Communication Power. Oxford University Press.

CNN Brasil. (2023). Em Paris, Lula Critica Desigualdade e Fala Sobre Acordo com UE e Desmatamento na Amazônia. [Video file]. Retrieved from https://www.youtube.com/watch?v=_PusBST5R_c&t=708s.

De Sartre, X. A., & Taravella, R. (2009). National sovereignty vs. sustainable development: Lessons from the narrative on the internationalization of the Brazilian Amazon. Political Geography, 28(7), 406–415.

Duchêne, F. (1972). Europe's role in world peace. In Europe Tomorrow: Sixteen Europeans Look Ahead (pp. 32–47).

Durán, G. M. (2020). EU External Environmental Policy. In R. Wesse (Ed.), EU External Relations Law: Text, Cases and Materials (2nd ed., pp. 462–501).

European Commission. (2019a). Communication From The Commission To The European Parliament, The European Economic and Social Committee and The Committee of The Regions:

stepping up the EU action to Protect and Restore the World's Forests. SWD(2019) 307 fina. Brussels.
European Commission. (2019b). The European Green Deal. COM/2019/640/Final. Brussels.
European Commission. (2019c). Protecting and restoring the world's forests: Stepping up EU action to halt deforestation and forest degradation. FactSheet.
European Commission. (2019d). The EU-Mercosur Agreement Explained. Retrieved from https://policy.trade.ec.europa.eu/eu-trade-relationships-country-and-region/countries-and-regions/mercosur/eu-mercosur-agreement/agreement-explained_en.
European Parliament. (2011). Climate Change: End US-China 'Ping-Pong Blame Game' and Agree Deal by 2015. Press Release. Retrieved from https://www.europarl.europa.eu/news/es/press-room/20111207IPR33461/climate-change-end-us-china-ping-pong-blame-game-and-agree-deal-by-2015.
European Parliament. (2019). Implementation of the Common Commercial Policy—Annual Report 2018. European Parliament resolution of 7 October 2020 on the implementation of the common commercial policy—annual report 2018 (2019/2197(INI)).
European Union. (2023). EU-Mercosur Joint Instrument, DRAFT-SENSITIVE. Retrieved from https://friendsoftheearth.eu/wp-content/uploads/2023/03/LEAK-joint-instrument-EU-Mercosur.pdf.
European Union. (2007). Treaty of Lisbon Amending the Treaty on European Union and the Treaty Establishing the European Community, 13 December 2007, 2007/C 306/01. Retrieved from https://www.refworld.org/docid/476258d32.html.
Grieger, G. (2020). Amazon deforestation and EU-Mercosur deal. European Parliamentary Research Service.
INPE. (2023). Desmatamento – Amazônia Legal. Retrieved from http://terrabrasilis.dpi.inpe.br/downloads.
Júnior, A. M. E. (2012). De Vargas a Geisel: As Estratégias da Política Externa Brasileira para a Criação do Tratado de Cooperação Amazônica -TCA (1940–1978). Cadernos do Tempo Presente, 10. KARATEPE, Ismail D.; SCHERRER, Christoph; TIZOT, Henrique. Mercosur-EU Agreement: Impact on Agriculture, Environemnt, and Consumers. ICDD Working Papers. 2020.
Karatepe, I. D., Scherrer, C., & Tizot, H. (2020). Mercosur-EU Agreement: Impact on Agriculture, Environment, and Consumers. ICDD Working Papers.
Karkkainen, B. C. (2004). New governance in legal thought and in the world: Some splitting as antidote to overzealous lumping. Minn. L. Rev., 89, 471.
Keukeleire, S., & Delreux, T. (2022). The foreign policy of the European Union. Bloomsbury Publishing.
Krzyżanowski, M. (2009). Europe in crisis? Discourses on crisis events in the European press 1956–2006. Journalism Studies, 10(1), 18–35.
Kulovesi, K. (2013). Climate change in EU external relations: Please follow my example (or I might force you to). In E. Morgera (Ed.), The External Environmental Policy of the European Union: EU and International Perspectives (pp. 76–91). Cambridge University Press.
Manners, I. (2002). Normative power Europe: A contradiction in terms? Journal of Common Market Studies, 40(2), 235–258.
Mata Diz, J. B. (2021). Por Uma Aliança Estratégica Mercosul e União Europeia: O papel da Europa para o fortalecimento da sustentabilidade. Latin American Journal of European Studies, 01(01), 356–374.
Mata Diz, J. B., & Paiva Araújo, H. E. de. (2021). Extraterritoriality and the impact of EU Regulatory Authority: Environmental protection as soft power. In Extraterritoriality of EU Economic Law: The application of EU Economic Law outside the territory of the EU (pp. 319–333).
Maull, H. W. (2007). Europe as a Global Power. University of Trier.
Malamud, A. (2022). Assessing the political dialogue and cooperation pillar of the EU-Mercosur Association Agreement: Towards a bi-regional strategic partnership? IN-DEPTH ANALYSIS Requested by the AFET committee.

Merson, J. (2000). Bio-prospecting or bio-piracy: Intellectual property rights and biodiversity in a colonial and postcolonial context. Osiris, 15, 282–296.
Miskimmon, A., O'Loughlin, B., & Roselle, L. (2017). Forging the World: Strategic Narratives and International Relations. University of Michigan Press.
Miskimmon, A., O'Loughlin, B., & Roselle, L. (2013). Strategic Narratives: Communication Power and the New World Order. Routledge.
Miskimmon, A., O'Loughlin, B., & Roselle, L. (2014). Strategic narratives: Communication power and the new world order. Routledge.
Niţoiu, C. (2014). The narrative construction of the European Union in external relations. Routledge, pp. 76–91.
Neves, E. G., et al. (Eds). (2021). Amazon Assessment Report 2021. United Nations Sustainable Development Solutions Network, New York, USA. Available from https://www.theamazonwewant.org/spa-reports/.
Nolte, D., & Ribeiro Neto, C. C. (2021). Mercosur and the EU: The false Mirror. Lua Nova: Revista de Cultura e Política, 87–122.
Prince, G. (1982). Narrative analysis and narratology. New Literary History, 13(2), 179–188.
Prodes. (2023). Portal Terra Brasilis: Mapa do Desmatamento. Retrieved from http://terrabrasilis.dpi.inpe.br/app/map/deforestation?hl=pt-br.
Rhodes, R. A. W. (1997). Understanding governance: Policy networks, governance, reflexivity and accountability. Open University.Roselle, L., Miskimmon, A., & O'Loughlin, B. (2014). Strategic narrative: A new means to understand soft power. Media, War & Conflict, 7(1), 70–84.
Rosenau, J. N., & Czempiel, E. O. (1992). Governance without government: Order and change in world politics. Cambridge University Press.
Roselle, L., Miskimmon, A., & O'loughlin, B. (2014). "Strategic narrative: A new means to understand soft power." Media, war & conflict 7, no. 1, 70–84.
Shenhav, S. R. (2006). Political narratives and political reality. International Political Science Review, 27(3), 245-262.
Simon, F. (2024). 'Blame China': EU Parliament's high-stakes gambit at COP28. EURACTIV. Retrieved from https://www.euractiv.com/section/energy-environment/news/blame-china-eu-parliaments-high-stakes-gambit-at-cop28/.
Souza, M. (2019). História da Amazônia: Do período pré-colombiano aos desafios do século XXI. Rio de Janeiro: Record.
Souza Costa, B., Costa Assis, C., & Sampaio, J. A. (2020). Dialética na Panamazônia: Linguagem, pluralidade e cooperação. In J. Miranda, C. Gomes, & A. Leão (Eds.), Diálogo Ambiental Constitucional Internacional: Desenvolvimento socioambiental e econômico: O diálogo para um planeta em crise (Vol. 14, pp. 13–34). Lumen Juris, Rio de Janeiro.
Toledo, A. de P. (2012). Amazônia: Soberania ou Internacionalização. Arraes, Belo Horizonte.
UNEP. (2022). Convention on Biological Diversity: Post-2020 Global Biodiversity Framework.CBD/WG2020/5/L.2. Montreal.
UNFCC. (2015). Paris Agreement to the United Nations Framework Convention on Climate Change. December 2015. No. 16-114.
United Nations (2015). Transforming Our World: The 2030 Agenda for Sustainable Development. Resolution Adopted by the General Assembly on 25 September 2015.
Van Vooren, B., & Wessel, R. A. (2014). EU external relations law: Text, cases and materials. Cambridge University Press.
Verdélio, A. (2023). Lula: Amazon countries should lead preservation efforts. Agência Brasil. Retrieved from https://agenciabrasil.ebc.com.br/en/politica/noticia/2023-01/lula-amazon-countries-should-lead-preservation-efforts.
Vogler, J. (2005). The European contribution to global environmental governance. International Affairs, 81(4), 835–850.
Wallström, M. (2008). Public Diplomacy and its role in the EU's external relations. SPPECH/08/494. Mortar Center for International Studies, Georgetown University, Washington DC, 2 October 2008.

Wessel, R. A. (2020). The European Union as a Global Actor. In B. Van Vooren & R. Wessel (Eds.), EU External Relations Law: Text, Cases and Materials (2nd ed.).

Wessel, R. A. (2021). Normative transformations in EU external relations: The phenomenon of 'soft' international agreements. West European Politics, 44(1),72–92.

Zhouri, A. (2004). Global-Local Amazon Politics: Conflicting Paradigms in the Rainforest Campaign. Theory, Culture & Society, 21.

Zhouri, A. (2006). "O ativismo transnacional pela Amazônia: entre a ecologia política e o ambientalismo de resultados." Horizontes Antropológicos 12, 139–169.

**Jamile Bergamaschine Mata Diz** is a Coordinator of the UFMG Jean Monnet Center of Excellence at the Federal University of Minas Gerais (UFMG) and Professor at the Law School in UFMG.

**Júlia Péret Tasende Tarsia** is a Ph.D. candidate at the Federal University of Minas Gerais (UFMG), has a LLM from Federal University of Minas Gerais (UFMG) in the research line European Studies (FAPEMIG scholarship holder). Master student in the University Santiago de Compostela (USC), and an Assistant Researcher at the UFMG Jean Monnet Center of Excellence. Graduated in Law from the Pontifical Catholic University of Minas Gerais. Graduated in State Sciences from UFMG.

# Chapter 16
# From the EU Timber Regulation to the EU Regulation on Deforestation-Free Products: Making a Difference?

**Wybe Th. Douma**

**Contents**



**Abstract** The European Union (EU) has employed various instruments to fight deforestation, notably to counter climate change. In 2003, it adopted an Action Plan on Forest Law Enforcement, Governance and Trade (FLEGT), followed by the 2005 FLEGT Regulation. Because FLEGT was not successful, in 2010, the EU Timber Regulation (EUTR) was adopted. It regulated the import of timber from non-FLEGT countries. EU traders had to demonstrate through a due diligence system that such timber was produced legally, in accordance with the law of the producing countries.

W. Th. Douma (✉)
European Environmental Law Consultancy and the Dutch Ministry of Social Affairs and Employment, The Hague, The Netherlands
e-mail: w.t.douma@gmail.com



R. A. Wessel et al. (eds.), *EU External Relations Law and Sustainability*, Global Europe: Legal and Policy Issues of the EU's External Action 4,
https://doi.org/10.1007/978-94-6265-655-0_16

Whether the production was sustainable did not play a role. The regulation was poorly enforced by the EU Member States and by the European Commission. In 2023, the EU Regulation on Deforestation-Free Products (RDFP) was adopted. This regulation, sometimes referred to as the EU Forests Law, is again based on a due diligence system, with some improvements compared to the EUTR. It covers not just timber, but also other commodities that cause deforestation. It demands that these commodities are produced legally (according to the law of the producing country) and do not cause deforestation (in line with EU norms). The new regulation forces those that market products in the Union to collect loads of information from producers in third countries. By examining details of the RDFP and comparing them with the old rules and their (non-)application in practice, this Chapter investigates whether the new rules stand a chance of not just becoming yet another law on the books, but also one that can achieve its objectives in practice. A key factor, based on past experiences, is whether the EU Member States and the Commission will apply and enforce the new rules better than the old EUTR rules. It also depends on producing countries and their producers: they would need to adjust their policies and actions in order to continue exporting to the European Union.

**Keywords** Deforestation · Commodities and derived products · Deforestation-free products · Climate change · Human rights · Indigenous peoples · Country benchmark system · Due diligence · Risk assessments · Supply chains · Legality and sustainability conditions · Monitoring · Traceability · Geolocation · Sovereignty · Liability · Investments · Inspections · Sanctions · Naming-and-shaming list

## 16.1 Introduction

The European Union (EU) has long argued that deforestation should be combated, especially because of its effect on climate change.[1] However, over the past twenty years, the instruments employed by the organization have had little to no effect. Deforestation continued to rise worldwide[2] and the EU's contribution hardly diminished. About 80% of tropical deforestation is caused by agricultural expansion to

[1] The Union, together with 40 countries (not including Brazil), signed the 2014 New York Declaration to Combat Deforestation, which agreed to end deforestation by 2030. The interim goal for 2020 was to halve deforestation. In reality, deforestation actually increased in the period 2014–2020. In 2021, the UNFCCC COP26 saw the Glasgow World Leaders Declaration on Forests and Land Use endorsed by 143 countries (including Brazil) representing about 90% of the world's forests. Once again it was agreed to halt deforestation by 2030, and yet again deforestation continues to increase. See Forest Declaration Assessment Partners, *2023 Forest Declaration Assessment: off track and falling behind* [https://www.forestdeclaration.org/resources/forest-declaration-assessment-2023].

[2] According to the FAO, the world has lost around 420 million hectares of forests in the last three decades, which is 10% of the total and more than the area of the entire European Union (point 2 preamble RDFP).

produce commodities such as soya, beef and palm oil. Because of its imports, the EU is estimated to be responsible for at least 10% of global deforestation.[3]

In 2003, the EU Action Plan on Forest Law Enforcement, Governance and Trade (FLEGT) was launched.[4] The FLEGT Action Plan proposed focusing on concluding bilateral agreements[5] with the most important timber producing countries and helping them to combat illegal deforestation. If they demonstrably had a working control system, their timber could be freely exported to the EU. This was further regulated in the 2005 FLEGT Regulation.[6] The FLEGT system relied on a bilateral approach and thus did not encroach on the sovereignty of third countries. By 2023, only Indonesia, from which a mere fraction of the total European Union timber imports comes, can issue FLEGT permits. Negotiations have been conducted without success for years with some fifteen other countries, but not with Brazil, for instance. While these negotiations might have had some effects on the ground in the countries that negotiations were carried out with, even the European Commission itself has admitted that the FLEGT system failed to have a clear effect on the protection of forests.

In 2010, the EU Timber Regulation (EUTR) was introduced to prevent the placing of illegally harvested timber and timber products on the EU market from non-FLEGT countries, with the legislation of the country of origin being decisive. The obligations are primarily aimed at companies that put timber and timber products on the EU market. Timber producers and exporters from all timber producing countries with the exception of FLEGT country Indonesia will need to supply information and documentation regarding the legality of their products to EU traders, bringing about some costs. The EUTR clearly is aimed at the promotion of sustainable development and the fight against deforestation outside the Union. While the EUTR rules form a unilateral approach, they hardly encroach on the sovereignty of non-EU countries as it is left up to them to decide what logging is legal and what is not.

The scope of timber products covered by the EUTR is limited. Seating furniture, for instance, is not included. Whether the timber was harvested in a sustainable manner is not a criterion. The EUTR provisions became applicable in March 2013.[7] Examining its functioning in 2014, it showed that the EU member states were not yet monitoring and enforcing the EUTR very well.[8] One of the challenges under the EUTR is the complicated due diligence system that is to be used to prove that the

[3] European Parliament, Resolution of 22 October 2020 with recommendations to the Commission on an EU legal framework to end and reverse EU-promoted deforestation at global level (2020/2006(INL)), point O.

[4] Communication from the Commission to the Council and the European Parliament—Forest Law Enforcement, Governance and Trade (FLEGT)—Proposal for an EU Action Plan, COM(2003)251 of 21 May 2003.

[5] These treaties are referred to as Voluntary Partnership Agreements (VPAs).

[6] Council Regulation (EC) No 2173/2005 of 20 December 2005 establishing a FLEGT licensing scheme for the import of timber into the European Community, OJ L 347, 30.12.2005, pp. 1–6.

[7] Regulation (EU) No 995/2010 of the European Parliament and of the Council laying down the obligations of operators who place timber and timber products on the market, OJ L 295, 12.11.2010, pp. 23–34.

[8] Douma and Van der Kooij 2014.

timber was legally logged in the country of origin. The low sanctions and the absence of a formal role of consumers, NGOs or indigenous people also did not help. Several years later, the application and enforcement of the EUTR still had many shortcomings.[9] The due diligence system turned out to be rather complicated for many traders, enforcers, public prosecutors and judges. This part of the EU deforestation legislation also did not function properly, in other words, so new legislation was developed.

In 2023, the Regulation on Deforestation-Free Products (RDFP)[10] was adopted, which will replace the EUTR from the end of 2024. Remarkably, the new regulation is again based on an (improved) system of due diligence requirements. Just like the EUTR, it imposes obligations on those that put commodities and products on the EU market. Indirectly, it introduces norms relevant to producers and traders in non-EU countries. Inter alia, they will need to provide detailed information on the geolocation of the plots where the products were produced, as well as adequately verifiable information that their production did not cause deforestation. Abiding by these norms brings about more costs for suppliers than was the case under the EUTR.

The RDFP, which is also referred to as the EU Forests Act, does not only apply to timber, but also to cattle, cocoa, coffee, palm oil, soya and rubber and products derived therefrom. Furthermore, it is no longer only the legality of the commodity and products that is decisive, but also their sustainability: the commodities and derived products must not have contributed to deforestation. That is an important change, as legally produced commodities and products can still cause deforestation. Of course, it also makes the system more complicated. Moreover, contrary to the legality condition that relies on the rules of the country of production of the goods, the sustainability condition relies on EU norms to decide whether commodities and products are sustainably produced or not, both inside and outside the European Union, which—for the latter area—raises the question whether the Union can impose such norms unilaterally.

This contribution first takes a closer look at what went wrong with the EU Timber Regulation, and (briefly) with the FLEGT Regulation, and how these instruments affected producing countries. It then discusses the improvements that have been made in the new Regulation on Deforestation-Free Products compared to the EUTR, especially in the area of the due diligence provisions. In addition, some of the new challenges that this law entails, notably as regards producing countries outside the Union are examined. Furthermore, a number of proposed improvements are discussed that will be decided upon at a later moment in time. In this way, it is assessed whether the RDFP could combat deforestation better than the EUTR and the FLEGT Regulation and whether the unilateral approach that the new regulation encompasses infringes on the sovereignty of non-EU countries.

[9] Tubbing and Douma 2017; Bazelmans and Douma 2020.

[10] Regulation (EU) 2023/1115 of the European Parliament and of the Council of 31 May 2023 concerning the making available on the Union market and the export from the Union of certain commodities and products associated with deforestation and forest degradation and repealing Regulation (EU) No 995/2010, OJ L 150, 9.6.2023, pp. 206–247.

More specifically, the following aspects are considered. Why was the scope of the EU legislation broadened? And why has the boat been held back, for the time being, on other points, such as the protection of other valuable nature reserves (that are equally important as forests to counter climate change), liability rules and the regulation of EU investments that contribute to deforestation? Will the RDFP make life easier or harder for companies and regulators? And will the RDFP, contrary to the EUTR and FLEGT, ensure that the European Union contributes less to deforestation? Or will we have to conclude in a couple of years that this law, like its predecessor, was also ineffective? Finally, attention is paid to the criticism voiced by countries like Brazil, notably as regards the costs that meeting the RDFP conditions will bring about for producers in these countries, and the alleged infringement on sovereignty. "The EU is the main destination for Brazilian coffee exports, the second-largest destination for Brazilian soybeans and oil palm and their derivatives, and the third-largest for Brazilian cattle and wood and their derivatives, as well as cocoa and rubber".[11] The new European regulation will thus have a significant impact on Brazilian agribusiness; hence the focus will be on that country when it comes to the effects of the RDFP on producing countries.

## 16.2 The EU Timber Regulation and FLEGT Regulation Evaluated

### *16.2.1 First Evaluations*

Initially, many Member States did not comply with formal obligations and did not allocate enough human and financial resources to competent authorities. Over time, the former problem was resolved, but the allocated resources remained substandard. The European Commission concluded that six years after the EUTR came into force, and despite the issuance of guidance providing further explanations,[12] continued efforts were still needed to ensure uniform and effective application of the EUTR in all countries.[13]

Remarkably, the Commission hardly took action against countries that did little or nothing on control and enforcement. In 2017, Belgium received a letter of formal

[11] Leme Lopes et al. 2023. By 2022, exports from Brazil to the EU of commodities and products covered by the RDFP amounted to some US$ 17.5 billion. Also see footnote 64 below.

[12] Commission Communication of 12 February 2016, *Guidelines for the EU Timber Regulation*, C(2016)755. Available at: https://www.nvwa.nl/documenten/import/hout/flegt/publicaties/hout-importeren-guidance-document-eutr. Accessed November 2023.

[13] European Commission, Regulation (EU) No 995/2010 of the European Parliament and of the Council of 20 October 2010 laying down the obligations of operators who place timber and timber products on the market (the EU Timber Regulation). Biennial report for the period March 2017–February 2019, COM(2020)629 final of 2 October 2020. Available at: https://eur-lex.europa.eu/legal-content/NL/TXT/?uri=CELEX:52020DC0629&qid=1692188251583. Accessed November 2023.

notice regarding the quantity and quality of controls carried out by its competent authorities. It is one of the largest timber importing countries in the Union, but employed less than one FTE who carried out a minimal number of checks annually.[14] Slovakia and Romania both received reasoned opinions due to their weak sanctions systems.[15] However, the Commission never actually instigated infringement proceedings: no Member State was called before the Court of Justice of the European Union. This represents a missed opportunity to improve the functioning of the EUTR.

Research confirmed that the EUTR and FLEGT worked poorly in practice.[16] It was also established that an EU approach which only focuses on wood and palm oil (the latter via the revised Renewable Energy Directive, RED II[17]) is not effective in combating deforestation. Deforestation is mainly caused by growing demand in the European Union and elsewhere for products such as meat and soya, and can therefore only be effectively combated by ensuring that many more products placed on the European and other markets do not contribute to deforestation, damage other ecosystems or cause human rights violations in the countries where they are produced. It was indicated that a mandatory due diligence system should be given teeth by clearer enforcement and liability provisions.

### 16.2.2 *Public Consultations*

In 2020, the Commission launched a one month public consultation on how the EUTR and the FLEGT Regulation worked globally, how effective, efficient, relevant and consistent these laws were and whether they usefully complemented national

[14] European Commission, *October infringement package: key decisions*, 4 October 2017. Available at https://ec.europa.eu/commission/presscorner/detail/EN/MEMO_17_3494. Accessed March 2024. Also see ClientEarth, *Belgium facing legal action for breaking illegal logging law*, 23 October 2017. Available at: https://www.clientearth.org/latest/news/belgium-facing-legal-action-for-breaking-illegal-logging-law/. Accessed March 2024.

[15] European Commission, *July infringement package: key decisions*, 2 July 2020. Available at https://ec.europa.eu/commission/presscorner/detail/en/inf_20_1212. Accessed March 2024. ClientEarth, *EUTR News—October to December 2016*, available at https://www.clientearth.org/media/bnjptwoq/eutr-news-october-to-december-2016-ce-en.pdf.

[16] See footnotes 9 and 10 above.

[17] Directive (EU) 2018/2001 of the European Parliament and of the Council of 11 December 2018 on the promotion of the use of energy from renewable sources, OJ L 328, 21.12.2018, p. 82 (*RED II*). Note that amended rules will need to be in place by 21 May 2025 under *RED III*: Directive (EU) 2023/2413 of the European Parliament and of the Council of 18 October 2023 amending Directive (EU) 2018/2001, Regulation (EU) 2018/1999 and Directive 98/70/EC as regards the promotion of energy from renewable sources, and repealing Council Directive (EU) 2015/652, OJ L, 2023/2413, 31.10.2023.

efforts.[18] A mere 39 contributions were received, most of which represented European interests. Another public consultation that year collected feedback regarding the concept impact assessment of regulatory and non-regulatory options for demand side measures to minimize the risk that deforestation linked products that are placed on the EU market, and on defining deforestation-free supply chains.[19] This consultation also was open for one month, but received 99 contributions. Some of these were from outside the European Union. For instance, scientists from Brazil and France warned against the emphasis placed on the use of private sustainability standards, certifications and audits that are carried out by the companies themselves or third parties hired by them, because of their serious flaws.[20]

Finally, a three month consultation was held in order to contribute to an impact assessment that would investigate the suitability of a range of different demand-side measures to address deforestation and forest degradation associated with EU consumption. The consultation objective was twofold: to ensure that all relevant stakeholders would be identified (including those from producing countries) and are given the opportunity to take part in the consultation, and to gather stakeholder opinions on the potential measures at the EU level. This time around, a staggering amount of feedback was received: over one million respondents from inside and outside of the European Union forwarded reactions. Over 30% of the reactions were from non-EU citizens; almost 38000 reactions came from Brazil.[21] A majority of the contributors did want the EU to take further action to reduce the Union's contribution to global deforestation and forest degradation. There was strong support for legally binding options rather than voluntary measures. According to the Commission, the vast majority of companies and NGOs were in favor of a system of binding due diligence requirements. Various suggestions were received about what that system should look like.

[18] Illegal logging—evaluation of EU rules (fitness check). Available at: https://ec.europa.eu/info/law/better-regulation/have-your-say/initiatives/11630-Illegal-logging-evaluation-of-EU-rules-fitness-check-_en. Last accessed November 2023. The 39 contributions are available at https://ec.europa.eu/info/law/better-regulation/have-your-say/initiatives/11630-Illegal-logging-evaluation-of-EU-rules-fitness-check-_en .

[19] Deforestation and forest degradation—reducing the impact of products placed on the EU market. Available at: https://ec.europa.eu/info/law/better-regulation/have-your-say/initiatives/12137-Deforestation-and-forest-degradation-reducing-the-impact-of-products-placed-on-the-EU-market_en. Accessed November 2023. The 99 contributions can be found at https://ec.europa.eu/info/law/better-regulation/have-your-say/initiatives/12137-Deforestation-and-forest-degradation-reducing-the-impact-of-products-placed-on-the-EU-market/feedback_en?p_id=6348580.

[20] Rajão et al. 2020.

[21] See https://ec.europa.eu/info/law/better-regulation/have-your-say/initiatives/12137-Deforestation-and-forest-degradation-reducing-the-impact-of-products-placed-on-the-EU-market/F_en

### 16.2.3 EP Resolution

On October 22, 2020, the European Parliament set out a series of recommendations what an EU legal framework for ending EU-promoted deforestation in the world should look like.[22] Among other things, calls were made for good rules on the protection of indigenous people and human rights, and for rules on the financial sector to prevent it from contributing to deforestation. Furthermore, the importance of the possibility for persons harmed by non-compliance with the rules on deforestation-free products to claim compensation was underlined. Market participants should be primarily liable for damage caused by human rights violations or damage to forests and natural ecosystems caused by entities under their control or economically dependent on them. Secondly, the European Parliament advised that they should be liable for such damage if it is directly related to their products, services or activities, unless they can demonstrate that they have exercised sufficient care.[23]

Third party certification schemes should only complement product origin determination and risk assessment if they are ensured to be transparent, impartial and reliable. The liability of economic operators should remain unaffected, according to the European Parliament.[24] As will be discussed below, many of these recommendations were not embraced in the RDFP, though some were placed on the backbench for potential future amendments.

### 16.2.4 Fitness Check

A so-called fitness check was carried out at the end of 2021,[25] which examined whether the EUTR and the FLEGT Regulation were *fit for purpose.* It was established that even the largest timber importing countries of the EU provide little manpower and other resources for controls and enforcement. The Netherlands, Belgium, Finland and Ireland, for example, only had two to three FTEs for the inspection of many thousands of loads of wood and wood products. Still, the EUTR is said to have led to between 12 and 29% less illegal timber being imported.[26] However, imports from countries with specific problems (such as Ukraine and Myanmar) continued and even increased.

---

[22] European Parliament resolution of 22 October 2020 with recommendations to the Commission on an EU legal framework to halt and reverse EU-driven global deforestation (2020/2006(INL)).

[23] Ibid., recommendation 5.2, p. 33.

[24] Ibid., recommendation 4.1 sub f, pp. 33–34.

[25] European Commission, Commission staff working document. Fitness check on Regulation (EU) No 995/2010 of the European Parliament and of the Council of 20 October 2010 laying down the obligations of operators who place timber and timber products on the market (the EU Timber Regulation) and on Regulation (EC) No 2173/2005 of 20 December 2005 on the establishment of a FLEGT licensing scheme for imports of timber into the European Community (FLEGT Regulation), SWD(2021)328 final of 17 November 2021.

[26] It was noted that this estimate was based on a relatively small number of data.

China also became a prominent player in the timber industry, complicating information gathering around longer supply chains. Furthermore, there were continued reports of significant illegal activity in some of the EU's main import sources, such as Brazil, Russia, Ukraine, Myanmar and a number of African countries.[27]

The competent authorities indicated that the EUTR provisions are insufficient for operators and authorities to effectively assess risks. Only the most serious violations of the EUTR lead to action, partly due to weak enforcement capacity and a lack of resources. It is difficult to meet the demands of the Public Prosecution Service, and it is difficult to prove violation of the EUTR's complex duty of care provisions in court.[28] At times, competent authorities could not cooperate with customs authorities. For example, customs data was sometimes refused to be shared with competent authorities due to the lack of specific authorization provisions.

As far as the FLEGT Regulation is concerned, the *fitness check* found that its most important instrument, the Voluntary Partnership Agreements (VPAs), were not producing results. The VPAs were described as complex agreements concluded for a single commodity and derivatives, that also cover labor, social and human rights aspects. Its negotiations take years, making the conclusion of VPAs slow and costly processes.[29]

As for the EUTR, it is noted that this law has clearly not had a significant impact on reducing illegal logging worldwide. A revealing annex shows that an earlier draft version made clear conclusions about how *fit for purpose* the EUTR was missing, as were take-always that are important for future decision-making. The Commission therefore supplemented the fitness check. While this was being done, work was already underway on the proposal for the new RDFP. That is far from an optimal order. It is also not very transparent. The RDFP proposal was ultimately presented on the same day as the modified fitness check.

## 16.3 The Proposal for the Regulation on Deforestation-Free Products

The RDFP proposal[30] focused not only on wood, but also on a range of other products that could cause deforestation, such as palm oil,[31] cattle, soya, coffee, cocoa and derivatives (such as beef, furniture and chocolate). Rubber and corn were missing: the Commission looked at these commodities but did not include them because

[27] Ibid., p. 24.

[28] Ibid., p. 21.

[29] Ibid., p. 45.

[30] European Commission, Proposal for a Regulation of the European Parliament and of the Council on the making available on the Union market and the export from the Union of certain commodities and products associated with deforestation and forest degradation and repealing Regulation (EU) No 95/2010, COM(2021)706 final of 17 November 2021.

[31] Where palm oil is placed on the EU market for use as biofuel, the rules of RED II apply.

they would contribute less to deforestation.[32] Furthermore, it encompassed a new condition. Besides the condition that commodities and relevant products have been produced in accordance with the relevant legislation of the country of production, they must also be deforestation free. A group of producing countries, including Brazil, complained about the proposal, regretting the EU has chosen "unilateral legislation instead of an international engagement" to deal with the shared objectives and urgent tasks of the fight against climate change and the conservation and sustainable management of forests.[33] The criticism was voiced again after the adoption of the RDFP[34] and will be returned to below when discussing the RDFP.

As explained above, the EUTR's due diligence system worked poorly. Nevertheless, it was decided to continue with such a system, with improvements. Besides the two conditions mentioned above, the third condition is that the commodities and products are covered by a due diligence statement.

Alternative policy options based on a different system were hardly explored, hence a variant of the existing system to combat illegal fishing was not considered.[35] The latter system is simpler than a due diligence system in many ways. Countries receive a yellow card from the European Union if insufficient action is being taken against illegal fishing, and if no improvements are subsequently made, they can be placed on a blacklist, which means they will no longer be allowed to export fish to the Union. That system appears to work satisfactorily,[36] raising the question why such a system was not even considered for products that come with a risk of deforestation.[37]

[32] About EUR 2.8 billion worth of corn and EUR 17.6 billion worth of rubber are imported annually, but the small contribution to deforestation would not outweigh the costs of bringing these two commodities under the scope of the RDFP. See the Commission's Impact Assessment SWD(2021)326, part 1, of 17 November 2021, p. 32.

[33] Argentina et al. 2022.

[34] Argentina et al. 2023.

[35] In particular Council Regulation (EU) 1005/2008 of 29 September 2008 establishing a Community system to prevent, deter and eliminate illegal, unreported and unregulated fishing, OJ L 286, 29.10.2008, pp. 1–32.

[36] See for example European Commission, *Report on the application of Regulation (EC) No 1005/ 2008 (the IUU Regulation)*, COM(2020)792 of 9 December 2020.

[37] See Douma 2018.

## 16.4 The New Deforestation-Free Products Regulation

### *16.4.1 General*

On May 31, 2023, the Regulation on Deforestation-Free Products was adopted.[38] The rules will largely apply from December 30, 2024[39] and replace those of the 2010 EU Timber Regulation (EUTR). Annex I to the RDFP contains the lists of relevant commodities and products, and Annex II indicates what information the due diligence declaration must include. A countries and regions benchmarking system is to be set up, encompassing low, standard and high risks of deforestation (see Sect. 16.4.2).

The main rule can be found in Article 3 RDFP: relevant commodities and products may not be placed or offered on the EU market or exported from the EU unless they are deforestation-free (see Sect. 16.4.5), were legally produced under the legislation of the production country (see Sect. 16.4.6) and are accompanied by a due diligence declaration about their sustainability (see Sect. 16.4.7). The latter declaration must include a risk assessment to determine whether there is a risk that the relevant products do not comply with the RDFP conditions. Only if that risk is absent or negligible,[40] economic operators may place them on the EU market. In line with one of the suggestions of the European Parliament, it is no longer allowed to rely solely on certification by third parties, as under the EUTR. Such information may still be used as part of the risk assessment, though.[41]

The EUTR mainly imposes obligations on companies that offer timber products on the market in the Union for the first time. The RDFP has more extensive obligations, namely for all companies that offer relevant products on the EU market, not just for companies that offer the products for the first time. However, exceptions apply for Small and Medium-sized Enterprise (SME) operators participants.[42] This group of entrepreneurs is very broad: it concerns market participants with a maximum of 250

[38] Regulation (EU) 2023/1115 of the European Parliament and of the Council of 31 May 2023 on the making available on the Union market and the export from the Union of certain commodities and products related to deforestation and forest degradation, OJ L 150, 9.6.2023, pp. 206–247.

[39] Under Article 38, para 3 RDFP, micro- and small operators (i.e. those who place relevant products on the Union market or export them) have until 30 June 2025 to comply with the main RDFP provisions, an extension of six months in other words.

[40] Article 2 sub 26 RDFP further defines negligible risk as the situation in which, after a full assessment of the product-specific and general information and, where necessary, the application of risk mitigation measures, there is no cause for concern about compliance with Article 3 sub a) or b) RDFP.

[41] This seems a wise move. See for instance Rajão et al. 2020, who discuss scientific research showing how unreliable certification by private companies is, and demonstrate that the certificate issued by a Norwegian company to JBS, the largest Brazilian meat processing company in the world, claiming that no deforestation had taken place in the JBS' supply chains, was contradicted by Brazilian Public Attorney's audits and independent investigations. The latter revealed large amounts of farms involved in deforestation selling meat to JBS.

[42] Article 2 sub 30 RDFP: micro, small and medium-sized enterprises within the meaning of Directive 2013/34/EU on annual financial statements.

employees and an annual turnover of 40 million euros. These SME's do not have to comply with the due diligence requirements for certain products. Similar exceptions do not exist for SME's in producing countries. Several developing countries pointed out that this forms a flaw in the RDFP.[43]

What is striking, are the very extensive evaluation obligations (Article 34), which require, among other things, to consider whether a large number of points still need to be added to the RDFP at a later moment in time (see Sect. 16.4.3). A full evaluation of the RDFP must be carried out no later than June 30, 2028.

### *16.4.2 Country and Regions Benchmarking*

In accordance with Article 29 of the RDFP, a list will be drawn up of countries and parts thereof, classified according to the risk of deforestation there. This benchmarking classifies countries and regions into a low, standard, or high risk category. The classification is carried out by the European Commission, in an "objective and transparent" manner. It is to be based primarily on the following assessment criteria: rate of deforestation and forest degradation, and rate of expansion of agriculture land and production trends of relevant commodities and products (Article 29(3) RDFP). While those criteria are fairly straightforward, the other criteria listed in Article 29(4) are not altogether unequivocal, because the Commission is not obliged to take them into account, but may do so. For instance, when it comes to "effective enforcement of laws protecting human rights [and] the rights of indigenous peoples" (Article 29(4)(d) RDFP), it is hard to follow why such information may or may not be taken into account and how that could be objective and transparent. The same holds true for the possibility to consider, or not, whether the country has relevant legislation in place and takes effective enforcement measures to prevent and punish deforestation and forest degradation, and applies sanctions that are strict enough to deprive benefits thereof (Article 29(4)(c)). This also applies to the existence of agreements and other instruments between the country concerned and the Union and/or its Member States to address deforestation and forest degradation and the compliance of relevant commodities and relevant products with Article 3 and their effective implementation to facilitate this (Article 29(4)(a)).[44] Finally, "the existence, compliance with, or effective enforcement of laws protecting human rights, the rights of indigenous peoples, local communities and other customary tenure rights holders" forms relevant information that *may* be taken into account (Article 29(4)(d)). In all of these cases, it would have made more sense to oblige the Commission to take all relevant information into account.

The risk assessment part of the due diligence declaration mentioned above is to take into account the risk category of the country or region in question. According to

---

[43] Argentina et al. 2023; Leme Lopes et al. 2023.

[44] This could contribute to the willingness of third countries to conclude such agreements. Compare Dobson 2022, p. 539.

a number of countries, the country benchmark system is intrinsically discriminatory and punitive, and may be inconsistent with the law of the World Trade Organization (WTO).[45] While aspects of the benchmark system might indeed be criticized, as also was done above, the system as such seems objectively necessary to protect humans (against climate change) and the environment (against deforestation and destruction of habitats). As the system ensures that products stemming from countries or regions that meet certain criteria and qualify as low risk countries can be put in the EU market relatively easily, and products from high risk countries bring about the duty to fulfill more conditions, the system seems proportionate as well. Rather than labeling the benchmarking system as punitive, it can thus also be described as encouraging producing countries and their companies to fight deforestation and aim for the status of low risk country.

The fact that the Commission has until 30 December 2024 to present the country benchmark list (Article 29 para 2) is to be considered a flaw. The RDFP rules apply as of that date and for all parties concerned it is important to know well in advance whether relevant products stem from a low, standard or high risk country. The Commission is to notify the country concerned of its intention to classify that country or a part thereof as a high risk, and invite it to provide any information deemed useful in that regard. An exact timeframe is lacking, but the country concerned needs to be granted 'sufficient time' to reply. Gracefully, the RDFP allows the country concerned to provide the Commission with information on measures taken by it to remedy the situation in its reply (Article 29 paras 6 and 7). Considering how high the stakes are, it would have been better to fix a minimum of deadlines. Could a country or region labeled a high risk address the Court of Justice of the European Union to fight such a decision? If it would have legal standing, the very broad discretion offered to the Commission might stand in the way of that Court concluding that a manifest error was made.

### *16.4.3 Wider Scope*

The scope of the RDFP is much broader than of the EUTR. It no longer covers only timber (wood) and timber products, but also the commodities cattle, cocoa, coffee, oil palm, rubber and soya, plus specific products listed in Annex I that contain these

---

[45] Argentina et al. 2022 claim that the country benchmarking system's 'most likely effect will be to generate trade distortion and diplomatic tensions, without benefits to the environment. Furthermore, it imposes additional controls, entails reputational risks for companies and is s likely to penalize producers in developing countries, especially smallholder farmers and SMEs.' These concerns overlook that the system also allows for *less* stringent controls of products from low risk countries. Compare Civillini 2023. The European Commission plans to postpone adopting the benchmarking list, in violation of the clear RDFP rules (see Bounds, Hancock and Beattle, *EU delays stricter rules on imports from deforested areas*, Financial Times 8 March 2024. Available at https://www.ft.com/content/8dab4dc6-197b-4a2f-86f0-d5e83ce00b09. Accessed March 2024).

commodities, or are fed, or manufactured with them.[46] Rubber was missing from the Commission proposal, but this commodity was added to the RDFP at the insistence of the EP and others.[47]

Remarkably, the RDFP proposal contained more or less the same list of timber products as the EUTR, while the EUTR had long been criticized for lacking a multitude of products, and research into broadening already existed.[48] There was therefore a strong insistence on adding more timber products to Annex I, for instance seating furniture.[49] With success: in the end, nine timber product groups were added compared to the EUTR.[50] By broadening the scope of the RDFP also in this respect, it could make a better contribution to combating climate change.

The scope of the rules was expanded in yet another way. They apply not only to commodities and products imported into the Union, but also to their exports. This could be important for EU countries such as Romania, where a lot of illegal logging still occurs.[51]

### 16.4.4 An Even Broader Scope in the Future?

Several proposals to broaden the scope of the RDFP even more had been pushed for without success, notably by the EP and the EESC.[52] For example, unlike rubber, maize was not included as a relevant commodity in the RDFP. It was stipulated that the Commission should submit an impact assessment and, if necessary, a legislative

---

[46] Article 1 sub 1 and 2 in conjunction with Annex I RDFP.

[47] Like the Forest Stewardship Council 2022, which pointed out that the Union is the world's second largest importer of rubber and that rising demand is leading to tropical forests being converted into rubber plantations.

[48] European Commission, *Impact assessment study for the revision of the product scope of the EU timber regulation: final report,* Publications Office, 2019 and WWF European Policy Office, Feedback on EUTR and FLEGT Fitness Check, 28 February 2020, Available at https://ec.europa.eu/info/law/better-regulation/have-your-say/initiatives/11630-Illegal-logging-evaluation-of-EU-rules-fitness-check-/F506744_en. Accessed November 2023.

[49] Łuszczek et al. 2021.

[50] Including seating furniture, tools, wood wool and charcoal.

[51] See for example Kuchlmayr et al. (2023). The Commission opened an infringement procedure against Romania in 2020 (NFR(2020)2033) for the lack of effective controls on economic operators and the failure to impose appropriate sanctions. This did not result in a case ending up at the Court of Justice of the European Union, see the answer to questions from a Member of European Parliament (E-000301/2023(ASW)). The country only had administrative sanctions where illegal wood was concerned. Criminal sanctions have now been introduced for specific serious cases.

[52] EESC 2022, p. 92, para 4.2, urged, among other things, to supplement the list of commodities and products with bananas, sugar cane and all meat from animals that were fed soya, such as pigs and poultry.

proposal by 30 June 2025 at the latest, indicating 'the need and the feasibility' to extend the scope of the RDFP 'to further commodities, including maize'.[53]

A widely expressed wish[54] that should be explored at the same time concerns expanding the scope of the RDFP to other valuable natural ecosystems, "including other land with high carbon stocks and with a high biodiversity value such as grasslands, peatlands and wetlands". This would be more consistent and coherent with the Renewable Energy Directive (RED), where the sustainability criteria concern not only the protection of forests but also of other valuable ecosystems.[55] It would encourage the Union to contribute less to the disappearance of natural ecosystems, which would provide further climate benefits. For example, peatlands play an important role in preventing and mitigating the effects of climate change.[56] It would also ensure that in a country such as Brazil, the RDFP would get more impact: for the time being, deforestation in large areas such as the Cerrado and Pantanal are not covered.[57] Therefore, it seems a good idea not to just focus on deforestation, as that is just one of the causes of climate change. This was also argued in the run-up to the RDFP, but the European Commission rejected expanding the scope of the RDFP in this respect. Here too, it is far from certain that this institution will deem it necessary and feasible to expand the scope of the RDFP in this respect by 2025.

The same impact assessment should also assess whether the list of relevant products set out in Annex I should be amended or extended. It must be ensured that the list includes the most relevant products containing, fed with or manufactured with relevant commodities. In particular, it should be examined whether biofuels should be added to the list.[58]

Finally, the 2025 impact assessment is to cover a highly relevant point: the role of financial institutions and investors in preventing financial flows that directly or indirectly contribute to deforestation and forest degradation. It should be assessed whether it is necessary to impose specific obligations under EU law on financial institutions in this regard.[59] The Commission had set out several reasons why the RDFP contained no rules regarding the financial and investment sectors.[60]

> "Existing initiatives in the area of sustainable finance, such as the implementation of the EU Taxonomy Regulation and the future Corporate Sustainability Reporting Directive, CSRD

---

[53] Article 34 para 2 RDFP. The Commission might very well use its discretionary space to maintain the view that it is neither necessary nor feasible to add maize to the RDFP.

[54] Including EESC 2022, p. 92, para 4.1.

[55] Article 29 para 3 REDD II includes sustainability criteria for biofuels based on land with a high biodiversity value, including, for example, grassland in addition to forests. Directive (EU) 2018/2001 of the European Parliament and of the Council of 11 December 2018 on the promotion of the use of energy from renewable sources, OJ L 328, 21.12.2018, p. 82.

[56] For the global dimension see IUCN 2021, and for the Dutch dimension Ekker 2017.

[57] Leme Lopes et al. 2023, p. 9.

[58] Article 34 para 3 RDFP.

[59] Article 34 para 4 RDFP.

[60] COM(2021)706 of 17 November 2021, Explanatory Memorandum, pp. 3–4.

(current Non-Financial Reporting Directive, NFRD) are well suited to address the deforestation impacts of the finance and investment sectors, thereby complementing and supporting this legislative initiative on deforestation.

The CSRD and taxonomy impose disclosure obligations also on non-financial undertakings: the CSRD foresees the publication of sectoral reporting standards by October 2023; under the EU Taxonomy Regulation, technical criteria have already been established under the delegated act for climate mitigation and adaptation for forestry, while the publication of criteria for agriculture has been delayed. Both economic activities can be covered under the delegated acts for the other four environmental objectives."

These arguments are not altogether convincing. The cited legislation mainly concerns reporting requirements that do not create obligations for financiers or investors to ensure that their financial services do not lead to deforestation or human rights violations.[61] The European Parliament had proposed introducing a duty of care for financial institutions operating in the EU to ensure that they comply with the requirements of the RDFP when providing financial services to companies engaged in harvesting, extracting, producing, processing or trading products and their derivatives with a risk to forests and ecosystems.[62] That is a sensible proposal, because strict RDFP requirements for products without stricter financing requirements amounts to mopping the floor while leaving the tap running. It is recommended to focus on financial flows that contribute directly or indirectly to deforestation and forest degradation. This could include, for example, financing infrastructure (roads, ports) for agriculture that poses a threat to forests.[63]

It is noteworthy that the impact assessment that must address all these points, and where necessary the associated legislative proposals, must be presented no later than June 30, 2025. The provisions of the RDFP will then only have applied for six months, which will make it difficult, at best, to assess whether its implementation and enforcement will get off the ground in practice. As long as this is not clear, the Commission may well be inclined to still label them as unnecessary and/or not feasible in 2025.

### *16.4.5 Deforestation-Free*

While the EUTR only requires that timber must be legal, the RDFP adds an additional requirement: the timber or other commodities and products must be deforestation-free. This is the case if they (a) contain relevant products, or are fed or manufactured

[61] See also Rice and Blumtritt 2021, p. 12.

[62] European Parliament resolution of 22 October 2020 with recommendations to the Commission on an EU legal framework to end and reverse EU-promoted deforestation at global level (2020/2005(INL)), para 24.

[63] Compare Kuipers 2018; Been and Groot 2023, discussing many cases of Dutch investors involved in deforestation in Brazil. In fact, by 2022, more than 23% of all investments in Brazil came from the Netherlands, more than from any other country, and more than half of all FDI from EU Member States, according to Santander Trade Markets 2023. Also see footnote 12 above.

with relevant commodities produced in areas where no deforestation has occurred after December 31, 2020, and (b) in the case of relevant products, contain or are manufactured from wood, the wood was felled in forests without causing forest degradation after December 31, 2020.[64]

It is positive that it must be demonstrated not only that products were legally produced, but also that they have not contributed to deforestation or forest degradation. The production of, for example, timber and meat, can be legal under the laws of the producing country but still cause large-scale deforestation or forest degradation, especially if producing countries prioritize economic development over forest protection.

The definitions of "forest", "forest degradation", "forest plantation" and "deforestation-free" are confusing. For example, it appears that transforming a natural forest into a forest plantation does not constitute deforestation under the RDFP, while of course it should.[65]

The RDFP requires that relevant products (such as cattle and beef) must be deforestation-free, which means, among other things, that they must not have been fed with relevant commodities (such as soya) that were produced in areas where deforestation took place after December 31, 2020.[66] This could indicate that derived products included in Annex I, such as cattle, may not be placed on the Union market. The preamble[67] states this in rather cryptic terms

> economic operators placing on the market relevant products (…) containing bovine animals (…) fed with relevant products containing or manufactured from other relevant commodities or products shall ensure, as part of their due diligence system, that the feed is deforestation-free.

This provision seems to indicate that imported beef may not contain soya that does not meet EU conditions. The situation is apparently different for other products that were produced with such soya (or palm oil, for example): these products do not appear on Annex I, and can therefore be put on the EU market.

However, the requirement for derived beef seems mainly symbolic. The operator only needs to follow the geolocation requirements for the places where the cattle themselves were kept. Article 9, para 1, sub (d) RDFP indeed states that for relevant products that have been fed with relevant products, the geolocation refers to all establishments where the cattle was kept. According to the preamble, no geolocation may be requested for the food they received. Only if a competent authority in an EU Member State obtains relevant information or is informed, for example by third parties, that there is a risk that the feed does not meet the requirements of the RDFP, then this authority must request detailed information about these feed requests.[68] In this way there is little or no guarantee that products do not contain animal feed that

---

[64] Article 1 sub 13 RDFP.

[65] Compare Łuszczek et al. 2021, p. 5; Forest Stewardship Council 2022, p. 1.

[66] Article 2 sub 13 RDFP.

[67] Point 39.

[68] Ibid.

causes deforestation. This point should be included by the Commission in the impact assessment in 2025, where necessary and feasible with proposals for legislation or delegated acts.[69]

## 16.4.6 Legality Requirement

In the EUTR, the only substantive requirement was that the wood had to be legally harvested, under the rules of the country where the wood came from. The proposal for the RDFP contained a rather limited definition of this legality requirement, one that was even more limited than the EUTR definition[70] of applicable law. This led to a lot of criticism. Fortunately, the final RDFP definition of 'relevant legislation of the country of production' became broader in scope:[71]

> "The laws applicable in the country of production regarding the legal status of the production area in terms of:
>
> (a) land use rights;
>
> b) environmental protection;
>
> (c) forest-related rules, including those relating to forest management and biodiversity conservation, if directly related to logging;
>
> d) rights of third parties;
>
> e) labor rights;
>
> f) human rights protected under international law;
>
> g) the principle of free, prior and informed consent (FPIC), as established, inter alia, in the UN Declaration on the Rights of Indigenous Peoples;
>
> (h) tax, anti-corruption, trade and customs regulations."

Since we are talking about the laws that 'apply' in the production countries, it seems that compliance with international human rights and the FPIC principle is only mandatory if the production country in question has also ratified these rights and principles and/or in national legislation has been implemented. In many production countries, this offers insufficient support, especially for the local population who are dependent on forests. Their rights are often not properly regulated in national legislation. Further research into the protection of human rights was therefore called for in the general review of mid-2028.[72]

---

[69] Article 34 paras 3 and 5 RDFP. See further Sect. 16.4.3 above.

[70] Article 2 sub h EUTR.

[71] Article 2 sub 40 RDFP. The quoted article contains a minor translation error in the Dutch version: in view of the French and English language versions, it should have stated 'the laws that apply'.

[72] Bosscher 2023 and Opinion of the European Economic and Social Committee (EESC) on the proposal for a regulation on the making available on the Union market and the export from the Union of certain commodities and products related to deforestation and forest degradation, OJ C 275, 18.7.2022, pp. 88–94, p. 93, para 4.4. The EESC had previously proposed setting up a public EU rating agency for human and environmental rights: see EESC, Own-initiative opinion on a binding UN treaty on business and human rights, para 1.15, OJ C 97, 24.3.2020, pp. 9–17, p. 10.

In the risk assessment (Article 10 RDFP), market participants are obliged to take into account the presence of indigenous peoples in the country of production, or parts thereof, consultation and cooperation in good faith with these peoples and their claims.[73] They must also take into account human rights violations in the production countries.[74]

Timber products that fall under the FLEGT Regulation and for which a FLEGT license has been issued by the production country are deemed to comply with the legality requirement of Article 3 sub b RDFP.[75] For the time being, this only concerns timber products from Indonesia, as explained above. The products must also be deforestation-free.

### *16.4.7 Due Diligence*

In order to place relevant products on the EU market or export them from the EU, economic operators must exercise *due diligence* on all relevant products supplied by each supplier (Article 8(2)). That due diligence includes, first of all, the collection of the necessary information, data and documents to comply with the information requirements of Article 9, and furthermore the risk assessment measures of Article 10 and the risk mitigation measures of Article 11.

It must be indicated from which country or parts thereof their products come (Article 9(1)(d)), and the geolocation of the plots where the relevant commodities containing or used to manufacture the relevant product were produced. Furthermore, the date or time period of production must be indicated (Article 9(1)(e)) and there must be adequately conclusive and verifiable information that the relevant products are deforestation-free (Article 9(1)(g)). Through a risk assessment, economic operators verify and analyze the collected information and other relevant documentation. On this basis, they have to determine whether there is a risk that the products do not comply with the RDFP. Operators shall not place the relevant products on the market or export them, except where the risk assessment reveals no or only a negligible risk that the relevant products are non-compliant (Article 10, para 1).

The country benchmarking list is to be taken into account (Article 10, para 2, sub a), as well as the prevalence of deforestation or forest degradation (idem, sub f), the degree of corruption, the prevalence of falsification of documents and data, the lack of law enforcement, violations of international human rights, armed conflicts or imposed sanctions (sub h). Account must also be taken of the presence of indigenous peoples in the country of production (sub c), the consultation and cooperation with these indigenous peoples (sub d), and "the existence of duly reasoned claims by

[73] Article 10 para 2 sub c, d and e RDFP. The latter claims must be substantiated on the basis of objective and verifiable information about use or ownership, which is too high a threshold for situations that are often based on custom.

[74] Article 10 para 2 sub h RDFP.

[75] Article 10 para 3 RDFP.

indigenous peoples based on objective and verifiable information regarding the use or ownership of the area used for the purpose of producing the relevant commodity" (sub e). The latter formulation seems to pose a fairly high threshold: after all, indigenous peoples living in forests will often not be able to produce formal property rights.

Finally, additional information on compliance with this Regulation should be taken into account, including information on certification or other third-party audited arrangements. These may be voluntary schemes recognized by the Commission in the context of Article 30(5) RED II, if the information meets the requirements laid down in Article 9 RDFP.

Market participants are required to document and evaluate risk assessments annually. They shall make these evaluations available to the competent authorities upon request. They must be able to demonstrate how the information collected has been tested against the risk assessment criteria and how they have determined the risk level (Article 10(4) RDFP).

If the risk assessment shows that there is no negligible risk of non-conformity of the relevant products, the economic operator shall establish appropriate risk mitigation procedures and measures before placing them on the market to reduce the risk to zero or negligible level (Article 11).

The RDFP's due diligence system and risk approach make it possible for market participants who mainly trade products from low-risk countries to identify and assess risks without too much effort. More efforts will be required for products from high-risk countries. This makes this approach par excellence a flexible and proportional process. The RDFP also includes a simplified due diligence procedure in Article 13 for products from low-risk countries.[76] As already explained in Sect. 16.4.1, there are also exceptions for SME companies; for example, they do not have to provide geolocation data for certain products (Article 4, para 8 and Article 5). Exporters from non-European countries that make it to the low-risk list thus have a definite advantage. That makes it even more important that the criteria used by the European Commission are clear and unequivocal, which they are not completely at the moment.

[76] Under the simplified procedure, operators are not required to fulfil the Articles 10 and 11 obligations where, after having assessed the complexity of the relevant supply chain and the risk of circumvention of the RDFP or the risk of mixing with products of unknown origin or origin in high-risk or standard-risk countries or parts thereof, they have ascertained that all relevant commodities and products have been produced in countries or parts thereof that were classified as low risk. Upon request, the operator shall make available relevant documentation demonstrating that there is a negligible risk of circumvention of this Regulation or of mixing with products of unknown origin or origin in high-risk or standard-risk countries or parts thereof. See also Rice and Kaja 2021, p. 6.

### 16.4.8 Enforcement and Controls

Whether the RDFP, unlike the EUTR and FLEGT, can reduce deforestation will largely depend on the EU member states and the Commission. Will they make sufficient resources available this time, and will the competent authorities monitor efficiently and effectively? And will the Commission this time put more pressure on Member States that fail to effectively enforce European rules?

Compared to the EUTR, the RDFP contains clearer and more detailed provisions on how Member States should enforce and monitor the legislation. Member States must first ensure that competent authorities have sufficient powers, functional independence and the resources to fulfill RDFP obligations.[77] Annually, they must inspect at least 1% of economic operators offering on the market or exporting products from low-risk countries, 3% for imports from standard-risk countries, and as many as 9% of economic operators if the products are coming from high-risk countries.[78] In principle, these checks must take place unannounced.[79] Such rules were necessary given the challenges under the EUTR, where Member States consistently provided insufficient resources in relation to the quantities of timber and timber products imported.[80] Although there are general rules and principles of European law that oblige Member States, among other things, to ensure that EU rules do not lose their effectiveness, in practice this only allows action to be taken against obvious cases—such as the situation in Belgium with less than 1 FTE for the EUTR discussed above in Sect. 16.2.1.

Article 20 RDFP provides that competent authorities may recover from economic operators or traders the costs of their activities related to cases of non-compliance. This includes the costs for storage and for carrying out DNA analyses and chemical tests.

The latter analyses and tests are included in Article 18, which contains further rules for economic operators and non- SME traders. However, these are formulated in too non-binding terms. Controls "may, where necessary" also include "on-site examination of products to verify their compliance with due diligence with the documentation used for due diligence", "all technical and scientific means"[81] appropriate to determine whether the products are deforestation-free, including anatomical, chemical DNA analysis and Earth observation data.[82] To really contribute to improved enforcement, this provision should have read: "if necessary, the checks include…". The final point, moreover, is at the end of Article 18: "if necessary, checks including audits may be carried out on the spot, where appropriate also in third countries,

[77] Article 14 para 4 RDFP.

[78] Article 16 paras 10, 8 and 9 RDFP.

[79] Article 16 para 13 RDFP.

[80] See Sect. 16.2.

[81] Article 18 para 2 sub a RDFP.

[82] Article 18 para 2 sub c and d RDFP.

subject to their consent, through cooperation with the authorities of those third countries."[83] This provision is also far too non-binding. There is also a lack of coherence here with the way in which the control of imports of other products, such as fish, is regulated in the Union. For fish, the European Commission's Directorate General for Health and Food Safety (DG SANTE) "regularly carries out audits in third countries to assess the control and certification system".[84] If this can be arranged for fish, then it is also possible for RDFP products. The desire expressed in Article 30 to arrive at a coordinated approach with producing countries, especially if they pose a high risk within the meaning of Article 29, does little to change this. Article 30 has once again been drafted in very vague terms, requiring the Commission to consider the use of relevant instruments.

Finally, the regulations laid down in Article 19 RDFP for the inspection of SME traders are very limited. These checks include the examination of documents and data proving that they have the details of the economic operators or traders who supplied them with the relevant products and the reference numbers of the due diligence declarations linked to those products, as well as the details of the economic operators or traders to whom they have supplied the relevant products. Furthermore, 'if necessary' checks on SME traders may also include on-the-spot checks, including on-the-spot audits.

### *16.4.9 Sanctions*

The main rules regarding sanctions are contained in Article 25 RDFP. Member States should lay down rules on the penalties applicable to infringements of the RDFP by economic operators and traders, and take all measures necessary to ensure their implementation. Sanctions must be effective, proportionate and dissuasive. These sanctions shall include (subpara (a)) fines proportionate to the environmental damage and to the value of the relevant commodities or relevant products concerned, the level of these fines being calculated in such a way as to ensure that the persons responsible receive the economic benefits they owed to their infringements, are effectively confiscated, and in the event of recurrence of a serious infringement, the fines are gradually increased.

So far, the RDFP and EUTR sanction provisions are still in agreement. What is new, is that it then provides that, in the case of a legal person, the maximum amount of such a fine shall be at least 4% of the total annual turnover throughout the Union of the economic operator or trader in the preceding financial year, and that, if necessary, that maximum amount should be increased to ensure that the fine exceeds the potential economic benefit. Subsection b, like the EUTR, prescribes that provision

[83] Article 18 para 2 sub e RDFP.

[84] NVWA, *Import fish from countries outside the EU 2016-2020.* Available at https://www.nvwa.nl/onderwerpen/visketen-in-beeld/toezicht-op-de-visketen/import-vis-uit-derde-landen. Accessed November 2023.

must be made for the possibility of seizure of the relevant products concerned from the market participant and/or trader. Subsection (c) adds confiscation of income that the market participant and/or trader has acquired from a transaction with the relevant products concerned; the EUTR does not provide for this. Sub d is also new: temporary exclusion for a maximum of twelve months from procurement procedures and from access to public financing, including procurement procedures, subsidies and concessions. Where the EUTR prescribed as a final sanction that provision should be made for "immediate suspension of the permit to engage in commercial activities",[85] the RDFP sub e) prescribes a temporary ban on placing relevant goods and products on the market; offer or implement, in the event of a serious infringement or repeated infringements and (f) a prohibition on exercising the simplified due diligence system of Article 13 in the case of a serious infringement or repeated infringements. The list of possible sanctions under the RDFP is therefore significantly broader than under the EUTR.

An interesting new provision can be found in Article 25 para 3 RDFP. It calls on the Member States to report to the Commission all final judgments against legal entities regarding RDFP infringements and imposed sanctions. It then publishes a list of these judgments on its website, the name of the legal entity; the date of the final judgment; a summary of the activities for which the Regulation was found to have been infringed and, if a financial penalty is involved, the amount thereof. A *naming-and-shaming* list, a public pillory that can contribute well to the effectiveness of the RDFP.

Article 25 RDFP applies without prejudice to the obligations arising from the Environmental Crime Directive 2008/99/EC. In its current form, that directive does not yet contain any rules regarding violations of the EUTR or the RDFP. However, there is a proposal for a new environmental crime directive.[86] It notes that there are major discrepancies between Member States' sanctions regimes applicable to breaches of the EUTR, and that sanctions are often too low to have a deterrent effect on illegal behavior. The differences between sanctions regimes and the lack of uniform implementation in the Member States affect the level playing field: economic operators who strictly comply with the rules are thus disadvantaged. That is why the new directive proposes harmonized criminalization and an approximation of sanction levels across the EU. In addition, according to the Commission, administrative sanctions are clearly inadequate in circumstances such as those that led to the murder of two forest rangers in one Member State who were investigating illegal logging.[87]

The proposal refers not only to the EUTR but also to the RDFP, and in particular regulates the introduction of criminal sanctions for offenses within the scope of Article 3 RDFP (commodities or products may only be placed on the market if they

---

[85] Article 19 para 2 sub c EUTR.

[86] European Commission, *Proposal for a Directive of the EP and of the Council on the protection of the environment through criminal law and replacing Directive 2008/99/EC*, COM(2021)851 of 15 December 2021.

[87] This concerned Romania. Note that in Brazil, hundreds of people who stood up against deforestation were killed (Human Rights Watch 2019 and 2022).

are deforestation-free and legally produced and be accompanied by a declaration of due care).

Among other things, a maximum prison sentence of at least six years should be introduced for this offense and at least ten years if the offense causes or threatens to cause death or serious injury to persons. If it concerns a legal entity, a maximum fine of at least 5% of the total worldwide turnover of the legal entity must be provided (slightly more than the 4% prescribed by the RDFP).

Finally, Article 23 RDFP stipulates that, in the event of violations found, provision must be made for the possibility of provisional measures (for example suspension of placing on the market) and corrective measures (such as withdrawal from the market). To prevent destruction, the RDFP stipulates that products that have been withdrawn from the market are donated to a charity or for purposes of general interest. Only if that is not possible can they be removed (Article 24(2)(c) and (d)).

## *16.4.10 Access to Justice*

If market participants could be held liable for violations of the RDFP, this could make a valuable contribution to the effectiveness of this regulation. The EP, among others,[88] had also insisted on this. However, those recommendations were largely ignored, as becomes clear when looking at the text of the RDFP. It lacks provisions on liability.

Even indigenous peoples who are chased off their land due to illegal deforestation have no possibility under the RDFP to seek compensation from companies that finance such actions and/or then obtain commodities and products from them to put on the EU market. One of the few things that is possible under the RDFP is the submission of "concrete indications" that traders or market participants are not complying with the regulation (Article 31 para 1 RDFP). It is unlikely that an indigenous tribe will be able to determine whether commodities and products are being marketed in the Union from their former habitat, so this provision is of little or no use to them. NGOs could perhaps do useful work in this regard. The competent authorities shall assess the validity of such claims "without delay, carefully and impartially" and take the necessary steps, such as provisional measures, to prevent the placing on the market, made available on the market or export of the relevant products under investigation (Article 31 para 2 in conjunction with article 23 RDFP). The reporters will be informed within 30 days of the steps taken and the reasons for them (Article 31, para 3, RDFP).

The only thing that the RDFP does regulate regarding access to justice is that natural and legal persons who have sufficient interest, including the persons who submitted concrete indications of violations under Article 31 RDFP, must be granted access, pursuant to Article 32 RDFP, to "administrative and judicial procedures to review the decisions, actions or negligence of the competent authorities." Under

[88] See Sect. 16.2.4 above.

the EUTR we saw cases in which the Dutch competent authority (NVWA) was not inclined to take action against importers of illegal timber from Brazil, until an involved NGO managed to persuade a Dutch court to order the NVWA to initiate criminal proceedings.[89] The option described here can therefore come in handy. In addition, it will be important that the European Commission ensures that competent authorities take effective action against violators under the RDFP better than under the EUTR.

Moreover, the Environmental Liability Directive does[90] not provide any relief for the indigenous population and others. Article 3(3) stipulates that private individuals are not entitled to compensation for environmental damage.

For the time being, access to justice and liability for damage to indigenous peoples and others due to deforestation are absent from the new regulation. This is very unfortunate, also because the possibility of submitting complaints about the (in)action of competent authorities of the Member States is not intended or suitable for compensating persons who suffer damage due to violations of the RDFP. The general evaluation of mid-2028 will have to address the effect of the RDFP on indigenous peoples. In the run-up to this, it will be important to put pressure on the European Commission to better protect the position of these peoples by proposing to introduce a liability regime. In this way, what is stated in the very first point of the RDFP preamble would also sound more sincere: "the destruction of forests has serious consequences for the livelihoods of the most vulnerable populations, including indigenous peoples and local communities, who are largely dependent on are of forest ecosystems."

## 16.5 Concluding Remarks and Final Observations

Around 2003, in order to fight deforestation, the EU set out to conclude bilateral agreements with major timber exporting countries under the FLEGT system. Despite years of negotiations with some fifteen timber exporting countries, merely one agreement between the Union and a producing country (Indonesia) exists that is being applied today. From the point of view of guarding the sovereignty of non-EU countries, the bilateral approach might be ideal, but in practice it barely functioned.

By 2013, the EUTR was added to the EU instruments aimed at fighting deforestation for timber from all countries except Indonesia. The regulation followed a more unilateral approach, yet it hardly encroached on the sovereignty of producer countries as the only criterion was the legality of logging—under the law of these producer countries. Producers and traders in non-EU countries did face limited additional costs when they were asked to provide documentation and other evidence of the legality of their timber. The EUTR thus did not meet with a lot of resistance

[89] See Tubbing and Douma.

[90] Directive 2004/35/EC of the European Parliament and of the Council of 21 April 2004 on environmental liability with regard to the prevention and remedying of environmental damage, OJ L 143, 30.4.2004, p. 56.

from non-EU countries. Another factor that could have played a role in this respect is that during the 10 years of its existence, the EUTR functioned very poorly. Export of illegally logged timber products to the Union thus remained possible. In the light of the vast quantities of timber and timber products being put on the EU market, the high price of timber and the number of importers and traders, the EU Member States simply did not make enough resources available to their executive and enforcement authorities to effectively enforce the regulation. The Commission noted this time and time again, and in a few cases even threatened infringement proceedings, but never actually brought an infringement case before the Court of Justice of the European Union. That was a missed opportunity to get EU Member States to better implement and enforce the EUTR. Another reason the EUTR did not work well was its complex due diligence system, aimed at demonstrating that the timber was legally logged. It posed a challenge in particular for importers and EU traders, but also for executive authorities, public prosecutors and judges in the European Union. If some EU member states started to implement and enforce the EUTR a bit better, importers moved their business to other EU member states where less stringent action was taken. Although it is estimated that between 12 and 29% less illegally logged timber was imported into the Union because of the EUTR, imports from countries where the risk of illegal logging is very high increased. All in all, even the European Commission had to agree that it functioned very poorly, hence a new instrument was developed and adopted by the EU legislators in 2023.

The Deforestation-Free Products Regulation (RDFP) will replace the EUTR by the end of 2024. It met with resistance from some countries. That might be partially because the new regulation applies to more commodities (namely cattle, cocoa, coffee, oil palm, rubber, soya and wood) and derived products that cause deforestation. The main reason for the resistance is the introduction of an additional criterion. Besides being legally produced under the laws of the producing countries, a condition that we already know from the EUTR, the commodities and products must not have contributed to deforestation after 31 December 2020. Companies that want to sell them in the European Union must follow due diligence rules and collect specific information from producers, like the geolocation of all plots of land where the relevant product was produced. They shall not place commodities and products covered by the RDFP on the EU market or export them, except where a risk assessment reveals that there is no or only a negligible risk that the relevant products are non-compliant. As these norms are unilaterally formulated by the European Union, the RDFP has been described as a unilateral encroachment on the sovereignty of producing countries. The use of the country benchmark system (dividing the world in countries and regions with low, standard and high risks of deforestation) has been described as discriminatory, but it was argued that with some tweaks, the system could work as a necessary and proportionate aspect of the RDFP. The same seems to hold true for the system as such, as it can contribute to the fight against deforestation and climate change in the interest of the European Union, the producing countries and others in a fairly proportionate manner. While also criticizing certain aspects of the new EU rules, Brazilian researchers have pointed out that an alignment exists between Brazilian environmental policies and the RDFP: "this regulation can be seen as an opportunity for sustainable Brazilian agricultural production to gain scale and a greater share of the European market". To this end, they recommend working toward the effective implementation of applicable Brazilian law, expand policies that help combat deforestation, and develop national systems for commodity chain monitoring and traceability.[91]

This research has also looked at some of the RDFP's shortcomings from the point of view of its scope. Some of these shortcomings were deferred to future decision-making by specifying that the Commission will prepare impact assessments and, if necessary, legislative or other proposals. This includes important topics such as rules for EU companies that finance deforestation (without which applying the RDFP will remain mopping with the tap open), a liability regime for people who suffer damages from deforestation (for instance indigenous people living in and/or using forests), and protecting not only forests but also other valuable ecosystems such as savannas, peatlands and wetlands (important also from the point of view of fighting climate change).

The RDFP includes several clear improvements compared to the EUTR, from the point of view of protection of forests and the fight against climate change. It is no longer just about the legality of logging, but also about the commodities and products being deforestation-free. Another improvement is that there are clearer rules on the minimum amount of controls in the European Union, and on various sanctions. However, the question remains whether the improvements will make the new law more effective than the EUTR. The deforestation-free norms mean that companies that want to sell specific commodities and derived products in the Union are obliged to ensure that these did not contribute to deforestation in the countries where they were produced. After having established whether these commodities and products stem from low, standard or high risk countries or regions through the benchmark system, the Union due diligence norms set out how companies that put them on the EU market need to carry out risk assessments demonstrating that these particular commodities and products in all likelihood did not contribute to deforestation.

A huge expansion of the scope of products has been chosen compared to the EUTR. The RDFP applies not only to wood and timber products, but also to numerous other commodities and products produced from them. The application and enforcement of the RDFP could pose many challenges in practice because of the enormous amounts of information that will need to be gathered. In addition, the RDFP has some counterproductive exceptions, such as the possibility of not having to comply with the due diligence and risk assessment provisions when it concerns products in low-risk countries, and the exceptions for SME companies. And an important factor in deforestation, namely the financial sector and investors, is not (yet) regulated, which amounts to mopping with the tap open. So, there is still a lot of room for improvement in this respect.

Twenty years of EU policy on deforestation has made little difference. The laws on paper did not function very well in practice. That is a rather sad conclusion.

[91] Leme Lopes et al. 2023, p. 12. The authors describe in detail which systems for tracking commodities are already in place in some parts of Brazil.

Now there is a new EU regulation. Whether it will be more effective in combating deforestation depends on the Member States and the Commission, on the one hand. They will need to take the RDFP more seriously than the old rules, and step up controls and enforcement. On the other hand, foreign producers and governments need to be willing and able to meet the demands of EU companies. Under the new rules, the latter are obliged to ensure they know how, where and when products were produced if they want to bring them on the EU market. In that manner, it is to be demonstrated that the risk that these products caused deforestation is either absent or negligible. If these actors pull out all the stops, the RDFP could stand a chance to be a success in combating deforestation worldwide and thus become not just another law on paper, but also one that delivers in practice. Aiding foreign producers and governments in their efforts to have laws protecting forests, its inhabitants and users in place, and effectively enforcing these laws, would be steps in the right direction. However, the experiences with the FLEGT and the EUTR might also form an indication that the 2028 evaluation of the new RDFP, some 25 years after the start of an EU policy against deforestation, will show that the new regulation, applicable to so many products, is not (yet) working properly. Hopefully that will not be the case: knock on wood.

## References


Argentina, Brazil, Colombia, Ghana, Guatemala, Indonesia, Ivory Coast, Nigeria, Paraguay, Peru, Honduras, Malaysia, Ecuador and Bolivia (2022), *Joint letter regarding European Union proposal for a regulation on deforestation-free products*, 28 November 2022 Available at: https://docs.wto.org/dol2fe/Pages/SS/directdoc.aspx?filename=q:/G/AG/GEN213.pdf&Open=True Accessed November 2023

Argentina, Brazil, Bolivia, Colombia, Dominican Republic, Ecuador, Ghana, Guatemala, Honduras, Indonesia, Ivory Coast, Malaysia, Mexico, Nigeria, Paraguay, Peru and Thailand (2023), *Carta de países em desenvolvimento a autoridades europeias sobre a entrada em vigor da chamada "lei antidesmatamento" da União Europeia (Letter of developing countries to the European authorities about the entry into force of the "anti-deforestation law" of the European Union)*, 7 September 2023 Available at: https://www.gov.br/mre/pt-br/canais_atendimento/imprensa/notas-a-imprensa/carta-de-paises-em-desenvolvimento-a-autoridades-europeias-sobre-a-entrada-em-vigor-da-chamada-201clei-antidesmatamento201d-da-uniao-europeia Accessed November 2023

Bazelmans JM and Douma WTh (2020), *Naar een effectievere aanpak van wereldwijde ontbossing. Een verkenning vanuit nationaal, Europees en internationaalrechtelijk perspectief*, Milieu & Recht 2020, nr. 10, p. 718-726, MenR 2020/101

Been J and Groot J (2023), *Rabobank en de ontbossers* (*Rabobank and the deforesters*), Financieel Dagblad, 3 November 2023

Bosscher A (2023), *Factsheet Europese Ontbossingsverordening (Factsheet European Deforestation Regulation)*, Greenpeace, 26 June 2023 Available at https://www.greenpeace.org/nl/klimaatverandering/59239/factsheet-europese-ontbossingsverordening/ Accessed November 2023

Civillini M (2023), *Amazon nations decry EU deforestation rules in thinly-veiled joint condemnation,* Climate Home News, 10 August 2023 Available at https://www.climatechangenews.com/2023/08/10/amazon-nations-decry-eu-deforestation-rules-in-thinly-veiled-joint-condemnation/ Accessed November 2023

Dobson N (2022), *De EU als bossenwaakhond: van vrijwillige partnerschappen naar autonome standaarden*, Ars Aequi July/August 2022, p. 534-543 Available at: https://arsaequi.nl/product/de-eu-als-bossenwaakhond-van-vrijwillige-partnerschappen-naar-autonome-standaarden/ Accessed November 2023.

Douma WTh (2018), *Squaring the circular economy: towards more coherence in the EU sustainable management of natural resources*, in: H.T. Anker, B.E. Olsen (eds), Sustainable management of natural resources, Intersentia, 2018, p. 29-48

Douma WTh and Kooij MM (2014), *Juridische aspecten van de strijd tegen illegaal gekapt hout op mondiaal niveau, in de EU en in de VS (Legal aspects of the fight against illegally harvested timber at a global level),* in: Bazelmans JM, Douma WTh et al. (eds.), Duurzame handel in juridisch perspectief (Sustainable trade in a legal perspective), VMR 2014-1, p. 181-202 Available at: https://www.milieurecht.nl/publicatie/49/duurzame-handel-in-juridisch-perspectief Accessed November 2023

Douma WTh and Tubbing AMCC, *Ontwikkelingen in de aanpak van fout hout: de actuele stand van zaken rond EUTR en FLEGT* (*Developments in tackling illegal wood: the current state of affairs regarding the EUTR and FLEGT*), Tijdschrift Milieu & Recht 2017, nr. 10, p. 900-911

Ekker H (2017), *Nederlands veen stoot steeds meer CO2 uit, bijna zo veel als een kolencentrale (Dutch peat emits more and more CO2, almost as much as a coal-fired power plant)*, NOS, 23 August 2017 Available at: https://nos.nl/artikel/2189420-nederlands-veen-stoot-steeds-meer-co2-uit-bijna-zo-veel-als-een-kolencentrale Accessed November 2023

European Commission (2019), *Impact assessment study for the revision of the product scope of the EU timber regulation: final report*, Publications Office, 2019 Available at https://op.europa.eu/en/publication-detail/-/publication/fd26ad03-9895-11e9-b2f2-01aa75ed71a1/language-en#:~:text=The%20EUTR%20defines%20its%20product,consequent%20impacts%20from%20such%20revisions Accessed November 2023

European Commission (2021), Commission Staff Working Document, *Impact Assessment minimising the risk of deforestation and forest degradation associated with products placed on the EU market*, SWD(2021)326, part 1, Brussel, 17 November 2021 Available at https://environment.ec.europa.eu/systexm/files/2021-11/SWD_2021_326_1_EN_impact_assessment_part1_v4.pdf Accessed November 2023

European Commission (2023), EU proposal for a Joint Instrument tabled to Mercosur in the context of discussions to finalise the EU Mercosur agreement, March 2023 Available at https://policy.trade.ec.europa.eu/eu-trade-relationships-country-and-region/countries-and-regions/mercosur/eu-mercosur-agreement/documents_en Accessed November 2023

Forest Stewardship Council (2022), *FSC's position and 10 recommendations to improve the new EU rules for deforestation-free products*, 26 April 2022 Available at: https://fsc.org/en/newscentre/general-news/fscs-position-and-10-recommendations-to-improve-the-new-eu-rules-for Accessed November 2023

Hedemann-Robinson M (2022), *Legislative pioneer on global forest protection? A commentary on the European Union's proposed Deforestation-Free Product Regulation*, 31 EEELR, 348-368

Human Rights Watch (2019), *Rainforest mafias : how violence and impunity fuel deforestation in Brazil's Amazon* Available at https://www.hrw.org/report/2019/09/17/rainforest-mafias/how-violence-and-impunity-fuel-deforestation-brazils-amazon Accessed November 2023

Human Right Watch (2022), *Crisis in the Brazilian Amazon*, Available at https://www.hrw.org/news/2022/04/19/crisis-brazilian-amazon#_edn20 Accessed November 2023

Human Rights Watch (2023), *EU: major step for 'deforestation-free' trade*, 19 April 2023 Available at: https://www.hrw.org/news/2023/04/19/eu-major-step-deforestation-free-trade Accessed November 2023

IUCN, *Peatland and climate change*, Issues letter November 2021 Available at https://www.iucn.org/sites/default/files/2022-04/iucn_issues_brief_peatlands_and_climate_change_final_nov21.pdf Accessed November 2023

Kuchlmayr F, Langhans K et al (2023), *Wie Holzräuber die ältesten Wälder Europas zerstören (How timber thieves are destroying the oldest European forests)*, Der Spiegel, nr. 10, March 1, 2023

Kuipers K (2018), *Duurzaamheid is slechts een verhaaltje* (*Sustainability is just a story*), De Groene Amsterdammer, nr. 17, 26 April 2018 Available at: https://www.groene.nl/artikel/duurzaamheid-is-slechts-een-verhaaltje Accessed November 2023

Leme Lopes C, Chiavari J and Segovia ME (2023), Brazilian environmental policies and the new European Union Regulation for Deforestation-Free Products: opportunities and challenges, Climate Policy Initiative (CPI) Available at: https://www.climatepolicyinitiative.org/publication/brazilian-environmental-policies-and-the-new-european-union-regulation-for-deforestation-free-products-opportunities-and-challenges/#_ftn13 Accessed November 2023

Łuszczek R, Mattioli F and Quirke J (2021), *The proposed EU law on deforestation-free products. What are the changes compared to the EUTR framework?, ClientEarth* Available at: https://www.clientearth.org/latest/documents/the-proposed-eu-law-on-deforestation-free-products-how-does-it-compare-to-the-eutr-framework/ Accessed November 2023

Marques da Silva I, *Why the global South is against the EU's anti-deforestation law*, Euronews, 20 September 2023 Available at: https://www.euronews.com/my-europe/2023/09/20/why-the-global-south-is-against-the-eus-anti-deforestation-law Accessed November 2023

Rice M and Blumtritt K (2021), *The proposed EU law on deforestation-free products. What is in the European Commission's proposal and what is left out?*, ClientEarth Available at: https://www.clientearth.org/latest/documents/the-proposed-eu-law-on-deforestation-free-products-what-does-it-include-and-what-is-left-out/ Accessed November 2023

Pontecorvo CM (2022), *The Proposed EU Regulation on Trade in Forest-Risk Commodities (FRCs): A First Assessment,* European Yearbook of International Economic Law, p. 507-549

Purnomo H, Puspitaloka B et al (2022), *Adapting global palm oil deforestation - free trade to benefit local economies and landscapes,* CIFOR-ICRAF Info Brief No. 369, August 2022 Available at: https://www.cifor.org/knowledge/publication/8636/ Accessed November 2023

Rajão R, Loconto A and Gautreau P (2020), *The limits of private certifications and the potential of state-led spatial data infrastructure in South America as to ensure deforestation-free exports to the EU*, feedback F509193, 18 March 2020 Available at: https://ec.europa.eu/info/law/better-regulation/have-your-say/initiatives/12137-Deforestation-and-forest-degradation-reducing-the-impact-of-products-placed-on-the-EU-market/F509193_en Accessed November 2023

Santander Trade Markets, *Brazil: foreign investment*, October 2023 Available at: https://santandertrade.com/en/portal/establish-overseas/brazil/foreign-investment#:~:text=The%20main%20investing%20countries %20in,France%2C%20Spain%2C%20and%20Canada. Accessed November 2023

Segara V (2022), *Unintended consequences of Europe's fight against deforestation*, EIAS, 30 September 2022 Available at: https://eias.org/publications/op-ed/unintended-consequences-of-europes-fight-against-deforestation/. Accessed November 2023

**Wybe Th. Douma** is an independent legal advisor at European Environmental Law Consultancy and senior legal advisor at the Dutch Ministry of Social Affairs and Employment. This contribution was written in his personal capacity.

# Part IV
# Case Studies in EU External Relations

# Chapter 17
# International Trade and the Mitigation of Environmental Damage: The Experience of the EU-Mercosur Free Trade Agreement

**Tiago Matsuoka Megale**

## Contents




**Abstract** The interdependence between trade and environment was concretized in Trade and Sustainable Development (TSD) chapters of European Free Trade Agreements (FTAs). This Chapter analyzes the consequences of these rules. The focus falls on the sustainability provisions of the EU-Mercosur FTA that was chosen due to current objections to the implementation of the agreement for environmental reasons and to its sustainability provisions' potential effects to alleviate the Brazilian environmental crisis. This crisis constitutes the identified research problem and is characterized by deforestation of the Amazon forest and fires in Cerrado and Pantanal. The main question addressed in this Chapter concerns whether the European concerns around the approval of the EU-Mercosur FTA and its possible implementation can contribute to mitigate the current environmental damages. The Chapter firstly identifies elements of the Brazilian environmental crisis and warnings in the sustainability impact assessment (SIA) of the agreement, a support material to EU decision-making, about the environmental risks regarding its ratification. Recently, the hindrance on



T. Matsuoka Megale (✉)
University of São Paulo, Ribeirão Preto Street, 506, São Paulo, Brazil
e-mail: tiago.megale@fgv.br; tiago.megale@usp.br





R. A. Wessel et al. (eds.), *EU External Relations Law and Sustainability*, Global Europe: Legal and Policy Issues of the EU's External Action 4,
https://doi.org/10.1007/978-94-6265-655-0_17

its approval procedure in the European Parliament, derived from the destruction of natural resources in Brazilian biomes, demonstrates that the European foreign policy is guided by ethical objectives. The Chapter then analyzes key general elements of the TSD chapter of the agreement that, when into force, limit the right to regulate, and protect environmental and social standards and can allow their effective implementation. Beyond their explanation, the analysis focuses on the issue of climate change that can be further mitigated with the compliance of obligations foreseen in the TSD chapter.

**Keywords** EU-Mercosur FTA · Deforestation · EU-Mercosur FTA SIA · EU-Mercosur FTA approval · TSD chapter · Climate change

## 17.1 Introduction

The bilateral Free Trade Agreements (FTAs) concluded by the EU during the last decade, aimed to deal with environmental concerns through the incorporation of dedicated Trade and Sustainable Development (TSD) chapters that address environmental and social issues. However, so far, no FTA seems to have provided the kind of integration of environmental issues that would be fit for purpose. These deficiencies have already engendered a significant discomfort in EU civil society concerning EU trade agreements and their role in promoting sustainable development.

Among the EU FTAs under negotiation or concluded, this paper analyzes the EU-Mercosur FTA. This agreement will allow the removal of trade barriers to both sides, deeper integration of value chains between the two regions and the projection of sustainable values. The agreement was particularly chosen as a case study in view of the European objections to the entry into force of the agreement for environmental reasons, and to potential effects of sustainability provisions of the agreement to alleviate the current Brazilian deep environmental crisis. This crisis constitutes the research problem that was identified and is characterized by the deforestation of the Amazon forest by miners, loggers and squatters, the invasion of Amazonian indigenous lands and the fires in Cerrado, Pantanal and Pampa. This Chapter aims then to answer the following question: can the European concerns around the approval of the EU-Mercosur FTA and its possible implementation contribute to mitigate the current environmental damages?

Questions raised by many EU actors such as the national parliaments of Member States, the European Parliament, as well as, by the EU civil society composed by business clusters, human rights activists and those concerned with the continuation after the treaty ratification of intrinsic economic inequalities, which have been a feature of economic relations between Europe and South America. They target the EU-Mercosur FTA due to felled forest in Brazil that enters the EU market and the non-compliance by Brazil of its commitments to reduce deforestation in the Amazon. As the main hurdle to the entry into force of the EU-Mercosur FTA are the European environmental concerns that arise from the Brazilian environmental crisis analyzed below, the focus of this chapter will fall on the environmental dimension of the concept of sustainable development. For the purposes of this chapter, the questioning by the EU civil society and the media of unfavorable environmental conditions in Brazil will be examined.

The Chapter starts with the identification of elements of the Brazilian environmental crisis and warnings that were released in the international civil society and in the sustainability impact assessment (SIA) of the agreement, and in support material underlying the EU's decision-making about the environmental risks regarding the agreement's ratification. At the moment of writing, the approval procedure of the agreement in the European Parliament faces grave hindrance that arises from the destruction of natural resources in the Amazon forest and in neighboring biomes, Cerrado and Pantanal. The aforementioned facts and factors demonstrate that EU trade diplomacy constitutes an instrument to stimulate environmental protection by making the ratification of the agreement conditional to the termination of unacceptable levels of deforestation in Brazil. The Chapter then moves to the analysis of key general elements of the TSD chapter in the draft agreement.

## 17.2 Warnings about the Ratification of the EU-Mercosur Free Trade Agreement

This section analyzes the main obstacles to the ratification of the agreement by all contracting parties. These are found in the Brazilian environmental crisis, in the 2020 SIA of the EU-Mercosur FTA, and in the EU's trade policy, which is increasingly founded on the central pillar of sustainability.

### *17.2.1 The Brazilian Environmental Crisis*

The deforestation probabilities in the Amazon rainforest (Amazon) correlate with biophysical factors, infrastructure, and policy decisions.[1] For instance, in areas of high precipitation, a lower risk of deforestation is verified. In the same sense, forests where the soils are classified as "severe constraints to agriculture" were considerably less prone to be deforested than areas that present soils with other features. According to the land use categories, most of the recent deforestation in the Amazon rainforest has occurred on non-designated public lands, private lands, and within rural settlements.[2] There was a higher propensity by landowners to deforest settlement projects as well as areas near roads than places distant from roads. Another

[1] Arima and Barreto 2020, p. 39.

[2] Soendergaard et al. 2021, p. 22.

fact to be highlighted is the higher degree of deforestation per unit of land in areas where government-backed land reform projects take place.[3]

The degree of deforestation on the aforementioned areas is provided by the most recent data gathered by the Brazilian National Institute for Space Research (INPE).[4] This institute revealed that approximately 17% of the vegetation in the Amazon biome has already been deforested. Almost half of the deforested spaces in the Brazilian Amazon were located in the state of Pará (46.3%), which also accounted for the largest portion in the period between 2008 and 2020 (44.2%). Mato Grosso is the state that faced the second highest deforestation rates, accounting for 17.5% in 2020 and 18.8% in the same period. Amazonas (13.3%) and Rondonia (12.6%) are the states that stand as the third and fourth main contributors to deforestation in 2020 and that, respectively, were responsible for 11.2% and 13.7% in the same period mentioned above. In terms of development in deforestation rates from 2008 to 2020, the highest growth is identified in Amazonas (6%) and Acre (5.1%). These numbers give rise to concerns as deforestation rates have been increasing again in the last years and reached a new peak in 2020. More recent data gathered by the INPE satellite Deter[5] reveal that, in October 2022, the deforested area was of 904 $km^2$, an area 3% higher than the one identified in the same month of 2021. Moreover, according to additional data collected by the satellite Deter, the accumulated deforestation from January 2022, to October 2022 was 9494 $km^2$, the highest area since the start of operation of a new version of the referred satellite in 2015.

Besides deforestation, the Amazon was punished with fires that also reached the Cerrado, the Pantanal and the Pampa. Data provided by INPE reveal that, in the period from January 2019 to August 19, 2019, a sharp growth in the number of fires was identified in Acre, Mato Grosso do Sul, Roraima and Rondonia. The data gathered by INPE are reflected in the fact of the "Day of the Fire"—a protest of farmers on August 10, 2019, which resulted in the increase of the number of fires in many municipalities. More recent data from MapBiomas[6] reveal that half of the burned area in 2022 was registered in September and corresponds to 5,825,520 ha. The most affected biome is the Cerrado, with 2,973,443 ha lost in the year, an area that is 13% higher than the one burned in the biome in the same month in 2021. The Amazon had the biggest increase in burned area compared to September last

---

[3] Arima and Barreto 2020, p. 39.

[4] INPE 2020.

[5] The INPE satellite Deter, as explained by the Brazilian National Institute for Space Research, provides a rapid survey of evidence of changes in forest cover in the Amazon. It consists on an alert system to support the inspection and control of deforestation and forest degradation carried out by the Brazilian Institute of the Environment and Renewable Natural Resources (IBAMA) and other bodies linked to this topic. The results of detections, by class, state, municipality and federal conservation units, are available via spatial query on the TerraBrasilis portal that can be accessed on the link http://terrabrasilis.dpi.inpe.br.

[6] MapBiomas constitutes a collaborative network formed by NGOs, universities and technology start-ups. The initiative produces annual mapping of land cover and use and monitors water surface and fire scars monthly with data from 1985. It also validates and prepares reports for each deforestation event detected in Brazil since January 2019.

year: 71% or additional 1,080,388 ha. In the first nine months of this year, a total of 11,749,938 ha, located mainly in the agricultural frontier, were burned in the country, an area 1% larger than that affected in the same period in 2021. In the last period, exponential growth rates were identified in the Amazon (40%) and in the Pampa (1500%). Pantanal, on the other hand, had the lowest area burned in the last four years, with an 83% reduction in the comparison between the period from January to September 2022 with the previous year.

Sanctions proved to be insufficient and contributed to the continuity of the environmental crisis. According to data provided by Ibama, in the period from January 1, 2019 until August 23, 2019, when considering all types of environmental infringement in the country, the institution reduced the assessments by 29.4%. More detailed figures reveal that, from the beginning of 2018 until August, 23 of the same year, Ibama applied 9771 fines of all kinds that are relative, for instance, to crimes against flora, illegal fishing, hunting and biopiracy. In the same period of 2019, the number of fines dropped to 6895. In crimes against flora, the reduction was of 38.7% in the same period: from 4138 fines in 2018 to 2535 in 2019. According to civil servants, former civil servants, authorities and environmentalists, the drop in the number of fines is linked to signals issued by the federal government since the beginning of the year against alleged excesses in inspections and changes in professionals at key Ibama posts. The attorney Luís de Camões Boaventura from the municipality of Santarém located in Para also claimed in a press interview that a preliminary survey carried out by his team shows that the few notices of infraction that have reached the Public Ministry concern administrative matters and that Ibama inspectors have not been in the field.[7]

The scenario of impunity relative to the illegal acts practiced against the Brazilian environment allows the increase of grave forms of criminality. One form was identified in the data from the NGO Global Witness that revealed that Brazil is the leading country in murders of environmentalists in the last decade from 2012 to 2021. According to the survey, more than 85% of the murders in the period happened in the Amazon and the majority of the victims were indigenous or black. Both are gravely under threat due to land dispute over territories that were previously occupied by them, as several cases collected over the years by the Pastoral Land Commission demonstrate.

The results of the Amazon deforestation, and the fires on different biomes, reach the national, regional and global level. The effects identified are not watertight and the ones originally identified in one level can provoke consequences on another level. In the first scenario that comprises mainly the national Brazilian territory, very negative consequences to the deforested environment and to the economic activities performed on it were identified. In a scenario in which there is a greater number of deforested areas, an initial effect is the impossibility of rainwater capture by vegetation, which leads to a reduction of the recycling of water in the region, the biome, and what remains of the forest. An additional effect is the change of most of the deforested Amazon biome, which has either been abandoned or turned into low productivity

---

[7] Shalders 2019.

pasture and unproductive economic activities, except for soybean planting, marked by a productivity level close to the national average.[8] The last scientifically demonstrated effect of the Brazilian Amazon deforestation is the reduction of the rainfall in the forest. Forest conversion to pasture and croplands affects moisture cycling and energy balance and may change rainfall patterns.[9] Hence, the regional climate could respond as much as or even stronger to land use and cover change than to global warming.[10] Forest loss impacts rainfall in different manners that vary according to the geographical scale. At the original spatial resolution of the Tropical Rainfall Measuring Mission (TRMM)[11] data, there is a dual effect of forest loss, with rainfall increasing with forest loss up to around 58% and then decreasing afterwards.[12] The gain before the critical threshold is of $+ 22.1 \pm 9.3$ mm of precipitation per 10% of additional forest loss.[13] However, beyond 57.5% of forest loss, precipitation reduces precipitously, with each 10% of additional forest loss inducing a net reduction in annual rainfall of $- 49.2 \pm 11.3$ mm.[14]

With regard to the second scenario that comprises the Amazon forest within the Brazilian territory and in eight other countries,[15] the literature identified the consequences of the aforementioned biophysical changes. They are regional warming and drying, changes in regional atmospheric circulation and moisture convergence and longer-range teleconnections.[16] Warming and drying promote wildfire drought, which greatly increases the risk of regional forest dieback and the associated loss of the aboveground biomass.[17] An ensuing climate-tipping point, once triggered, may cause local ecosystems to move toward an alternate stable state, in which grass plant functional types are dominant, wildfires are prevalent, and carbon stocks are considerably reduced.[18]

Regarding the third scenario in which Brazil is placed alongside other countries, the factual results explained above imply an increase of Brazilian GHG emissions that contribute to the worsening of climate change. Comparing the Brazilian GHG emissions profile to the main global emission sources, agriculture and land use changes represent a much larger share than the global average, which is close to 20%.[19] In

---

[8] Soendergaard et al. 2021, p. 22.

[9] Leite-Filho et al. 2021, p. 2.

[10] Ibid.

[11] The Tropical Rainfall Measuring Mission (TRMM) constitutes a joint space mission between NASA and Japan's National Space Development Agency created to monitor and analyze tropical and subtropical precipitation and the associated release of energy.

[12] Leite-Filho et al. 2021, pp. 2–3.

[13] Leite-Filho et al. 2021, p. 3.

[14] Ibid.

[15] For the sake of clarity, the eight other countries are Bolivia, Colombia, Ecuador, French Guiana, Guiana, Peru, Suriname and Venezuela.

[16] Li et al. 2022, p. 2.

[17] Ibid.

[18] Ibid.

[19] Soendergaard et al. 2021, p. 23.

2019, Brazil was the 6th largest GHG emitter, with an amount of 2.2 gigatons of $CO_2$ equivalent annually, behind China, USA, India, Russia and Indonesia and its emissions were mainly due to harmful land-use practices and outdated production models in parts of its agricultural and livestock sector.[20]

In a prospective analysis of the Brazilian environmental crisis, the literature also identified a scenario with the areas of the Brazilian Amazon more susceptible to deforestation that arises from the implementation of the EU-Mercosur FTA. The states where deforestation is more likely to occur are Pará (39.9%), Rondonia (32.6%) and Mato Grosso (25.2%) and the remaining areas of predicted deforestation are in Amazonas and Maranhão.[21] Small isolated points that deserve to be mentioned are the new deforestation border in southern Amazonas state east of Humaitá close to the Transamazon road. The EU-Mercosur FTA would also add the risk of deforestation in the vicinity of indigenous lands and conservation units.[22]

### *17.2.2 The Small Pros and Dangerous Cons of the Agreement According to the 2020 SIA*

For the sake of clarity, SIAs are activities conducted by external consultants in a neutral and unbiased manner. They begin as soon as the Council of the EU formally permits the Commission to participate in trade negotiations, allowing the SIA to contribute meaningfully to the negotiation process at key moments. Their central objective is to guarantee that the resulting trade agreements respect high social, economic and environmental standards as well as human rights. They also aim to contribute to the transparency of the analysis and of the ongoing trade negotiations by providing stakeholders with comprehensive information on the possible impacts of the agreement.[23] Stakeholders are also allowed to comment on the information received and to provide their input. The negotiators are also targeted as they are informed in due time of consultation activities and then have sufficient time to provide their input and contributions.

The SIA, relative to the EU-Mercosur FTA, contains a specific chapter that provides a baseline analysis of environmental policies in Mercosur countries, and an analysis of the environmental impacts arising from the association agreement (AA). This chapter was partly built on the features of the Brazilian environmental crisis previously analyzed. The baseline analysis reveals that Brazil outperforms other Mercosur and upper-middle-income countries in terms of adoption of climate change policies.[24] Deforestation, however, remains a concern in Mercosur countries,

---

[20] Ibid.

[21] Arima and Barreto 2020, p. 39.

[22] Ibid.

[23] European Commission 2016.

[24] London School of Economics and Political Science 2020, p. 103.

with the exception of Uruguay.[25] While the situation improved during the first part of the twenty-first century, recent trends suggest a resurgence of deforestation, which is what happens in Brazil, where, from 2014, Brazilian deforestation rates began to rise due to a combination of worse economic conditions and deteriorating commitments to environmental regulation and enforcement.[26]

The analysis of the environmental impacts of the AA, in its turn, shows a negligible impact on global greenhouse gasses (GHG) emissions.[27] Two issues of moderate concern are also emphasized. First, the expected expansion of the agricultural and animal sectors poses some moderate concerns regarding the increased use and contamination of water resources, if appropriate management practices are not implemented. This is particularly concerning given the observed rise in the use of pesticides and the absence of adequate price incentives to encourage efficient use of pesticides, fertilizers and water in agriculture.[28] Secondly, some moderate concerns are raised in terms of the impact of the AA on deforestation, particularly in Brazil, if the policy environment that allowed past reductions in deforestation is not maintained and any expansion of the agriculture and animal sectors are met by an increase in forest clearing instead of by increases in productivity and the conversion of existing low-efficiency meadows and pasturelands.[29] On the other hand, there is the expectation of some positive effects since the AA is likely to strengthen the parties' commitments in the Paris Agreement, to contribute to increasing trade in environmental goods and services, and stimulate international cooperation for the development of green technology and the protection of natural resources, such as fisheries.[30]

In order to address the two concerns above explained, the SIA makes some policy recommendations. Mercosur countries should transform existing degraded pasture areas into land designed for sustainable agriculture to avoid the destruction and degradation of forest land to attain the expected increase of agricultural production. Brazil should ameliorate anti-deforestation legal measures and law implementation activities to detect illegal logging and increase monitoring along the supply chain by means of the renovation of the policy environment that led to the decrease in deforestation verified up to 2012. Fruitful measures that were implemented in the past encompass the "Soy Moratorium", initiative that aims to ensure that soy, produced in the Amazon biome, is free from deforestation that occurred after July 22, 2008, as well as the wider anti-deforestation policies performed by the Ministry of the

[25] Ibid., p. 103.
[26] Ibid., p. 103.
[27] Ibid., p. 103.
[28] Ibid., pp. 103–104.
[29] Ibid., p. 104.
[30] Ibid., p. 104.

Environment in the first ten years of the twenty-first century.[31] Brazil should stimulate private sector agents to lengthen the Soy Moratorium to the Cerrado and to increase the effectiveness of the Beef Moratorium, a set of agreements between the Federal Public Ministry and more than 100 farm and slaughterhouse owners to the environmental regularization of the sector, by, for instance, enlarging monitoring to every property in the supply chain. The government should increase its investments in Ibama to restock its labor force and reaffirm its authority over auditing that takes the form of inspections. The government should, moreover, apply the available data on illegal logging, regularly gathered through satellite imagery, to target law enforcement measures.

The SIA also provides policy recommendations on environmental themes foreseen in the TSD chapter of the EU-Mercosur FTA targeted at Mercosur countries. On the issue of climate change, Mercosur and the EU should accomplish their Paris Agreement commitments and attain their GHG emissions targets, as foreseen on their nationally determined contributions. In regards to biodiversity, Mercosur countries should adopt National Diversity Strategies and Action Plans beneath the Convention of Biological Diversity and guarantee the protection of land rights mainly among indigenous communities to transform local stakeholders into a full member that acts for the biodiversity protection. On the topic of sustainable management of fisheries and aquaculture, the EU, Brazil and Argentina should maintain the engagement in the All-Atlantic Ocean Research Community in order to foment the sustainable management of the Atlantic Ocean. Key areas of cooperation within the community include responsible and sustainable fisheries management, improvements in aquaculture, and addressing pollutants that are becoming increasingly apparent or prominent.

However, the timing of the release of the SIA related to the EU-Mercosur FTA was subject to criticism. The first draft of the SIA draft interim report of the Mercosur agreement was not published until four months after the official announcement of the end of the negotiations, resulting in a formal complaint in 2020 by five civil society organizations.[32] These five civil society organizations revealed concern that the Commission undertook trade negotiations that lacked an up-to-date assessment of its possible environmental impact. After this complaint, an inquiry by European Ombudsman Emily O'Reilly has identified that the European Commission should have finished an updated SIA before the EU-Mercosur trade agreement was concluded and that its inability to act in this way was due to poor management. On

[31] Examples of such policies are the Action Plan for Prevention and Control of Deforestation in the Legal Amazon and the Sustainable Amazon Plan. The first one was launched in 2004 when Marina Silva was Minister of the Environment of the first government of President Luiz Inácio Lula da Silva. The action plan sought to continuously reduce deforestation and create the conditions for the transition to a sustainable development model in the Legal Amazon. The second one was launched in 2008 when Marina Silva was Minister of the Environment of the second government of President Luiz Inácio Lula da Silva. The plan gathers a set of strategic guidelines, drawn up based on an updated diagnosis of the contemporary Amazon and its challenges, to reference government policies, private investments and meeting social demands. The plan aimed to become an institutional locus capable of providing long-term answers to the challenges in the region.

[32] Lamy et al. 2021, p. 12.

the other hand, despite the unavailability of the final report of the SIA at the time of the conclusion of the negotiations, the Commission services considered that the SIA process offered a solid platform for engagement with civil society and has duly and extensively informed the negotiators.[33]

### *17.2.3 The Features of European Foreign Policy*

The facts of the Brazilian environmental crisis and the aforementioned warnings raised against the implementation of the EU-Mercosur FTA due to poor environmental performance of Brazil contradict core aspects of European foreign policy. As established in Article 21(2) TEU and further explained below, the Union shall define and pursue common policies and actions, and shall work for a high degree of cooperation in order to ensure sustainable development. On the economic field, in its "Trade for All" strategy, the EU expressed its will to incorporate sustainable development considerations in the TSD chapter of trade negotiating texts and in all relevant areas of FTAs such as energy, raw materials or public procurement provisions.[34]

The legal foundations of European foreign policy are found on the Treaty on European Union (TEU) and the Treaty on the Functioning of the European Union (TFEU). The first treaty establishes on its Article 21 the EU commitment to protect and promote human rights during the development and implementation of its foreign policy. The second treaty, on its Article 207, requires the implementation of the EU external trade policy, also known as common commercial policy, in accordance with Article 21 TEU and the support and promotion of EU's values and standards concerning human rights, environmental health and consumer protection and sustainable development.

Besides the legal text, the European Commission, through its trade policy communications, emphasizes the need to weigh the search of concrete trade interests by means of trade instruments with EU values. The challenge of operationalizing these principles is significant, given that the performance of EU trade partners on the various values listed in Articles 21 TEU and 207 TFEU will vary and be context specific.[35]

The content of the legal provisions and of the European Commission legal communications were reflected on the paths taken by EU external trade policy. The EU pursues a balanced and progressive trade policy to harness globalization, premised on the idea that global trade is a key contributor to a competitive and prosperous EU but needs to be proactively shaped and managed to ensure that it is fair, projects values and remains firmly anchored in a rule-based system.[36] In other words, the EU's approach is that international trade and investment can be used "as a tool to

[33] European Commission 2021, p. 3.

[34] London School of Economics and Political Science 2020, p. 66.

[35] Hoekman and Rojas-Romagosa 2022, p. 46.

[36] World Trade Organization 2020, p. 9.

achieve genuine global sustainable development" and that "trade has never been just about goods—it is about values too".[37]

The first main objective of its policy is to conclude new trade partnerships to build progressive rules for global trade and to use trade policy to promote universal values regarding environmental, social and labor protection and fundamental rights, and preserve the right to regulate in the public interest.[38] In this sense, the EU has sought, since 2011, when the FTA with South Korea was concluded, to insert systematic rules on sustainable development in its FTAs. This approach has led to the conclusion of twelve such agreements until December 2021. This can be considered as a significant achievement, given that this period corresponds to an intensive period of bilateralism on the part of the EU and reflects a determination to achieve a meaningful consideration of sustainable development with each trading partner that the EU was negotiating with.[39] These agreements are known as new-generation FTAs, which include provisions on sustainable development and typically go beyond provisions on liberalization of trade in goods.[40]

With regard to institutions, Domestic Advisory Groups (DAGs) established by the European Commission occupy a central place. DAGs are bodies established under trade agreements, composed of independent representatives from civil society groups, including non-governmental organizations, businesses, and trade unions, ensuring a balanced representation of various economic, social, human rights, environmental, and other interests.[41] DAGs and joint civil society meetings were conceived as mechanisms to further the goals of the trade and sustainable development dimensions of EU trade agreements, by encouraging civil society to monitor progress in promoting sustainability objectives.[42] In other words, DAGs act as instruments of participatory democracy allowing civil society actors to contribute to the monitoring and implementation of EU trade agreements.[43] In concrete terms, DAGs make available advice and opinions to the contracting parties. More recently, the European Commission, in its communication about the strengthening of the implementation and enforcement of TSD chapters of the EU's FTAs, clarified that DAGs can file collective complaints on violations of TSD commitments and an EU-based complainant can represent the interests of a party located in a partner country.[44]

In reality, nevertheless, civil society schemes in EU FTAs have been subject to criticism due to the absence of added value and effectiveness.[45] Firstly, the time dedicated to the deliberation of the administrative organization of the DAG appears to be disproportionate to the debate of questions directly connected to the FTA. Secondly,

[37] Malmstrom 2018.

[38] World Trade Organization 2020, p. 9.

[39] Lamy et al. 2021, p. 3.

[40] World Trade Organization 2020, p. 40.

[41] European Commission 2022, p. 8.

[42] Hoekman and Rojas-Romagosa 2022, p. 58.

[43] Lamy et al. 2021, p. 10.

[44] European Commission 2022, pp. 9–10.

[45] Driegue et al. 2022, p. 591.

the membership of the DAGs is criticized in the sense that inside them there is no proportional representation of dissonant interests that reflect the concerns of civil society. This situation is due in part to the insufficient representation of environmental stakeholders. While environmental stakeholders experiment with expertise or capability limitation, other stakeholders are reluctant to engage themselves as they believe that their interests will not be considered in a timely and appropriate manner.

Besides, the scope or mandate of these civil society schemes is constantly being questioned. Under the EU FTAs, DAGs can only debate sustainability aspects relative to trade and sustainable development chapters of the agreements. However, a recurrent criticism points that the coverage of DAGs' actions should be broadened to cover sustainability aspects of the whole trade agreement.

Considering specific EU foreign policy measures, the EU presented a legislative proposal on the issue of deforestation that can significantly attenuate the Brazilian environmental crisis. As a central economy that gathers consumers of commodities whose harvest takes place after deforestation and forest degradation, the EU is partially responsible for this issue and it wishes to be a leader on the paths to settle it. The goals derived from its proposal that it seeks to achieve are the diminishment of carbon emissions provoked by consumption and production inside the EU of the pertinent commodities in an amount of at least 32 million metric tonnes per year and the fight of all deforestation that arises from the increase of agricultural activities that encompass the production of commodities within the scope of the proposed regulation and forest degradation.

The EU legislative proposal was approved and became Regulation (EU) 2023/1115, known as the EU Deforestation Regulation. The commodities consumed in the EU that are reached by the Regulation whose production leads to carbon emissions are cattle, cocoa, coffee oil palm, rubber, soya and wood. On a temporal dimension, the EUDR reaches goods obtained on June 29, 2023 or after this date, with the exception for timber and timber products that were produced before that date and inserted on the EU market since December 31, 2027. A key measure that allows the achievement of the regulation's goals is the practice of extensive diligence to guarantee that the products derive from recent deforestation or forest degradation. An innovation found in the regulation is that it targets legal deforestation in conformity with the laws of the country of production. A basis to identify legal deforestation is a report by the Forest Policy Trade and Finance Initiative quoted on the regulation that found out the legality of approximately 30% of deforestation aimed at the practice of commercial agriculture in tropical regions from 2013 to 2019.

The Regulation in force aims to allow only the entry of deforestation free products in EU markets. In more detailed terms, Article 3 of Regulation (EU) 2023/1115 provides that relevant commodities and products may be placed or made available on the Union market, or exported from the Union market only if all the following conditions are fulfilled: they are deforestation-free; they have been produced in accordance with the relevant legislation of the country of production; and they are covered by a due diligence statement. "Deforestation-free" means that the relevant commodities and products, including those used for or contained in relevant products, were produced on land that has not been subject to deforestation after December 31,

2020 and that the wood has been harvested from the forest without inducing forest degradation after December 31, 2020.

## 17.3 Understanding the Essential Elements of the TSD Chapter

Inside the global governance framework, the content of TSD obligations reveals a connection between the regional EU level and the multilateral level. The firm anchoring of bilateral TSD commitments in the multilateral context makes the EU's TSD approach legitimate, credible and effective.[46] It also seeks to avoid fragmentation and instead ensures that TSD commitments reinforce global efforts.[47] By being founded on international standards, TSD commitments can also shield against protectionist reflexes.[48]

The rules on TSD mirror ethical values that are accommodated with trade practice. The primary motivation for including provisions on values in reciprocal trade agreements is not because of the potential impacts of trade on the realization of these provisions or that non-respect by a trading partner may affect trade, but that these provisions represent values that the EU seeks to promulgate and protect.[49]

These values reflect European concerns about the environment and core labor rights and are addressed in a manner that goes beyond the enforcement of Multilateral Environmental Agreements (MEAs) and International Labor Organization (ILO) conventions. TSD chapters address the domestic environmental policies in more general terms, trying to set a balance between the competing interests of supporting policies pursuing high levels of environmental protection, avoiding that these policies disguise protectionist and/or discriminatory intents, and condemning a 'race to the bottom' whereby more favorable competitive conditions on international markets are achieved by deregulation.[50]

A variety of environmental and trade-related environmental issues are foreseen in TSD chapters. Examples of the first ones are biodiversity, endangered species, sustainable use of marine resources, climate change mitigation and forest management. Examples of the second ones are fishing subsidies, renewable energy and trade in environmental goods as timber and fish products. TSD chapters also address public participation in environmental decision-making and the monitoring of the implementation of the FTAs themselves.

---

[46] European Commission 2022, p. 5.

[47] Ibid., p. 5.

[48] Ibid., p. 5.

[49] Hoekman and Rojas-Romagosa 2022, p. 46.

[50] Adinolfi 2020, p. 32.

### 17.3.1 The Effectiveness of the Minimum-level Clause, the High-level Clause and the Non-regression Clause

In broad terms, the EU-Mercosur FTA confirms initially each party's right to regulate labor and environmental aspects, subject to the condition that such domestic laws and policies must be in conformity with that party's international commitments. Limitations on the sovereign right to regulate in environmental and labor issues are found in the following general provisions: the minimum-level clause, the high-level clause and the non-regression clause. Their content and the comments found in the literature to their limits and effects are explained below.

The minimum-level clause[51] consists of the obligation to respect "each party's commitment to the international agreements and labor standards", which provides the minimum level of environmental and labor protection to be fulfilled by each of the FTA parties. In what concerns social protection, parties undertake to respect, promote and effectively implement the core labor standards foreseen on the ILO Constitution and on the ILO Declaration on Fundamental Principles and Rights at Work, namely freedom of association and the right to collective bargaining, the elimination of all forms of forced or compulsory labor, the effective abolition of child labor and the elimination of discrimination with respect to employment and occupation. Moreover, parties further reiterate their engagement towards ratifying the fundamental ILO Conventions, Protocols and other relevant ILO Conventions to which it is not yet a Party and that are classified as up-to-date by the ILO.[52] As for the environmental standards, each party reaffirms its commitments to promote and effectively implement MEAs, protocols and their amendments to which it is a party.[53] The agreements, nonetheless, are not specified.

The limitations and potentialities of the aforementioned clause were analyzed in the literature. This first limitation on the right to regulate adds nothing substantially new, since it merely incorporates a bilateral/regional context of international environmental and labor standards that are already binding on the FTA parties under the relevant multilateral agreements.[54] On the other hand, the effect of the minimum-level clause is to make pre-existing multilateral commitments applicable between the EU and the third country concerned on the basis of the FTA itself and, hence, enforceable in accordance with the FTA dispute settlement procedures in addition to multilateral compliance mechanisms.[55] It is worth clarifying that TSD chapters are commonly excluded from the general dispute settlement mechanism (DSM) of the FTA and have their own DSM.

---

[51] A deep analysis of the clause is found in *What is There to Enforce? Provisions regarding Multilateral Agreements and Standards in the TSD chapters of EU FTAs,* written by Saide Esra Akdogan which is Chap. 7 of this book.

[52] Article 4(4) of the TSD Chapter of EU-Mercosur FTA.

[53] Article 5(3) of the TSD Chapter of EU-Mercosur FTA.

[54] Durán 2020, p. 1037.

[55] Durán 2020, p. 1038.

The high-level clause, the second limitation on the right to regulate, is the effort of parties to guarantee that their laws and policies establish and stimulate high and effective levels of environmental and labor protection. The broadness of the clause can be inferred from the non-confinement to internal legislation that meets international standards and its application even when bilateral trade is not affected. However, this clause is legally weaker, being couched in best-endeavors and imprecise terms, which allow for different levels of compliance.[56]

The aspirational legal nature of the high-level clause also implies that the content of TSD chapters of EU trade agreements are partly pursued by means of trade policy instruments that are non-binding enforceable commitments. However, there is no evidence that this broader set of aspirations in the EU's FTAs has produced better, more meaningful results than the supposedly narrower set of enforceable norms in the FTAs concluded by the USA and Canada.[57] In that regard, many lawyers and civil society organizations have called for non-trade policy objectives-related provisions in EU trade agreements to be made binding.[58] Besides this claim, the literature understands that, if the presence of weaker provisions is thought to affect the enforceability of more robust norms in the sustainability chapter, the former are best put elsewhere as in an annex or political declaration accompanying the FTA.[59] In case they have not really been found useful, these weaker provisions might also be removed entirely from the FTAs.[60]

The non-regression clause, the third limitation on the right to regulate, is the prohibition on parties to waive or derogate from their environmental or labor laws "in order to encourage trade or investment" between them. In comparison with the EU-Singapore FTA, which provides in Article 12.12(1) that domestic environmental and labor laws cannot affect trade or investment between the parties, the EU-Mercosur FTA provides in Article 2(4) of the TSD chapter a softer formulation and generates a more intent-based test. But in either case, the specific link to trade sheds light on the purpose of the non-regression obligation in the sense that it is primarily aimed at preventing competitive pressures arising from greater economic liberalization leading to a "race-to-the-bottom" in environmental and labor regulations, and thereby preserve a "level-playing field" in bilateral and investment relations.[61]

A comparative analysis allows the identification of the similarities and differences between the clauses explained above. A common aspect is that the three clauses constitute interpretative guidelines to the right of the parties to the agreement to determine the levels of environmental and labor protection. In comparison to the minimum-level clause, the scope of application of the non-regression obligation is broader since it extends to all domestic environmental and labor laws, and not only those that implement international environmental and labor standards as is the case of

---

[56] Ibid.

[57] Bronckers and Gruni 2021, p. 33.

[58] Hoekman and Rojas-Romagosa 2022, p. 57.

[59] Bronckers and Gruni 2021, p. 33.

[60] Ibid.

[61] Durán 2020, p. 1039.

the minimum-level obligation.[62] Despite this apparent overlap, the trade-effect condition in the non-regression clause is likely to help in delimiting the respective scope of application of these two obligations in practice.[63] Due to its higher trade-related evidentiary burden, the non-regression obligation seems mostly relevant for domestic environmental and labor laws that go above and beyond international standards, and which are not covered by the minimum-level clause.[64]

### 17.3.2 The Obligations of the TSD Chapter and the Mitigation of Climate Change

Among the specific obligations foreseen in the TSD chapter, the obligation of the parties to effectively implement the Paris Agreement allows one step beyond the handling of nationally determined contributions (NDCs) as foreseen in the Paris Agreement. In regards to this obligation, it is worth firstly clarifying that the Paris Agreement on Climate Change is generally understood as lacking an obligation of result beyond a requirement on States to report the measures taken to fulfill NDCs. Article 4, para 2 of the Paris Agreement establishes that each party shall prepare, communicate and maintain successive NDCs that it intends to achieve and shall pursue domestic mitigation measures, with the aim of achieving the objectives of such contributions. In addition to this core obligation, the Paris Agreement places normative expectations on parties in relation to their contributions—expectations that each successive contribution will reflect a progression and that each party will undertake its highest possible ambition.[65]

The EU achieves positive results through the inclusion of an obligation in its FTAs to 'effectively implement' NDCs under the Paris Agreement. The language of the EU's proposed FTA with Mercosur, according to which Brazil assumed an obligation to effectively implement the Paris Agreement, adds to the intentions that Brazil manifested under the Paris Agreement. By virtue of the FTA, any doubt has been removed that Brazil's NDC might be considered voluntary, as the commitment in the Paris Agreement has been firmed up by the FTA.[66] Before the EU at least, Brazil would clearly recognize that its NDC consists of a binding obligation (and, vice versa, of course). While the nature of that commitment is probably not an obligation of result, the obligation to effectively implement its NDC would at least require Brazil to make its best efforts to achieve its goals under the Paris Agreement.[67] Brazil continues with the choice of the means of implementation to accomplish the goals inserted on its NDC, but the best-efforts obligation plausibly imposes a continuous obligation

[62] Durán 2020, p. 1039.
[63] Ibid.
[64] Ibid.
[65] Rajamani 2019, p. 167.
[66] Bronckers and Gruni 2021, p. 29.
[67] Ibid.

on Brazil to prove that it is on the path towards the compliance of its stated goal within the stipulated time frame foreseen on its NDC. Moreover, independently of the understanding of Brazil's commitment to implement the Paris Agreement before the EU, it does not depend on whether or not Brazil's environmental measures have an impact on trade with the EU.

It is plausible that the EU, once the FTA is ratified, could question Brazil's policies under the FTA's bilateral dispute settlement mechanism specific to the TSD chapter,[68] arguing that the country is not acting towards its stated goal and thus not complying with its best-effort obligation. The FTA's sustainability provisions are not, however, ideal, as a specific discipline on the implementation of the standards is not foreseen. Furthermore, dispute settlement decisions issued by the panelists that are part of the mechanism specific to the sustainability chapter lack binding force. Consequently, if, in a hypothetical situation, Brazil would not comply with an arbitration award, the FTA does not establish that the EU could adopt sanctions. On the other hand, the Commission proposed in 2022 the possibility of trade sanctions as a matter of last resort in instances of serious violations of the Paris Agreement and of ILO fundamental principles and rights at work.[69]

On the other hand, part of the literature interprets the TSD chapter of the EU-Mercosur FTA as lacking binding force and silent on certain sustainable issues.[70] The sustainability clauses, which are there to be used as a due diligence mechanism, a precondition for the trade pillar, are not legally enforceable and therefore they are not enough to make the TSD chapter be in conformity with the Paris Agreement rules either.[71] Citing mere *de jure* proviso (unfit to invoke thicker regulation needed for *de facto* compliance) does not lend authority to either Mercosur's sustainability agenda or the AA.[72] The repercussions are inherently local and especially brutal for Brazil.[73] As such, the lack of a legally enforceable TSD chapter shall be the basis for the perpetuation of the agribusiness' ruthless expansion model (entailing, that is, human rights infringements), and with it the further endangerment of indigenous peoples' rights.[74] In fact, the TSD chapter callously disregards the displacement of these peoples (rife in this expansion model), as there is no provision safeguarding against this.[75]

---

68 Article 17 of the TSD chapter of EU-Mercosur FTA.

69 European Commission 2022, p. 5.

70 Correa and Lehmen 2023; Kramer, 2021.

71 Diz 2022, p. 16.

72 Ibid.

73 Ibid.

74 Ibid.

75 Ibid.

## 17.4 Conclusion

This chapter systematizes the identified dimensions of the Brazilian environmental crisis and, following the sequence of the topics of the Chapter, the possible answers to the main question posed in the Introduction. A high degree of deforestation in the Amazon and the recurrent fires in the Cerrado, the Pantanal and the Pampa, which were detected by INPE, are the key features of the grave environmental situation in Brazil. The insufficient sanctions contribute to perpetuating this crisis as they create a scenario of impunity, which creates the expectation of grave forms of criminality. Consequences are felt on the national level, the regional level and the international level. Examples of them on the first level are the impossibility of capture of the rainwater by vegetation and low productivity pasture. Examples of them on the second level are warming and changes in the atmospheric circulation and moisture convergence. The main effect on the third level is the increase of GHG emissions that contribute to the worsening of climate change.

The policy recommendations of the SIA relative to the EU-Mercosur FTA and the values and institutions of European foreign policy can contribute to alleviate the environmental damages recently experienced in Brazil. The policy recommendations formulated after the identification of issues of moderate concern previously explained regard, for instance, the transformation of existing degraded pasture areas into land designed to sustainable agriculture to avoid the destruction and degradation of forest land to attain the expected increase of agricultural production and the attainment of the GHG emissions targets as foreseen on the NDCs established in the Paris Agreement. The values of the European foreign policy that deserve to be highlighted are human rights, environmental health and consumer protection and sustainable development. The main EU institution is the DAG that deepens the objectives of the trade and sustainable development aspects of EU trade agreements through the encouragement of monitoring by civil society of improvement in promoting sustainability objectives.

Inside the TSD chapter of the EU-Mercosur FTA, the minimum-level clause, the high-level clause and the non-regression clause limit the sovereign right to regulate, protect environmental standards and make possible their effective implementation. The first one aims to deepen the acceptance and effective implementation of the international environmental regime in force, instead of creating a system of parallel normative standards. The second one, despite its aspirational legal nature, can be interpreted as a clear weakening of environmental laws in force inconsistent with the strive to the improvement of these standards. The third one constitutes a guarantee against regression in current degrees of legislative protection in environmental and social standards conditional to the intent of encouraging trade or investment. The obligation to effectively implement the Paris Agreement orients the manner by which Brazil must act, that is being bound by a continuous best-effort obligation to prove that it seeks to comply with its stated goal within the stipulated period of time established on its NDC.

Beyond the content analyzed in this paper, the research agenda on environmental rules inside FTAs reveals the existence of a further proposal that can increase the effective compliance of the referred rules towards greater environmental protection. In face of the reduced effectiveness of dispute settlement mechanisms, there is the suggestion of the addition of remedies to ensure an effective enforcement, by introducing a formal complaint mechanism for business and civil society to lodge their assessments of the implementation of the FTA in relation to sustainable development and adopting a more proactive attitude toward consultations under the TSD dispute settlement mechanism.[76]

## References

Adinolfi G (2020) A Cross-Cutting Legal Analysis of the European Union Preferential Trade Agreements' Chapter on Sustainable Development. In: Beverelli C et al (eds), International Trade, Investment and the Sustainable Development Goals. Cambridge University Press, Cambridge, pp 15–49

Arima E, Barreto, P (2020) The EU-Mercosur trade agreement: where is the greatest risk of deforestation in Brazil? In: Amazon Institute of People and the Environment, Is the EU-Mercosur trade agreement deforestation-proof? Instituto do Homem e Meio Ambiente da Amazônia, Belém, pp. 36–58

Bronckers M, Gruni G (2021) Retooling the Sustainability Standards in EU Free Trade Agreements, Journal of International Economic Law 24:25–51

Correa F, Lehmen A (2023) Trade, Sustainable Development and Climate Change: How Can Free Trade Agreements Be Leveraged for Increased Climate Action? Perspectives on the EU-MERCOSUR Agreement, Legal Issues of Economic Integration 50:287–306

Diz J (2022) The Mercosur and European Union relationship: an analysis of the incorporation of the Association Agreement in Mercosur, Europe and the World: A law review. https://doi.org/10.14324/111.444.ewlj.2022.01

Driegue L, Orbie J, Potjomkina D, Shahin J (2022) Participation of Civil Society in EU Trade Policy Making: How Inclusive is Inclusion?, New Political Economy 27: 581–596

Durán G (2020), Sustainable development chapters in EU free trade agreements: emerging compliance issues, Common Market Law Review 57:1031–1068

European Commission (2016) Handbook for Trade Sustainability Impact Assessment. https://op.europa.eu/en/publication-detail/-/publication/8b3a2b37-1028-11e6-ba9a-01aa75ed71a1. Accessed: 3 October 2022

European Commission (2021) European Commission Services' Position Paper on the Sustainability Impact Assessment in Support of Negotiations for the Trade Part of the European Union-Mercosur Association Agreement. https://circabc.europa.eu/ui/group/09242a36-a438-40fd-a7af-fe32e36cbd0e/library/a974649b-a1ef-4436-9dc4-c80492a64b70/details. Accessed: 26 November 2022

European Commission (2022) The power of trade partnerships: together for green and just economic growth. https://circabc.europa.eu/ui/group/8a31feb6-d901-421f-a607-ebbdd7d59ca0/library/8c5821b3-2b18-43a1-b791-2df56b673900/details. Accessed: 26 November 2022

Hoekman B, Rojas-Romagosa H (2022) EU Trade Sustainability Impact Assessments: Revisiting the Consultation Process, Journal of International Economic Law 25:45–60

[76] Lamy et al. 2021, p 2.

Instituto Nacional de Pesquisas Espaciais [National Institute for Space Research] (2020) Prodes: Projeto de Monitoramento do Desmatamento da Floresta Amazônica Brasileira por Satélite [Project for Monitoring the Deforestation of the Brazilian Amazon Forest by Satellite]. http://terrabrasilis.dpi.inpe.br/app/dashboard/deforestation/biomes/legal_amazon/rates. Accessed: 25 November 2022

Instituto Nacional de Pesquisas Espaciais [National Institute for Space Research] (2023) TerraBrasilis [TerraBrasilis]. http://terrabrasilis.dpi.inpe.br/. Accessed: 10 October 2023

Kramer L (2021) A Lost Opportunity? The Environment and the EU-Mercosur Trade Agreement, Journal for European Environmental & Planning Law 18:143–163

Lamy P, Pons G, Garzon I, Kauffmann L (2021) Sustainable development in EU trade agreements. Much ado about nothing? https://www.europejacquesdelors.eu/publications/gt7-sustainable-development-in-eu-trade-agreements. Accessed: 24 September 2022

Leite-Filho A, Soares-Filho B, Davis J, Abrahão G, Borner J (2021) Deforestation reduces rainfall and agricultural revenues in the Brazilian Amazon, Nature Communications 12:1–7

Li Y, Brando P, Morton D, Lawrence D, Yang H, Randerson J (2022) Deforestation-induced climate change reduces carbon storage in remaining tropical forests, Nature Communications 13:1–13

London School of Economics and Political Science (2020) Sustainability Impact Assessment in Support of the Association Agreement Negotiations between the European Union and Mercosur. https://www.lse.ac.uk/business/consulting/assets/documents/SIA-in-Support-of-the-Association-Agreement-Negotiations-between-the-EU-and-Mercosur-Final-Report.pdf. Accessed: 26 November 2022

Malmstrom C (2018) From earning profits to earning trust. European Union Directorate-General for Trade, Brussels

Rajamani L (2019) Innovation and Experimentation in the International Climate Change Regime, Collected Courses of the Hague Academy of International Law 404:9–234

Shalders A (2019) Queimadas disparam, mas multas do IBAMA despencam sob Bolsonaro [Fires soar, but IBAMA fines plummet under Bolsonaro]. https://www.bbc.com/portuguese/brasil-49430376. Acessed: 26 May 2024.

Soendergaard N, Sá C, Jank M, Gilio, L (2021) Decoupling Soy and Beef from Illegal Amazon Deforestation: Brazilian Private Sector Initiatives. Brazilian Center for International Relations, Rio de Janeiro

World Trade Organization (2020) Trade Policy Review. Report by the Secretariat. European Union. Revision. https://docs.wto.org/dol2fe/Pages/SS/directdoc.aspx?filename=q:/WT/TPR/S395R1.pdf&Open=True. Accessed: 26 November 2022

**Tiago Matsuoka Megale** is a PhD candidate at University of São Paulo, Ribeirão Preto Street 506, São Paulo, Brazil.

# Chapter 18
# Sustainable Development in Public Procurement: The EU-Mercosur Trade Agreement

**Bruno Fontenelle and Maria Tereza Dias**

**Contents**



**Abstract** It is expected that the open market sought by the Agreement between the European Union (EU) and Mercosur will have great repercussions for sustainable development. Therefore, this Chapter seeks to provide a study on how the Agreement enables the promotion of sustainable development, specifically in public procurement, as the two blocs understand this matter from very different perspectives. First, the EU has substantially improved its legal norms related to sustainable development in public bidding. In 2014, it introduced "the most economically advantageous tender" as an award criterion that may promote social, environmental, and other objectives contributing to sustainable development. This new criterion is not necessarily connected to the traditional goals of public procurement, such as obtaining the lowest price. However, this broad understanding of sustainability is not present within Mercosur. The term "sustainable development" does not exist in the southern bloc's most important charters, such as the Treaty of Asunción (1991) and the Protocol of Ouro Preto (1994), and can only be found briefly in the Mercosur Public Procurement Protocol (2017). Nevertheless, the Agreement establishes that the two blocs

B. Fontenelle (✉)
Universidade Federal de Minas Gerais, Av. João Pinheiro 100, Belo Horizonte, Minas Gerais, Brazil
e-mail: brfontenelle@hotmail.com

M. T. Dias
Faculdade de Direito, Universidade Federal de Minas Gerais, Avenida João Pinheiro, Nº 100, Bairro Centro, CEP 30130-180 Belo Horizonte, Minas Gerais, Brazil
e-mail: terezadias@ufmg.br



R. A. Wessel et al. (eds.), *EU External Relations Law and Sustainability*, Global Europe: Legal and Policy Issues of the EU's External Action 4,
https://doi.org/10.1007/978-94-6265-655-0_18

shall recognize the contribution of transparent, competitive, and open tendering to economic development and set as their objective the effective opening of their respective public procurement markets. Furthermore, it includes the necessity of the inclusion of sustainable development. Therefore, it is clear that the Trade Agreement could face many difficulties in its application, especially related to sustainable development in public procurement, due to the undeniable differences between the two blocs.

**Keywords** Tenders · Public procurement · Sustainable development · Mercosur · European Union · Agreement

## 18.1 Introduction: Overview of the Understanding of Sustainable Development

The concept of sustainable development has been crafted over the years, after the establishment of several international and national conventions that discussed ways to reconcile environmental protection, social development, and economic growth. According to Mata Diz, the 1972 Stockholm Conference is considered a milestone for discussions on sustainability, and it was the first step to coordinate efforts directed at protecting the environment, including sustainable development as a reference. As she argued, this Conference clearly connected environmental sustainability, economic growth and social development established for the first time.[1] In the words of the author:

> [The concept of sustainable development] was essential to generate a consensus between antagonistic positions: those who defended the complete stagnation of economic growth, predicting a catastrophic future for humanity due to environmental degradation, and those who defended growth at all costs, claiming the issue of environmental protection served the interests of developed countries, which intended to go against the industrialization of peripheral countries.[2]

Subsequently, in 1987, the World Commission on Environment and Development released the Brundtland Report, also called "Our common future", which revolutionized the perspective of sustainable development by being the first international document to define the term, defining it as the development which "[...] ensure that it meets the needs of the present without compromising the ability of future generations to meet their own needs."[3] Furthermore, according to the document, it is necessary to understand the concept of sustainability not only from the environmental standpoint, but also to include the social and economic perspectives. According to the Report:

> The satisfaction of human needs and aspirations in the major objective of development. The essential needs of vast numbers of people in developing countries for food, clothing,

[1] Mata Diz 2019, p. 84.

[2] Ibid., p. 86.

[3] Brundtland 1987.

> shelter, jobs—are not being met, and beyond their basic needs these people have legitimate aspirations for an improved quality of life. A world in which poverty and inequity are endemic will always be prone to ecological and other crises. Sustainable development requires meeting the basic needs of all and extending to all the opportunity to satisfy their aspirations for a better life.[4]

This has revealed that environmental challenges and the pursuit of sustainable development are intrinsically linked to the eradication of poverty, the provision of basic housing, health, and food, enabling discussions about new energy sources that prioritize renewables and technological innovation. It is clear, therefore, that the predominant focus on the ecological aspect does not imply a greater predominance of this aspect, but rather a growing urgency to address the challenges it faces. Thus, sustainable development has come to be understood as the type of development that meets current needs without compromising the ability of future generations to fulfill their own aspirations.[5]

In this way, as stated in a passage from the Report itself, this conception of development is: "[...] a process of change in which the exploitation of resources, the direction of investments, the orientation of technological development, and institutional change are made consistent with future as well as present needs."[6] In this sense, Larejo and Armeriso understand sustainable development as follows:

> The concept of sustainable development has been incorporated by the Report of the World Commission on Environment and Development (Brundtland Report). In this sense, sustainability can be seen as a simple corollary of the dependence of the human community on its environment. The environment is both a source of the resources necessary for economic and social survival, a sink for the waste generated by production and consumption activities, and a support for a wide range of free environmental services provided to the human population.[7]

Regarding this expanded view of sustainable development, the sociologist Elkington coined the term "triple bottom line", in which the economic, environmental, and social dimensions should be interconnected within the context of sustainable development, aiming to reconcile the well-being of people, the planet, and profits in a balanced manner. This way of thinking represents a shift in business mentality, emphasizing that organizations should measure the value they create or destroy in the social, environmental, and economic dimensions.[8] Therefore, sustainability should be perceived as a business management approach that seeks to provide a financial return to shareholders while encompassing economic progress, social advancement, and the preservation of global natural resources.[9]

However, according to the Freitas, the concept proposed by the Report, despite having been essential for the promotion of the idea of sustainable development, says little about the multidimensional nature of sustainability, which should include, in

---

[4] Ibid.

[5] Ibid.

[6] Ibid.

[7] Laredo and Armeriso 2001, pp. 405–406.

[8] Brasil 2020, p. 79.

[9] Ipiranga et al. 2011, p. 14.

addition to the economic, social and environmental aspects, the legal-political and ethical dimensions. According to the author, post-Brundtland sustainability must take on "[...] demands concerning physical and psychological well-being, and not only meet material needs, but do so without increasing the risks posed, on an industrial scale, by human beings themselves", and must be politically and ethically consistent.[10] Thus, the author defines sustainability, at least in the scope of Brazilian law, as "a constitutional principle that determines the promotion of social, economic, environmental, ethical and juridical-political development in order to ensure favorable conditions for the well-being of future and present generations."[11] In Freitas' view, in order to have real conditions for effective well-being in the future, it is up to sustainability to shape development in the present—and not the other way around. Therefore, based on this multidimensional view of sustainability, it is necessary to understand the different facets of this concept, namely the (a) social, (b) ethical, (c) environmental, (d) economic, and (e) legal-political ones.

The social dimension (a), according to Freitas, is the one that harbors the fundamental social rights, with the purpose of helping the disadvantaged, through positive actions and effectively rewarding interventions, which allow facing poverty, not admitting an excluding development model. Thus, the social dimension of sustainability calls for an increase in intra and inter-generational equity and favorable conditions for human potential to flower through the consolidation of healthy habits.[12]

The ethical dimension (b), on the other hand, consists in "[...] acting in such a way that the homeostatic production of lasting well-being can be universalized, in the intimate and in the interaction with nature".[13] It requires recognizing the interconnectedness of all living beings beyond strict anthropocentrism, as well as the moral demand for honesty in light of human dignity.[14]

Furthermore, the environmental dimension of sustainability (c) is linked to the right of current generations, without prejudice to future generations, to a clean environment and environmental protection.[15] From this perspective, it is crucial to combat hyper-consumerism, especially in affluent countries, as a high quality of life and sustainable existence cannot be achieved in a context where healthy planetary relationships are absent.[16] In this regard, the author explains that "[...] there can be no human life without the vigilant safeguarding of environmental sustainability in a timely manner. Indeed, either we protect socio-environmental justice, or there simply will be no future for our species."[17]

The economic dimension of sustainability (d) is related to the appropriate balance between efficiency and equity, considering all direct and indirect benefits and costs

---

[10] Freitas 2019, p. 52.

[11] Ibid., p. 55.

[12] Ibid., pp. 65–67.

[13] Ibid., p. 69.

[14] Ibid., pp. 70–71.

[15] Ibid., p. 72.

[16] Ibid., p. 341.

[17] Ibid., pp. 341–342.

of trade relations. In the author's view, "economic viability implies the prohibition of waste in the broad sense, as well as strengthens fiscal responsibility and encourages responsible innovation, especially in view of the so-called 'fourth industrial revolution."[18] Additionally, it should work with justified market regulation to correct failures such as abuse of dominant power, asymmetric information, and negative externalities[19].

Finally, the legal-political dimension of sustainability (e) is related to the duty to protect the freedom of each person, in this status, in the process of inter-subjective stipulation of the inter-temporal content of fundamental rights and duties of present and future generations, whenever directly possible.[20] According to the author, this dimension can also be understood as a norm that establishes the intertemporal effectiveness of fundamental rights, not only those of the third dimension but also serves as a criterion for determining the illegality of behaviors that cause harm related to condemnable practices of patrimonialism and other vices.[21]

Based on this multidimensional nature of sustainability, the United Nations (UN) in 2015 adopted it famous global action plan called "2030 Agenda" that brings together 17 Sustainable Development Goals (SDGs), designed to eradicate poverty, and promote a decent life for all, within the conditions that our planet offers and without compromising the quality of life of the next generations. Specifically about goal 12, which is to ensure sustainable production and consumption patterns, it is important to highlight sub-goal 12.7, which establishes the need to "promote public procurement practices that are sustainable in accordance with national policies and priorities".[22]

From this perspective of sustainable public procurement and its impact on the economic chain of production of goods and services, the present Chapter aims to analyze the different proposals for sustainable development in bidding in Mercosur and the European Union (EU), in addition to its application in the Trade Agreement between the two blocs, signed in 2019.[23] It thus seeks to analyze the standards related to sustainable development and Mercosur and EU public procurement rules, with a view to the provisions of the Agreement, to understand how the blocs seek to promote sustainable public procurement together.

---

[18] Ibid., p. 342.

[19] Ibid., p. 76.

[20] Ibid., pp. 76–77.

[21] Ibid., p. 342.

[22] UN (United Nations) [2022]. The 17 goals. https://sdgs.un.org/goals. Accessed 13 November 2022.

[23] Ghiotto and Echaide 2019, p. 2.

## 18.2 Sustainable Development in European Union Public Procurement

The Treaty on the Functioning of the European Union (TFEU) states in Article 11 that "Environmental protection requirements must be integrated into the definition and implementation of the Union's policies and activities, in particular with a view to promoting sustainable development".[24] In the same vein, the Treaty on European Union (TEU) provided, in its Article 3(3), that sustainable development is an objective of the Union, making it a guiding principle in the EU's actions.[25]

In the context of public procurement, the EU adopted Directive 2014/24/EU, which seeks to promote sustainability in bidding, reconciling the concern with public accounts and sustainable development through the criterion of the most economically advantageous tender. This criterion, called in this article the "European criterion", is an awarding method found in EU bidding law, referring to the possibility of using other parameters besides the lowest price, to include qualitative aspects—both environmental and social—to determine the winner in a competition.[26]

According to Article 67(2) of the Directive, such tender must be identified on the basis of price or cost, and may also include the best price-quality ratio, which must be evaluated on the basis of criteria that include qualitative aspects, both environmental and social, always linked to the subject matter of the contract. Such criteria may include—but are not limited to—quality, aesthetic and functional characteristics, accessibility, and characteristics perceived as social, environmental and innovative. Recitals (98 and 99) of the Directive provide, for example, measures to promote equality between women and men at work, to protect the environment or animal welfare, to protect the health of personnel involved in the production process and to promote the social integration of disadvantaged persons or members of vulnerable groups.

Furthermore, according to Article 67(5), there will be a relative weighting between the price of each proposal and the social and/or environmental criteria stipulated in the invitation to bid. Thus, the criterion of the most economically advantageous proposal will work with a weighted average between the lowest value of the proposal and the sustainable criteria, giving equal importance, within the award criteria, between the lowest value and sustainability.

European literature understands that public bids have other social values to be achieved besides the lowest economic cost to the public coffers. This social function of bidding is named by Arrowsmith a "horizontal policy", and can be described

---

[24] Europe 1958, The Treaty on the Functioning of the European Union.

[25] The Article reads as follows: "Article 3 [...] 3. The Union shall establish an internal market. It shall work for the sustainable development of Europe based on balanced economic growth and price stability, a highly competitive social market economy, aiming at full employment and social progress, and a high level of protection and improvement of the quality of the environment. It shall promote scientific and technological advance."

[26] Europe [2014] Directive 2014/24/EU of the European Parliament And Of The Council of 26 February 2014; BOVIS 2007, p. 79.

as "[...] one facet of public procurement, namely its use to promote social, environmental and other societal objectives that are not necessarily connected with the procurement's functional objective[...]".[27]

Feliu also believes that European public procurement should pursue social and environmental objectives, as an instrument at the service of public policies, in order to face global challenges, such as climate change, and for the sake of sustainable development in the European Union. According to the author, public procurement is not only a way to supply raw materials or services in the most advantageous conditions for the State, but also a policy of intervention in economy, social and public life, capable of guiding certain behaviors.[28] In his words:

> **What has just been explained explains that** public procurement can—and must, we insist—be a technique that promotes the achievement of social, environmental or research objectives, in the conviction that these entail a proper understanding of how public funds should be allocated**. It can therefore be a major lever for activating economic change. An economic model that must be aligned with the SDGs (target 12.7 focuses specifically on promoting "promote public procurement practices that are sustainable in accordance with national policies and** priorities"**).**
>
> **The new legal and economic reality of public procurement requires a strategic vision in a globalized economic context.** A correct use of public procurement, as an instrument at the service of public policies, must make it possible to reinforce the principles inherent to the European social model and guarantee its sustainability **in a geopolitical situation that is becoming increasingly tense from the Eastern markets, which makes it necessary to rethink and reinforce the strategy of the European internal market.**[29]

It is in this sense that recital 2 of Directive 2014/24 states that public procurement plays a key role in the economy, being one of the market instruments used to achieve smart, sustainable and inclusive growth, while ensuring the most efficient use of public funds. In the same line of reasoning about the social function of procurement, the Directive states in Recital 91 that:

> Article 11 TFEU requires that environmental protection requirements be integrated into the definition and implementation of the Union policies and activities, in particular with a view to promoting sustainable development. This Directive clarifies how the contracting authorities can contribute to the protection of the environment and the promotion of sustainable development, whilst ensuring that they can obtain the best value for money for their contracts.[30]

From this perspective on public procurement, the European Commission,[31] in conjunction with the Court of Justice of the European Union (CJEU), over time and fromulated what is currently understood as the criterion of the most economically advantageous tender. It is in this sense that Bovis understands that such a

[27] Arrowsmith and Kunzlik 2009.

[28] Feliu 2021, pp. 71–72.

[29] Ibid., pp. 70–71.

[30] Europe [2014] DIRECTIVE 2014/24/EU OF THE EUROPEAN PARLIAMENT AND OF THE COUNCIL of 26 February 2014.

[31] Europe [2020] Comissão europeia.

criterion "[...] has provided the Court with the opportunity to balance the economic considerations of public procurement with policy choices",[32] as will be shown below.

The first case related to this matter was Case C31/87, in which the parties were Beentjes BV and the Kingdom of the Netherlands. The litigation began as a result of the decision of the contracting authority of the Ministry of Agriculture and Fisheries, which expressed the view that Beentjes' proposal "had insufficient specific experience for the work to be performed, [...] and that it did not appear to have the capacity to employ the long-term unemployed."[33] Beentjes, on the other hand, stated that even having presented a proposal with the lowest value, another company won, breaching the legal dispositions at the time.[34]

The CJEU held that the legislation establishes the discretion of the contracting authorities to determine the most economically advantageous tender, provided that it is based on objective criteria, mentioned hierarchically, and previously established in the call for tenders.[35] By this, "tenderers and interested parties can clearly ascertain the relative weight of factors other than price in the [bidings] evaluation process".[36] Moreover, specifically about the criterion of hiring the long-term unemployed, the Court understood that it was recurrently used by the Netherlands as a form of social inclusion, understanding its conformity with the European provisions.[37]

Later, in Case C-513/99 (*Concordia Bus Finland vs HKL-Bussiliikenne*), the Court held that the preservation of the environment may be linked to the criterion of the most economically advantageous tender. This, however, must be related to the object of the contract, be expressed in the call for tenders and not violate the fundamental principles of European Union law, such as non-discrimination, making it impossible for the contracting authority to have an unrestricted freedom of choice. As it can be seen:

> It follows from the above considerations that, where the contracting authority decides to award a contract to the tenderer who submits the economically most advantageous tender, in accordance with Article 36(1)(a) of Directive 92/50, it may take criteria relating to the preservation of the environment into consideration, provided that they are linked to the subject-matter of the contract, do not confer an unrestricted freedom of choice on the authority, are expressly mentioned in the contract documents or the tender notice, and comply with all the fundamental principles of Community law, in particular the principle of non-discrimination.[38]

It is important to underline, however, that the contracting authority has the autonomy to determine the weighting of the economic and social criteria to be used in the bidding, in order to identify the most economically advantageous proposal, as can be seen in Case C-448/01 (*EVN AG vs. Republic of Austria*). According to the Court,

[32] Bovis 2007, p. 274.

[33] European Court of Justice [1988]. Case C-31/87. para 5.

[34] Ibid., para 2.

[35] Ibid., para 27.

[36] Bovis 2007, p. 274.

[37] Caranta and Trybus 2010, p. 19.

[38] European Court of Justice [2002]. Case C-513/99. para 64.

as long as contracting authorities adhere to Union law requirements, they have the freedom not only to select the criteria for awarding a contract but also to determine how important each criterion is. However, these weightings should allow for an overall assessment of the criteria to identify the most economically advantageous bid. Regarding the specific award criterion in question, the Court has previously ruled that using renewable energy sources to generate electricity is beneficial for environmental protection. It contributes to reducing greenhouse gas emissions, a major factor in the climate change challenge that the European Union and its Member States have committed to addressing.[39]

Finally, it is important to highlight case C-368/10 (*European Commission vs Kingdom of the Netherlands*), in which it was established that not only economic award criteria may be considered, but also qualitative ones, provided that they are linked to the subject matter of the contract, have an objective evaluation and with effective competition conditions. On such criteria, see the excerpt from the decision:

> [...] As the Advocate General points out in point 103 of her Opinion, the fourth paragraph of that recital also indicates that 'a contracting authority may use criteria aiming to meet social requirements, in response in particular to the needs—defined in the specifications of the contract—of particularly disadvantaged groups of people to which those receiving/using the works, supplies or services which are the object of the contract belong'. It must therefore be accepted that contracting authorities are also authorised to choose the award criteria based on considerations of a social nature, which may concern the persons using or receiving the works, supplies or services which are the object of the contract, but also other persons.[40]

It is on the basis of this jurisprudential framework that Directive 2014/24 established the criterion of the most economically advantageous tender, "[...] considering it best to accommodate considerations related to issues of innovation, the environment and the social sphere."[41]

## 18.3 Sustainable Development in Mercosur Public Procurement

Unlike the European Union—a political and economic organization with supranational status and legal personality, comprising 27 countries and with executive, legislative and jurisdictional powers—Mercosur is merely a regional intergovernmental organization between Brazil, Argentina, Uruguay, Paraguay and Venezuela (currently suspended), with no executive capacity. Furthermore, Mercosur does not have an executive body like the European Commission, capable of proposing legislative acts for the bloc autonomously in relation to its Member States.

Thus, in the case of Mercosur, the proposition of rules for the Common Market will be the responsibility of the countries themselves, requiring unanimity among

[39] European Court of Justice [2003]. Case C-448/01. paras 39–40.

[40] European Court of Justice [2012] Case C-368/10. paras 85–87.

[41] Faustino 2014, p. 133.

the members for legislative approval—which is not necessary in the European Union system. The fact is that such unanimity ends up making it impossible to establish sustainable development standards more often in the bloc, as will be shown.

Regarding the bloc's economic matters, according to data from GEO Mercosur, 60.5% of exports between the states parties to the Common Market are related to the trade of primary goods, such as minerals, oil, soybeans, coffee, poultry, beef, among others.[42] In this way, it is notorious that this increase in the trade of agricultural and livestock products linked to the expansion of cultivable areas has culminated in the loss of green areas, serious deforestation, attacks on ecosystems and a reduction in soil quality.[43] Faced with the imminent natural resource crisis, the Mercosur member states have sought several times to create agreements and norms capable of dealing with the problem of deforestation, negotiating texts relating to the bloc's sustainable policy.

The Treaty of Asuncion, the agreement constituting the Southern Common Market, established in its preamble the need to preserve the environment and improve the lives of the peoples of the bloc:

> Considering that the expansion of the current dimensions of their national markets, through integration, is a fundamental condition for accelerating their processes of economic development with social justice
>
> Understanding that this objective should be achieved through the more effective use of available resources, the preservation of the environment, the improvement of physical interconnections, the coordination of macroeconomic policies, the complementation of the different sectors of the economy, based on the principles of gradualness, flexibility and balance;[44]

According to the Larejo and Armeriso, such a preamble assists in establishing adequate conditions of environmental competition among the States Parties and with third parties, making the regional integration process aimed at both strengthening human capital and eradicating differences in social welfare and disparity of opportunities among the participating countries.[45] However, Paixão Silva Oliveira emphasized that the concern with sustainable development and the environment is only established in the preamble of the Treaty, which, in her view, corroborates the thesis that Mercosur member states have relaxed their commitment to the subject.[46] However, she points out that even with its non-binding legal character, the preamble of the Asuncion Treaty functions as "[...] a fundamental precept to guide its interpretation; thus, the inclusion of environmental protection in it can be interpreted as a way chosen by the negotiators to include the environmental theme as a guiding axis in the interpretation of the entire text."[47]

Thus, it can be noted that, with regard to the Treaty of Asuncion, unlike what happened in the founding acts of the European Union, the Mercosur Member States

---

[42] GEO Mercosur 2008.

[43] Food and Agriculture Organization of the United Nations, 2021.

[44] MERCOSUL [1991] Tratado de Assunção.

[45] Laredo and Armeriso 2001, p. 8.

[46] Oliveira 2016, p. 408.

[47] Ibid., p. 9.

have not demonstrated a direct concern with the issue of sustainable development. However, the preamble, even if simple, contributes in some way to the proposition of a sustainable agenda.

Throughout Mercosur's history, it has become clear that the issue of sustainability has not been high on thebloc's agenda. First, it is important to note that the Ouro Preto Protocol, responsible for developing institutional bases for Mercosur and for creating permanent organs of the bloc, such as the Mercosur Administrative Secretariat, the Common Market Council and the Economic and Social Consultative Forum,[48] did not make any other reference to the subject of sustainable development or any kind of environmental protection.

Furthermore, since April 1996, Mercosur's Working Subgroup on the Environment (SGT6) began negotiating the creation of the Additional Protocol to the Treaty of Asunción on the Environment, with the aim of consolidating issues related to sustainable development, which did not yet exist in the Common Market's legislation. According to Paixão Silva Oliveira, the negotiated protocol:

> "[...] had the ambition of making a positive contribution to the internal legislation in force in each Member State, in order to improve the mechanisms for the implementation of environmental legislation and encourage the formation of a 'sustainable Mercosur'. It could be said that the Protocol would represent the materialization and recognition of environmental protection in the Southern Cone ."[49]

The protocol proposal, initiated through the initiative of the Brazilian delegation, included the implementation of sustainable development, the harmonization of laws among the bloc members, the incentive to reduce pollutant gas emissions, the adoption of inspection policies for activities that impact the environment, and the strengthening of environmental management institutions.[50]

In 1997, the project was approved by SGT6 and presented to the Common Market Group (GMC) for further discussions. However, in the GCM, the Argentine delegation vetoed the project, returning to the Subgroup for new discussions and changes in the project's content. Nevertheless, in 1999, after a new submission to the GCM, the Argentine delegation again did not approve the proposal.[51]

As Paixão Silva Oliveira explains, the gap between the position of SGT6 (essentially made up of environmentalists) and the position of the Common Market Group (composed by governmental representatives linked to trade liberalization) made the proposal that elaborated on environmental protection and sustainable development issues in the bloc unfeasible.[52] Furthermore, according to Oliveira, the Argentinean delegation twice refused to approve of the protocol once it feared that the Protocol could make the country's agricultural trade unviable because of the use of genetically modified soybeans. According to the author:

---

[48] MERCOSUL [1994]. Protocolo de Ouro Preto. 1994.

[49] Oliveira 2016, p. 3.

[50] Ibid.

[51] Ibid.

[52] Ibid., pp. 8–9.

> Articles 60 to 62 establish the rules for biosafety. Under the Protocol, each State Party shall establish regulatory mechanisms on biosafety, based on scientific knowledge and biotechnology risk management. The mechanisms shall establish legal instruments for the safety and oversight of the use of biotechnology techniques in the cultivation, handling, transport, marketing, consumption and release of genetically modified organisms, with the aim of protecting the environment and the health of the population. The release of genetically modified organisms into the environment must be preceded by a risk assessment to avoid possible damage.
>
> [...]
>
> in Argentina, in 1999, approximately 90% of soybean exports were of transgenic species, [...] the surface of the area cultivated with genetically modified soybeans increased from 1,756,000 hectares from 1997-1998 to 8,555,000 between 2000-2001. In short, the Argentine blockade is directly linked to the impact on the agricultural market for fear that its adoption would lead to increased production costs and loss of competitiveness.[53]

Considering this difficulty of promoting sustainable development, after years of continuous negotiations, in March 2021 the Mercosur Framework Agreement on the Environment (AQMAM) was signed in Florianopolis, Brazil. According to Larejo and Armeriso, the Framework Agreement sought to establish the need for environmental protection and the most efficient use of available resources, the incorporation of environmental issues in the decisions adopted within the Common Market, and the promotion of sustainable development through reciprocal support between the environmental and economic sectors.[54] However, after a detailed analysis of the Agreement, it can be observed that it establishes ambitious rules, but that effective compliance is more difficult:

> Article 6 The States Parties shall deepen the analysis of the environmental problems of the subregion, with the participation of the competent national agencies and civil society organizations, and shall implement, among others, the following actions: [...]
>
> b) encourage national policies and instruments in environmental matters, seeking to optimize the management of the environment; [...]
>
> e) to contribute to the promotion of environmentally healthy and safe working conditions in order to, within the framework of sustainable development, enable the improvement of the quality of life, social welfare and the generation of employment;
>
> f) To contribute so that the other Mercosur forums and instances consider the relevant environmental aspects in an adequate and timely manner; [...]
>
> h) to encourage scientific research and the development of clean technologies; [...]
>
> j) To encourage the harmonization of legal and institutional guidelines with the objective of preventing, controlling and mitigating environmental impacts in the States Parties, with special attention to border areas; [...]
>
> m) Consider cultural aspects, where relevant, in environmental decision-making processes; and [...][55]

It can be observed that the use of verbs such as "encourage" and "contribute" do not establish a concrete obligation to the Member States, but rather objectives to

[53] Ibid., p. 9.

[54] Laredo and Armeriso 2001, p. 410.

[55] MERCOSUL [2001]. Acordo-Quadro sobre o Meio Ambiente do Mercosul.

be met, without the stipulation of goals or procedures to be followed. In Oliveira's interpretation, the Agreement "[...] looks more like a political declaration than a binding text. Due to the resistance of the State Parties, the agreement sought to encourage rather than impose the preservation of the environment and the practice of sustainable development."[56]

Finally, with regard to sustainable development in public procurement, once again Mercosur's lack of interest in furthering its work on the subject can be observed. As can be seen in the Mercosur Protocol on Public Procurement (MERCOSUR/CMC/P. DEC. No. 37/17), the Common Market has only established in its general principles the need to promote sustainable development, with no further mention of the subject in the other 32 Articles of the Protocol.[57]

This reveals a clear disparity between the proposals related to sustainable development in public procurement in the European Union and Mercosur; something that will certainly cause difficulties in the eventual application of the agreement between the two blocs.

## 18.4 The Agreement between Mercosur and the European Union and Sustainable Development in Public Procurement

On June 28, 2019, Mercosur and the European Union announced the end of a 20-year long negotiation on a free trade agreement between the two blocs. The so-called "Association Agreement between Mercosur and the European Union" includes three essential pillars: political dialogue, cooperation and free trade. Specifically related to the trade part, it addresses issues such as antitrust, customs cooperation, government procurement (access commitments) and sustainable development.[58] With the agreement, it is expected that 92% of Mercosur imports and 95% of tariff lines will enter the European Union tariff-free. The South American market, on the other hand, intends to free 91% of imports from the European Union and 91% of tariff lines.[59]

It turns out that—also according to MEPs Luciana Ghiotto and Javier Echaide—the agreement concluded is extremely asymmetric, since European industry is much more competitive than that of the southern bloc: while in 2018 Mercosur's GDP per capita was $10,600, the European bloc's reached $41,890. According to the Observatory of International Conjuncture and Foreign Policy (OCIPEX), Germany alone has a GDP almost equal to that of Argentina, Brazil, Uruguay, and Paraguay combined.[60]

[56] Oliveira 2016, p. 5.

[57] Article 3—GENERAL PRINCIPLES: [...] 2. Public procurement processes for goods and services shall be geared towards promoting the sustainable development of States Parties.

[58] Brazil 2019; **Acordo de associação Mercosul-União Europeia** 2019, p. 2.

[59] Ibid., p. 3.

[60] Ghiotto and Echaide 2019, p. 17.

Furthermore, the South American market is much more specialized in the production and export of agricultural materials, while the European Union has expertise in the production of technology, making the agreement extremely disadvantageous for Mercosur's industry. Highlighting the discrepancy in production, Ghiotto and Echaide report that:

> During the period 2014-2016, the top twelve products that Mercosur imported from the EU were industrial goods. For example, the first six products were as follows: 1) medicine, vaccines, and immunological products; 2) airplanes and airplane parts; 3) parts and accessories for motor vehicles, motor cars, and vehicles; 4) light and medium oils and preparations; 5) fungicides; 6) light vessels and floating cranes. Mercosur, on the other hand, in the same period, exported products related to agricultural activities and mining to the EU: 1) products related to soybeans (soya-beans and oils); 2) wood pulp; 3) copper and iron ores and concentrates; 4) coffee; 5) petroleum oils; 6) orange juice; 7) bovine meat.[61]

Thus, according to these MEPs, the European Union will import much more meat and agricultural products, which will lead to higher emissions of polluting gasses, deforestation, soil contamination, and expansion of cultivable areas in South American territory, for an example, worsening the situation of the Amazon forest fires. Besides this, the Mercosur bloc will import more cars, chemical products and machinery coming from Europe, risking the industrial production chain of the southern bloc countries.[62]

Regarding the chapter on sustainable development, the agreement establishes principles and commitments of both blocs on environmental issues in the context of trade and investment. In this sense, it reiterates the need to follow the sustainable development goals of the 2030 Agenda, indicating that this will occur "in accordance with the national capacities of the parties".[63]

In addition, this chapter addresses the issues of climate change, making reference to the sustainable management of fisheries and forests, the protection of biodiversity, and the need to comply with the Paris Agreement.[64] Thus, under this agreement, Brazil, the largest Mercosur country, would be committed to act against illegal deforestation and to deliver 12 million hectares of reforestation in the Amazon forest, which plays a crucial role in regulating the planet's climate.[65]

However, according to Abdenur, the inclusion of these clauses in the European Union and Mercosur Agreement does not mean that Brazil would be bound by global warming mitigation commitments, since this Agreement does not establish mechanisms or sanctions in case of non-compliance with the obligations of the Paris Agreement.[66] Moreover, the language used in the clauses weakens the relevance of climate in the agreement and gives member states ample scope to avoid fulfilling

---

[61] Ibid., pp. 16–17.

[62] Ibid., p. 6.

[63] Brazil 2019; Acordo de associação Mercosul-União Europeia, p. 15.

[64] Ibid.

[65] Ghiotto and Echaide 2019, p. 63.

[66] Abdenur 2019.

concrete commitments.[67] For example we can cite Article 6 of the Agreement (on Trade and Climate Change): "The Parties shall cooperate, as appropriate, on issues of climate change linked to trade bilaterally, regionally and in international fora such as the UNFCCC." According to Ghiotto and Echaide, the phrase "as appropriate" removes the obligation for Member States to cooperate effectively on climate change issues.[68] Moreover, according to these European authors, despite the reference to the Paris Agreement, "the impact of trade liberalisation on the climate involves multiple negative effects, including deforestation, emissions from methane gas, the increase in pollution from maritime transport, and an increase in the use of pesticides."[69]

Finally, regarding the clauses related to bidding, the agreement will open the public procurement market for goods, construction and infrastructure services at all levels to European and Mercosur companies under equal conditions. By this, Brazilian companies, for example, will have equivalent access to the European procurement market.[70]

According to the Brazilian government, the agreement will allow an increase in competition in public bids and will provide a more efficient use of public resources, besides ensuring an international standard of transparency rules. It emphasizes, however, that the access of Brazilian suppliers to the European market will be broader than the access of European companies to the Brazilian market. Additionally, public policies in technological development, public health, promotion of micro and small companies and food safety "are still safeguarded".[71]

However, specifically with regard to sustainable development in public procurement, the agreement is silent, not establishing any norm for the promotion of sustainability in light of goal 12.7 of the SDGs, which establishes the need to "promote public procurement practices that are sustainable in accordance with national policies and priorities". Thus, one can observe the great difficulty of the member blocs in effectively contributing to the UN agenda, given the lack of provisions on the matter, in addition to the great difference between the concern that the two blocs give in relation to the sustainability.

## 18.5 Conclusion

After analyzing the legislation of Mercosur and the European Union on sustainable public procurement, it can be observed that only the European standards have effectively contributed to achieving goal 12 of the 2030 Agenda, which is to ensure sustainable production and consumption patterns. Besides, the agreement between

---

67 Ibid.

68 Ghiotto and Echaide 2019, p. 66.

69 Ibid.

70 Ibid., p. 92.

71 Brazil [2019]. Acordo de associação Mercosul-União Europeia, pp. 7–8.

the two blocs does not clearly establish how it will be possible to reconcile public procurement with sustainable development.

In the case of Mercosur, the vagueness of the rules related to the theme, in addition to the lack of procedural discipline, either in the constituent agreements or in the sparse legislation, makes it impractical for any real fulfillment of the sustainable development objective, in the scope of bids in the South American bloc. European Union legislation, on the other hand, is much more consistent in promoting sustainability in public procurement, since it manages to reconcile the concern with public accounts and sustainable development, based on the criterion of the most economically advantageous tender (Article 67 of Directive 2014/24).

It is therefore to be expected that the EU-Mercosur Trade Agreement may encounter many difficulties in its application—especially related to sustainable development in public procurement—due to the undeniable differences between the two blocs. Additionally, the lack of discipline on the subject makes a real improvement of the matter unfeasible within the scope of the legislation of the Southern Common Market. Nevertheless, nothing prevents the South American countries from reconfiguring the rules of the bloc in this regard, and they may even be based on the European criterion of the most economically advantageous tender—found in the Directive 2014/24. At this moment, however, Mercosur does not show the same level of commitment to sustainability with regard to public procurement, which will make it difficult to reach a level-playing field.

## References

Abdenur A (2019) A double-edged sword: The climate politics of the Mercosur-EU Agreement. https://climate-diplomacy.org/magazine/cooperation/double-edged-sword-climate-politics-mercosur-eu-agreement. Acessed in 10 october 2022.

Arrowsmith S, Kunzlik, P (2009) Social and environmental policies in EC procurement law - New Directives and New Directions. Cambridge University Press, New York 2009.

Bovis C (2007) EU Public Procurement **Law**. Cheltenham: Edward Elgar PublishingLimited, Cheltenham.

Brasil, Bárbara Dayana. A contratação pública como mecanismo de fomento para o desenvolvimento sustentável. *In*: SCHIER, Adriana da Costa Ricardo; BITENCOURT, Carolina Müller. Direito administrativo, políticas públicas e estado sustentável. Curitiba, Íthala, 2020. p. 75–96.

Brazil (2019) Acordo de associação Mercosul-União Europeia, 2019. https://www.gov.br/mre/pt-br/arquivos/documentos/politica-externa-comercial-e-economica/2019_10_24__Resumo_Acordo_Mercosul_UE_CGNCE.pdf

Brundtland, G (1987). Our Common Future – The World Commission on Environment and Development. Oxford University, Oxford University Press, New York.

Caranta R, Trybus M (2010) The Law of Green and Social Procurement in Europe v. II. DJØF Publishing Copenhagen, Aarhus.

European Court of Justice (2012) Case C-368/10.

European Court of Justice (2003) Case C-448/01.

European Court of Justice (2002). Case C-513/99.

European Court of Justice (1988). Case C-31/87.

Europe (1958) Consolidated Version Of The Treaty On The Functioning Of The European Union.

Europe (2009) Treaty of Lisbon.
Europe (2014) DIRECTIVE 2014/24/EU OF THE EUROPEAN PARLIAMENT AND OF THE COUNCIL of 26 February 2014.
Europe (2020) **Comissão europeia**: https://europa.eu/european-union/about-eu/institutions-bodies/european-commission_pt. Accessed in 29 july 2021.
FAO Food and Agriculture Organization of the United Nations, (2021). COP26: Agricultural expansion drives almost 90 percent of global deforestation. https://www.fao.org/newsroom/detail/cop26-agricultural-expansion-drives-almost-90-percent-of-global-deforestation/en Acessed in 28 february 2024.
Faustino, P (2014) Award criteria in the new EU Directive on public procurement. Public Procurement Law Review, London.
Feliu, J (2021) Los objetivos de sostenibilidad e inclusividad de la Agenda de Naciones Unidas y su incidência en la contratación pública – De las ideas a la acción .In: Esteve Pardo, J (eds.). La Agenda 2030. Implicaciones y retos para las administraciones locales. Madrid: Fundación Democracia y Gobierno Local, Madrid, p. 67–100.
Freitas, J (2019) Sustentabilidade: direito ao futuro Fórum, Belo Horizonte.
GEO Mercosur. PROGRAMA DE LAS NACIONES UNIDAS PARA EL MEDIO AMBIENTE (PNUMA). GEO Mercosur. Gráfica Mosca, Montevideo.
Ipiranga, Ana Silva Rocha; Godoy, Arilda Schmidt; Brunstein, Janette. Introdução. **RAM. Revista de Administração Mackenzie** (*Online*), São Paulo, v. 12, n. 3, jun. 2011. Disponível em: https://www.scielo.br/j/ram/a/Xv3r9ypsxNsjLtTqtPCBnJP/?format=pdf&lang=pt. Acessed in 16 june 2023.
Laredo, I. M. and Armeriso C. Et al (2021) Sistema de indicadores ambientales para e desarrollo sustentable del Mercosur. https://rephip.unr.edu.ar/bitstream/handle/2133/8132/Laredo%2Cameriso%2Ccicare_sistema%20de%20indicadores%20ambientales.pdf?sequence=3&isAllowed=y Acesses in 17 october 2022.
Mata Diz, J (2019) Os objetivos do Desenvolvimento Sustentável e sua Incorporação pela União Europeia e pelo Brasil. In: MATA DINIZ, J and GAIO, D Desenvolvimento Sustentável na Contemporaneidade. Belo Horizonte: Arraes, Belo Hozionte. P. 84–103.
Mercosul (1991), Tratado de Assunção. 1991 https://www.mercosur.int/documento/tratado-de-assuncao-para-a-constituicao-de-um-mercado-comum/?wpdmdl=4616&masterkey=5b8c5dce93746
Mercosul (1994) . Protocolo de Ouro Preto. https://www.mercosur.int/pt-br/documento/cmc-1994-protocolo-ouro-preto/
Mercosul (2017) Protocolo de Contratações Públicas do Mercosul. https://www.gov.br/siscomex/pt-br/arquivos-e-imagens/2020/12/dec-37-2017-protocolo-de-contratacoes-publicas-cleaned-docx.pdf
Oliveira, L (2016). Direito ambiental do Mercosul: da negociação do protocolo adicional à ratificação do Acordo-Quadro sobre o meio ambiente do Mercosul. In: Revista de Direito Ambiental. Vol. 83 (july - september 2016). http://www.mpsp.mp.br/portal/page/portal/documentacao_e_divulgacao/doc_biblioteca/bibli_servicos_produtos/bibli_boletim/bibli_bol_2006/RDAmb_n.83.13.PDF
UN (United Nations) - Department of Economic and Social Affairs Sustainable Development. (2022) The 17 goals: https://sdgs.un.org/goals

**Bruno Fontenelle** is a Lawyer and holds a Bachelor's and Master's degree in Law and Public Administration from UFMG. Academic exchange student at Radboud University (Netherlands) in 2019.

**Maria Tereza Dias** holds a Ph.D. and Master's degree in Administrative Law from UFMG. Associate Professor at the Department of Public Law at UFMG and a Visiting Professor at King's College London through the UFMG/CAPES/PRINT program. CNPq Researcher. Lawyer.

# Chapter 19
# Sustainability as a Financial Matter in the European Union and Competitive Impacts in Third Countries

**Marina Leonel Pereira Santana Paiva and Rafaela Moraes Cardoso**

**Contents**



**Abstract** This Chapter focuses on analysing the key legislative acts of the European Union pertaining to environmental issues, their impact on corporate governance in third countries, and the global economic and competitive consequences of managing sustainability as a potential financial matter. In particular, this study presents the premises of the Action Plan: Financing Sustainable Growth (2018), the Green Deal (2019), as well as the provisions of the new Corporate Sustainability Reporting Directive (CSRD) and the Corporate Sustainability Due Diligence Directive (CSDD). In this analysis, the mentioned documents guiding sustainability management will be scrutinised with a view to their effects on third countries, including the effects on forest degradation.

**Keywords** Sustainability · European Union · Legislative acts · Financial matters · Environmental issues · Green Deal

---

M. L. P. S. Paiva (✉)
KPMG, São Paulo, Brazil
e-mail: marinalpsp@gmail.com

R. M. Cardoso
Alvarez & Marsal, São Paulo, Brazil
e-mail: mcardoso.rafaela@gmail.com



R. A. Wessel et al. (eds.), *EU External Relations Law and Sustainability*, Global Europe: Legal and Policy Issues of the EU's External Action 4,
https://doi.org/10.1007/978-94-6265-655-0_19

## 19.1 Introduction

Ever since the term Environmental, Social and Governance (ESG) was first coined by the United Nations Environment Programme Finance Initiative (UNEP FI) in its study 'Who cares wins',[1] environmental risks have been understood as non-financial risks. This is because traditional financial risk management has been considered non-financial "all quantitative and qualitative data on the policy pursued, the business operations, and the results of this policy in terms of output or outcome, without a direct link with a financial registration system".[2] Therefore, sustainability was imputed as a minor aspect within a broader and less specific realm, representing all non-financial issues.

Until November 2022, the Non-Financial Reporting Directive (NFRD) outlined the European Union's overarching objectives regarding the issue of social and environmental performance reports. However, in January 2023, aligning with the Green Deal objectives, a new directive called the Corporate Sustainability Reporting Directive, was published. This piece of legislation will affect reports to be presented in 2025, covering the 2024 financial year. CSRD is a consequence of the Green Deal and aims to support the achievement of its goals. It also aligns with the provisions outlined in Articles 11 and 191 of the Treaty on the Functioning of the European Union (TFEU), which call for integrating environmental issues across European Union policies. In essence, the mentioned directives seek to (i) enhance transparency in the disclosure of social and environmental information provided by companies, as well as (ii) encourage the adoption of corporate practices that align with ESG principles.

The European Union (EU) was able to impact the world's commercial and competitive manner of coping with sustainability. This influence outside EU's borders contains three main elements: (i) *arguing*, which involves persuasion, (ii) *bargaining*, which can involve offering and granting rewards and threatening punishment, and (iii) *coercion*, which entails inflicting punishment.[3] In other words, the EU has the potential to address the need for regulation on sustainability issues beyond the borders of the European Union, "commanding change, controlling agendas, and establishing preferences",[4] performing the so-called 'hard' and 'soft power'[5] especially in relation to developing countries. In this context, the European Union faces the challenge of promoting sustainable economic growth while upholding the 'no one is left behind' Green Deal principle. This principle mandates preserving the security of supply and competitiveness, even in situations where others are unwilling to take action.

This Chapter focuses on analysing the key European Union legislative acts pertaining to environmental issues with a view to assessing their impact on corporate governance and the global economic and competitive consequences of managing

---

[1] United Nations 2004.

[2] Haller 2017, pp. 407–429.

[3] Schunz 2014.

[4] Nye Junior 2011.

[5] Nye Junior 2004.

sustainability as a potential financial matter. To achieve that, the Chapter will present the foundations of (i) the 'Action Plan: Financing Sustainable Growth' (2018), (ii) the Green Deal (2019), as well as the provisions of the new sustainability reporting directive, (iii) the CSRD, and, from a target perspective, the (iv) Regulation on Deforestation-free Products, which is a recent regulation that has entered into force as part of a broad strategy to address deforestation and forest degradation. Subsequently, an imminent European Union legislative act will be discussed: the CSDD, which mandates specific due diligence requirements concerning environmental issues.

## 19.2 Sustainability-related European Legal Milestones

Under EU law, a 'Directive', such as the CSRD, is a type of legislative act that establishes specific objectives that all European Union Member States are required to meet by a set date.[6] How these objectives are implemented, is left to the discretion of each Member State, which will be responsible for developing (or adapting) its own laws and measures to meet these goals. 'Regulation', such as the Regulation on Deforestation-free Products, is a legislative act that applies directly and uniformly throughout the European Union.[7]

In order to fulfil the objectives of sustainable economic growth, the European Commission has proposed new Directives and Regulations regarding the treatment and reporting of sustainability by corporations.

### *19.2.1 The Action Plan on Financing Sustainable Growth' and 'The Green Deal'*

A Communication entitled 'Action Plan: Financing Sustainable Growth'[8] was published by the European Commission in 2018. This initiative is a milestone in establishing the relationship between financial/economic factors and environmental matters, especially regarding integration between ESG concepts and investment decisions. The Plan set the following main objectives:

1. reorient capital flows towards sustainable investment in order to achieve sustainable and inclusive growth;
2. manage financial risks stemming from climate change, resource depletion, environmental degradation and social issues; and
3. foster transparency and long-termism in financial and economic activity.

---

[6] European Union.

[7] Ibid.

[8] European Commission 2018.

Furthermore, the document provides comprehensive information on the outlined objectives, corresponding initiatives and deadlines. Consequently, it becomes imperative to establish regulations mandating companies to engage in long-term planning pertaining to environmental risk management (objective 2) and to transparently report their endeavours in this regard (objective 3). This scenario implies the disclosure of information to the market, enabling capital allocation towards sustainable investments (objective 1).

The year following the publication of the 'Action Plan: Sustainable Growth', the Green Deal was published, proposing (i) zero greenhouse gas emissions by 2050 and (ii) promoting economic growth decoupled from the depletion of natural resources. In addition, the pact goes beyond the mere establishment of goals and institutes concrete actions to effectively achieve the proposed objectives, accompanied by a schedule for the completion of each stage. To ensure proper implementation of actions and compliance with the schedule, the Green Deal itself stipulates that:

> in addition to launching new initiatives, the Commission will work with the Member States to step up the EU's efforts to ensure that current legislation and policies relevant to the Green Deal are enforced and effectively implemented (EUROPEAN UNION, 2019).

In line with the premises of the Action Plan, the Pact established that:

> Sustainability should be further embedded into the corporate governance framework, as many companies still focus too much on short-term financial performance compared to their long-term development and sustainability aspects. At the same time, companies and financial institutions will need to increase their disclosure on climate and environmental data so that investors are fully informed about the sustainability of their investments

That being said, among the planned measures to achieve the goals set, the Green Deal also determines that 'the Commission will review the Non-Financial Reporting Directive'.[9]

### *19.2.2 The Corporate Sustainability Reporting Directive (CSRD)*

As a result of the revision determined by the Green Deal, Directive 2022/2464, known as the CSRD, was published on December 14, 2022. In force since January 5, 2023, the directive introduced changes to the sustainability reporting requirements for companies, replacing the rules of the previous Directive 2014/95, the Non-Financial Reporting Directive (NFDR).

The NFDR was applicable to large undertakings categorised as 'public-interest entities exceeding on their balance sheet dates the criterion of the average number of 500 employees during the financial year'.[10] This included financial institutions,

---

[9] European Union 2019.

[10] European Union 2014a, b.

companies listed on the stock exchange, and insurance companies, among others, totalling approximately 11,700 companies and groups in the European Union.[11]

In contrast, the CSRD extends its applicability to a broader range of entities, including small and medium-sized public-interest enterprises and companies that meet at least two of the following three criteria: (i) have more than 250 employees, (ii) turnover greater than EUR 40 million and (iii) more than EUR 20 million total assets listed.[12] This implies that approximately 50,000 companies will be required to publish periodic reports on their impacts on social and environmental issues.[13] This expansion in reporting requirements signifies the increased scope of reporting obligations.

In terms of content, there were also significant changes from the NFRD[14] to the CSRD.[15] The NFRD addressed environmental aspects as a non-financial issue alongside social matters such as treatment of employees, respect for human rights, anti-corruption and bribery and diversity in the boards of directors of companies (in terms of age, gender, education, and professional training). the requirements listed by the CSRD, on the otehr hand, include (i) a description of the company's model and strategy, (ii) the preparation and reporting of a schedule of objectives related to long-term sustainability issues and their monitoring, role and performance of governance in the defined objectives, (iii) a due diligence process applicable to the operation and supply chain to ensure compliance with the established environmental goals, and (iv) the description of the main risks associated with the company's activities and the mitigating measures and controls implemented.[16]

Besides that, the CSRD[17] considers that:

> There has been a very significant increase in demand for corporate sustainability information in recent years, especially on the part of the investment community. That increase in demand is driven by the changing nature of risks to undertakings and growing investor awareness of the financial implications of those risks. That is especially the case for climate-related financial risks.

In this sense, it is evident that the CSRD can be understood as a tool to mitigate addressing environmental issues as both an ethical or moral dilemma and a potential financial matter.[18] It is important to note that the CSRD does not explicitly state that environmental issues are financial issues but rather recognises the practical impacts of dealing with the matter—prioritising substance over form—which aligns with the financial agenda.

Additionally, it is worth mentioning that, with the expectation of strengthening the Green Deal guidelines, the CSDR emerged in a 'context of increased demand

---

[11] European Parliament 2022.

[12] European Union 2022.

[13] European Parliament 2022.

[14] European Union 2014a, b.

[15] European Union 2022.

[16] Ibid.

[17] Ibid.

[18] Ibid.

for topics related to corporate sustainability, due to inherent risks associated with the topic, and global mobilisation in response to climate and cultural changes'.[19] Among those modernising and strengthening expectations, the inaccuracy of the 'non-financial' term, used as a premise of the NFDR, is mentioned:

> Many stakeholders consider the term 'non-financial' to be inaccurate, in particular because it implies that the information in question has no financial relevance. Increasingly, however, such information does have financial relevance. Many organisations, initiatives and practitioners in the field of sustainability reporting refer to 'sustainability information'. It is therefore preferable to use the term 'sustainability information' in place of 'non-financial information'.[20]

Along the same line, even if indirectly, the Action Plan and the Green Deal establish economic investment premises and goals regarding sustainable growth, leading to a consolidation of sustainability as a financial aspect. In conclusion, regardless of how sustainability issues are referred to, they have now been set as potential financial matters.

### *19.2.3 The Corporate Sustainability Due Diligence Directive (CSDD)*

Corroborating with the ideas presented above, on Dezember 2023, "the Council and the European Parliament reached a provisional deal on the corporate sustainability due diligence directive".[21] For now, the following categories fall under the scope of the CSDD[22]:

| Companies | Generated global (net) turnover | Employees |
|---|---|---|
| EU-based companies | Exceeding EUR 40 million | More than 250 employees |
| EU-based parent companies | Exceeding EUR 150 million | More than 500 employees |
| Non-EU companies | Exceeding EUR 150 million, provided that at least EUR 40 million is generated from within the EU | N/A |
| Non-EU parent- companies | Exceeding EUR 150 million, from which at least EUR 40 million was generated in the EU | More than 500 employees |

Regarding the CSDD, it is mentioned that

> the behaviour of companies across all sectors of the economy is key to succeed in the Union's transition to a climate-neutral and green economy in line with the European Green Deal and

[19] Ibid.

[20] Ibid.

[21] European Council 2023.

[22] Bosselaar et al. 2023.

> in delivering on the UN Sustainable Development Goals. [...] This requires implementing comprehensive mitigation processes for adverse human rights and environmental impacts in their value chains, integrating sustainability into corporate governance and management systems, and framing business decisions in terms of human rights, climate and environmental impact, as well as in terms of the company's resilience in the longer term.[23]

That said, in order to achieve the goals, the Directive provides that corporations will have 'to identify and prevent, bring to an end, or mitigate the actual and potential impacts of their activities on the environment and on human rights abuses'.[24] Therefore, due diligence regarding sustainability and social matters must be conducted for the company's operations and third parties in their value chains with whom it has business relationships—directly or indirectly.

The Directive will be enforced both at the Member State level and through the company's governance. According to Article 22(1), Member States:

> shall ensure that companies are liable for damages if: (a) they failed to comply with the obligations laid down in the directive and; (b) as a result of this failure an adverse impact that should have been identified, prevented, mitigated, brought to an end or its extent minimised through the appropriate measures mentioned in the document occurred and led to damage.[25]

Further, Article 25(1) stipulates that Directors of corporations bound by the CSDD should 'take into account the consequences for sustainability matters when fulfilling their duty to act in the best interest of the company'.[26]

The substantive content complements what was established by the CSDR, expanding the company's responsibility to mitigate environmental risk and determine action plans not only for its own business but also for its suppliers. These predictions represent a major turning point in how sustainability is managed and reported—involving all agents of the supply chain—and reinforce the financial aspect of environmental matters developed in the aforementioned previous legislation.

### *19.2.4 The Regulation of Deforestation-Free Products*

In a more targeted fashion than the previously mentioned instruments, on June 9, 2023, the European Parliament formally approved Regulation (EU) 2023/1115, that substitutes the Timber Regulation,[27] published in 2010, which also aimed to mitigate deforestation, with a specific target on timber and timber products. In a broader scope, the new Regulation (i) prohibits the importation of relevant commodities and relevant products linked to deforestation and demands that they (ii) are produced according

[23] European Commission 2022.

[24] KPMG Internacional 2022.

[25] European Commission 2022.

[26] Ibid.

[27] European Union 2010.

to its origin country's laws and (iii) have undergone due diligence.[28] The main goods affected are wood, cattle, cocoa, coffee, palm oil, soy and their derivatives.

Regarding third countries, Article 7 determines that the first natural or legal person established in the Union who makes relevant products from a person established outside the Union available on the market shall be deemed to exercise the due diligence processes,'[29] which includes the collection of data. The materials must indicate the product's name, type and quantity, a description with all relevant products or commodities contained in the imported goods. Additionally, the documentation should confirm the country of production and viable information that the product was produced in compliance with local legislation and is unlinked to deforestation.

Subsequently, with all this information, the operator[30] must perform a risk assessment to verify 'whether there is a risk that the relevant products intended to be placed on the market or exported are non-compliant'.[31]

These standards are directly linked to the global economy, especially because the mentioned products are mostly exported by developing countries, which tend to have less strict regulations regarding environmental matters. That said, Regulation 2023/1115 can create competition disadvantages and segregation of products, as will be discussed on the next topic, regarding the practical impacts of the legislative acts presented.

## 19.3 Governance and Economic Impacts

Even though all entities, either governmental or non-governmental, coexist in the same market and may have some common goals, their core purpose, risks and roles in society are often very different, sometimes even antagonistic. For this reason, their perspective on the importance and the manner of developing a more environmentally friendly society is quite distinct. However, governments benefit from the privilege of legal power to lead changes towards environmental progress through sustainability agendas, as well as to dictate regulatory requirements applicable to firms.

Regarding the European Union, 'because of its particular historical evolution, its hybrid polity, and its constitutional configuration, can be said that it has a normatively different basis for its relations with the world'.[32] This introduces the concept of normative power, that, in the eyes of Manners, suggests that 'not only is the EU

---

[28] European Union 2023a, b.

[29] Ibid.

[30] Operator means any natural or legal person who, in the course of a commercial activity, places relevant products on the market or exports them. European Union 2023a, b, p. 16. https://eur-lex.europa.eu/legal-content/EN/TXT/PDF/?uri=CELEX:32023R1115. Accessed 13 June 2023.

[31] Ibid.

[32] Manners 2002.

constructed on a normative basis, but importantly that this predisposes it to act in a normative way in world politics'.[33]

In a normative sense, this action can take different forms, but for the present study the concept of transference is worth highlighting: diffusion takes place when the EU exchanges goods, trade, aid or technical assistance with third parties through largely substantive or financial means. Such transference may be the result of the exportation of community norms and standards[34] or the 'carrot and stickism' of financial rewards and economic sanctions'.[35]

Hence through transference, the European Union guarantees a central role in terms of adjusting environmental agendas using commerce as an instrument to achieve its goals.[36] Seen in these terms as a competitive tendency, compliance with EU Directives, Regulations, decisions, recommendations, and opinions represents a potential commercial advantage. This applies even to companies that are not directly regulated by these legislative acts, which brings the discussion to the 'soft power' theme. First mentioned by Nye Junior, 'soft power' has been described as the manner in which a country gets the outcomes without tangible threats of payoffs, in other words, without inducements or threats. 'Soft power' is used as a tool to co-opt countries rather than coerce them.[37] It is important to point out that the EU legal framework has been interfering rather indirectly with private companies, as well as entire countries, as will be exemplified below. In what follows, some 'soft power' interactions permeated by the aforementioned legislative acts of the European Union will be presented.

First of all, the Green Deal has influenced the reform of comprehensive fiscal policies by setting the removal of 'subsidies for fossil fuels, shifting the tax burden from labour to pollution, and taking into account social considerations'.[38] It is important to highlight that fossil fuel is part of a historically structural power tool in the interaction between countries, as well as a financial possibility to enable development, especially in the context of developing countries.[39]

Therefore, the plans described in the Green Deal are tangible not only for the European Union countries but also, sensibly, for developing countries. This is because if they often have no choice but to follow the EU's lead—which, as explained, has a tendency of happening because of Europe's normative power—changing the main source of financing welfare requires investments and considerable effort from companies and countries to adapt to the new reality.

Khor (2017), an activist in civil society, former member of the United Nations Committee on Development Policy, and Executive Director of the South Center (an intergovernmental organisation of developing countries)—explained that:

---

[33] Ibid.

[34] Cremona 1998.

[35] Manners 2002.

[36] European Commission 2024.

[37] Nye Junior 2004.

[38] European Commission 2018.

[39] Wang H et al. 2022.

> some developing countries have also benefited from the proliferation of global value chains in which developing countries often produce an intermediate product that are then shipped crossborder to other locations; levies on bunker fuel and other transportation costs could reduce the extent of such trade to the disadvantage of developing countries. Where developing countries currently enjoy competitive advantage, there is a direct impact on livelihoods and employment and a longer term impact on the decline in the availability of resources developing countries could use for investing in new sectors and economic diversification.[40]

Besides that, developing countries often face limited resources and fragile infrastructures, as well as challenges resulting from climate change regarding increasing vulnerabilities to natural disasters.[41] Briefly, 'soft power' co-opt developing countries, implying they rely on constrained resources to mould their preferences in alignment with EU regulations. Meanwhile, they also face inequalities regarding structures to deal with climate change.[42]

Accordingly, the well-known scenario of potential competitive and social imbalances arises from the dichotomy represented by developed and developing countries. Developed countries seem to have various advantages since their governmental agenda has greater capacity to invest resources in environmental sectors. In contrast, in developing countries, social issues still occupy a large portion of budgets and require greater efforts than in developed countries. In those cases, resources available to invest in adapting sustainable standards and/or proposing subsidy programs for such adjustments are much restricted.

The CSDR tends to incorporate ESG issues as a potential financial matter, indirectly helping to set the order for developing countries. As previously seen, one of the significant differences between the NFRD and the CSRD is the range of companies required to publish periodical reports on the impact of ESG issues. The new Directive includes some third-country companies, which reaffirms the concept of European influence as "commanding change, controlling agendas, and establishing preferences".[43] In other words, 'soft powering' developing countries with ESG legal acts may result in the construction of market opportunities while simultaneously bolstering the competitive advantage held by developed countries in the realm of global trade.[44] Additionally, the CSDD reiterates the CSRD provisions concerning due diligence reporting. In the first directive, due diligence is mentioned in a more general manner. However, the CSRD's proposed provision provides mandatory and specific requirements. In the same vein, the CSDD will be applicable to both inside and outside companies of the European Union.

This situation constitutes a paradox: at the same time, the European Union (i) leads the way towards a sustainable economy but (ii) can compromise the hegemony and competitiveness of the global economic environment. In order to elucidate that, it is worth resuming the Regulation on Deforestation-free Products prohibiting the

---

[40] Martin et al. 2017.

[41] Chinowsky et al. 2011.

[42] Nye Junior 2004.

[43] Nye Junior 2011.

[44] Martin et al. 2017.

importation of goods linked to deforestation. Initially, it is known that most of the deforestation in the world occurs in developing continents, especially South America and Africa. More specifically, in Brazil, deforestation is mostly related to the formation of pasture for cattle farming and the cultivation of soybeans, which are the country's main exports.[45]

The EU is one of the major importers of the aforementioned Brazilian products, which are now classified as relevant commodities, falling under the scope of Regulation 2023/1115. A study funded by the NGO German Watch[46] has revealed that during the early 2000s, the EU was the biggest external contributor to deforestation in Brazil due to the importation of meat and soybean products. The new regulation comes as a reinforcement of Europe's late goal to decrease its participation in deforestation regarding its imports. Still, it also constitutes a hard hit on countries such as Brazil, which throughout the years had increased their production to support the European market, many times sacrificing environmental protection to achieve that. A researcher from the Norwegian University of Science and Technology analyses that 'developed countries are driving deforestation overseas through consumer choices, since the farmers, foresters and countries where deforestation is occurring are responding to signals from the global market'.[47]

For example, the Brazilian government is concerned that the Regulation does not differentiate between legal and illegal deforestation, which is incompatible with the current Brazilian environmental legislation. The legislation stipulates that the 'original vegetation must be preserved in 80% of the property in the Amazon, 35% in the cerrado (savannah), and 20% in other areas. In other words, the remaining percentages are legally authorised for exploitation,[48]' but the goods produced there do not comply with Europe's new standards.

On the other hand, the 'Ambassador-Chief of the Delegation of the European Union in Brazil, Ybáñez, believes that the law focuses on combating forest degradation and Brazil can turn it into a competitive advantage'.[49] Considering Mr. Ybáñez's opinion, it is undeniable that Regulation 2023/1115, as well as the other directives, present an opportunity to turn sustainability into a competitive advantage. However, this only applies when discussing countries with the same level of development. As mentioned, developed countries already have the advantage of addressing social issues more effectively, leading to a more accessible agenda and budget to promote changes in the environmental sector compared to others.

In summary, our study confirms that through its legislations and 'soft power', the European Union sets an important standard that motivates third countries to adapt and become more 'eco-friendly'. However, that change is much more challenging for developing countries due to the higher opportunity cost of investing in environmental matters compared to developed ones.

---

[45] Zell-Ziegler 2017.

[46] Ibid.

[47] Tandon 2021.

[48] Brasil 2012.

[49] Loures 2023.

According to the KPMG Global Survey of Sustainability Reporting 2022, regional N100,[50] reporting rates continue to trend upward in North America and Europe, whereas Latin America dropped 12% over the past 5 years. To be more specific, the research shows that

- in North America, 88% of the companies reported on the sustainability of ESG, while 95% in 2020 and 97% in 2022, representing an increase of 9% in the past 5 years;
- in Western Europe, 82% of the companies reported on the sustainability of ESG, while 83% in 2020 and 85% in 2022, representing an increase of 3% in the past 5 years;
- in Eastern Europe, 65% of the companies reported on the sustainability of ESG, while 74% in 2020 and 72% in 2022, representing an increase of 7% in the past 5 years;
- in Latin America, 81% of the companies reported on the sustainability of ESG, while 72% in 2020 and 69% in 2022, representing a decrease of 12% in the past 5 years.

Besides that, there are 20 countries the sustainability reporting rate scores of which were higher than 90% in 2022. Among them, there are 2 North American countries (Canada, United States), 9 European (Finland, Germany, Iceland, Ireland, Italy, Norway, Spain, Sweden, United Kingdom), 1 African (South Africa) and 7 Asian (Japan, Malaysia, Pakistan, Singapore, South Korea, Taiwan, Thailand), with three of them being part of the Asian Tigers, a group of Asian economies that have experienced rapid industrialisation followed by a continuous growth rate of around 7%.[51] It is clear attempts toa homogenise international commercial interactions have are intervened by historical structures, 'shaped by vast disparities in wealth, income, and bargaining power of different states, corporations, and consumers'.[52]

Fostering good ESG practices is necessary, not only in developed countries but also in developing countries. Especially with regard to sustainability, it is worth importing the principle formalised in the United Nations Framework Convention on Climate Change (UNFCCC): 'Common But Differentiated Responsibilities'. According to this principle, all countries bear responsibility for climate change, but not to the same degree, considering the countries' industrialisation history. Therefore, 'developed country Parties are to provide financial resources to assist developing country Parties in implementing the objectives of the UNFCCC'.[53] Along the same line, the Green Deal established the 'no one is left behind principle, recognising the need to maintain the security of supply and competitiveness even when others are unwilling to act'.[54]

[50] N100 are the worldwide sample of the top 100 companies by revenue in 58 countries, territories and jurisdictions. KPMG Internacional 2022. https://assets.kpmg.com/content/dam/kpmg/se/pdf/komm/2022/Global-Survey-of-Sustainability-Reporting-2022.pdf. Accessed 10 July 2023.

[51] Ibid.

[52] Christoff et al. 2013.

[53] United Nations.

[54] European Commission 2019.

That said, it is worth highlighting that the predictions resulting from the EU instruments mentioned throughout this Chapter, induces all agents in the economic market to comply with the terms and stipulated deadlines. In this logic, adaptation to the legislation and compliance with the ESG indicators may result in changes in the competitive dynamics of a given market. This is due to to the fact that the costs incurred for adapting the production chain to a sustainable model,

> are not necessarily incorporated into the price of products and service, companies that are more environmentally proactive and have the capacity to produce and offer environmentally friendly products on the market, due to their higher costs, may face greater challenges in competing with their rivals.[55]

In view of this, in order to harmonize the benefits derived from adhering to social norms and expectations—resulting in goodwill—with potential decreases in comparative advantage. Comparative advantage refers to a producer's ability to manufacture a product with a lower opportunity cost compared to another producer, influenced by transformation costs. Additionally, companies have aimed to comply with legislative norms and market expectations. As a result, they have collaborated on initiatives for the advancement of green technology and the integration of sustainable practices.

Given this scenario, companies have endeavored to *balance* the advantages obtained by complying with social norms and expectations—translated into goodwill[56]—with eventual reductions in comparative advantage. Comparative advantage refers to the capacity of a producer to manufacture a product with a lower opportunity cost when compared to another producer,[57] brought by the costs of transformations. Additionally, companies have aimed to *comply* with the norms established by the legislator and expected from the market. As a result, they have sought to act in collaboration for the development of green technology and harmonisation of sustainable practices.

While states traditionally took the first steps in multilateral environmental negotiations and summits, they now find themselves drawing on the participation of a much broader range of actors.[58] With regard to the European Union, following the initial actions facilitated by the Green Deal, both businesses and non-governmental institutions have devised strategies to adapt to the ongoing and forthcoming shifts. Yet, even though the new policies of the European Union are indispensable tools to lead the world to sustainable growth and preservation of the environment, they may imply a potential increase in inequalities between countries.

---

[55] Lameira 2018.

[56] "Goodwill can be described as the sum of those intangible attributes of a business that contribute to its success, such as a favorable location, a good reputation, the skill and expertise of its employees and managers, and its enduring relationship with creditors, suppliers, and customers." GLAUTIER, M. W. E.; UNDERDOWN, B. Accounting theory and practice. 7., Essex: Prentice Hall, 2001 apud MARTINS, E.; ALMEIDA, D. L.; MARTINS, E. A.; COSTA, P. S. Goodwill: uma análise dos conceitos utilizados em trabalhos científicos. Revista Contabilidade & Finanças - USP, v. 21, n. 52, art. 5, pp. 1–25, 2010.

[57] Mankiw 2014. Our translation.

[58] Ibid.

## 19.4 Conclusion

Since its conception, ESG has been understood as a non-financial issue from a traditional financial risk management perspective. However, the European Union has been modernising its legislation concerning ESG and has incorporated the idea that those issues can also be understood as potential financial matters. Besides that, there is an expansion of companies that are regulated by the mentioned legislations, now including third-country companies.

This modernisation includes (i) the new Corporate Sustainability Reporting Directive (CSRD), which supports the Green Deal's goals and emphasises transparency in disclosing social and environmental information; (ii) the Regulation on Deforestation-free Products—Regulation (EU) 2023/1115, which is expected to minimise greenhouse gas emission and global biodiversity loss; and (iii) the still not adopted Corporate Sustainability Due Diligence Directive CSDD, that aims to establish a corporate due diligence duty.

These legislative instruments can potentially influence global corporations and governments to view sustainability as a financial issue. They also address sustainability regulation needs beyond the EU, considering it's function asa role model with 'soft power' influence. The challenge now is to promote sustainable economic growth without compromising the Green Deal's 'no one is left behind' principle.

## References

Bosselaar M et al. (2023) The Corporate Sustainability Due Diligence Directive (CSDDD): the position of the European Parliament and potential impact. https://www.loyensloeff.com/insights/news--events/news/the-corporate-sustainability-due-diligence-directive-csddd-the-position-of-the-european-parliament-and-the-impact/. Accessed 13 July 2023.

Brasil (2012) Lei nº 12.651. https://www.planalto.gov.br/ccivil_03/_ato2011-2014/2012/lei/l12651.htm. Accessed 30 January 2023.

Christoff P and Eckersley R (2013) Globalisation and the Environment. Cambridge, Polity Press.

European Commission (2018) Action Plan: Financing Sustainable Growth. https://eur-lex.europa.eu/legal-content/EN/TXT/?uri=CELEX:52018DC0097. Accessed 1 July 2023.

European Union (2012) Consolidated version of the Treaty on the Functioning of the European Union. https://eurlex.europa.eu/LexUriServ/LexUriServ.do?uri=CELEX:12012E/TXT:en:PDF. Accessed 20 May 2023.

European Union (2023) Regulation (EU) 2023/1115. https://eur-lex.europa.eu/legal-content/EN/TXT/PDF/?uri=CELEX:32023R1115. Accessed 14 June 2023.

European Union (2023) Types of legislation. https://european-union.europa.eu/institutions-law-budget/law/types-legislation_en. Accessed 13 July 2023.

European Union (2023) Corporate sustainability reporting. https://finance.ec.europa.eu/capital-markets-union-and-financial-markets/company-reporting-and auditing/companyreporting/corporate-sustainability-reporting_en. Accessed 10 July 2023).

European Union (2019) Green Deal. https://eurlex.europa.eu/resource.html?uri=cellar:b828d1651c2211ea8c1f01aa75ed71a1.0002.02/DOC_1&format=PDF. Accessed 27 September 2023.

European Union (2014) Directive 2014/95. https://eur-lex.europa.eu/eli/dir/2014/95/oj>. Accessed 5 March 2024

European Union (2022) Directive 2022/2464. https://eur-lex.europa.eu/legal-content/EN/TXT/PDF/?uri=CELEX:32022L2464. Accessed 27 September 2023.

European Commission (2022) Proposal for a DIRECTIVE OF THE EUROPEAN PARLIAMENT AND OF THE COUNCIL on Corporate Sustainability Due Diligence and amending Directive (EU) 2019/1937. https://eur-lex.europa.eu/resource.html? uri=cellar:bc4dcea4-9584 11ecb4e401aa75ed71a1.0001.02/DOC_1&format=PDF. Accessed 27 September 2023.

European Union (2014) Directive 2014/95. https://eur-lex.europa.eu/legal-content/EN/TXT/PDF/?uri=CELEX:32014L0095. Accessed 1 May 2023.

Haller A et al. (2017) "The term 'non-financial information' – a semantic analysis of a key feature of current and future corporate reporting". Accounting in Europe 14/3: 407-429.

KPMG Internacional (2022) Big shifts, small steps. https://assets.kpmg.com/content/dam/kpmg/se/pdf/komm/2022/Global-Survey-ofSustainability-Reporting-2022.pdf. Accessed 10 July 2023.

Lameira E (2018). Os Acordos Ambientais entre Empresas no Direito da Concorrência da União Europeia. Tese (Mestrado em Direito) – Especialização em Ciências Jurídico-Empresariais, Menção em Direito Comercial, Universidade de Coimbra.

Loures V (2023) Diante de impactos comerciais, Brasil pode recorrer à OMC contra lei europeia sobre desmatamento. https://www.camara.leg.br/noticias/979331-diante-de-impactos-comerciais-brasil-pode-recorrer-a-omc-contra-lei-europeia-sobre-desmatamento/. Accessed 11 2023.

Mankiw G (2014) Introdução à Economia. São Paulo, Cenage Learning.

Martin K et al. (2017) Promoting Sustainable Development by Addressing the Impacts of Climate Change Response Measures on Developing Countries. https://martinkhor.org/wp-content/uploads/2019/10/RP81_Promoting-Sustainable-Development-by-Addressing-the-Impacts-of-Climate-Change-Response-Measures-on-Developing-Countries_EN-1.pdf. Accessed 3 July 2023.

Manners I (2002) Normative Power Europe: A Contradiction in Terms?. JCMS 40: 235-58

Nye Junior J (2004) Soft Power: the means to success in world politics. New York, Public- Affairs.

Nye Junior J (2011) The future of Power. New York, Public- Affair

Schunz S (2014) European Union Foreign Policy and the Global Climate Regime. Peter Lang S.A. International, Academic Publishers Brussels.

Tandon A (2021) Scientists calculate trade-related 'deforestation footprint' of rich countries. https://www.carbonbrief.org/scientists-calculate-trade-related-deforestation-footprint-of-rich-countries/. Accessed 20 Jun 2023.

Wang H et al. (2022) Fossil Energy Demand and Economic Development in BRICS Countries. https://doi.org/10.3389/fenrg.2022.842793/full. Accessed 27 September 2023.

Zell-Ziegler C (2017) Is the EU a Major Driver of Deforestation in Brazil? Quantification of $CO_2$-emissions for Cattle Meat and Soya Imports. https://www.germanwatch.org/sites/default/files/publication/19810.pdf. Accessed 12 July 2023.

**Marina Leonel Pereira Santana Paiva** is a senior consultant at KPMG Brazil. MBA in Management and Digital Transformation at the University of São Paulo (USP). Bachelor in Law from the Federal University of Minas Gerais (UFMG).

**Rafaela Moraes Cardoso** is a senior consultant at Alvarez & Marsal specialized in complex corporate investigations and intelligence background check. Bachelor in Law from the Federal University of Minas Gerais (UFMG).

# Chapter 20
# The Environmental Legal Framework Governing the Relations Between the EU And Serbia as a Candidate Country

**Dunja Duić and Uroš Ćemalović**

## Contents




**Abstract** The Republic of Serbia is officially a candidate for the European Union (EU) accession for more than ten years, as membership negotiations started in 2014. However, the negotiations on Cluster 4 (Green Agenda and Sustainable Connectivity) were opened only in December 2021, eight years after the first EU-Serbia Intergovernmental Conference. As was the case for all former candidate countries in recent EU enlargement rounds, the legal requirements on environmental protection and sustainability have occupied an important place in the accession negotiations, reflecting also some of the Union's major orientations in its external relations taken *lato sensu*. The objective of this Chapter is to offer a case study of the complex and often tense relations between, on the one hand, the EU's legal requirements on environmental protection, and, on the other, its enlargement policy and membership negotiations with, in this case, Serbia. The first part of the analysis defines the place of environmental sustainability in the EU's external action and examines how the Union's actions influenced sustainability and environmental policies. The second part of the analysis examines how and to which extent the existing EU's environmental



D. Duić (✉)
Faculty of Law, University of Osijek, Stjepana Radića 13, 31000 Osijek, Croatia
e-mail: dduic@pravos.hr

U. Ćemalović
Institute of European Studies, Trg Nikole Pašića 11, 11000 Belgrade, Serbia
e-mail: uros@ies.rs





R. A. Wessel et al. (eds.), *EU External Relations Law and Sustainability*, Global Europe: Legal and Policy Issues of the EU's External Action 4,
https://doi.org/10.1007/978-94-6265-655-0_20

legal framework influences its relations with Serbia as a candidate country. In conclusion, the Chapter argues that—in spite of its growing importance—the EU's legal framework on sustainability and environmental protection encounters ever-growing challenges regarding its implementation in the candidate countries.

**Keywords** EU external relations law · Environmental law · Sustainable development · EU enlargement · Law approximation · Serbia

## 20.1 Introduction

In the European Union's (EU) external relations taken *lato sensu*, the weight and importance of its sustainability-related policies and legislation are progressively increasing. The same holds true for the EU's enlargement policy, especially with regard to the Western Balkans candidate countries, one of those being Serbia. However, while the membership incentive was a given in the previous EU enlargement rounds, the current candidates find that "the conditions are much less conducive for this incentive to result in strong EU influence".[1] Thus, the integration of environmental protection requirements into the definition and implementation of Union's policies and activities (Articles 11 and 191 TFEU) appears to be progressively demanding, especially in terms of the EU's enlargement policy.

In December 2021, nearly eight years after the first EU-Serbia Intergovernmental Conference (January 2014), negotiation Cluster 4 (Green Agenda and Sustainable Connectivity) was finally opened. Serbia's path to EU membership is burdened by high domestic adjustment costs to meet the EU's legal and political conditionality in numerous areas. Moreover, this adjustment cost is not only economic, but also societal and political, whereby the most challenging issues are environmental protection and sustainable development. Serbia's energy mix is still mainly based on fossil fuels, its green energy transition is at a stalemate, and the level of air pollution in many urban areas remains alarmingly high. It is then hardly surprising that the cost of a full alignment of Serbia's national legislation with the existing EU acquis in the field of environmental protection is significant—by conservative estimates, in the vicinity of at least 10 billion euros "for environmental infrastructure alone".[2] Through that lens, the dimension of environmental sustainability in the future negotiations between the EU and Serbia will be crucial, not only for negotiation Cluster 4, but also for overall relations between the two parties, and, consequently, for the entire EU enlargement process.

This Chapter offers a case study of the complex and often tense relations between, on the one hand, the EU's legal requirements on environmental protection, and, on the other, its enlargement policy and membership negotiations with Serbia, as part of

---

[1] Epstein and Sedelmeier.

[2] European Commission, Annex II to the Commission Implementing Decision on the financing of the annual action plan in favour of the Republic of Serbia for 2021, p. 4.

its external relations. The first part of the analysis defines the place of environmental sustainability in the EU's external action and accession policies—with a focus on the two previous enlargement rounds (2007 and 2013)—and examines how the EU's actions influenced said sustainability and environmental policies. The second part of the analysis examines how and to which extent the existing EU's environmental legal framework could influence its relations with Serbia as a candidate country. In conclusion, the authors argue that—in spite of its growing importance—the EU's legal framework on sustainability often remains subject to various exceptions and different interpretations.

## 20.2 EU Environmental Law in the Current Geopolitical Context

Environmental protection was not a political priority in the initial European integration period. Between 1957 and 1972, only a small number of European Economic Community measures could have been classified as environmental protection measures.[3] The term "environment" was regulated in the Treaties since the 1987 Single European Act. However, the Single European Act cannot be interpreted as the starting point of EU environmental law, given that environmental issues existed prior to the formal inclusion of the environment in the Treaty.[4] Since 1987, environmental protection has been aimed at preservation, protection and improvement of the quality of the environment, the protection of human health, and the prudent and rational use of natural resources. Promotion of measures at the international level to solve regional, i.e., global environmental issues, and especially the fight against climate change, were added as objectives by the Treaty of Lisbon in 2009.[5]

The present-day legal basis for the adoption of environmental protection legislation is laid down in Articles 4, 11, 114 and Chapter XX (Articles 191–193) of the Treaty on the Functioning of the European Union (TFEU), whereby the basic articles for the adoption of legislation are Article 192 of the TFEU (legal basis for environmental protection) and Article 114 TFEU (legal basis for the internal market). Since its inception in the Single European Act, EU environmental legislation has grown steadily; today, it covers a range of different measures and instruments. It can, nevertheless, be roughly divided, i.e. differentiated into environmental-policy legislation and environmental issues embedded in other EU policies. This matter was addressed by the Court of Justice of the European Union (CJEU) within the framework of multiple policies, including environmental disputes.[6] As its case-law shows, the CJEU gives preference to the use of a single legal basis for enacting environmental

[3] Davies Peter 2004, p. 2.

[4] Von Zeben 2020.

[5] Article 191 TFEU.

[6] *Commission v Council,* Case C-300/89 [1991] ECR 1-2867.

protection legislation.[7] Article 192 TFEU prescribes the basic competence within the framework of the EU's environmental policy in accordance with the objectives specified in Article 191 TFEU. However, the objectives in Article 191 TFEU are defined in negative terms; specifically, the measures adopted pursuant to Article 192 TFEU can only be adopted to achieve the objectives from Article 191 TFEU. Consequently, for any environmental objective not prescribed in Article 191 TFEU, the EU must use a different legal basis.

In December 2019, the European Commission announced the European Green Deal—the EU's attempt to encourage the transition towards the common objective of a climate-neutral economy through reducing carbon emissions by at least 50% by 2030 and carbon neutrality by 2050.[8] The main elements of the European Green Deal are climate action, clean energy and sustainable industry.[9]

Although the EU aims to achieve the green transition and to become the world's first climate-neutral region by 2050, accomplishing this objective is far from straightforward and will require a radical transformation of the European economy and society. Since the adoption of the European Green Deal, the EU, as well as the whole world, had to face the Covid-19 pandemic, and in 2022, the geopolitical crisis caused by the war in Ukraine. The two global issues certainly impede the adoption of legislation in the field of environmental policy. For example, the set of energy amendments has been only partially implemented and experienced a delay, as the initial intention was to implement all the modifications by the start of 2023 but up today not all energy legislation planned has been adopted. The adoption of the European Climate Law[10] is nevertheless an important cornerstone for the necessary changes in EU legislation. However, it is only a first step towards overall European environmental legislation.[11] It remains to be seen whether the EU's plans for recovery from Covid-19 together with the implementation of the European Green Deal is sturdy enough groundwork for stimulating the development of legislation in the field of environmental protection and energy policy.

In terms of the enlargement process, it should be emphasized that Article 11 TFEU stipulates that "environmental protection requirements must be integrated into the definition and implementation of the Union's policies and activities, in particular with a view to promoting sustainable development,"[12] which clearly shows that environmental issues should be included in the enlargement process.

To date, there have been seven stages of EU enlargement: Denmark, Ireland and Great Britain (1973), Greece (1981), Portugal and Spain (1986), Austria, Finland

[7] Peeters and Eliantonio 2020, pp. 11–12.

[8] Sikora 2021, p. 681.

[9] Communication from the Commission to the European Parliament, the European Council, the Council, the European Economic and Social Committee and the Committee of the Regions (2019), The European Green Deal. COM. 640 final.

[10] Regulation (EU) 2021/1119 of the European Parliament and of the Council of 30 June 2021 establishing the framework for achieving climate neutrality and amending Regulations (EC) No 401/2009 and (EU) 2018/1999 ('European Climate Law').

[11] Duić 2022, pp. 7–25.

[12] Article 11 TFEU.

and Sweden (1995), Cyprus, Czech Republic, Estonia, Hungary, Latvia, Lithuania, Malta, Poland, Slovakia and Slovenia (2004), Bulgaria and Romania (2007) and Croatia (2013). The legal basis for enlargement is laid down in Article 49 of the Treaty on European Union (TEU), which is supplemented with the so-called Copenhagen criteria. The latter criteria were formalized only after the first three enlargement stages, at the 1993 EU summit in Copenhagen.[13] In the European Communities era, an applicant country only needed to be an "European State".[14] The Lisbon Treaty limited the membership to European states that hold the values referred to in Article 2 TEU and are committed to promoting them.[15] At the time, candidate countries had to satisfy three overlapping categories: the Copenhagen criteria, neighbourly and regional cooperation that was introduced for the Western Balkan countries, and transposition of the EU *acquis*.[16] Given that it is affected by internal, external, economic, legal and social factors, the issue of EU enlargement is a highly complex one.[17] Nonetheless, as a procedure that allows the EU to transform third states into Member States, enlargement should be considered as the most successful EU foreign policy.[18]

To unpack this claim, we first turn to analysing how the EU influenced environmental sustainability in candidate countries during the accession process. For some, especially central and eastern European countries, the adjustment to a democratic society, free market and improved environmental rules took place concurrently with the EU accession process.[19] As it is clear from the abundant discussion on the EU environmental policy implementation,[20] the considerable challenges regarding the implementation of the environmental *acquis*—that existed already in the 1980s enlargement—only grew more acute with the environmental *acquis* extension resulting from the 2004 and 2007 enlargement.[21] With regard to Serbia, this chapter will take into comparison only its on-going accession process (and not the yet-to-come Accession Act).

[13] Hillion 2004, pp. 1–22.

[14] Cremona 2007.

[15] Article 49 TEU.

[16] Bogojevic and Drenovak-Ivanovic 2019, p. 956.

[17] Mrkša 2001, p. 34.

[18] Hillion 2016, p. 6.

[19] Buzogány and Cotta 2021, p. 2.

[20] Knill and Lenschow 2000, pp. 251–282; Börzel 2003; Bondarouk and Mastenbroek 2018, pp. 15–27.

[21] Skjærseth and Wettestad 2007, pp. 263–280; Bogojevic and Drenovak-Ivanovic 2019, pp. 949–978; Börzel and Buzogány 2019, p. 318.

## 20.3 EU Environmental Law in the Enlargement Process

The focus of this Chapter is the role of EU environmental requirements as just one aspect of the complex EU accession process, as well as the accession conditions, observed through the lens of only the 2007 and 2013 EU enlargement stages. Accession agreements and Stabilisation and Association Agreements (SAAs) were international agreements concluded between Accession Country, Member States and European Community—by nature, mixed agreements[22] based on what was then Article 310 of the (Treaty establishing the European Community (TEC).[23] Accession agreements contain provisions on environmental requirements that are designed to achieve a variety of objectives, related both to environmental protection and sustainability.

Accession partnerships were introduced in 1998 for the ten central and eastern European countries[24] that were in the process of accession. This is considered as a turning point in the EU enlargement process, considering that it introduced conditionality for a wide range of policy areas.[25] In regard to environmental law, all candidate countries were required to continue the transposition of the *acquis* and to inaugurate detailed programmes for approximation of their national legislation with EU norms, as well as implementation strategies.[26] The priorities and intermediate objectives were divided into short- and medium-term.

The short- and medium-term objectives for the 2007 enlargement countries (Bulgaria and Romania) were identical. Short-term objectives included the continuing of transposition of framework and horizontal legislation, establishment of detailed approximation programmes and implementation strategies related to individual legislative acts.[27] Long-term objectives, on the other hand, entailed the development of monitoring and implementation control structures and capacities, and continuous planning and implementation of approximation programmes related to individual legislative acts. A particular emphasis was to be placed on air pollution, the waste and water sectors, including their institutional requirements. Environmental protection requirements and the need for sustainable development was required to be integrated into the definition and implementation of national sectoral policies.

The European Commission published regular reports on Romania's and Bulgaria's progress towards accession from 1998 till 2005. In the first such report on Bulgaria, the Commission concluded that some progress had been made toward meeting the

---

[22] *Mixed agreements* is the term used for agreements that are concluded jointly by European Union and Member States with a third party or third parties.

[23] Rodin 2003, p. 592.

[24] Czech Republic, Estonia, Hungary, Poland, Slovenia, Bulgaria, Latvia, Lithuania, Romania, Slovakia.

[25] Grabbe 1999, p. 9.

[26] Grabbe 1999, p. 19.

[27] Council Decision 98/266/EC of 30 March 1998 on the principles, priorities, intermediate objectives and conditions contained in the accession partnership with the Republic of Bulgaria and 98/261/EC: Council Decision of 30 March 1998 on the principles, priorities, intermediate objectives and conditions contained in the accession partnership with Romania.

requirements of the short-term priorities within the Accession Partnership, but that significant additional efforts were required before Bulgaria could assume all membership obligations in the environmental sector. In particular, this entailed full compliance with EU environmental regulations and enhancing administrative capability of relevant institutions, such as monitoring laboratories and implementation and enforcement mechanisms. Additional efforts were found needed in order to establish framework and horizontal legislation transposition, as well as approximation programmes and implementation strategies, particularly in the water sector, waste management, and air pollution.[28]

In each of the said pre-accession Reports, the Commission issued opinions on environmental issues. In the 2005 (and last such) Report on Bulgaria, the Commission found that Bulgaria largely satisfied the air quality, noise, and environmental protection commitments and requirements established in the accession negotiations. The Commission also forecasted that with the then pace of change, Bulgaria was on track with successfully implementing the *acquis* in those areas. The Commission also found that Bulgaria still needed to strengthen air quality management and monitoring, as well as to create a regulatory mechanism to address exceptional circumstances in compliance with fuel regulations. Nature protection-wise, the Commission found that Bulgaria needed to concentrate on finalising its lists of proposed nature protection sites and special bird protection zones.

The Commission also took the view that Bulgaria needed to put forth more effort to complete the transposition process and ensure the implementation and enforcement of transposed legislation with particular regard to horizontal legislation, waste management, water quality, industrial pollution and risk management, chemicals and genetically modified organisms, nuclear safety, and radiation protection. In these areas, the Commission found, administrative capacity needed significantly more strengthening, especially in the area of regional and local environmental agencies. Waste management-wise, the establishing of waste collection systems and recovery and disposal facilities were set as key priorities (with an emphasis on linking the necessary waste disposal permit procedures with public access to relevant information).

Concerning water quality, the Commission underlined the need to complete the regional drinking water monitoring systems before the accession, as well as to develop the necessary action plans and programmes. In regard to industrial pollution, permits for installations covered by the Integrated Pollution Prevention and Control Directive (IPPC) still needed to be issued and adhered to by the *acquis* deadlines.[29] It was found critical that the industry follow the issued permits, that the administrative capacity of the enforcing authorities be made able to rise to the immensely complex and demanding activities that they needed to complete, and that the existing rate of granting permits be maintained. Administrative capability

[28] Regular report from the Commission on Bulgaria's progress towards accession 1998. COM (98) 707 final, 17 December 1998. [Bulletin of the European Union Supplement 12/98].

[29] Carius et al 2001, p. 73.

regarding chemicals and genetically modified organisms was found to need improvement, and biocide approval processes to need devising. The Commission also found that noise, action plans and strategic noise maps still needed to be developed.

Overall, Bulgaria was found to need to accelerate its efforts to complete the alignment process and increase the administrative capability of environmental authorities at all levels and to increase the respective personnel and training. Found in need of enhancement was coordination among the many environmental institutions and authorities.[30]

In its first Report on Romania, the Commission assessed the overall state of Romania's approximation efforts as low. Much remained to be done in order to satisfy the Accession Partnership's short-term priorities. Major challenges were found to be surrounding Romania's air pollution, water management, and waste management. The legal approximation and institution building pace needed to be increased, and the Ministry of Waters, Forests, and Environmental Protection restructured. The development of a coherent environmental strategy, as well as directive-specific implementation programmes, were found indispensable for continued progress in European integration.[31]

In its 2005 (and last such) Report on Romania, the Commission found that the candidate generally met the membership requirements and, predicted that, given the strong progress in both legislation and administrative capacity alignment, it would be able to implement the *acquis* prior to accession in the areas of air quality, nature protection, chemicals and genetically modified organisms, noise, nuclear safety, and radiation protection. Romanian legislation had yet to be fully aligned in these areas. The Commission also found that Romania needed to strengthen air quality management and monitoring, as well as to finalize air quality plans and programmes.

In terms of nature protection, Romania had yet to ensure that essential safeguards are in place prior to accession. More efforts were found to be required in the areas of horizontal legislation, waste management, and water quality to complete the transposition process and guarantee that the transposed legislation is implemented. Waste management plans had yet to be adopted and recycling and recovery rates increased; the monitoring of water quality required further improvement and action plans needed to be adopted. Serious concerns were found to exist in relation to industrial pollution.[32] The Commission found that, prior to accession, significant efforts would be necessary to guarantee that relevant permits are provided at the municipal and regional levels and that they are followed for all relevant IPPC facilities. It was also found that Romania needed to considerably accelerate its efforts to boost the administrative capability of the environmental authorities in this field at all levels.

Overall, while Romania had achieved progress in terms of administrative capability, both in terms of personnel recruitment and training, more strengthening was

---

[30] Comprehensive monitoring report Bulgaria 2005a. SEC (2005) 1352 final, 25 October 2005. [EU Commission—SEC Document].

[31] Regular report from the Commission on Romania's progress towards accession 1998. COM (98) 702 final, 17 December 1998. [Bulletin of the European Union Supplement 9/98].

[32] Buzogány 2009, p. 188.

found to be needed, particularly at the regional and municipal levels. The lack of adequate coordination between national, municipal, and the relatively newly formed regional environmental agencies were found to have remained a major source of concern, and to require addressing to allow a proper implementation of the environmental *acquis*.[33] Already in 2007, at the very outset, Romania's membership was marked by a large number of Formal Notices from the Commission. Nearly one fourth of the Notices were launched in the field of environmental policy.[34]

What the EU enlargement process clearly demonstrates is the EU's undeniable ability to act externally through the accession process and to influence environmental and sustainability issues. However, as indicative from the above-presented contrast between Romania and Bulgaria's accession progress, that EU's ability is not absolute in its power.

For the Western Balkan countries (Albania, Bosnia and Herzegovina, Croatia, Serbia and North Macedonia (previously: Former Yugoslav Republic of Macedonia), the EU introduced a new tailor-made category of pre-accession contractual relations: the Stabilization and Association Agreements (SAAs). The first SAAs were signed in 2001, with North Macedonia and Croatia.[35] Croatia's SAA entered into force in March 2002. The signing of an SAA demonstrates a country's efforts to satisfy the relevant political and economic criteria, and is a sign that the country is regarded as credible for full EU membership. A special review process—the first of its kind—was developed for SAAs, which includes an annual assessment of the Stabilization and Association Process and its various instruments, as well as of a country's progress.[36]

With regard to the environment, Croatia's SAA in Article 103 prescribes that the Parties are to expand and deepen their collaboration in the critical mission of combating environmental degradation in order to promote environmental sustainability.[37] Article 103 also sets a number of priorities: water quality, including wastewater treatment, particularly in transboundary watercourses; combating local, regional, and transboundary air and water pollution (including drinking water); effective monitoring of pollution levels and emissions; development of global and climate strategies; efficient, sustainable, and clean energy production and use; classification and safe handling of chemicals; industrial plant safety; waste reduction, recycling and safe disposal, and the implementation of the Basel Convention on the control of transboundary movements of hazardous wastes and their disposal (Basel 1989); the environmental impact of agriculture; soil erosion and pollution from agricultural chemicals; the protection of flora and fauna, including forests, and biodiversity conservation; town and country planning, including construction and urban planning; the use of economic and fiscal instruments to improve the environment; carrying out

[33] Comprehensive monitoring report Bulgaria 2005b. SEC (2005) 534 final, 25 October 2005. [EU Commission—SEC Document].

[34] Buzogány 2020, p. 8.

[35] For the text of the SAA Macedonia, see COM(2001) 90 final, 19 February 2001; for the SAA Croatia, see COM(2001) 371 final, 9 July 2001.

[36] Pippan 2004, p. 239.

[37] Vlašić and Vlašić 2006, pp. 321–346.

impact assessment and strategic environmental assessment; continuous approximation of laws and regulations to Community standards; international environmental Conventions to which the Community is a Party; regional and international cooperation, as well as education and information on environmental issues and sustainable development.[38]

The European Commission issued Regular reports on Croatia's progress towards accession from 2005 until its accession in 2013. In its first such Report, the Commission found that sound progress was made in the fields of air quality and waste management. While some progress was found to have been made in terms of water quality, there were only minimal advances in other sectors. Overall progress was found to be slower than planned under Croatia's 2004 National Programme for EU Integration. A general timeframe for transposing the environmental *acquis* had yet to be developed, and the implementation and enforcement of new legislation was found lagging.

With regard to the issues raised in its Opinion on Croatia's application for EU membership, the Commission found that most had persisted and continued to represent a danger to the *acquis*'s successful transposition. The Commission also found that Croatia had yet to make more substantial and continuous efforts in the environmental sector to show that the Croatian government gives necessary weight to environmental protection. The Ministry of Environmental Protection, Physical Planning and Construction was found to have been given great responsibility for environmental protection, but not the authority necessary to deliver on it. The Ministry's comparative weakness, coupled with the distribution of responsibilities for operational aspects of environmental protection, was found to continue to stymie approximation, implementation, and enforcement of legislation.[39]

From 2004 until 2012, the Commission issued opinions on environmental issues in each Report. Per the Commission's final pre-accession report from 2005, Croatia was found to be generally meeting the commitments and requirements arising from the accession negotiations and was deemed able to implement the environmental *acquis* by accession. Nevertheless, additional efforts were found to be required to adequately implement and enforce legislation. To be able to implement the *acquis* prior to the accession, urgent actions were yet required, the Commission found, particularly in the area of climate change. Some industries were found to require increased administrative capability, particularly in the area of climate change, but also industrial pollution control and risk management, nature protection, and chemicals. Investments were found to be required across sectors, but particularly in air quality, industrial pollution control and risk management, water, and waste, to ensure that Croatia completes its membership preparations by accession. Special emphasis was to be placed on timely delivering of commitments in the sectors of waste and IPPC.

---

38 The States in the process of accession are required to demonstrate their ability for applying and enforcing the Union rules and laws while balancing the needs of their own national, social and economic developments. L 26 (28/01/2005).

39 Republic of Croatia 2004 Pre-Accession Economic Programme 2005-2007. [EU Commission—Working Document].

The quality of environmental impact assessment for projects was found to need significant improvements.[40]

In December 2012, the Croatian government announced the 2013 Programme for Transposition and Implementation of the EU *Acquis*. Croatia has fulfilled most of the obligations arising from the EU accession process, particularly in regard to the process of aligning national legislation with the *acquis*. However, as the *acquis* is continually developing, so Croatia as a Member State is obliged to continually align its national legislation.[41] That environmental issues were monitored more closely compared to the 2007 accession process, is a testament to the EU's continued ability to act externally through the accession process and to influence environmental and sustainability issues. Countries in the process of EU accession are required to demonstrate their ability for applying and enforcing EU rules and laws, while balancing the needs of their own national, social and economic developments.[42] To date, no candidate country has been able to fully adopt the *acquis* prior to accession.[43] Accession countries in general benefit from a variety of transitional periods that allow them to gradually comply with complex EU environmental legislation, while observing country-specific challenges.[44]

## 20.4 The Place of EU Environmental Legislation in the Process of Serbia's Accession to the Union

Sixteen years after the signature of the SAA between the European Communities and Serbia, and ten years from the moment of the official opening of accession negotiations with the EU, Serbia's perspective of joining the Union is still uncertain and its path towards the membership considerably slowed down by various external and internal factors. The latest formal step recently undertaken in the process of Serbia's accession was the opening (December 2021) of negotiating Cluster 4 (Green Agenda and Sustainable Connectivity). In spite of broadly positive changes this opening could generate in the mid-term perspective, the lack of broad pro-European consensus in Serbia, some related economic, societal and political challenges, as well as country's high domestic adjustment costs to meet the EU's criteria in the field of environmental protection are considerably burdening the perspective of a swift transformation of national legislation and policies in this field. Moreover, the

[40] Comprehensive monitoring report on Croatia's state of preparedness for EU membership. Commission staff working document accompanying the communication. SWD (2012) 338 final, 10 October 2012. [EU Commission—SEC Document].

[41] Vlada Republike Hrvatske, Program Vlade Republike Hrvatske za preuzimanje i provedbu pravne stečevine Europske unije za 2013. godinu, http://www.sabor.hr/Default.aspx?sec=1207. Accessed 12 October 2022.

[42] Finckenberg-Broman 2022, p. 86.

[43] Kapios 2002, p. 6.

[44] Čavoški 2019, p. 117.

aggression of the Russian Federation on Ukraine has significantly impacted not only some major energy policies of the EU, but has also put in question its plans in the field of climate change, significantly compromising its aptitude to impose the environmental conditionality to candidate countries. Therefore, before examining the general framework of Serbia's law approximation policy (Sect. 20.4.2), the role of EU *acquis* in transformation of Serbian national environmental legislation and some major challenges in this process (Sect. 20.4.3), this chapter will first strive to answer whether there is an "environmental conditionality" in the EU accession process (Sect. 20.4.1).

### *20.4.1 Is There Still an "Environmental Conditionality" in the EU Accession Process?*

On the political level, the process of European integration of Serbia and other counties of the region has been launched at the EU-Western Balkans Summit, which was held in Thessaloniki on 21 June 2003. Unfortunately, almost two decades later, the EU-Western Balkans Zagreb Summit, held via video conference on 6 May 2020, has not gone much further, avoiding to mention the notion of the membership in the EU, and, therefore, keeping alive the consecrated formulation "European perspective of the Western Balkans countries," in theory often referred to as "political declaration usually devoid of important consequences".[45]

In any case, the EU accession criteria (defined at the European Council held in Copenhagen back in 1993) have remained predominantly focused on political and economic conditions, putting the issue of "conditionality" into the focus of the interest of both academics and practitioners. However, while the accession criteria taken *stricto sensu* have not evolved significantly over time, the Union's acquis—with which the candidate countries are obliged to harmonise their internal national legal systems—is in the process of constant development and mutation, especially in the field of environmental protection and climate change.

Unfortunately, the Russian aggression on Ukraine is a significant (and far from being only political) game-changer, influencing numerous EU's ambitious goals in environmental protection and, consequently, its aptitude to impose the same objectives to potential future Member States. Moreover, some recent decisions of national constitutional courts of the Member States are "challenging not only the Union's legal order, but also its institutional viability".[46] We will, therefore, first examine the quintessence of the notions of political and economic conditionality, before turning to the question of whether the same concept of conditionality can (still) be applied to the EU accession criteria related to environment and climate change.

In spite of the fact that, in recent decades, the notion of political conditionality is most often associated to the process of EU enlargement, various "international

45 Ćemalović 2020, p. 180.

46 Ćemalović 2021, p. 71.

institutional lenders, such as the IMF, World Bank and ECDR have frequently practised conditionality in the pursuit of neo-liberal fiscal policy during regime change" while, applied to the candidates for accession to the EU, it necessarily includes a "monitoring system embodied primarily in the political dialogue set up in the association agreements and in the European Commission's annual evaluations of the candidates' progress".[47] Given the fact that the principle of political conditionality can be resumed in the mechanism of "carrots and sticks," it is not surprising some recent studies have revealed that its implementation "resulted in limitations in rule adoptions due to weaker incentives,"[48] while, more generally, the political conditionality presupposes the weakness of the state to which it is applied, thus allowing the economic and/or strategic interests to "trump the norm of conditionality".[49] Moreover, while the substance of the economic conditionality is different than it is the case of its political counterpart, its mechanisms are practically the same. In conclusion, the quintessence of the principle of conditionality relies on two fundamental elements; first, there is a weak, potentially dysfunctional and economically less developed candidate country and, second, the entire process depends on the volume and nature of the incentives which—at least from the perspective of the "object" of conditionality—have to be higher than the adjustment costs of the required reforms.

With ever-growing EU acquis dedicated to the environment, energy transition and climate change, it has long seemed appropriate to complete the array of EU accession "conditionalities" with the one denominated "environmental conditionality". Notwithstanding the fact that the bibliography dedicated to this issue is significantly less voluminous than those dedicated to political and economic, the notion of EU's environmental conditionality is constantly present in theory for more than two decades, while the treated topics are ranging from the relationship between global competition and EU environmental policy[50] to the issue of the impact of trade on environmental law.[51] However, various EU policies—particularly including those related to energy transition[52] and response to the threat of climate change—have been significantly impacted by the consequences of the Russian aggression on Ukraine. It is, therefore, most likely that, the Union's relations with accession countries, the environmental conditionality (if any) will, on the one hand, remain hostage of broader mechanisms of economic and political conditionality while, on the other, it will still critically depend on the volume of the incentives (on the EU side) and the ever-growing environmental adjustment costs (on the side of candidate countries).

[47] Malová and Rybář 2003, p. 99.

[48] Şemşit 2016, p. 229.

[49] Smith 1998, p. 255.

[50] Golub 1998.

[51] Delimatsis and Reins 2021.

[52] Osička and Černoch 2022.

## 20.4.2 The General Framework of Serbia's Law Approximation Policy

The membership in the EU necessarily implies the obligation of the future Member States to gradually harmonise its legal system with the Union's *acquis*. This obligation does not end on the day of the accession, given that all Member States of the EU also have to transpose into their internal legal systems every provision of the EU law that requires the normative intervention of national legislators (most often, EU directives). However, when it comes to the process of law approximation, there are no universal general rules applicable to all candidate countries and, consequently, all former candidates have had their own policies in this field, mainly defined by bilateral treaties signed with the EU. When it comes to Serbia, the main characteristics of its law approximation policy are defined by Article 72 of the SAA, but the general framework of this policy has also some other country-specific features, mainly due to the specificities of its internal legal and political systems. Before turning to the issue of Serbia's general law approximation policy (Sect. 3.2.2), it is first necessary to clearly delimit the process of law approximation, on the one hand, and the procedure of membership negotiation, on the other (Sect. 3.2.1). As technical as it may seem at first, the difference between those two notions is fundamental for the understanding of the entire relation EU—candidate country and, therefore, necessary in order to define the position of Serbia.

### 20.4.2.1 Law Approximation versus Membership Negotiation

The complex relation between those two notions (and processes) can be summarised in three main features: legal, political and methodological. First, Serbia accepted the obligation to harmonise its national legal system with the EU *acquis* (in 2008, by signing and ratifying the SAA) well before it was granted EU candidate status (2012). Moreover, according to the national constitutional provisions, the SAA is a ratified international treaty and, therefore, has a greater legal force than domestic laws. Second, different national instances (and, more broadly speaking, different branches of power) are in charge of the process: in the process of membership negotiation, the executive branch has more emphasised role, while the adoption of laws that are (supposedly) harmonised with the EU acquis is the exclusive competence of the Parliament (the Government, most often, prepares and formally submits draft laws to the legislator). Finally, law approximation and membership negotiation are substantially different from ontological and methodological points of view; while the first one has primarily legal substance, the second is quintessentially political and economic, largely defined by interests and often subject to broader strategic, geo-political and societal considerations.

#### 20.4.2.2 Specificities of the National Law Approximation Policy

Serbia's general policy in this field—therefore applicable to the adoption of EU's environmental legislation—can be broadly defined by the following three aspects: graduality, unspecified time framework and dependency on enforcement policies.

Graduality of the law approximation process is the direct consequence of Article 72-1 of the SAA, according to which "Serbia shall endeavour to ensure that its existing laws and future legislation will be gradually made compatible with the Community acquis". Notwithstanding the fact that this prerogative should be applied *bona fide* by Serbian national authorities, both textual and teleological interpretation of Article 72-1 unequivocally implies that neither the executive branch of power (when it proposes laws or when it enacts by-laws), nor the national legislator in the Republic of Serbia is obliged to provide that every proposed and adopted legal act is necessarily, immediately and fully harmonized with the Union's *acquis* in the moment of adoption of the national legislation. In other words, the compliance of national legislation with EU law can be partial, or even non-existent, if this is "justified by country-specific needs, legitimate interests or by simple non-transferability of a specific provision of EU law at this particular moment".[53]

Unspecified time framework is, only partially and indirectly, the consequence of the graduality of law approximation process. More substantially, it derives from the fact that law approximation is influenced by progress in membership negotiation between Serbia and the EU, on the one side, but also by internal national legislative and political dynamics, on the other. The latest, fourth revised National Program for the Adoption of the EU Acquis (NPAA), has been adopted in July 2022 and it covers the period until 2025. However, the ambition of previous versions of the NPAA—as well as of a far predecessor of this document, adopted in 2008 and entitled the National Program of Integration—was to achieve the highest possible level of compatibility with the EU acquis in the period of up to four years from its adoption. This objective has been systematically postponed for several times and it is clear that—as it is the case of membership negotiation—both Serbia and the EU are either unwilling or incapable to define a clear time framework in this respect.

Dependency on enforcement policies means that even the most successful and complete law approximation remains largely pointless if it is not followed by implementation and enforcement of harmonised national legislation. In spite of the fact that both Article 72-1 of SAA and internal constitutional and legal provisions in Serbia insist on proper, full and effective enforcement of adopted national laws, in numerous cases, covering different policy areas, the country either hasn't adopted the necessary by-laws or simply does not (fully) enforce national legislation. Unfortunately, one of these policy areas is also the one dedicated to the environmental policy taken *lato sensu.*

[53] Ćemalović 2018, p. 301.

### 20.4.2.3 The Role of EU Acquis in Transformation of Serbian National Environmental Legislation and Some Major Challenges in this Process

Due to the large amount of both EU and national acts in the field of environmental protection and climate change, the focus of this section will be on the following three issues: protection of air (Sect. 3.1.1), protection of water (Sect. 3.3.2) and waste management (Sect. 3.3.3). In each subsection, we will first analyse the state of Serbian national legislation in the respective field, before turning to the issue of main challenges regarding its implementation and enforcement.

**Protection of Air**

In Serbia, the systemic (general) law on Environmental Protection (LEP) was adopted in 2004, bringing in a series of provisions on the general framework of protection, and aiming to guarantee the constitutional rights "to a healthy environment and [...] to timely and full information about the state of environment".[54] Apart from defining the air as "natural value" and "public natural good", the LEP dedicates only one Article (24) exclusively to the protection of air, while the other mentions (like, for example, in Articles 39, 40, 81, 90v) of this notion are made together with other protected values. The Law on Air Protection (LAP), as a *lex specialis* in relation to the LEP, has been adopted in 2009, and amended in 2013 and 2021. Before the amendments adopted in 2021, the LAP transposed only some of the provisions of the Directive of the European Parliament and of the Council 2008/50/EC of 21 May 2008 on ambient air quality and cleaner air for Europe. Moreover, back in 2013, in preparatory and explanatory acts accompanying the draft version of the LAP, the government omitted to refer to the Directive 2008/50, even if its provisions were transposing some of its formulations. The amendments to the LAP adopted in 2021 have brought some more substantial alignment with the EU law, allowing the European Commission to conclude, in its 2021 and 2022 Progress Reports dedicated to Serbia, that "in the field of air quality, Serbia has a good level of alignment with the EU acquis".[55] However, the major problems related to ambient air quality in Serbia have remained the same for years: not fully developed and functional system of air monitoring, non-implementation or partial implementation of local air quality plans (in agglomerations that adopted it) and non-alignment of national legislation with EU acquis on national emission ceilings.

**Protection of Water**

In this field, the situation is substantially different—both on legislative and administrative level—that the one related to the protection of air. First, the nomotechnical approach in Serbia and in the EU differ when it comes to the number and scope of legal acts related to the protection of water. While the EU has a whole range of various sources of law dedicated to particular and precise issues (drinking water,

[54] Article 74-1 LEP.

[55] Serbia Progress Report 2022, p. 123; 2021 Serbia Progress Report, p. 114.

bathing water, urban waste water treatment, nitrates in water), in Serbia there is one systemic Water Law dedicated to a wide and complex set of issues. Second, the level of alignment of the national legislation with the EU acquis is significantly lower than in the field of air protection, especially when it comes to water quality standards, dangerous pollutants and wastewater infrastructure. Third, in spite of the fact that the Action plan for implementing the Water management strategy was adopted, the three biggest Serbian cities (Belgrade, Novi Sad and Niš) are still discharging the quasi-totality of their wastewater directly into rivers, while the administrative capacity (both on national and local level) to effectively implement the Action plan is either weak or practically non-existent. For example, according to the latest report of the Serbian Institute for Public Health, more than 30% of the population covered by the public water supply networks gets the water that is unsafe for drinking,[56] while the local authorities do not have functional monitoring and reporting mechanisms.

**Waste Management**

As it is the case of air quality, the level of alignment of the national legislation on waste management with the EU acquis is estimated as "good" by the European Commission's latest 2022 Report. However, the level of effective implementation of this legislation is particularly low. In spite of the recently adopted National waste management programme (2022–2031) and Waste action plan (2022–2024), the global national recycling rate is 15.5%, while it falls to only 2% regarding the waste generated by households. In spite of the generally well-transposed EU regulatory framework regarding the waste movement control system, the Law on Waste Management (LWM) remains largely non-implemented. For example, Article 45 of the LWM provides that every movement of all kinds of waste (excluding the one generated by households) has to be duly documented, as it is required by the EU Directive 2008/98. Even when various operators formally respect this obligation, the national authorities do not have sufficient administrative capacity to monitor and control the process.

According to the latest accessible data, the total number of all active environmental inspectors working in all municipalities is 216, while the majority of municipalities (112 of 145) had only one inspector and some of them (10) none. It is, therefore, not surprising why both academics and practitioners estimated that, in the field of waste management, the enforcement of the national legal provisions is "either non-existent or uneven, often provoking confusion and discontent of operators and citizens."[57] Therefore, before whichever new step towards a further alignment of Serbian national legislation with the EU *acquis* is made, the country should first endeavour to implement the existing regulatory framework.

[56] Serbian National Institute for Public Health "Dr Milan Jovanović Batut", 2021 Report on Drinking Water from Public Water Supply, pp. 17–20.

[57] Protić 2022, p. 225.

## 20.5 Conclusion

In the process of EU accession, all candidate countries have to balance their own national, social and economic situation with application of EU rules and laws. In accordance with that, no candidate country has fulfilled all the accession criteria prior to accession. Moreover, accession countries often benefit from a variety of transitional periods that allow them to gradually comply with complex EU legislation, while observing country-specific challenges. What the EU enlargement process clearly demonstrates is the Union's undeniable ability to act externally through the accession process and to influence environmental and sustainability issues. The importance of environmental sustainability and the EU's influence thereon has increased with the introduction of SAAs, which is clear from the case of Croatia's accession. As the first country to accede to the EU within the SAA framework, it was obliged to fulfil more detailed and complex requirements. Moreover, the ostensible differences between analysed countries' scope and quality of the fulfilment of obligations point to political compromises that the EU makes in deciding on the finalisation of the pre-accession process. As the case of Serbia clearly demonstrates, the role of EU *acquis* in the transformation of national environmental legislation is significant, but very often faced with various obstacles, such as high domestic adjustment cost, unspecified time framework and dependency on national enforcement policies. As the analysis regarding the issues of protection of air, protection of water and waste management has shown, even when the level of formal adjustment of national legislation with the EU acquis is satisfactory, its enforcement is often non-existent or uneven, potentially jeopardizing not only Union's enlargement, but also affecting its sustainability-related policies.

## References

Annex II to the Commission Implementing Decision on the financing of the annual action plan in favour of the Republic of Serbia for 2021 [European Commission, Action Document "EU for Connectivity and Green Agenda"]

Bogojevic S, Drenovak-Ivanovic M (2019) Environmental Protection Through the Prism Of Enlargement: Time For Reflection. Common Market Law Review 56:949–978

Bondarouk E, Mastenbroek E (2018) Reconsidering EU compliance: implementation performance in the field of environmental policy. Environmental Policy and Governance, 28 (1)

Börzel T A (2003) Environmental leaders and laggards in Europe. Why there is (not) a "Southern Problem". Hampshire: Ashgate

Börzel T A, Buzogány A (2019) Compliance with EU environmental law. The iceberg is melting, Environmental Politics, 28:2, 315–341

Buzogány A (2009) Romania: Environmental Governance — Form without Substance. In: Börzel, T.A. (eds) Coping with Accession to the European Union. Palgrave Studies in European Union Politics. Palgrave Macmillan, London

Buzogány A (2020) Beyond Balkan exceptionalism. Assessing compliance with EU law in Bulgaria and Romania, European Politics and Society, 22:2, 185–202

Buzogány A, Cotta B (2021) Post-accession backsliding and European Union environmental policies, Post-Communist Economies, 34:5, 647–665
Carius A et al (2001) Environmental Policy and Law in Bulgaria—Towards EU-Accession. BoD—Books on Demand, Germany
Čavoški A (2019) EU environmental compliance assurance. Environmental Law Review 21(2): 111–118
Ćemalović U (2018) Usklađivanje domaćeg pravnog Sistema sa pravnim tekovinama Evropske unije – opšti teorijski okvir i neka praktična pitanja. In: Vasić R (ed) Nomotehnika i pravničko rasuđivanje, str. 281–316, Pravni fakultet Univerziteta u Beogradu i Program Ujedinjenih nacija za razvoj u Srbiji, Belgrade, pp 281–316
Ćemalović U (2020) One Step Forward, Two Steps Back: The EU and the Western Balkans After the Adoption of the New Enlargement Methodology and the Conclusions of the Zagreb Summit, Croatian Yearbook of European Law and Policy 15:179–196
Ćemalović U (2021) Supremacy of EU Law Over National Legislation and Supreme Jurisdictions of the Member States—a Quest for a New Balance. In: Zakic K, Demirtas B (eds) Europe in Changes: The Old Continent at a New Crossroads, 63–78
Commission v Council, Case C-300/89 [1991] ECR 1-2867
Communication from the Commission to the European Parliament, the European Council, the Council, the European Economic and Social Committee and the Committee of the Regions (2019), The European Green Deal. COM. 640 final
Comprehensive monitoring report Bulgaria 2005a. SEC (2005) 1352 final, 25 October 2005. [EU Commission—SEC Document]
Comprehensive monitoring report Bulgaria 2005b. SEC (2005) 534 final, 25 October 2005. [EU Commission—SEC Document]
Comprehensive monitoring report on Croatia's state of preparedness for EU membership. Commission staff working document accompanying the communication. SWD (2012) 338 final, 10 October 2012. [EU Commission—SEC Document]
Council Decision 98/261/EC of 30 March 1998 on the principles, priorities intermediate objectives and conditions contained in the accession partnership with Romania
Council Decision 98/266/EC of 30 March 1998 on the principles, priorities intermediate objectives and conditions contained in the accession partnership with the Republic of Bulgaria
Cremona M (2007) Enlargement of the European Union, Oxford University Press, Oxford
Davies Peter G G (2004) European Union Environmental Law: An Introduction to Key Selected Issues. Routledge, New York
Declaration of EU-Western Balkans 2020 Summit https://www.consilium.europa.eu/media/43776/zagreb-declaration-en-06052020.pdf [Accessed 15 September 2022]
Delimatsis P, Reins L (eds) (2021) Trade and Environmental Law, Edward Elgar Publishing, Cheltenham
Duić D (2022) Europska politika zaštite okoliša i Europski zeleni plan. In: Duić D and Ćemalović U (eds.), Zakonodavstvo zaštite okoliša i održivi ekonomski razvoj u EU. Pravni fakultet Osijek, Osijek
European Commission (2021) Serbia 2021 Report, Accompanying the Communication on Eu Enlargement Policy (2021) https://neighbourhood-enlargement.ec.europa.eu/serbia-report-2021_en [Accessed 29 September 2022]
European Commission (2022) Serbia 2022 Report, Accompanying the 2021 Communication on EU Enlargement Policy (2021). https://neighbourhood-enlargement.ec.europa.eu/serbia-report-2022_en [Accessed 29 September 2022]
Finckenberg-Broman P (2022) Weaponizing EU State Aid Law to Impact the Future of EU Investment Policy in the Global Context, Springer, Cham
Golub J (ed) (1998) Global Competition and EU Environmental Policy, Routledge
Government of the Republic of Serbia (2022) Forth Revised National Program for the Adoption of the EU Acquis (2022–2025) https://www.mei.gov.rs/upload/documents/nacionalna_dokumenta/npaa/NPAA_2022-2025_EN.pdf [Accessed 10 October 2022]

Grabbe H (1999) A partnership for accession? The implications of EU conditionality for the Central and East European applicants. EUI Working Paper RSC No. 99/12, http://hdl.handle.net/1814/1617

Hillion C (2004) The Copenhagen Criteria and Their Progeny. In: Hillion, Christophe (eds.), EU enlargement. Oxford: Hart Publishing, Oxford

Hillion C (2016) The Creeping Nationalisation of the EU Enlargement Policy, Swedish Institute for European Policy Studies, Stockholm

Kapios (2002) Environmental enlargement in the European Union: Approximation of the acquis communautaire and the challenges that it presents for the applicant countries. International and Comparative Environmental Law 4:21

Knill C and Lenschow A (2000) Do new brooms really sweep cleaner? Implementation of new instruments in EU environmental policy. In: C Knill and A Lenschow, eds. Implementing EU environmental policy: new directions and old problem. Manchester: Manchester University Press

Malová D, Rybář M (2003) The European Union's policies towards Slovakia – Carrots and sticks of political conditionality. In: Jacques Rupnik and Jan Zielonka (eds.), The road to the European Union—Volume 1: The Czech and Slovak Republics, Manchester University Press, Manchester

Mrkša S (2001) Politika proširenja Europske unije (European Union enlargement policy), Pravo i porezi 9: 34–56

Osička J, Černocha F (2022) European energy politics after Ukraine: The road ahead, Energy Research & Social Science 91:102

Peeters M and Eliantonio M (2020) Research Handbook on EU Environmental Law. Edward Elgar, Massachusetts

Pippan C (2004) The Rocky Road to Europe: The EU's Stabilisation and Association Process for the Western Balkans and the Principle of Conditionality. European Foreign Affairs Review 9: 219–245

Protić, J (2022) Conditions and obstacles for effective attainment of individual and collective labour rights of workers in waste management". Doctoral dissertation, Faculty of Law, University of Belgrade

Regular report from the Commission on Romania's progress towards accession 1998. COM (98) 702 final, 17 December 1998. [Bulletin of the European Union Supplement 9/98]

Regulation (EU) 2021/1119 of the European Parliament and of the Council of 30 June 2021 establishing the framework for achieving climate neutrality and amending Regulations (EC) No 401/2009 end (EU) 2018/1999 ('European Climate law')

Republic of Croatia 2004 Pre-Accession Economic Programme 2005–2007. [EU Commission—Working Document]

Rodin S (2003) Sporazum o stabilizaciji i pridruživanju u pravnom poretku Europske zajednice i Republike Hrvatske (Agreement on stabilization and association in the legal order of the European Community and the Republic of Croatia). Zbornik Pravnog fakulteta u Zagrebu, 53 3–4; 591–613

Şemşit S (2016) The EU's Enlargement and Neighbourhood Policy Strategies: The Role of Political Conditionality. In Sieglinde G (ed.), The European Neighbourhood Policy in a Comparative Perspective Models, Challenges, Lessons, Routledge, Oxon.

Serbian National Institute for Public Health "Dr Milan Jovanović Batut", 2021 Report on Drinking Water from Public Water Supply https://www.batut.org.rs/download/izvestaji/higijena/Godisnji%20vode%20za%20pice%202021.pdf [Accessed 11 Noveber 2022]

Sikora A (2021) European Green Deal—legal and financial challenges of the climate change. ERA Forum 21

Skjærseth, J B and Wettestad J (2007) Is EU enlargement bad for environmental policy? Confronting gloomy expectations with evidence. International Environmental Agreements: Politics, Law and Economics, 7 (3), https://doi.org/10.1007/s10784-007-9033-7

Smith K (1998) The Use of Political Conditionality in the EU's relations with Third Countries: How Effective?, European University Institute Working Paper 1997/07

Stahl B (2013) Another 'strategic accession'? The EU and Serbia (2000–2010). Nationalities Papers, Special Issue: Europeanization, State-Building and Democratization in the Western Balkans, 41–3:447–468

UN Environment Programme (2020) Review of national legislation. https://leap.unep.org/countries/rs/national-legislation/water-law [Accessed October 2022]

Vlašić I and Vlašić Feketija M (2006) The importance of environmental protection: Croatia in the European Union accession process. In K. Ott (ed.), Croatian accession to the European Union. Vol. 4, The challenges of participation. Zagreb: Institute of Public Finance, Zagreb

Zeben von J (2020) Environmental Law. The Oxford Encyclopaedia of EU Law, University of California Press

**Dunja Duić** is an Associate Professor at the Faculty of Law of University of Osijek, Stjepana Radića 13, 31000 Osijek, Croatia. (Parts of this paper written by Dr. Duić are the result of work that has been fully supported by the Faculty of Law Osijek—Josip Juraj Strossmayer University of Osijek, under the project nr. IP-PRAVOS-6. Project title: Implementation of EU Law in Croatian Legal System).

**Uroš Ćemalović** is a Doctor of the University of Strasbourg, and a Senior Research Associate, Institute of European Studies, Trg Nikole Pašića 11, 11000 Belgrade, Serbia.

The manufacturer's authorised representative in the EU is Springer
Nature Customer Service Centre GmbH, Europaplatz 3, 69115 Heidelberg,
Germany. If you have any concerns regarding our products, please
contact ProductSafety@springernature.com

Printed and bound by CPI Group (UK) Ltd, Croydon, CR0 4YY
23/09/2026
02221534-0003